Computation and Comparison of Efficient Turbulence Models for Aronautics – European Research Project ETMA

Edited by
Alain Dervieux
Marianna Braza
and Jean-Paul Dussauge

Notes on Numerical Fluid Mechanics (NNFM)   Volume 65

Computation and Comparison of Efficient Turbulence Models for Aeronautics – European Research Project ETMA

Edited by
Alain Dervieux
Marianna Braza
and Jean-Paul Dussauge

vieweg

Die Deutsche Bibliothek – CIP-Einheitsaufnahme

**Computation and comparison of efficient turbulence models
for aeronautics – European Research Project ETMA** / ed. by
Alain Dervieux ... – Braunschweig; Wiesbaden: Vieweg, 1998
   (Notes on numerical fluid mechanics; Vol. 65)

http://www.vieweg.de

Produced by Geronimo GmbH, Rosenheim
Printed on acid-free paper

ISSN 0179-9614
ISBN 978-3-322-89861-6   ISBN 978-3-322-89859-3 (eBook)
DOI: 10.1007/ 978-3-322-89859-3

FOREWORD AND INTRODUCTION

The computation of complex turbulent flows by statistical modelling has already a long history. The most popular two–equation models today were introduced in the early seventies.

However these models have been generally tested in rather academic cases. The developement of computers has led to more and more acurate numerical methods. The interactions between numerical and modelling techniques are generally not well mastered. Moreover, computation of real life cases, including 3D effects, complex geometries and pressure gradients based on two–equation models with low–Reynolds treatment at the proximity of walls are not really of common use. A large number of models has been proposed; this is perhaps the sign that none of them is really satisfactory, and then the assessment of their generality is not an easy task: it requires a lot of understanding of the physics and a lot of work for testing the large number of relevant cases in order to assess their limits of validity which is a condition for an improved confidence in engineering applications.

This is probably why workshops and working groups are frequent and the ETMA consortium has choosen to build a state of the art in theoretical and numerical statistical turbulence modelling for real life computations by taking some marks with respect to previous workshops such as the Stanford meetings (1980,1981); some problems are kept or updated by new experiments, some problems are discarded, some new problems are introduced; the focus is kept on flows with 2D geometries.

The ETMA Project

The project on Efficient Turbulence Models for Aeronautics (ETMA) was established to study the statistical turbulence models, from the points of view of building new models, making new experiments, deriving new numerical methods, computing new cases or better computing existing cases. The aims of ETMA involved also knowledge dissemination through a data base involving experimental and numerical results together with numerical kernels.

The ETMA project was co–financed by the Industrial and Materials Technology Programme Area 3 Aeronautics under the Third Research Framework Programme of the European Union (EU).

The ETMA Consortium

Participants to the consortium were twenty research organisations and universities from eleven European countries and five industrial endorsers:

ETMA Partners

INRIA, Sophia–Antipolis/Rocquencourt (F)	LSTM, Erlangen (D)
IST, Lisbon (P)	Ecole Centrale de Paris (F)
Imperial College, London (UK)	AFM, Lyngby (DK)
Aero–Delft (NL)	LAT, Patras (GR)
NTUA–LTT, Athens (GR)	NTUA–LA, Athens, (GR)
University of Sevilla (E)	UMIST, Manchester (UK)

VUB, Brussels (B)
IMF, Toulouse (F)
Ecole Centrale de Lyon (F)
FFA, Bromma (S)

Politecnico, Torino (I)
INSA, Rouen (F)
IMST, Marseille (F)
KTH, Stockholm (S)

Industrial Endorsers

Daimler–Benz Aerospace Airbus (D)
Dassault Aviation, Saint–Cloud (F)
SNECMA, Villaroche (F)

Alenia, Torino (I)
Hispano–Suiza, Le Havre (F)

Coordination

The responsibility for the coordination of ETMA was placed on A. Dervieux (INRIA, Sophia–Antipolis), F. Durst and I. Lekakis (LSTM, Erlangen)

ETMA Project Overview

The ETMA project involved the following activities:

Model Testing and Computations

Such as testing of different second moment models (wall treatment: wall functions or integration to the wall) in a large range of incompressible flows: homogeneous flows, free turbulent flows, plane channel flow, boundary layers, sink flows(including laminarization), oscillating boundary layers and channel flows, 3–D turbulent boundary layers, round impinging jets, backward facing step, tube bundle.

There were also studied second–order closure predictions of separated flows with and without combustion, flow around a disk with central orifice, the prediction of a turbulent premixed V–shape flame stabilized behind a cylindrical rod by the k–ε and k–λ models for turbulence and the coherent flame model for combustion, the prediction of incompressible sudden expansion, including near–wall effects, heat and particle transport, using the k–ε and Reynolds stress models.

Furthermore, k–ε, k–τ and RSM predictions were also investigated for a wall mounted obstacle in duct (2–D) and fence–on–wall, bump–on–wall flow geometries.

The prediction of incompressible turbulent wake flow behind a flat plate and a boundary layer both under adverse pressure gradient was studied by using the k–ε, k–λ and Reynolds–stress models and k–ε, k–τ and RSM predictions were applied to low and transonic Mach numbers flows over airfoils.

At last, spatial DNS was performed on a zero–pressure gradient boundary layer and computations of the two–point correlations needed for model developments at LSTM–Erlangen (k–λ etc.).

A special focus was put on *compressible models of industrial interest*; new models, variants, and assessments have been considered for the improvements for zero– and two–equation and RSM steady models.

Four important aspects of this domain have been specially addressed, the effects of compressibility, the introduction of anisotropy in two–equation models, the low–Reynolds treatment of compressible boundary layers, the application of statistical models to unsteady flows by phase averaging and homogenisation.

Compressibility effects are important on free shear flows; from the Stanford meeting, errors in mixing layers have been pointed out; shock–boundary–layer interactions result in separation difficult to predict.

The anisotropic and nonlinear two–equation models constitute a compromise between basic two–equation models and second moment models.

Low–Reynolds boundary treatment (without multi–layer models) was an important issue because of the impact of accurate modelling in separation predictions.

Experimental Test Cases

A series of experimental test cases has been conducted. They are classified in three subgroups: internal and external separated flows (sudden pipe expansion, wall–mounted prismatic obstacle, sudden expansion with transport across streamline, PTV–studies of sudden expansion with water, fence–on–wall and bump–on–wall), 2–D and 3–D boundary layers (2–D boundary layer under pressure gradient, pipe flow and near–wall measurements, 3–D boundary layer, swept concave/convex surface, premixed propane/air flame in rectangular duct), free shear flows and wake/boundary layer interactions (mixing layer/boundary layer interaction, two–point correlations in round free jets, effects of pressure gradient on wakes, compressible/supersonic free shear layers).

New Numerical Methods

The project has contributed to the building, validation, and assessment of new numerical methods involving both approximation and solution aspects.

For approximation, new schemes applying unstructured triangulations were built and applied to most of the compressible flows cases considered, as will be observed by the reader of this book.

The solution of high speed turbulent flows necessitate the application of sophisticated algorithms; crucial issues are robustness and efficiency. The ETMA consortium proposed three families of algorithms, fast explicit algorithms, implicit ones, multigrid ones.

Conclusive comments on these numerical aspects will be found in the global synthesis in the last chapter of this book.

The ETMA Workshop

The ETMA Workshop on turbulent modeling for compressible flow arising in aeronautics was held at UMIST, Manchester, U.K. in November 14–16, 1994 .

The workshop was open to any people proposing a contribution. About 60 contributions were selected from extended abstracts from criteria related to the novelty of models or of numerical methods, the accuracy of the results (authors were strongly encouraged to perform mesh convergence).

For each test case, one or two partners of ETMA took the responsibility not only of test case and output formats definition but also of meshes availability, assistance to contributors, synthesis of the contributions, and pre–edition of finally written contributions. They are mentioned in the following short description of test cases.

Test Cases for Analysis

Test Case 1: Supersonic mixing layer
(Resp. M. Braza (IMF–Toulouse), T. Chacon (U. of Sevilla)).

Steady evolution from supersonic boundary layer to a mixing layer (Experiment : Barre, Quine, Dussauge 1994, $M_c=0.62$) unsteady; in the same flow information on the frequency of the energetic turbulent structures and their convection velocity.

Test Case 2: Supersonic rearward facing step
(Resp. K. Giannakoglou (NTU–Athens))

The experiments refer to axisymmetric configurations (Roshko, Thomke 1966, $2 \geq M \geq 4$) to avoid three dimensional perturbations due to side walls.

Test Case 2C: Incompressible flows with separation
(Resp. P. Larsen (AFM, Lyngby))

This involves a series of cases for which the ETMA project has obtained new measurements.

Test Case 3: Turbulent boundary layers
(Resp. Ch. Hirsch, E. Shang, (VU–Brussels))

It was proposed to compute several cases : the reference cases of the subsonic zero pressure gradient flat plate boundary layer (Klebanoff 1954), an accelerated boundary layer (sink flow direct simulation, Spalart 1986), and a boundary—layer in an adverse pressure gradient (Samuel et Joubert 1974). The calculations were extended to supersonic zero–pressure gradient boundary layers on adiabatic wall ($M \leq 4$).

Test Case TC4: Compression ramp flow
(Resp K. Giannakoglou (NTU–Athens))

In the retained experiment (Princeton data), the Mach number is 2.84; several ramp angles from 8 to 24 degrees are explored, which cover the range of non–separated interactions, incipient separation, and full separation.

Test Case TC5: Two dimensional transonic bump
(Resp. A. Loyau, D. Vandromme (INSA–Rouen))

This flow is one of the classical cases for shock/turbulent boundary layer interactions as described by the experiments conducted at ONERA by Délery and his group.

Test Case TC6: Shock reflection on a flat plate
(Resp. R. Arina (Poli–Torino))

In this flow at Mach 2 (Délery 1992), the incident shock wave is imposed from the external flow, and hence it does not depend on the incoming turbulence.

Test Case TC8: Flow around airfoil (steady)
(Resp. L.J. Johnston (UMIST, Manchester))

The selected experiments are those performed by Cook, Mc Donald and Firmin (1979) for the RAE 2822 aerofoil, and by Bucciantini, Oggiano, Onorato (1979) for a MBB–A3 aerofoil at Mach numbers around 0.75.

Test case TC8 bis: Flow around airfoil (unsteady)
(Resp. M. Braza (IMF–Toulouse), S. Tsangaris (NTU–Athens))

In a limited range of Reynolds and Mach numbers in the transonic regime, there is a competition between the trailing edge and the shock induced separation, leading to periodic flow oscillations. The present case, based on the experiments of Mc Devitt, Levy Jr. and Deiwert (1976), was proposed to test the ability of numerical methods to predict this unsteadiness.

Finally discarded from the ETMA workshop were two extra test cases relying on (1) a curved mixing layer experimented by Castro I.P., and Bradshaw P., and (2) a compressible circular jet.

General Organisation of the Book

The book is organised with respect to test cases, a test case being either one single flow or a set of similar flows.

For each test case, a set of contributions was proposed. For each contribution a short scope of the model and numerics is given; emphasis is then put on discussing the computation conditions and results. Besides the different contributions, an accurate description of each test case is given, with some motivation for its choice, and the outputs expected from contributors are specified, in order to allow to use this book for the validation of new codes. Also, a synthesis yields to the reader the main keys for the interpretation of results and the identification of the best ones.

Acknowledgements

The scientific editors wish to express their warmest thanks to: Catherine Juncker and Monique Simonetti, Bureau des Relations Extérieures de l'INRIA for their vigourous and efficient organisation of the workshop and Leslie L. Johnston for his contribution to the local organisation at UMIST.

Our thanks also go to Françoise Trucas–Martin for taking care of the assembling of this book, and to Christian Olivier who was responsible of the ETMA Data Base.

The test cases responsibles and in particular P. Larsen for the set of incompressible test cases, made important pre–edition work and we express our gratitude to them.

Peter Bradshaw and Jean Délery brought important help in test case definition and information.

The work of the ETMA project was supported by EU through the Industrial and Materials Technology Programme. Particular thanks go to Dietrich Knoerzer, who was a very open–minded and helpful partner at the European Commission and contributed to the success of the ETMA program and workshop. The workshop meeting was also supported by the EU–programme COMETT II through UETP ERCOFTAC, which is gratefully acknowledged.

Sophia–Antipolis, November 1997

Alain DERVIEUX
Marianna BRAZA
Jean–Paul DUSSAUGE

CONTENTS

CONTENTS (continued)

CONTENTS (continued)

CONTENTS (continued)

CHAPTER 1 : MIXING LAYERS

Test Case 1: Supersonic mixing layers

J.-P. Dussauge
I.R.P.H.E., 12, Avenue Général Leclerc
F-13003 Marseille

Summary

Mixing layer flows are briefly presented, with emphasis on the influence of compressibility on the turbulence of such free shear flows. It is proposed to consider three problems found in the analysis of compressible mixing layers. The first one is the determination of asymptotic properties of fully developed mixing layers. The second one is the evolution from a boundary layer flow at the trailing edge of a plate towards an asymptotic mixing layer, and the last one is the determination of the development of large scale structures. Pertinent parameters of the problems are recalled. An assessment of the accuracy of the experimental data is given. Finally the instructions for computing these flows are given.

1. Presentation of the test case

The mixing layer is the flow which develops between two parallel streams of unequal velocities. Between the two streams, large eddies are formed, and in the subsonic regime, these eddies prove to be very efficient to mix the properties of the two flows. At high speeds, the experimental evidence is that there is an inhibition of the mixing, probably due to a reduction of the importance of the coherent structures found in subsonic flows: the effect of compressibility has macroscopic consequences (Sirieix, Solignac 1968, Papamoschou Roshko 1988). The mixing layers are therefore a very good test case for turbulence models far from walls, and also to calibrate models for compressible turbulence.

Several cases can be considered. The simpler one is the case of fully developed, self similar mixing layers in which conditions are well identified and for which compressibility effects are unambiguous. Some models can be efficient for mixing layers in which compressibility effects are strong, but not for boundary layers where the influence of compressible

3

turbulence is not felt for M<5. Therefore, a second test case, more general than the previous one, is the development of a mixing layer from the trailing edge of a separating plate: in this flow, turbulence evolves from a boundary layer (weakly compressible turbulence) to a mixing layer (effective compressibility effects). A last point of interest is the determination of large scale ("coherent") structures. It is known that such eddies are present in mixing layers, and contribute efficiently to the mixing. The influence of compressibility is to alter their growth. The simulation of these large scale structures requires at least for their validation to be compared to experimental statistical information on these structures, such as the spectra, their characteristic length scales or the coherence function of the fluctuations and their group velocity, and is a first step for the validation of two-point turbulence closures.

The studies on compressible mixing layers show that the properties of the supersonic mixing layers are different from the properties of the layers with variable density at subsonic speeds. It is suggested that these properties can be represented by a single parameter, the convective Mach number M_c. This Mach number is based as follows: the large eddies are supposed to be convected at the speed U_c. An observer moving with the large eddies would feel only the compressibility effects associated with the velocity difference between the eddies and the external flows. Two convective Mach numbers, relative to each of the external flows can be defined In simple cases, when the turbulent structures do not induce shock waves in the external flows, these two Mach numbers are equal. The resulting convective Mach number is defined as $M_c = (a_1 U_2 + a_2 U_1) / (a_1 + a_2)$. Experiments have shown that this parameter is useful for correlating the data in compressible mixing layers, for example the rate of spread (Papamoschou & Roshko 1988) and the level of turbulence (Barre, Quine, Dussauge 1994). It will be used in the present specifications

The assessment of uncertainties is generally an uneasy task in such flows. Essentially, mean quantities are measured with an accuracy of the order of ±1%. Note that in classical measurements with Pitot- and pressure probes to determine the Mach number and thermocouple probes to measure total temperature, if these three quantities are determined with ±1% accuracy, for M=2, the accuracy on mean temperature and mean velocity is about ±2%. Derived quantities such as spreading rate are generally determined with an accuracy not better than ±10%. The accuracy on Reynolds stresses (or on turbulent kinetic energy) is certainly not better than ±20%, whatever the method.

2. Instructions for the computation of Test Case 1

2.1 Definition of the sub- test cases:

As suggested in the presentation, this Test Case is divided into three sub-test cases: TC011 *Asymptotic supersonic mixing layer*, TC012 *Evolution from an initial mixing layer to an*

4

fully developed mixing layer (steady) and TC013 Mixing layer (unsteady) .TC013 is the same flow as TC012, but only the large scale structure and their unsteady aspects are considered. In the present document, the precise conditions for computing these flows are described.

2.2 Definition of the variables:

A sketch of the flow is given in figure 1.

ΔU is the velocity difference across the flow: $\Delta U = U_1 - U_2$.

δ is the mixing layer thickness. It may be defined as in the 1980 Stanford Conference, $\delta = y_1 - y_2$.
  y_1 is the ordinate where $U = U_2 + (0.9)^{\frac{1}{2}} \Delta U = U_2 + 0.95 \Delta U$
  y_2 is the ordinate where $U = U_2 + (0.1)^{\frac{1}{2}} \Delta U = U_2 + 0.316 \Delta U$.

δ' is its longitudinal derivative (spreading rate) $\delta' = d\delta/dx$.

δ'_0 is the rate of spread of the subsonic layer with the same velocity- and density ratios. An empirical correlation for δ'_0 is

$$\delta'_0 = (\delta'_{ref}/2)\Delta U/U_c = (\delta'_{ref}/2)(1-r)(1+s^{1/2})/(1+rs^{1/2}) \qquad (1)$$

where $r = U_2/U_1$ and $s = \rho_2/\rho_1$. δ'_{ref} is the value of the rate of spread of the low speed half-jet with constant density, i.e. for $M=0$, $U_2=0$ and $s=1$. The consensus value recommended by the Stanford Conference for δ'_{ref} is $\delta'_{ref} = 0.115$ (Kline et al. 1980).

$F = \delta'/\delta'_0$ is the normalized rate of spread. F is supposed to be a function of the convective Mach number M_c only.

$-u'v' / \Delta U^-$ is the dimensionless friction. $-u'v'$ is here the maximum friction in a section.

M_c is the convective Mach number. $M_c = \Delta U/(a_1 + a_2)$ when the same gas is used in both of the external streams. a_1 and a_2 are the sound speeds in flows 1 and 2. Otherwise,

$M_c = (U_1 - U_c)/a_1 = (U_c - U_2)/a_2$ with $U_c = (a_1 U_2 + a_2 U_1)/(a_1 + a_2)$.

$y^* = (y - y_1)/\delta$ is the normalized ordinate.

$U^* = (U - U_1)/(U_1 - U_2)$ is the normalized velocity.

3. Computation of TC011

In TC011; it is proposed to determine the rate of spread of an asymptotic mixing layer; here, asymptotic properties mean that the influence of the upstream conditions is no more felt, and that self-similar profiles are obtained. It is assumed that the influence of compressibility on the rate of spread can be observed on mean quantities through the function F(Mc) defined in the previous paragraph. This function for the thickness δ defined by the Stanford Conference is given in figure 2:

5

The first goal is to check that the computations match correctly these data. Of course, it will be interesting to have many computations describing F in great detail. It was felt however that precise comparisons for correctly chosen values of M_c would be the only firm basis for further discussion. Three values of M_c have been retained: Mc= 0.45 which should be practically the same as for $M_c = 0$; M_c=0.65 and 1.0. The values are:

M_c	$F(M_c)$
0.45	1
0.65	0.77
1	0.5

For each case, the computation of the subsonic mixing layer with the same velocity- and density ratio r=U_2/U_1 and s=ρ_2/ρ_1 should be performed, in order to check the results of the code at low speeds, and to compare them to formula 1 quoted hereabove.

The conditions on r and s to obtain the prescribed values of M_c is let to the choice of the computor. However it may be remarked that the simplest way to perform such calculations is to run the codes in the case when ρ_2=ρ_1=cst. As pressure is constant throughout the layer, T_1=T_2, and a_2=a_1. In this case, Uc=(U_1+U_2)/2 and M_c =(U_1-U_2)/2a. It should be checked however that the codes give the same function $F(M_c)$ when the same value of the convective Mach number is obtained in a different way, for example, when total enthalpy is constant, as in many experiments. Note that the condition of constant total enthalpy implies that s<1, and determines M_1 and M_2 for prescribed s and M_c. The relations are:

Constant total enthalpy
$$s = \frac{1+(\gamma-1)M_2^2/2}{1+(\gamma-1)M_1^2/2}$$

Mach numbers
$$M_1 = M_c \frac{1+s^{1/2}}{s^{1/2}(1-r)}$$
$$M_2 = M_1 r s^{1/2}$$

If two other values are known, for example the value of velocity in one of the external flows and of pressure, all the conditions in the outer flows are determined.

Other ways to get the values of M_c (non constant total enthalpy or gases with different γ in flows 1 and 2) are of course possible and interesting. They make it possible to check if M_c is a similarity parameter of the problem, and the control of this point will be welcome. They lead however to more complicated computations, and therefore are not mandatory. The choice of combination supersonic/ supersonic flows and supersonic/ subsonic flows is to be decided by the computors. However their choice should be indicated.

In these flows, the mean velocity profiles will be given and compared to the data shown in figure 3. Note that this velocity profile (Ikawa, Kubota 1975) whose shape is in agreement with subsonic velocity profiles, has been obtained in a case of co-gradient, s<1. In such case, the velocity and density gradients have the same sign. In the case of cross-gradients, s>1, the shape of the velocity profiles seems somewhat altered. However, this case is only partially documented, even in subsonic flows, and is not retained here for the shape of the mean velocity profiles. For this reason, it is recommended to test the codes for s<1, which corresponds to the majority of the experiments in supersonic flows. When available, the level of turbulent friction and of k will be compared to the compilation presented in figure 4, and discussed in the next section.

A last condition on Reynolds number is given: the Reynolds number $\delta\Delta U/\nu$,where ν is the average value of the kinematic viscosity between flows 1 and 2, should be larger than 10^5.

4. Initial profiles of mean velocity, k , ε and Reynolds stresses

Some reasonable guesses have to be made on the initial profiles of mean velocity, k, ε and possibly Reynolds stresses. The shape of the mean velocity is not very sensitive to Mach number and can be taken as in subsonic flow. A possible example is given in figure 3.

There is no real evidence that the shape of the profiles of k and ε should be very different at low and high speeds. At the opposite, their level is significantly altered. A compilation of maximum shear stress data is given in figure 4. An analysis was proposed (Barre et al. 1994) which suggested that, as a first approximation, the maximum of shear stress varies with M_c like the spreading rate. As little is known on the Reynolds stress anisotropy, it may be assumed that k, and the other Reynolds stresses assume the same variations. The same analysis suggested also that this level of turbulent kinetic energy can also depend on s. However, even if there are some experimental evidence of that in subsonic flows, the accuracy of measurements is not good enough to derive reliable empirical laws. It may be remarked that the data supporting the proportionality of k (or of the shear stress $-\overline{u'v'}$) to the rate of spread have been obtained in experiments where s is around 0.6. Checking the level of k (or of ') should be made for values of s in this range. $-\overline{u'v'}/\Delta U^2 = 0.0095$ is suggested as a consensus value for low speed constant density mixing layers.

Finally, the initial profiles of ε can be determined by assuming that production practically balances dissipation. P. Bradshaw recommends to use production=ε as an approximation for starting the computation and getting a reasonably rapid convergence. From an empirical basis referring to low speed mixing layer studies it may be preferred ε= 70% of production which is essentially the same approximation, but perhaps closer to the expected solution.

References

Barre S., Quine C., Dussauge J.P. Compressibility effects on the structure of supersonic mixing layers. *Journal of Fluid Mechanics.* 259: 47-78, 1994.

Ikawa H., Kubota T. Investigation of supersonic turbulent mixing layer with zero pressure gradient. *AIAA Journal*, 13 (5):566-572, 1975.

Kline S.J., Cantwell B.J., Lilley G.M. Ed., "Complex turbulent flows, Computations, Experiment", Vol.1, p. 367, Stanford University, 1981.

Papamoschou D., Roshko A.. The compressible turbulent shear layer: an experimental study. *Journal of Fluid Mechanics.* 197:453-477, 1988.

Sirieix M., Solignac J.L., "Contribution à l'étude expérimentale de la couche de mélange turbulente isobare d'un écoulement supersonique", Symp. on Separated Flows, AGARD Conf., Proc. 4, 241-270, 1968.

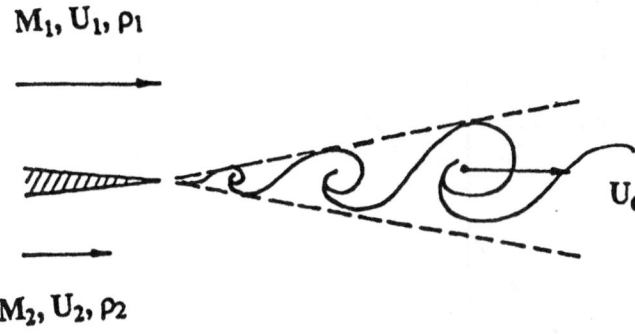

Fig. 1: Sketch of the mixing layer flow.

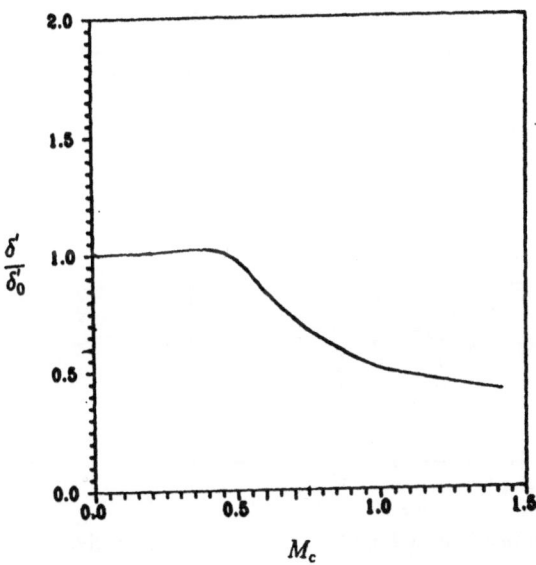

Fig. 2 Normalized spreading rate as a function of convective Mach number

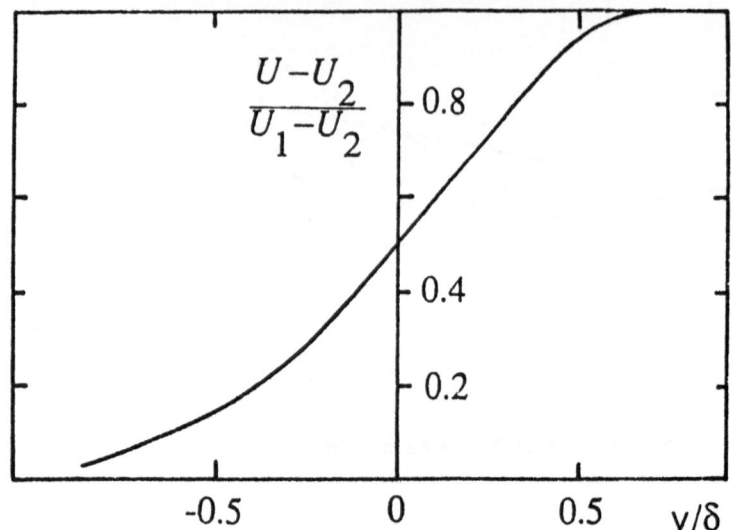

Fig. 3: Example of mean velocity profile, for s<1. Adapted from Ikawa & Kubota (1975).

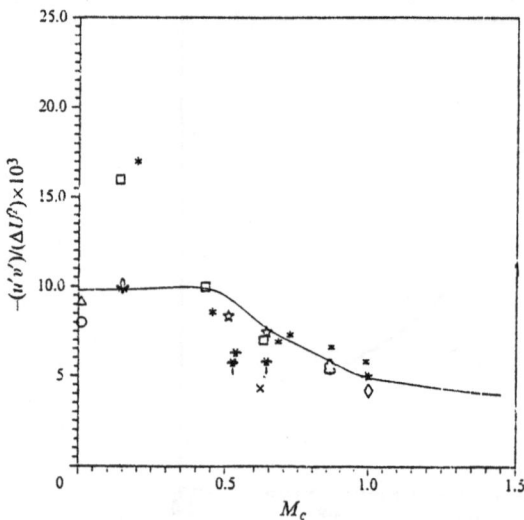

Fig. 4 : Compilation of the dimensionless turbulent friction vs. convective Mach number. Reproduced from Barre et al. 1994.

NUMERICAL SIMULATION AND MODELING OF AN UNSTEADY SUPERSONIC MIXING-LAYER FLOW

M. Braza, F. Hanine

Institut de Mécanique des Fluides de Toulouse, Unité Mixte de Recherche
C.N.R.S n° 5502, .Av. du Prof. Camille Soula
31400 Toulouse Cedex, France

Summary

In the present work we perform the prediction of the inherently unsteady turbulent flow in a supersonic mixing layer, by using zero-equation turbulence models, especially modified to allow the appearance of the coherent structures in the mixing layer. It is shown that a good agreement with the experiment can be obtained by this category of models, concerning the mean flow parameters, the growth rate of the mixing layer and the frequency of the coherent eddies.

1. Introduction

The general context of this work is the development of turbulence models for high-speed flows, characterized by the appearance of inherently unsteady phenomena. In fact, a wide category of strongly shear flows favorize the onset and the amplification of instabilities, owing to the inflexional shape of the streamwise velocity profiles and to a sufficiently high level of the Reynolds number, beyond a critical value. Owing to a high vorticity level, the amplification of instability waves lead to the appearance of organized coherent structures, having *distinctive* characteristic length and time scales, comparing to the co-existing fine-scale turbulent motion. Owing to this, the energy spectrum of these flows presents predominant frequency peaks interspersed in continuous frequency windows and has an overall slope which may be very different from the one of flows characterized by a statistical equilibrium. We remind that the majority of turbulence models have been conceived for flows in statistical equilibrium, according to the statistical theory of Kolmogorov. For this reason, there is a need to reconsider the hypotheses of these models when they are used in the case of flows in non-equilibrium. In this context, we consider firstly the algebraïc models of turbulence. Despite the lack of this category of models to take into account the physics of the anisotropic properties of the fine-scale turbulence in the category of strongly shear flows, it is worthwhile to apply them as a first step to the prediction of high-speed flows with coherent structures, thanks to the simplicity in their implementation. Hence, in this study we analyse the performances of the algebraïc model suggested by Cebeci & Smith [1], [2] and by Baldwin & Lomax [3]. In respect to the above discussion, we study in the present paper the performances of a modification of the length scale of this model that we introduce by using a local vorticity function, in order to reduce the excessively high level of the turbulent viscosity, produced usually by the Baldwin-Lomax model. These aspects are based in the work by Braza & Noguès [4], and are discussed in a next paragraph.

Among the high-speed flows with coherent structures, we have chosen the supersonic mixing-layer flow as defined in the experiment by Barre, Quine & Dussauge [5]. We recall here the thermodynamic parameters governing this experiment:

Table 1 Upstream thermodynamic parameters

	M	U(m/sec)	Tt(K)	T(K)	ρ(Kgm-3)
Flow 1:	1.79	481	296	180	0.208
Flow 2:	0.30	101	291	286	0.130

The experimental visualizations and the spectral analysis of the above studies report the appearance of unsteady coherent structures, whose the frequency does not remain constant downstream. Hence, the objectives of the present numerical simulation are to predict the natural unsteadiness of the flow, the predominant frequencies and the mean flow characteristics (mean velocities, spreading rate of the mixing layer) by using the zero-equation turbulence model.

We discuss in the following the outlines of the numerical method and of the turbulence modeling, implemented in the numerical code, as well as the choice of the optimum numerical parameters in order to ensure a *time-accurate* computation. In the last paragraph we present the results obtained in this study.

2. The governing equations and the numerical method

$$\frac{\partial \rho}{\partial t} + \frac{\partial (\rho u_k)}{\partial x_k} = 0 \tag{1}$$

$$\frac{\partial (\rho u_i)}{\partial t} + \frac{\partial (\rho u_i u_k - \sigma_{ik})}{\partial x_k} = 0 \tag{2}$$

$$\frac{\partial (\rho E)}{\partial t} + \frac{\partial (\rho E + p)}{\partial x_k} u_k = \frac{\partial (\tau_{ik} u_k - q_k)}{\partial x_k}. \tag{3}$$

σ_{ik}, τ_{ik} et q_k represent respectively the dynamic tensor, the tensor of viscous stresses and of the heat fluxes:

$$\sigma_{ik} = - p\delta_{ik} + \tau_{ik}$$

$$\tau_{ik} = \mu S_{ik} = \mu \left[\left(\frac{\partial u_i}{\partial x_k} + \frac{\partial u_k}{\partial x_i} \right) - \frac{2}{3} \delta_{ik} \left(\frac{\partial u_l}{\partial x_l} \right) \right]$$

$$q_k = - \lambda \, \partial T / \partial x_k$$

with μ the molecular viscosity, λ the thermal conductivity of the fluid and the deformation tensor, S_{ik}.

2.1 FAVRE AVERAGING

According Favre [8], [9], every unknown function F is decomposed as follows:
$$F = \widetilde{F} + f''.$$

$\widetilde{F}$ designates the mean of the function F, compensated by the density, f'' designates the fluctuation. These two quantities verify the relations:

$$\overline{\rho F} = \overline{\rho} \, \widetilde{F} \text{ et } \overline{\rho f''} = 0.$$

2.2 EQUATIONS OF MOTION ACCORDING TO THE FAVRE AVERAGING

By expressing the different quantities in the equations of motion (1)-(3), by the decomposition of Reynolds [6], [7] for the density ρ and for the pressure p and by using the Favre decomposition for the velocity u_k and the total energy E. Hence, we obtain the following system of equations where the fluctuations of the molecular viscosity and of the thermal conductivity are neglected:

$$\frac{\partial \overline{\rho}}{\partial t} + \frac{\partial (\overline{\rho} \, \widetilde{u_k})}{\partial x_k} = 0 \tag{4}$$

12

$$\frac{\partial(\overline{\rho}\ \widetilde{u}_i)}{\partial t} + \frac{\partial(\overline{\rho}\widetilde{u}_i\ \widetilde{u}_k + \overline{\rho u"_i u"_k} - \overline{\sigma_{ik}})}{\partial x_k} = 0 \tag{5}$$

$$\frac{\partial(\overline{\rho E})}{\partial t} + \frac{\partial\left(\overline{\rho}\ \widetilde{E}\ \widetilde{u}_k + \overline{\rho\ E"\ u"_k} + \overline{\rho}\widetilde{u}_k + \overline{\rho u"_k}\right)}{\partial x_k} = \frac{\partial\left(\overline{\mu\left(S_{ik}\ \widetilde{u}_i + \overline{S_{ik}u"}_i + \lambda\frac{\partial\widetilde{T}}{\partial x_k}\right)}\right)}{\partial x_k}. \tag{6}$$

We remark that equation (4) has a form which is analogue to that of equation (1), because the density-velocity correlation $\overline{\rho\ u"}_k$, are zero.

In equation (5) the non-linearity of the convection terms introduce the Reynolds stress correlation (5), $\overline{\rho\ u"_i u"_k}$..

By defining the tensor of the stresses: $\overline{\Sigma_{ik}} = \overline{\sigma_{ik}} - \overline{\rho\ u"_i u"_k}$, equation (5) is written:

$$\frac{\partial(\overline{\rho}\ \widetilde{u}_i)}{\partial t} + \frac{\partial(\overline{\rho}\widetilde{u}_i\ \widetilde{u}_k)}{\partial x_k} = \frac{\partial\overline{\Sigma_{ik}}}{\partial x_k}. \tag{7}$$

In the same way, equation of the total energy (6), makes appear the terms: $\overline{\rho\ E"u"}_k$ et $\overline{pu"}_k$ that have to be explicited. The ideal gaz approximation allows writing :

$$p = (\gamma-1)\,\rho\,c_v\,T = (\gamma-1)\,\rho\,c_v(\widetilde{T} + \theta"). $$

The term $\overline{pu"}_k$ is then written :

$$\overline{pu"}_k = (\gamma-1)\,c_v\,\overline{\rho\ \theta"u"}_k. \tag{8}$$

We have also :

$$\rho\ E = \rho\ c_v T + \rho\ \frac{u_i u_i}{2} = \rho\ \widetilde{E} + \rho\ E" \tag{9}$$

and $\rho\ E\ u_k = \rho\ c_v T u_k + \rho\ \frac{u_i u_i}{2}\,u_k. \tag{10}$

By taking the mean of the equations (9) et (10), we obtain :

$$\overline{\rho}\ \widetilde{E} = \overline{\rho}\ c_v\widetilde{T} + \overline{\rho}\ \frac{\widetilde{u}_i\widetilde{u}_i}{2} + \rho\ \frac{\overline{u"_i u"_i}}{2}$$

$$\overline{\rho}\ \widetilde{E}\ \widetilde{u}_k + \overline{\rho E"u"}_k = \overline{\rho}\ c_v\widetilde{T}\widetilde{u}_k + \overline{\rho c_v\theta"u"}_k + \overline{\rho}\ \frac{\widetilde{u}_i\widetilde{u}_i}{2}\ \widetilde{u}_k + \overline{\rho\ u"_i u"_k}\ \widetilde{u}_i + \rho\ \frac{\overline{u"_i u"_i}}{2}\ \widetilde{u}_k + \rho\ \frac{\overline{u"_i u"_i u"_k}}{2}.$$

By combining the above two relations we have :

$$\overline{\rho E"u"}_k = \overline{\rho c_v\theta"u"}_k + \overline{\rho\ u"_i u"_k}\ \widetilde{u}_i + \rho\ \frac{\overline{u"_i u"_i u"_k}}{2}. \tag{11}$$

The equation of averaged total energy (6) is written in respect to relations (8) and (11) :

$$\frac{\partial(\overline{\rho E})}{\partial t} + \frac{\partial\left(\overline{\rho}(\widetilde{E}+\overline{p})\widetilde{u}_k + \overline{\rho u"_i u"_k}\widetilde{u}_i + \frac{\overline{\rho u"_i u"_i u"_k}}{2} + \gamma c_v\overline{\rho\theta"u"}_k\right)}{\partial x_k} = \frac{\partial\left(\overline{\mu(\overline{S_{ik}}\widetilde{u}_i + \overline{S_{ik}u"}_i + \lambda\frac{\partial\widetilde{T}}{\partial x_k})}\right)}{\partial x_k}. \tag{12}$$

2.3   *TERMES TO BE MODELED AND DIFFERENT CLOSURE LEVELS*

The understanding of the additional terms which appear in the equations (5, 12), especially the correlation $\overline{\rho\ u"_i u"_k}$, $\overline{\rho\ \theta"u"}_k$ and $\overline{\rho\ u"_i u"_i u"_k}$, is essential for determining the characteristic quantities of the mean flow. The presence of these terms needs the use of phenomenological laws or of additional transport equations, which have as a purpose to express these unknown quantities as a function of the principal unknown quantities to be solved and therefore to close the system of the averaged equations for the turbulent flow.

2.4   PHASE-AVERAGING

The statistical modeling of turbulence, with different closure levels, based either on turbulent viscosity concept or second order modeling gives relative satisfaction only in the case of fully developed turbulence, under Kolmogorov type spectral equilibrium assumptions. Now, in many industrial situations, for inner as well as for outer flows, the onset of turbulence motion occurs simultaneously with organized structures. These structures are linked with geometrical conditions of the flow domain. In order to take into account this kind of structures, we suggest the use of the *phase-average decomposition* of every unknown quantity [12], [14], [23], [4]. A more detailed discussion is given in our paper concerning the test-case TC8bis. It is briefly reminded here that the methodology based on the phase-averaged Navier-Stokes equations makes the distinction between the structures to be predicted and those to be modelled upon the physical criterion of the organized or random character of the structures and not on their size. Furthermore this approach, called Organized Eddy Simulation (OES), is not inherently 3-D comparing to the LES approach.

3. The turbulence model

We recall briefly that in order to close the system of equations, two categories of turbulent models are currently used:

i) Models using the concept of turbulent viscosity μ_t (first - order closure).

ii) Models using the transport equations of the Reynolds stresses (second order closures). A detailed numerical study of compressible, turbulent boundary-layer flows can be found in the doctoral thesis of Hanine [18]. In this work, we detail the first-order closures and we focuss our interest to the zero-equation models.

According to the eddy-viscosity assumption, the Reynolds stresses are expressed as a function of the tensor of mean deformation, $\widetilde{S}_{ij}$, through the coefficient of the eddy-viscosity, μ_t . This relation, given by (13) was originally suggested by Boussinesq [19]. It is conceptually based on an analogy between the behaviour of the turbulent stresses and of the viscous stresses.

$$ -\overline{\rho\, u''_i u''_j} = \mu_t\left[\left(\frac{\partial \widetilde{u}_i}{\partial x_j} + \frac{\partial \widetilde{u}_j}{\partial x_i} - \frac{2}{3}\,\delta_{ij}\frac{\partial \widetilde{u}_l}{\partial x_l}\right)\right] - \frac{2}{3}\,\delta_{ij}\,\overline{\rho}\,\widetilde{k}\,. \tag{13}$$

The modeling of μ_t passes through the introduction of a velocity and a length scale, u and l. Hence, $\mu_t = \overline{\rho}\, u\, l$. Different closure levels are then established according the way of determining these two scales. Therefore, three main categories of models have been developed: Zero equation models, one equation and two equation models.

3.1   THE ALGEBRAÏC MODEL

In this category of models, the turbulent viscosity is given by a phenomenological law, based on theoretical developments and on turbulent experiments. Although the approaches available for algebraïc models suppose a conditionning of the variations of the turbulent stresses on those of the mean velocity gradients, these models merit to be examined owing to their simplicity in implementation in flows of complex geometry and in three-dimensional applications, interesting the industrial domain.

14

The turbulent viscosity is expressed as a function of mean velocity gradients, through a length scale l_m which is the mixing length, which is defined by phenomenological laws. The mixing length in its physical meaning represents the characteristic length between two shear layers of the turbulent flow, along which the exchange of energy through the turbulent motion takes place.

In the seventies, more sophisticated algebraïc models have been developed, taking into account two different expressions of the mixing length and of the eddy viscosity for internal and external regions of boundary layer configurations. We present in more details in the paper discussing the test-case TC8bis the algebraïc model by Cebeci & Smith and Baldwin & Lomax. We comment here the modification introduced by Braza & Noguès, suggesting to modify the length scale by a dimensionles vorticity function:

$$\mu_{te}(y) = \alpha \, \bar{\rho} \, C_{cp} \, F_{sillage} \, F_{Kleb}(y)/\omega/. \tag{14}$$

The modification of the length scale

In the case of the modeling of unsteady flows with coherent structures, there is a need to model differently the length scale in the outer region, owing to a different energy cascade from the external flow to the different classes of vortices, due to the fact that a considerable amount of the externally supplied energy is devoted to sustain the motion of the coherent structure and not only to produce the energy cascade towards the fine scale motion.

The classic law suggested by Cebeci and Smith [1] in this region is based on physical experiments of Klebanoff [1954] for a flat plate boundary layer with zero pressure gradient, which is not the case in our study. In that work it is shown that the dimensionless eddy-viscosity becomes nearly constant across the main outer part. This allowed to set the following relation:

$$v_t = a_1 u_\tau \delta . \tag{15}$$

because the mixing length is approximately constant, $1 / \delta = $ const.

with :   * δ, the boundary-layer thickness.

   * a_1, an experimental constant between 0.06 and 0.075.

   * u_τ, the shear stress velocity.

In that work, using experimental data for determining the constants, the authors deduced the classic law . It is also recalled that l is the distance that the turbulence eddies travel mainly in the direction perpendicular to the wall. However this theory, based on Prandtl's mixing layer assumption, is valid under the hypothesis that the fluctuations of v velocity component are analog to the ones of u component. This is valid in flows without unsteady organized coherent structures.

In our case, we have to reconsider the scale of this law, because the unsteady vorticity generated by the inlet velocity profile, makes the alternating vortices and other organized vortices of the shear layer to be convected downstream, and to interact non-linearly with the small-scale turbulence. On the other hand, Cebeci and Smith pointed out that the length and the velocity scales used to normalize the eddy-viscosity in that region *are not the only possible characteristic scales*. For these reasons we suggest another length scale, $l|\omega|$, in the law of the phase-averaged eddy-viscosity. This represents a mixing length that the smaller-scale turbulent eddies travel as they are swept by the larger-scale organized eddies: in fact, as it is shown in numerous experimental studies (see collected experimental visualisation works reported by Van Dyke [22]) and also as discussed in the numerical study by Braza &

Noguès, smaller-scale eddies are mixed progressively with the adjacent organized eddies. This mixing process is modelled in our work by setting the above mixing length, conditioned by the unsteady phase-averaged vorticity. This scaling leads to the relation (14) and takes into account the non-linear character among the organized vortices and the smaller-scale ones. The new length scale seems to represent more physically the interactions of the fine scale vortices with the organized large-scale ones, as it involves the vorticity, quantity which dominates the birth and the evolution of all the above structures. The classic model gives prohibitively high values for v_t, which are not physically justified for the present case of a strongly shear flow. In fact, the energy supplied to the flow is attributed to sustain not only the random turbulence, but also the organized coherent structures. For this reason, we need a less dissipative model for v_t for the outer region, achieved by the present modifications. We have preliminarily studied the evolution of the unsteady vorticity in the shear layers, in the objective of diminishing v_t, as it is shown in the results (figure 1). By introducing the length scale, $1|\omega|$, in the law of v_t, we achieve also that the vortex motion in the separation region is taken into account, whereas the classic models of this kind do not allow this possibility.

In this study, we have implemented the averaged unsteady equations for a turbulent compressible flow, as explicited above. We have implemented the algebraïc models of Baldwin & Lomax and the one including the modification of Braza & Noguès, in the numerical code using the MUSCL scheme (van Leer [20], [21]) in a second-order of accuracy and in a flux-vector-splitting formulation.

We detail in the following the assumptions that we use for the turbulence model closure, concerning the correlation due to the compressibility effects: Concerning the total energy equation, three terms need to be modelled: The triple correlation velocity $\overline{\rho u''_i u''_i u''_k}$ and the terms deriving from deformation tensor $\overline{S_{ik} u''_i}$ and $\overline{S_{ik} \tilde{U}_i}$. These terms are neglected when using the algebraïc models. The two first terms are taken into account for higher order modeling whereas the third one is usually neglected.

4. The numerical parameters and boundary conditions

The Navier-Stokes equations are solved by using the flux vector splitting scheme of van Leer on a cartesian grid. The method is second order accurate in space and time. The boundary conditions are zero-gradient for the velocity and energy at the outlet boundary. On the upper wall, slip boundary conditions are adopted whereas the boundary layer is taken into account for the lower wall. According to the physical experiment, a boundary layer thickness of 1mm is assumed to match a laminar boundary-layer velocity profile with the measured profile at x=1mm. The numerical grid is an equidistant one (512x151) points. The numerical domain has a length of 600 mm and a width of 145 mm. The horizontal and vertical grid spacings are 0.0075 and 0.00662 made non-dimensionles by the height (145 mm). The reference thermodynamic quantities are taken those of the supersonic stream, according to the paper by Barre, Quine, Dussauge. We have performed numerous tests for the optimum time-step to use. We have found that values of the time step higher than 0.05 (dimensionles) do not allow the appearance of the inherent unsteadiness.

5. The results

The U mean velocity profiles are plotted for the sections x=8mm, 32 mm and 100 mm and are compared with the experimental results (figure 2). A general good agreement is obtained. The expansion of the mixing layer is clearly obtained, in accordance with the physical experiment.

We have assessed a predominant frequency for the organized motion, by considering the V velocity component at the point: x=74 mm and y=0, (figure 3). Its value is 1.05 KHz at this point, which is in good agreement with the results obtained by a different method of the present workshop proceedings by S. Tsangaris et.al. and in the same range as suggested by the physical experiment [5].

6. Conclusions

The modified zero-equation turbulence model presented in this paper, coupled with the Organized Eddy Simulation (OES) approach allows the appearance of the inherent unsteadiness of the supersonic mixing layer flow, due to the development of a Kelvin-Helmholtz instability. This methodology provides a good agreement of the mean velocity profiles with the physical experiment. However, the numerical scheme used needs fine grids and consequently small time-steps, (but larger than the ones needed in explicit formulations), in order to achieve a satisfactory time-accurate procedure. For these reasons, the restitution of results is considerably long. Furthermore, a general remark on the behavior of the OES approach including the present turbulence model for unsteady shear flows is that even with relatively simple models of turbulence, the main organized characteristics and mean properties of the high-speed flow are predictable, because they are included in the phase-averaged quantities. Of course, it is worthwhile to couple the OES approach with higher order of complexity turbulence models, including compressibility corrections, in order to assess the relative benefits and weaknesses of the different levels of modeling, especially in the far field region.

Acknowledgements

We are thankful to Dr J.P. Dussauge, to Prof. S. Tsangaris and to Mr A. Pentaris for their valuable discussions.

References

[1] CEBECI, T. & SMITH, A. M. O. 1974 'Analysis of turbulent boundary layers'. Applied Mathematics and Mechanics. Ed. Frankiel F. N. and Temple, G. Academic Press, New-York.

[2] CEBECI T., SMITH A. M. O., MOSINSKI S. G.1970 "Calculations of compressible adiabatic turbulent boundary layer" AIAA Journal, N°8, pp 1974-1982,.

[3] BALDWIN B. S., LOMAX H. 1978 "Thin layer approximation and algebraïc model for separated turbulent flows." AIAA Paper, 78-257, January 1978.

[4] BRAZA M., NOGUÈS, P 1991 "Numerical simulation and modeling of the transition past a rectangular afterbody". Proceedings of the VIIIth Turbulent Shear Flow Conference, Munich, 9-11 Sept. 1991., pp. I-2-1, I-2-2.

[5] BARRE, QUINE, DUSSAUGE 1994 "Compressibility effects on the structure of supersonic mixing layers: experimental results". J. Fluid Mech., vol. 259, pp.47-78.

[6] REYNOLDS O. 1883 "An experimental investigation of the circumstances which determine whether the motion of water shall be direct of sinuous and the law of resistance in parallel channels. "Phys. Trans. Roy. Soc. London, 174, pp. 935-982.

[7] REYNOLDS O.1884 "On the dynamically theory of incompressible viscous fluid and the determination of the criterion." Phys. Trans. Roy. Soc. London, 186, pp. 123-161.

[8] FAVRE A. 1965 "Equations des gaz turbulents compressibles". Formes générales, Journal de Mécanique, Vol. 4, N°3, pp 361-390, Septembre 1965.

[9] FAVRE A 1965 "Méthode des vitesses moyennées, méthode des vitesses macroscopiques pondérées par la masse volumique". Journal de Mécanique, Vol. 4, N°4, pp 390-421, Décembre 1965.

[10] HINZE, O. 1959 "Turbulence" McGraw-Hill and second edition 1975. New York.

[11] HUSSAIN, A. K. M. F. & REYNOLDS, W. C. 1975 J. Fluid Eng. Vol. 97, 568.

[12] CANTWELL, B. J., & COLES, D. 1984 J. Fluid Mech. Vol. 136, 321-374.

[13] FRANKE, R. & RODI, W. 1991 Proceedings of the Eighth Symposium on Turbulent Shear Flows, Munich, Germany, Sept. 9-11. Vol. 2, 20-1.

[14] HA MINH, H., CHASSAING, P., BRAZA, M., KOURTA, A. & SEVRAIN, A. 1987 Rapport final, convention DRET / INPT n°84/190, Mars 1987, Institut de Mécanique des Fluides de Toulouse..

[15] HA MINH, H. & CHASSAING, P.1987 Rapport de convention DRET/INPT 87/131, Institut de Mécanique des Fluides de Toulouse.

[16] HA MINH, H., VIEGAS , J. R., RUBESIN, M. W., VANDROMME, D. D. & SPALART, P. 1989 Proceedings of the Seventh Symposium on Turbulent Shear Flows, Stanford, CA, USA, August 21-23.

[17] G. JIN & M. BRAZA 1994 "A two-equation turbulence model for unsteady separated flows around airfoils". AIAA J., Vol. 32, N° 11, Nov. 94, pp. 2316, 2320.

[18] HANINE, F. "Physique, Modélisation et Simulation des couches limites turbulentes compressibles". Thèse de Doctorat, I.N.P. Toulouse, 1992.

[19] BOUSSINESQ J.,1897 "Théorie de l'écoulement tourbillonnant et tumultueux des liquides dans les lits rectilignes à grande section." I-II- Gauthiers-Villars, Paris.

[20] VAN LEER B., THOMAS J. L., ROE P. L., NEWSOME R. W., 1987 "A comparison of numerical flux formulas for the Euler and Navier-Stokes equations". AIAA Paper 87-1184.

[21] VAN LEER B. 1979 "Towards the ultimate conservative difference scheme. A second order sequel to Godunov's method". J. Comp. Phys. , Vol. 32, 101-136.

[22] van DYKE, M. 1982 "An Album of Fluid Motion", Ed. Parabolic Press.

[23] M. BRAZA 1986 "Analyse physique du comportement dynamique d'un écoulement externe, décollé, instationnaire, en transition laminaire-turbulente. Application: Cylindre circulaire". Thèse de Doctorat d'Etat-ès-Sciences, I.N.P.T., Décembre 1986.

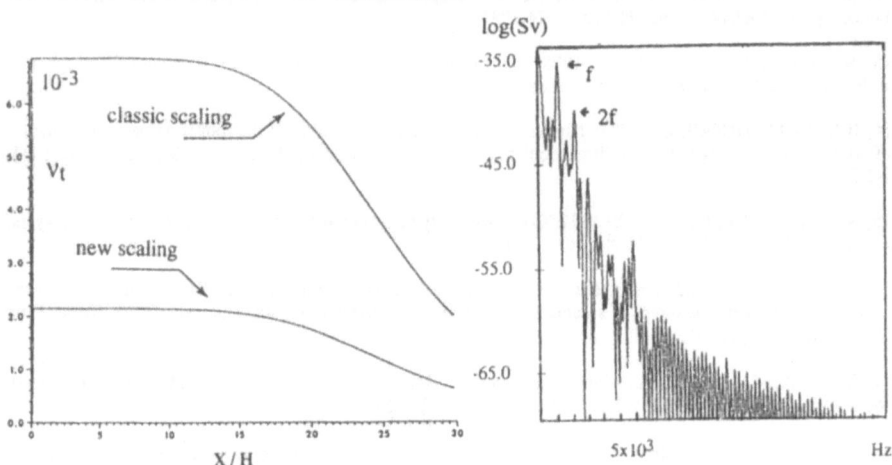

Fig. 1 Reduction of the level of the unsteady turbulent viscosity. Comparison with the Baldwin-Lomax model, [4].

Fig. 3

V component spectrum at x=74mm, y=0.

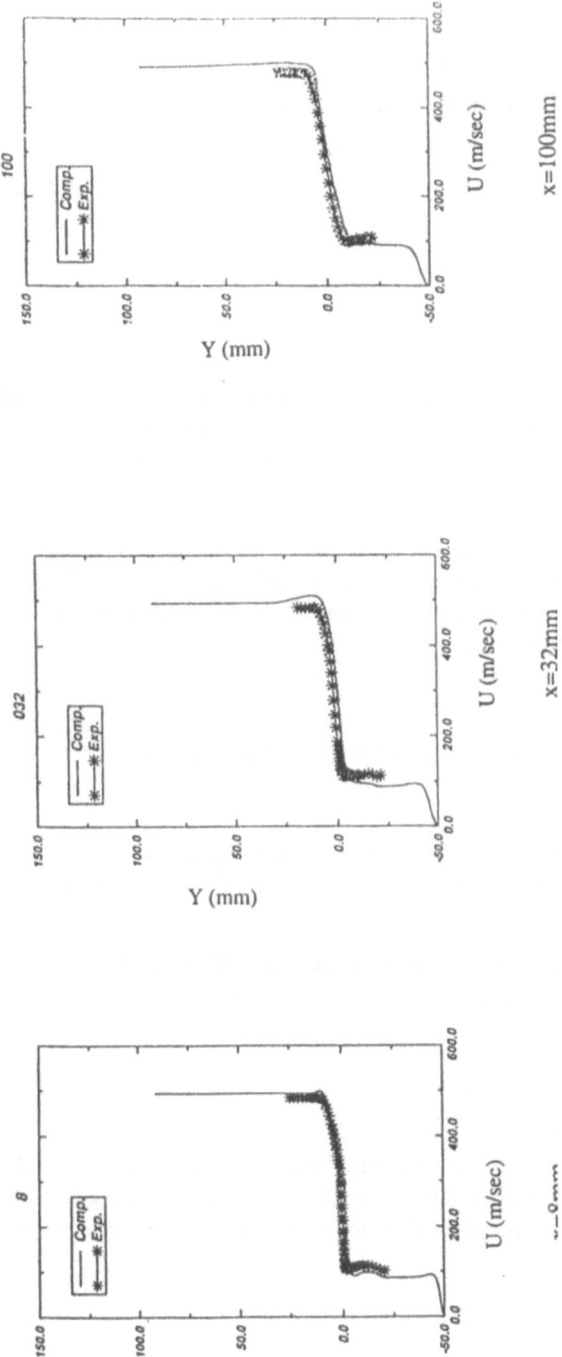

Figure 2. Mean velocity profiles (U component) along the supersonic mixing layer

A modified k-ε model derived by homogenization techniques

T. Chacón, D. Franco, F. Ortegón, I. Sánchez

Dpto. Ecuaciones Diferenciales y Análisis Numérico

Universidad of Sevilla. Facultad de Matemáticas. 41080 Sevilla, SPAIN

SUMMARY

We introduce a nonlinear k-ε model whose main feature is that the Reynolds stress tensor is modelled by means of mathematical homogenization techniques. An important characteristic of it is that the closure terms are computed from the solution of a PDE system that governs the turbulent perturbation. This system is coupled to the mean flow PDE system. We solve the model by a mixed finite volumes–finite elements technique with upwinding. We test it for mixing layer with different convective Mach numbers. The model is shown to be rather numerically unstable for a certain kind of closure terms, and to be close to the standard k-ε model for some other kind of closure terms.

DERIVATION OF THE MODEL

We consider flow of perfect gas with spectral gap and periodic microscale behaviour. A formal homogenization argument leads to a set of *microstructure* equations governing the turbulent perturbation w (cf. Chacón [2]):

$$\left.\begin{array}{l} \tilde{w}_{,\tau} + (\tilde{w} \cdot \nabla_y)\tilde{w} + C\nabla_y\pi = 0, \quad \nabla_y \cdot \tilde{w} = 0, \\ \tilde{w} \quad \text{odd and periodic in the cell } Y =]0, 2\pi[^3, \\ \langle\tilde{w}\rangle = 0, \\ \dfrac{1}{2}\langle\tilde{w}C^{-1}\tilde{w}\rangle = 1. \end{array}\right\} \tag{1}$$

Here, τ and y are the microstructure time and space variables. The symbol $\langle\cdot\rangle$ denotes mean in a period cell. Also, C is a 3×3 symmetric matrix defined through the Inverse Lagrangian co-ordinates $a(x,t)$ associated to the mean velocity field u:

$$C = G^t G, \quad \text{with} \quad G = \nabla a \tag{2}$$

whereas a is solution to

$$\frac{\partial(\rho a)}{\partial t} + \nabla \cdot (\rho a \otimes u) = 0, \qquad a(x, 0) = 1. \tag{3}$$

Once the *canonical* perturbation $\tilde{w}$ is known, the perturbation w is then given by

$$w = \sqrt{k}\, G^{-T} \tilde{w}\,.$$

As it was pointed out by Bègue et al. in [1], for incompressible flows the Reynolds stress modelling provided by the above approach is able to take into account some specific transient effects that usual turbulence models do not simulate properly. However, it is unable to take into account eddy diffusions effects. Instead, it is proposed to perform a mixed modelling, using the usual Boussinesq's hypothesis. In the case of compressible flows, this suggests the following modelling of the Reynolds stress tensor:

$$\overline{\rho u' \otimes u'} \simeq R = \langle w \otimes w \rangle - \mu_t S(u) = \rho k G^{-t} \langle \tilde{w} \otimes \tilde{w} \rangle G^{-1} - \mu_t S(u), \qquad (4)$$

where $S(u)$ is the Cauchy stress tensor, $S(u) = (\nabla u + \nabla u^T) - \dfrac{2}{3}(\nabla \cdot u)I$, and μ_t is the turbulent viscosity coefficient.

Now, any standard turbulence model may be used to model the turbulent viscosity coefficient. We shall consider here the k-ε model, for which $\mu_t = c_\mu \dfrac{k^2}{\varepsilon}$.

STRUCTURE OF CLOSURE TERMS

Let us consider 2D mean flows. Then,

$$u(x,t) = \begin{pmatrix} u_1(x_1, x_2, t) \\ u_2(x_1, x_2, t) \\ 0 \end{pmatrix} \quad \text{and, under suitable b. c.,} \quad a(x,t) = \begin{pmatrix} a_1(x_1, x_2, t) \\ a_2(x_1, x_2, t) \\ x_3 \end{pmatrix}.$$

Some algebra yields

$$G = \nabla a = \left[\begin{array}{cc|c} M_{2\times 2} & & 0 \\ & & 0 \\ \hline 0 & 0 & 1 \end{array}\right], \quad C = \left[\begin{array}{cc|c} D & & 0 \\ & & 0 \\ \hline 0 & 0 & 1 \end{array}\right], \quad \text{where } D = M^T M\,.$$

Frame invariance analysis allows to prove that the tensor $T = \langle w \otimes w \rangle$ depends only on $\hat{D} = MM^t$, as follows:

$$T = \left[\begin{array}{cc|c} \Phi(\hat{D}) & & 0 \\ & & 0 \\ \hline 0 & 0 & \beta(\hat{D}) \end{array}\right]. \qquad (5)$$

Here, Φ is a 2x2 matrix function, and β is an scalar function. Both of them depend only on the invariants $j_1 = \text{trace}\,\hat{D}$, $j_2 = \det \hat{D}$ of $\hat{D}$. However, only Φ is relevant to model tensor R. The actual structure of Φ is

$$\Phi(\hat{D}) = \beta_0 I + \beta_1 \hat{B}, \quad \text{where } \hat{B} = \hat{D} - \frac{1}{2} j_1 I. \qquad (6)$$

Coefficients β_0 and β_1 are computed from the solution $\tilde{w}(C)$ of problem (1), as follows:

$$\beta_0 = 1 - \frac{\mu_2}{2}, \quad \beta_1 = \frac{2}{j_1^2 - 4j_2}\,(\mu_1 - j_1 \beta_0), \text{ with } \mu_1 = \langle |\tilde{w}_1|^2\rangle + \langle |\tilde{w}_2|^2\rangle, \ \mu_2 = \langle |\tilde{w}_3|^2\rangle. \quad (7)$$

Although β_1 is not defined if $j_1^2 - 4j_2 = 0$, both β_0 and $\beta_1 \hat{B}$ are continuous functions of j_1 and j_2 if $\tilde{w}(C)$ depends continuously on matrix C.

As a conclusion, in the case of 2D mean flow, we can replace the modelling of the Reynolds stress tensor in (4) by

$$\overline{\rho u' \otimes u'} \simeq R = \rho k \Phi(\hat{D}(u)) - \mu_t S(u) \qquad \text{(in 2D).} \qquad (8)$$

COMPUTATION OF CLOSURE TERMS

The Reynolds' tensor T issued from homogenization depends only on the two invariants j_1 and j_2 of matrix D. This suggests to compute β_0 and β_1 as functions of these invariants, and just for diagonal matrices. We also must compute $\dfrac{\partial \beta_0}{\partial j_1}$ and $\dfrac{\partial \beta_0}{\partial j_2}$, since these quantities appear as closure terms in the equations for the energy in our model.

We thus consider the matrices

$$C = \left[\begin{array}{c|c} D & \\ \hline & 1 \end{array} \right], \quad \text{with } D = \left[\begin{array}{cc} \lambda_1 & \\ & \lambda_2 \end{array} \right]. \qquad (9)$$

Notice that as C is symmetric positive-definite, we must take only $\lambda_1, \lambda_2 \in \mathbf{R}_*^+$. As $j_1 = \lambda_1 + \lambda_2$, $j_2 = \lambda_1 \lambda_2$, then λ_1 and λ_2 are the (real) solutions to $\lambda^2 - j_1 \lambda + j_2 = 0$.

Consequently, we must only consider j_1, j_2 such that the discriminant of this equation is non negative,

$$\Delta = j_1^2 - 4 j_2 \geq 0. \qquad (10)$$

In order to compute the closure coefficients μ_1 and μ_2, we have solved the steady-state version of the system verified by the canonical microstructures, following the technique developed by Chacón and Ortegón in [3] and [5]. The solution technique is based upon a least square formulation of (1): A solution of this problem is characterized as minimizing a certain cost functional J.

Figure 1 represents the evolution of the normalized cost J versus the iterations of the conjugate gradient process. In almost all cases, the initial cost J is divided by 1000 in approximately 20 iterations. This takes about 90 seconds of CPU per iteration on a CRAY-YMP machine. Note that the larger the quotient j_1 / j_2 is, the faster the convergence of the conjugate gradient process results.

This technique of computation of closure terms that we have developed allows to compute these terms in a relatively small region of $\hat{D}$ around 2×2 identity matrix. However, in practice the computing times may be very long, and we need to know the asymptotic behaviour of $\Phi(\hat{D})$ for large values of j_1 and j_2. An asymptotic analysis of the behaviour of $\tilde{w}$ in terms of $\sigma = \dfrac{1}{\sqrt{j_1}}$ yields $\tilde{w} = O(\sigma)$ and so $R = O(\dfrac{1}{j_1})$ as $j_1 \longrightarrow \infty$. This result has been confirmed numerically by solving (1) for large values of j_1 and diagonal matrices.

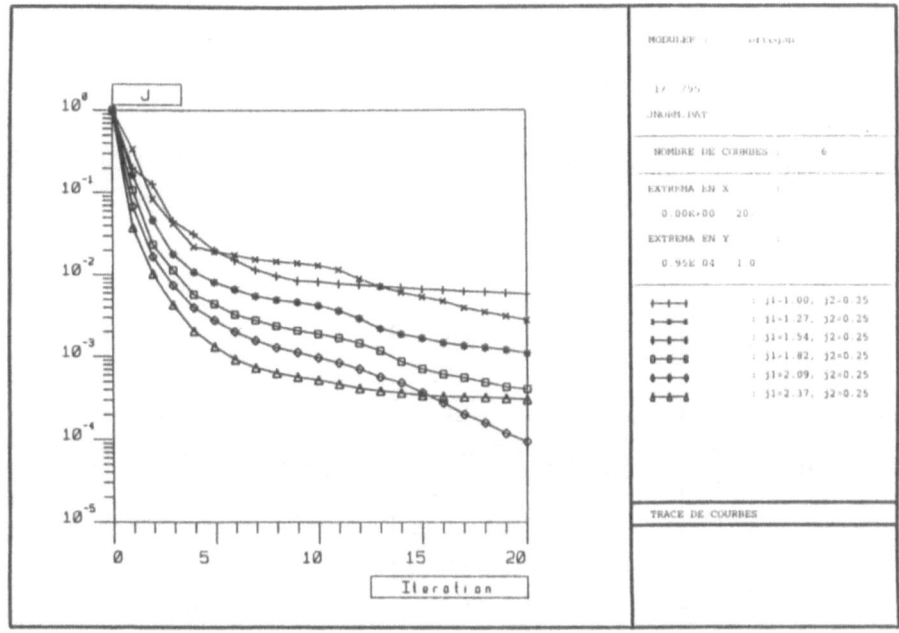

Fig. 1: Normalized cost J for the computation of $\tilde{w}$ for different values
of j_1 and $j_2 = 0.25$.

Another possible asymptotic behaviour is obtained numerically if matrix C in (1) is taken with the following structure, which corresponds to 1D mean flows (see [2]):

$$
C = \left[\begin{array}{cc|c}
1 & \alpha & \\
\alpha & 1+\alpha^2 & \\
\hline
& & 1
\end{array}\right].
$$

These tabulations show that parameters μ_1 and μ_2 tend to constant values as $j_2 \to \infty$.

CHANGEMENT OF VARIABLES

We have adapted the numerical method introduced by LeRibault in [4] to the solution of our model for 2D flows. This is a mixed finite elements–finite volumes method based upon a laminar version. It needs, in particular, a re-formulation of the model equations in order to have the same law of state as in the laminar case (Law of perfect gases). To do this, we have introduced the following variables:

$$q = p + \beta_0 \rho k, \qquad E' = E + \beta \rho k, \quad \text{with } \beta = -1 + \beta_0/(\gamma - 1). \tag{11}$$

Then, E' and q are indeed related by the Law of perfect gases: $q = (\gamma - 1)\left(E' - \frac{1}{2}\rho|u|^2\right)$.

The equation for E' is obtained from that for E in the k-ε model equations. This yields the definitive 2D formulation of our model. Under conservative form, it is the

23

following:

$$\frac{\partial}{\partial t}W + \nabla \cdot F(W) = \frac{1}{Re}\nabla \cdot R(W, \nabla W) + \frac{1}{R_t}\nabla \cdot \tilde{R}(W, \nabla W) + S(W) \qquad (12)$$

with

$$W = \begin{bmatrix} \rho \\ \rho a_1 \\ \rho a_2 \\ \rho u_1 \\ \rho u_2 \\ F \\ \rho k \\ \rho \varepsilon \end{bmatrix}, \quad F_1(W) = \begin{bmatrix} \rho u_1 \\ \rho a_1 u_1 \\ \rho a_2 u_1 \\ \rho u_1 u_1 + q \\ \rho u_1 u_2 \\ (F+q)u_1 \\ \rho k u_1 \\ \rho \varepsilon u_1 \end{bmatrix}, \quad F_2(W) = \begin{bmatrix} \rho u_2 \\ \rho a_1 u_2 \\ \rho a_2 u_2 \\ \rho u_1 u_2 \\ \rho u_2 u_2 + q \\ (F+q)u_2 \\ \rho k u_2 \\ \rho \varepsilon u_2 \end{bmatrix};$$

$R(W, \nabla W) = (R_1(W, \nabla W), R_2(W, \nabla W))$, with

$$R = \begin{bmatrix} 0 \\ 0 \\ 0 \\ \tau \\ \tau u + \dfrac{\gamma}{P_r}\nabla e \\ \nabla k \\ \nabla \varepsilon \end{bmatrix};$$

$\tilde{R}(W, \nabla W) = (\tilde{R}_1(W, \nabla W), \tilde{R}_2(W, \nabla W))$, with

$$\tilde{R} = \begin{bmatrix} 0 \\ 0 \\ 0 \\ \tau - \hat{\beta}_1 \rho k \hat{B} \\ (\tau - \hat{\beta}_1 \rho k \hat{B})u + \dfrac{\gamma}{P_{rt}}\nabla e + \dfrac{1}{\sigma_k}\nabla k \\ \dfrac{1}{\sigma_k}\nabla k \\ \dfrac{1}{\sigma_\varepsilon}\nabla \varepsilon \end{bmatrix};$$

24

where, $\hat{\beta}_1 = \beta_1 R_t$ and $\tau = \nabla u + \nabla u^t$;

$$S(W) = \begin{bmatrix} 0 \\ 0 \\ 0 \\ 0 \\ 0 \\ \psi_k \\ -P + \rho\varepsilon \\ C_{\varepsilon 1}\dfrac{\varepsilon}{k}P - C_{\varepsilon 2}\rho\dfrac{\varepsilon^2}{k} \end{bmatrix};$$

where

$$\psi_k = \beta\left(\frac{\partial(\rho k)}{\partial t} + \nabla \cdot (\rho k u)\right) - \frac{2}{\gamma - 1}\rho k\left(\hat{D} : \nabla u\frac{\partial \beta_0}{\partial j_1} + j_2\nabla \cdot u\frac{\partial \beta_0}{\partial j_2}\right).$$

We have adapted Le Ribault's k-ε solver to perform a first order in time–second order in space solver of our model.

NUMERICAL TESTS

We have tested our numerical code for mixing layers. Low Mach mixing layers is a good test case for our model. Indeed, the results furnished by the k-ε model are well known, and reasonably accurate with respect to experimental measurements. Since our model may be considered as a perturbation of the k-ε model, including memory terms, we may use this model as the reference to test ours.

We have considered a test case with Mach numbers of $M_1 = 0.1$ and $M_2 = 0.2$ in the upper and lower layers, respectively. The convective Mach number of the flow is approximately $M_c = 0.1$.

Our results present large qualitative differencies, depending essentially on the asymptotic behaviour of the closure terms that have been used.

If the first kind of closure terms, obtained by Chacón in [2] are used, the results are quite close to those given by the k-ε model. For instance, the steady states obtained for k and mean velocity profiles are quite close. This probably happens because in this case the additional term $\beta_1 \hat{B}$ introduced in the modelization of R is almost zero, for both small or large values of j_1.

If the other kind of closure terms are used, the code turns out to be somehow unstable and some unphysical features appear in the results obtained. In particular, the accuracy of the computed spreading rate of the layer decreases largely as the longitudinal distance to the inflow boundary increases. Also, the isolines of k and ε also show a progressive unphysical expansion of the layer.

We have also tested supersonic flows corresponding to convective Mach numbers of $M_c = 0.45$, $M_c = 0.65$ and $M_c = 1$. We have used the tabulations obtained by Chacón, with results close to those furnished by the k-ε model. Figures 2 and 3 show the iso-lines of ρk and $\rho\varepsilon$ when $M_c = 0.45$. The spreading rate of the layer is in good agreement with experiments. Table 1 shows the normalised spreading rates δ'/δ'_0 versus M_c.

25

Table 1: Normalised spreading rates δ'/δ_0' versus M_c, where $\delta_0' = 0.02447$ has been obtained with our model in the subsonic mixing layer experiment.

	$M_c = 0.45$	$M_c = 0.65$	$M_c = 1$
δ'/δ_0'	1.032	1.032	1.032

Our conclusion is that the model is highly sensitive to the actual closure terms used. The results given by Chacón's closure terms are quite close to those furnished by the standard k-ε turbulent model. In particular, the effect of compressibility on the spreading rate of the layer is not well taken into account. Also, when the new closure terms are used the model turns numerically unstable, yielding unphysical solutions.

We think that these unphysical results are, at least partially, due to the inaccuracy in the computations of closure terms. Probably, this may be improved by solving the transient Euler equations (1) instead of the steady ones.

REFERENCES

[1] C. BÈGUE, B. CARDOT, C. PARÉS, O. PIRONNEAU: *Simulation of turbulence with transient mean.*

Int. J. Num. Meth. in Fluids, Vol. 11, 1990.

[2] T. CHACÓN: *Étude d'un modèle pour la convection des microstructures.*

Thèse 3ème Cycle, Université Paris VI, 1985.

[3] T. CHACÓN, D. FRANCO, F. ORTEGÓN: *Homogenization of incompressible flows with helical microstructures.*

Adv. in Math. Sciences and Appl. Vol. 1, No. 2, pp. 251-300, 1992.

[4] C. LE RIBAULT: *Simulation des écoulements turbulents compressibles par une méthode mixte éiements finis-volumes finis.*

Thèse de Doctorat, Université Lyon, 1991.

[5] F. ORTEGÓN: *Modélisation des écoulements turbulents à deux échelles par méthode d'homogénéisation.*

Thèse de Doctorat, Université Paris VI, 1989.

```
ISOVALEURS :

20 ____ 0.4832E-02
19 ____ 0.4578E-02
18 ____ 0.4324E-02
17 ____ 0.4070E-02
16 ____ 0.3816E-02
15 ____ 0.3562E-02
14 ____ 0.3308E-02
13 ____ 0.3054E-02
12 ____ 0.2800E-02
11 ____ 0.2546E-02
10 ____ 0.2292E-02
 9 ____ 0.2038E-02
 8 ____ 0.1784E-02
 7 ____ 0.1530E-02
 6 ____ 0.1276E-02
 5 ____ 0.1022E-02
 4 ____ 0.7685E-03
 3 ____ 0.5145E-03
 2 ____ 0.2605E-03
 1 ____ 0.6572E-05
```

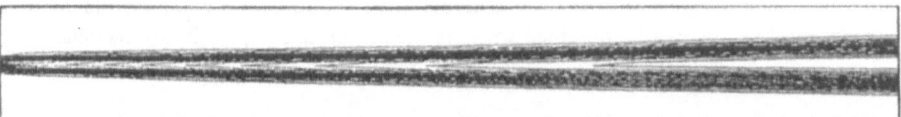

Fig. 2: Iso-lines of ρk for $M_c = 0.45$, mixed k-ε-MPP model.

```
ISOVALEURS :

20 ____ 0.1933E-02
19 ____ 0.1831E-02
18 ____ 0.1729E-02
17 ____ 0.1628E-02
16 ____ 0.1526E-02
15 ____ 0.1424E-02
14 ____ 0.1322E-02
13 ____ 0.1221E-02
12 ____ 0.1119E-02
11 ____ 0.1017E-02
10 ____ 0.9157E-03
 9 ____ 0.8140E-03
 8 ____ 0.7123E-03
 7 ____ 0.6106E-03
 6 ____ 0.5089E-03
 5 ____ 0.4072E-03
 4 ____ 0.3055E-03
 3 ____ 0.2039E-03
 2 ____ 0.1022E-03
 1 ____ 0.4673E-06
```

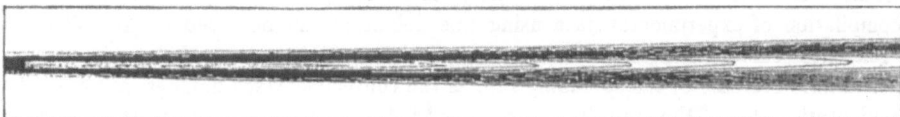

Fig. 3: Iso-lines of $\rho\varepsilon$ for $M_c = 0.45$, mixed k-ε-MPP model.

COMPRESSIBILITY MODELS APPLIED TO SUPERSONIC MIXING LAYERS

D. Guézengar[*,†] – H. Guillard[†]

*IMST, 12 Av Gnl Leclerc , 13003 Marseille, France.
†INRIA, BP 93, 06902 Sophia-Antipolis Cedex, France.

SUMMARY

In this paper, we test the use of several compressibility models for the computation of supersonic mixing layers with the $k - \epsilon$ model. It is shown that a good agreement with experimental results can be obtained with simple algebraic compressibility models.

INTRODUCTION

It is known that at high convective Mach numbers, compressibility has a very strong influence on the behavior of mixing layers. In particular the spreading rate is significantly reduced. Following the works of Sarkar & al [14] and Zeman [16] several compressibility models have been proposed to take into account these effects in one-point closures. This paper examines the performance of some of these models.

EXPERIMENTAL DATA

The fundamental parameter allowing the comparison of the numerical results with experimental data is the normalized spreading rate δ'/δ_0' as function of the convective Mach numbers. A recent compilation of these data is to be found in [1]. The definition of the mixing layer thickness used in this work is the one proposed at the Stanford conference:
$\delta = Y1 - Y2$, where $Y1$ is the ordinate where $U = U2 + \sqrt{0.9} \times (U1 - U2)$
$\qquad Y2$ is the ordinate where $U = U2 + \sqrt{0.1} \times (U1 - U2)$.
Another common definition is the vorticity thickness $\delta_W = (U1 - U2)/(\partial U/\partial y)_{max}$. A compilation of experimental data using this definition can be found in [2]. The two "reference" curves extracted from these compilations exhibit the same behavior and show a dramatic decrease of the spreading rate as the convective Mach number increases. An asymptotic value of the spreading rate around $0.4 - 0.5 \, \delta_0'$ is reached for $M_c > 1$. Note that the scatter of the experimental data is rather large and makes difficult to assess the performance of the different tested models. Some limited data do exist on the maximum values of the shear stress. They exhibit the same type of behavior than the spreading rate and can be used for comparison with the numerical results. However, in this case,

we also note that the scatter of the experimental data prevents a definite assessment of the model accuracy.

COMPRESSIBILITY MODELS

We consider the standard $k - \varepsilon$ model, supplemented by some modifications of it proposed by Sarkar and his co-workers and Zeman. With respect to the incompressible case, additional terms expressing the explicit effects of compressibility appear in the turbulent kinetic energy equation. These terms : the dilatation dissipation ε_c and the pressure-dilatation $\overline{p'd'}$ have to be modeled.

THE DILATATION DISSIPATION MODELS

Sarkar and his co-workers and Zeman have a similar approach to model this term. Both, they express ε_c as a function of the solenoidal dissipation ε_s and of the turbulent Mach number $M_t = \sqrt{2k}/\sqrt{\gamma RT}$.

Sarkar and co-workers models : With the help of an asymptotic analysis [14] and the results of Direct Numerical Simulation (DNS) [6], Sarkar & al have proposed the following algebraic model:

$$\varepsilon_c = \alpha_1.M_t^2.\varepsilon_s \tag{1}$$

They suggested [14] firstly that the constant, calibrated with DNS results for the case of decaying isotropic turbulence, was $\alpha_1 = 1$. But, from DNS results of homogeneous shear flows, it was proposed [15] $\alpha_1 = 0.5$ as an alternate value.

Zeman's models : DNS of compressible mixing layers as well as the homogeneous turbulence computations of Passot and Pouquet [12] have shown that structural changes occur in the flow when the Mach number increases: the turbulent field, initially solenoidal, exhibits "shocklet"-like structures. Zeman [16] uses this fact and introduces the concept of dilatation dissipation caused by these turbulent shocklets (called "anomalous" dissipation). He then proposed a model based on a certain threshold, specified by the apparition of shocklets and thus of ε_c:

$$\varepsilon_c = C_d \times F\left(M_t\right) \times \varepsilon_s$$

$$\begin{cases} F\left(M_t\right) &= 0 & \text{if } M_t < M_{t0} \\ F\left(M_t\right) &= 1 - e^{-\left(\frac{M_t - M_{t_0}}{\sigma_0}\right)^2} & \text{otherwise} . \end{cases}$$

The first model proposed by Zeman [16] is characterized by:

$$\begin{aligned} C_d &= 0.75 \\ M_{t_0} &= 0.1 \text{ and } \sigma_0 = 0.6 . \end{aligned} \tag{2}$$

For shear layers, Zeman & al. [19] have proposed other values for these parameters:

$$\begin{aligned} C_d &= 0.75 \\ M_{t_0} &= 0.25 \text{ and } \sigma_0 = 0.8 . \end{aligned} \tag{3}$$

But this model must be associated with a model for the pressure-dilatation correlation.

THE PRESSURE-DILATATION MODELS

Sarkar's model: The objective of Sarkar [13] was to propose an algebraic expression to evaluate the correlation $\overline{p'd'}$. His model depends on the production term and on the solenoidal dissipation [15]:

$$\overline{p'd'} = -\alpha_2.\bar{\rho}.\mathcal{P}.M_t + \alpha_3.\bar{\rho}.\varepsilon_s.M_t^2 . \tag{4}$$

$\alpha_2 = 0.15$ if $0.2 < M_t < 0.6$ and $\alpha_3 = 0.2$
This model must be employed with the compressible dissipation model where $\alpha_1 = 0.5$.
Zeman's models: The $\overline{p'd'}$ correlation can be decomposed in the following way :

$$\overline{p'd'} = (\overline{p'd'})_e + (\overline{p'd'})_{\nabla.U} + (\overline{p'd'})_\rho$$

where each term requires a different modeling:
$(\overline{p'd'})_e$: expresses that the pressure fluctuations tend to relax to an equilibrium value (see [17], [19]).
$(\overline{p'd'})_{\nabla.U}$: results from a directional rapid mean compression (see [20], [4]).
$(\overline{p'd'})_\rho$: is due to the interaction between mean density gradient and pressure fluctuations (see [18]).
These models have not been tested in this work.

APPLICATION TO THE $k - \varepsilon$ MODEL

The above compressibility models have been developed for second-order turbulence closure in which the Reynolds stresses are obtained by transport equations. In the $k-\varepsilon$ model, the Reynolds stresses are defined by the eddy viscosity μ_t. In this framework k/ϵ represents the turbulent time scale T_t. The most natural definition of this time scale is $T_t = k/\epsilon_t$ where ϵ_t is the total dissipation. This also implies the following definition of the eddy viscosity $\mu_t = C_\mu \dfrac{\rho k^2}{\varepsilon_s + \varepsilon_c}$. However, it seems that in some works [10], [3], the incompressible definition $T_t = k/\epsilon_s$ and thus $\mu_t = C_\mu \dfrac{\rho k^2}{\varepsilon_s}$ has also been used. In addition, we have also considered a variant defined by the use of the turbulent time scale k/ϵ_t in the turbulent kinetic energy equation (implying $\mu_t = C_\mu \dfrac{\rho k^2}{\varepsilon_s + \varepsilon_c}$) but the use of the "incompressible" time scale k/ϵ_s in the solenoidal dissipation transport equation (with $\mu_t = C_\mu \dfrac{\rho k^2}{\varepsilon_s}$).

SUMMARY

All the models tested in this work can be put in the following form :

$$\frac{\partial \bar{\rho}\tilde{k}}{\partial t} + \frac{\partial(\bar{\rho}\tilde{u}_i\tilde{k})}{\partial x_i} = \frac{\partial}{\partial x_i}\left[\left(\mu + \frac{\mu_t}{\sigma_k}\right)\frac{\partial \tilde{k}}{\partial x_i}\right] + (1 - \alpha_2 M_t)\bar{\rho}\mathcal{P} - \bar{\rho}((1 - \alpha_3 M_t^2)\tilde{\varepsilon}_s + \tilde{\varepsilon}_c)$$

$$\frac{\partial \bar{\rho}\tilde{\varepsilon}_s}{\partial t} + \frac{\partial(\bar{\rho}\tilde{u}_i\tilde{\varepsilon}_s)}{\partial x_i} = \frac{\partial}{\partial x_i}\left[\left(\mu + \frac{\mu_t}{\sigma_{\varepsilon_s}}\right)\frac{\partial \tilde{\varepsilon}_s}{\partial x_i}\right] + \bar{\rho}\frac{1}{T_t}(C_{\varepsilon_1}\mathcal{P} - C_{\varepsilon_2}\tilde{\varepsilon}_s)$$

where $\mathcal{P} = -\tilde{u}_{i,j}\widetilde{u_i''u_j''}$ is the production term and T_t is the turbulent time scale equal either to k/ε_t or k/ε_s. Table 3 summarizes the tested models. k/ε_s and k/ε_t signifies:

k/ε_s in the ε-equation and k/ε_t in the others.

DEFINITION OF THE TEST-CASES

The three test-cases calculated: $M_c = 0.45$ - 0.65 - 1.00 are characterized by a velocity ratio $r = 0.58$ and a density ratio $s = 1.56$. With these values of r and s the computation of a subsonic mixing layer gives a spreading rate $\delta'_0 = 0.0270$ that compares favorably with the value of 0.0315 given by the empirical correlation $\delta'_0 = (0.115/2)(1-r)(1+\sqrt{s})/(1+r\sqrt{s})$ ([1]).

The compressible computations correspond to supersonic/supersonic flows. The value $Mc = 0.45$ represents the experience of Goebel & al. [8] . The pressure has a constant value of $49kPa$ in the two free streams.

Mc	0.45	0.65	1.00
U1 (m/s)	700.0	1 021.0	1570.8
U2 (m/s)	406.0	592.2	911.1
M1	1.91	2.79	4.29
M2	1.37	2.02	3.10

NUMERICAL METHODOLOGY

The numerical method used is a mixed finite element – finite volume method described in detail in [7], [11] or [9] A time-advancing scheme is used and the computation advances in time until a steady state defined by a normalized residual of 10^{-6} is reached. Explicit or implicit time integration have been tested. With an explicit method, we found that the use of compressibility models improves the convergence, as can be seen on Figure 4. With implicit time integration, no differences has been noticed in the convergence rate. The computations reported here have used an implicit time scheme and the steady state is reached after ~ 200 iterations.

The computational domain is a 48 mm x 576 mm rectangle with a refined mesh, at the two streams interface, and is composed of 2469 nodes. The dimensions were chosen in reference to the test section of Goebel's experiment. However we found that the height of the experimental domain was too small which induces a compression of the flowfield. Therefore, we use far-field boundary conditions on the lateral walls instead of slip conditions. The domain length corresponds approximately to 100 time the initial thickness of the mixing layer equal to 5.1 mm. This value was determined in order to reach an asymptotic state in downwind sections of the computational domain. Actually, we found numerically that an asymptotic state is reached for the mean quantities after $35 \times \delta(x = 0)$ whereas, for the turbulent variables, $70 \times \delta(x = 0)$ are necessary. These values are coherent with the values noted in experiments [5].

We have checked through mesh refinement studies that this mesh is fine enough to get mesh independent solutions. It must be noted that mesh independence as well as the obtention of similarity profiles are much easier to obtain for the mean quantities than for

31

the turbulent ones. For instance, at least 15 nodes in the interfacial zone of the mixing layer are necessary to get similarity profiles for the turbulent kinetic energy while 7 nodes suffice to obtain velocities profiles in similarity.

The upwind boundary conditions for the mean variables are determined from experimental data. For the turbulent variables (k and ε) we chose to express ε as being 90 per cent of the production term. It then remains to specify the value of the turbulent kinetic energy. This was done with reference to Figure 21 of [1], that gives $(k/\Delta U^2)_{max} \sim 0.03\Phi(M_c)$ where $\Phi(M_c)$ denotes the Langley curve. Using these boundary conditions, we found that similarity profiles for the turbulent variables are obtained only in the downwind sections of the computational domain. Moreover, the asymptotic turbulent kinetic energy level depends on the used models. Therefore, we found useful to apply the following strategy to specify the upwind boundary condition : a first computation is realized using the above boundary conditions, then after convergence, the profiles obtained in the downwind sections are scaled using similarity transformation and used as boundary conditions for a second computation. This process can of course be repeated as many times as necessary, but for the present test cases we found that only one re-computation was necessary. Figure 3 displays the maximum value of k with respect to the abscissa after convergence of the first computation and of the second one. It can be checked that k is almost constant after the second computation indicating that an equilibrium state has been reached in the whole computational domain.

NUMERICAL RESULTS

For all the tested models, all the Mach numbers and all sections of the computational domain (using the strategy described above to specify the boundary condition) we found that the variables cluster on the same similarity profiles. This can be seen for instance on Figures 2 - 1 that display respectively the velocity and turbulent kinetic profiles.

We next examine the influence of the different definitions of T_t. The discussion is made here for Zeman's model, but similar results are obtained for the other models. The results are rather close to each others and in the limits of the experimental scatter of the data. From Figure 5 it can be seen that the ZEMAN1 and ZEMAN2 models (see table 3 for the definition of the models) tend to overestimate the spreading rate while ZEMAN3 tends to underestimate it. However, it seems that the slope of the ZEMAN3 curve follows more closely the decrease of the Langley consensus curve. Moreover, looking at table 1 displaying the maximum shear stress, it can be seen that the results obtained with the ZEMAN3 model are closer to the Langley curve. Therefore, in the sequel, we will use the two definitions of the turbulent time scales as explained before.

On Figure 6, are plotted the results obtained by the different models of Sarkar and his co-workers. It can be seen that the absence of a threshold value in these models induces a clear underestimation of the spreading rate at low convective Mach numbers. Clearly, the value $\alpha_1 = 1$ (SEHK3-1) in the dilatation dissipation modeling is too large. The value $\alpha_1 = 0.5$ (SEHK3-05) seems more appropriate and gives results within the limits of the experimental dispersion. A good improvement of the results especially at hight convective Mach numbers can be noticed with the use of the pressure dilatation modeling. On Figure 7 and tables 0-2 are also displayed the results obtained without any compressibility modeling. It is clear that the STANDARD model does not reproduce the decrease of the spreading rate and of the maximum shear stress.

CONCLUSION

The models, based on the dilatation dissipation, proposed by Sarkar & al. and Zeman, are very simple but can take in account the compressibility effects and reproduce the decrease of the spreading rate and of the maximum shear stress. At low convective Mach numbers, the absence of a threshold value in the Sarkar and co-workers models induces a certain underestimation of the spreading rate. Certainly due to the clipping effect of the threshold value, the Zeman's models have a better behavior at these low Mach numbers. At high convective Mach number, on the contrary, the models show an overestimation of the experimental results. This is especially true for Zeman's models while the modeling of the pressure dilatation term in Sarkar's model improves substantially the results for this range of Mach numbers. We also note that the use of a different time scale in the solenoidal dissipation transport equation yields a large improvement of the results for high Mach numbers. This point has yet to be fully understood. With respect to Figure 7 that shows the comparison of the "best" models, it can be seen that they give results in reasonable agreement with the experimental data. A more definite assessment of the performance of the models would require other test cases or more accurate experimental results.

Acknowledgments

A large part of this work results from many discussion with J.P. Dussauge that we thank for his collaboration.

REFERENCES

[1] BARRE S., QUINE C., DUSSAUGE J. P. - *J. Fluid Mech., Vol. 259, p.47-78, 1994.*

[2] BOGDANOFF D. W. - *AIAA Journal, Vol.21, No.6, p.926-927.*

[3] COLEMAN G. N., MANSOUR N. N. - *8th Symposium on turbulent shear flows 1993 p.283-296.*

[4] DURBIN P. A., ZEMAN O. - *J. Fluid Mech., Vol. 242, p.349-370, 1992.*

[5] DUSSAUGE J. P. : private communication.

[6] ERLEBACHER G., HUSSAINI M. Y., KREISS H. O.,SARKAR S. - *ICASE report 90-15, 1990.*

[7] FEZOUI L., LANTERI S., LARROUTUROU B., OLIVIER C. - *INRIA Report 1033, 1989.*

[8] GOEBEL G., DUTTON JC., KRIER H., RENIE JP. - *Exper. in Fluids Vol.8 p. 263-272, 1990.*

[9] INRIA-SIMULOG - Manuel theorique du logiciel NSTC3D V 2.1.

[10] LE RIBAULT C., BUFFAT M., JEANDEL D. - *Comp. Fluid Dynamics'92, Vol.1 .*

[11] OLIVIER C., B. LARROUTUROU - *INRIA Report 1526, 1991.*

[12] PASSOT T., POUQUET A. - *J. Fluid Mech., Vol.187, p.441-466, 1986.*

[13] SARKAR S. - *ICASE report 91-42, 1991.*

[14] SARKAR S., ERLEBACHER G., HUSSAINI M. Y., KREISS H. O. - *ICASE report 89-1789, 1989.*

[15] SARKAR S., ERLEBACHER G., HUSSAINI M. Y. - *ICASE report 92-6, 1992.*

[16] ZEMAN O. - *Phys. Fluids Vol.2, p.178-188, 1990.*

[17] ZEMAN O. - *Phys. Fluids Vol.3 p.951-955, 1991.*

[18] ZEMAN O. - *AIAA Paper 93-0897, Reno, Nevada, Jan. 11-14, 1993.*

[19] ZEMAN O., BLAISDELL G. A. - *Adv. in turbulence 3, Eds A.V Johansson P.H Alfredsson, 1991.*

[20] ZEMAN O., COLEMAN G. N. - *8th Symposium on turbulent shear flows 1993 p.283-296.*

FIGURES AND TABLES

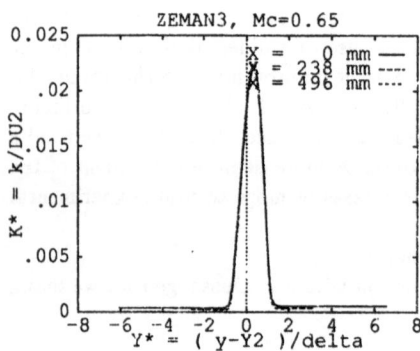

Figure 1: Kinetic turbulent energy profiles in similarity variables

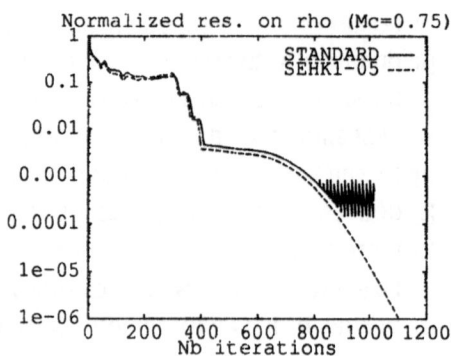

Figure 2: Mean velocity profiles in similarity variables

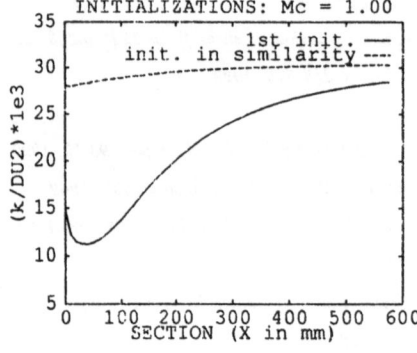

Figure 3: Effect of Boundary Conditions on the k behavior

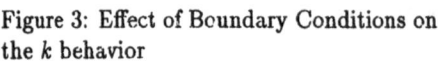

Figure 4: Explicit convergence rate

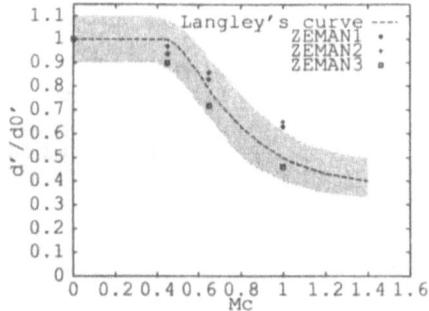

Figure 5: Comparison of the different adaptations to the $k - \varepsilon$ model

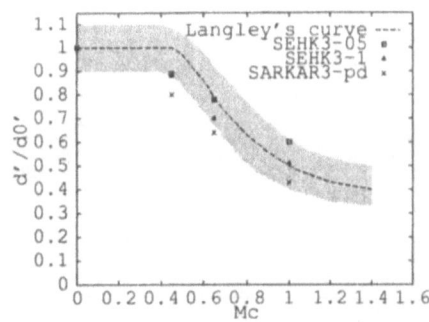

Figure 6: The different variants of the SEHK's models

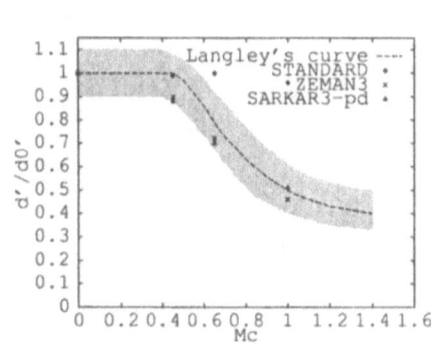

Figure 7: Comparison of the different models

Table 0: $\dfrac{\delta'}{\delta_0}$ versus M_c

M_c	0.45	0.65	1.00
STANDARD	0.99	0.99	0.99
SEHK1-1	0.89	0.79	0.61
SEHK2-1	0.90	0.81	0.65
SEHK3-1	0.80	0.64	0.43
ZEMAN1	0.94	0.83	0.63
ZEMAN2	0.97	0.86	0.65
ZEMAN3	0.90	0.72	0.46
SEHK3-05	0.89	0.78	0.60
SARKAR3-pd	0.88	0.70	0.51
LANGLEY	1.00	0.77	0.50

Table 1: $\left(-\dfrac{\widetilde{u'v'}}{\Delta U^2} \right)_{max} \times 10^3$ versus M_c

M_c	0.45	0.65	1.00
STANDARD	10.4	10.4	10.0
SEHK1-1	9.5	8.7	7.1
SEHK2-1	9.3	8.4	6.6
SEHK3-1	8.4	7.0	4.8
ZEMAN1	10.0	9.1	7.1
ZEMAN2	9.9	8.8	6.7
ZEMAN3	9.5	7.8	5.1
SEHK3-05	9.3	8.4	6.5
SARKAR3-pd	9.25	7.3	5.4
LANGLEY	**9.5**	**7.5**	**5.5**

Table 2: $\left(\dfrac{k}{\Delta U^2} \right)_{max} \times 10^3$ versus M_c

M_c	0.45	0.65	1.00
STANDARD	29.5	29	28.5
SEHK1-1	27	24	19
SEHK2-1	27	24	19
SEHK3-1	24.5	18	11.5
ZEMAN1	28.5	25.5	19.5
ZEMAN2	28.5	25.5	19
ZEMAN3	27.5	22.5	12.5
SEHK3-05	27	24	16.5
SARKAR3-pd	26.5	21	13.5

Table 3: : definition of the models

Model	ε_c	T_t	α_2	α_3
STANDARD	0	k/ε_s	0	0
SEHK1-1	formula (1), $\alpha_1 = 1$	k/ε_s	0	0
SEHK2-1	formula (1), $\alpha_1 = 1$	k/ε_t	0	0
SEHK3-1	formula (1), $\alpha_1 = 1$	k/ε_s and k/ε_t	0	0
ZEMAN1	formula (2)	k/ε_s	0	0
ZEMAN2	formula (2)	k/ε_t	0	0
ZEMAN3	formula (2)	k/ε_s and k/ε_t	0	0
SEHK3-05	formula (1), $\alpha_1 = 0.5$	k/ε_s and k/ε_t	0	0
SARKAR3-pd	formula (1), $\alpha_1 = 0.5$	k/ε_s and k/ε_t	0.15	0.2

Numerical simulation of supersonic mixing layers at different convective Mach numbers with a $k - \epsilon$ model

C. Le Ribault

LMFA, CNRS URA 263,ECL, UCB Lyon 1
36, avenue Guy de Collongue
69131 Ecully cedex FRANCE

SUMMARY

Supersonic boundary layers for different convective Mach numbers ($M_c = 0.45, M_c = 0.65, M_c = 1$) have been computed with a $k - \epsilon$ model. Additional terms have been introduced in the kinetic energy equation to take in account compressibility effects (compressible dissipation, pressure-dilatation correlations). The first difficultie was the computation of inlet profiles for the turbulent quantities which give a mixing layer in equilibrium. Then, similarity properties for the velocity and kinetic energy profiles have been checked. The normalised spreading rate in function of the convective Mach number is compared with the experimental curve. The standard $k - \epsilon$ model fails to predict the decrease of the spreading rate in function of the convective Mach number. The models which takes in account compressibility effects improve the results.

Introduction

Turbulent mixing layers are important flow fields which are encountered in gas combustors and engine exhausts. A detailed understanding of their physics is essential for the development of turbulence models. The comportment of a supersonic mixing layer is different from the comportment of an incompressible mixing layer. Several experimental studies have confirmed that the spreading rate of a supersonic mixing layer relative to its low-speed counterpart is reduced. The intrinsic compressibility of a mixing layer is measured by the convective Mach number. Experimental curves plotting the growth rate relative to the convective Mach number have been constructed.

The objective of this test case is to see if the turbulence models are able to predict the reduced growth rate and predict the experimental curve.

Numerical method

A mixed finite element/finite volume method on non structured meshes adapted for compressible flows, initially developed by Partner 1 (INRIA), is used (Le Ribault[5], Dervieux[1]). The convective fluxes are integrated by a finite volume method with a Roe solver to catch the discontinuities and a 2nd order interpolation and the source and viscous terms are integrated by a finite element method. The time integration uses an implicit scheme, with performant preconditioned linear solvers (Hallo[4]). The resulting code, "NaturNG", developed at the LMFA, is a powerfull tool to simulate 2D and 3D compressible turbulent flows in complex geometries.

Turbulence models

The standard $k - \epsilon$ model does not include compressibility effects. To take into account an increase of the dissipation with the turbulent Mach number $Ma_t = \frac{\sqrt{2k}}{c}$, an additional compressible dissipation ϵ_c is added into the k equation. Different models, based on direct simulation results, have been proposed for ϵ_c (Sarkar [7], Zeman [9], Wilcox [8]) in term of the incompressible dissipation ϵ_s and the turbulent Mach number Ma_t:

$$\epsilon_c = F(Ma_t)\epsilon_s .\tag{1}$$

The difference between the models are the expression of the dependence on the turbulent Mach number $F(Ma_t)$:

$$F(Ma_t) = 0.5 * Ma_t^2 \text{ (Sarkar)}\tag{2}$$

$$F(Ma_t) = \begin{cases} 1.5 * (Ma_t^2 - 0.25^2) & \text{if } Ma_t \geq 0.25 \\ 0 & \text{if } Ma_t \leq 0.25 \end{cases} \text{ (Wilcox)}\tag{3}$$

$$F(Ma_t) = \begin{cases} 1 - e^{-\frac{(Ma_t - 0.25)^2}{0.66^2}} & \text{if } Ma_t \geq 0.25 \\ 0 & \text{if } Ma_t \leq 0.25 . \end{cases} \text{ (Zeman)}\tag{4}$$

$$\tag{5}$$

For compressible flow, the pressure fluctuation induces a decrease of the production and the dissipation through the pressure-dilatation correlation. Sarkar proposed the following model for the pressure-dilatation term in the turbulent kinetic energy equation:

$$\overline{p'\frac{\partial u_i'}{\partial x_i}} = \alpha_1 Ma_t^2 \overline{\rho u_i' u_j'}\frac{\partial \tilde{u}_i}{\partial x_j} + \alpha_2 Ma_t^2 C_\mu \bar{\rho}\epsilon\tag{6}$$

with $\alpha_1 = 0.4$ and $\alpha_2 = 0.2$.$\tag{7}$

When the compressible dissipation of Sarkar is used alone, the model is refered as Sarkar 1. When the compressible dissipation and the pressure-dissipation correlation are used together the model is refered as Sarkar 2.

Description of the test case and inflow conditions

This test-case has been defined by Dussauge [2]. The computations begin in a section where the mixing layer is fully developed and the flow field has similarity properties. Three different values of the convective Mach number have been chosen ($Mc = 0.45$, $Mc = 0.65$, $Mc = 1$) by Dussauge. In the first case ($Mc = 0.45$) the spreading rate should be almost the same as in the incompressible case whereas in the last case (Mc=1.0) the asymptotic value is reached. To normalise the compressible growth rate, we have performed subsonic computations with the same mean freestream velocity and density ratio.

The choice between supersonic/supersonic flows or supersonic/subsonic flows and the choice of the mean freestream velocity and density ratio are let free. For our mixing layer computations, two supersonic flows corresponding to an experiment realized by Goebel [3] have been chosen. The mean freestream velocity ratio $r = U_1/U_2$ is equal to 0.58 and the mean density ratio $s = \rho_1/\rho_2$ is equal to 1.36. In the experiment of Goebel, the Mach numbers of the two incoming flows are respectively equal to 1.91 and 1.37 and the convective Mach number is equal to 0.45. To check if the results don't depend on the freestream velocity and density ratio, additional computations have also been performed with the same freestream velocity ratio but with a lower density ratio equal to 1. The two cases are referenced in the text as case 1 (s=1.36) and case 2 (s=1).

In the inlet section, the profiles for the velocity and the density are deduced from experimental similarity profiles. The experiment gives also the value of the maximum of the turbulent kinetic energy k. However for the dissipation, no data are available. A first approximation of the dissipation is obtained by assuming that the turbulent viscosity is constant and 100 times greater than the molecular viscosity. But the mixing layer predicted by those boundaries conditions is not in equilibrium. The kinetic energy profiles should have similarity properties and the maximum of k should remain approximatively constant. In our computations, important longitudinal variations of k appear. To find more suitable values for the turbulent quantities, the outflow turbulent profiles are used as inflow conditions for a new computation and this operation is repeat until the k level remains approximatively constant in the middle of the mixing layer. Only two or three computations

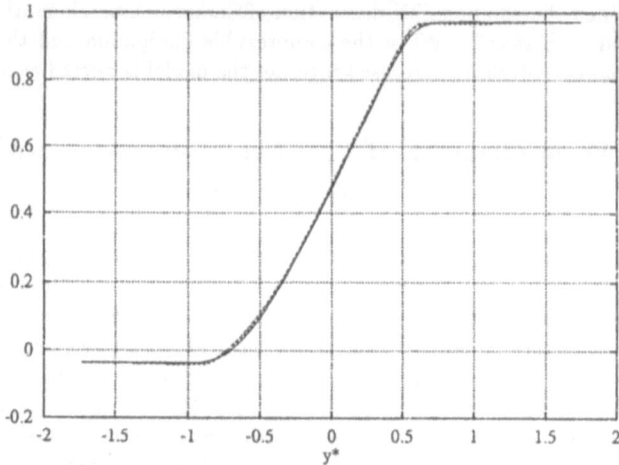

Figure 1: Mean velocity similarity profiles Mc=1.

are needed for each case. This operation must be repeated for each convective Mach number and each turbulence model.

Computational results

To ensure that the results are not dependant on the mesh, preliminary computations have been performed on three different meshes. From those tests we conclude that it is not necessery to use very refined meshes because in the asymptotic mixing layer, the gradients are not very strong. A mesh with 1800 nodes has been chosen for all computations (30 nodes in the x direction, 60 nodes in the y direction).

Similarity profiles for the mean velocity, the turbulent kinetic energy and the dissipation have then been plotted. The profiles are plotted in three sections, the beginning, the middle and the end of the mesh. The first plot presents $U^* = f(y^*)$, the second $k^* = f(y^*)$, the third $\epsilon^* = f(y^*)$ where $U^* = \frac{(U-U_1)}{\Delta U}$, $k^* = \frac{k}{\Delta U^2}$, $\epsilon^* = \frac{\epsilon L}{\Delta U^3}$, $y^* = \frac{(y-y_1)}{\delta}$, and y_1 is the coordinate where $U = U_2 + 0.95\delta U$. The profiles have been plotted for a mixing layer at a convective Mach number of 1. computed with the Sarkar 2 model. The similarity properties are respected.

The velocity thickness is defined by the section $\delta = y_1 - y_2$ where y_1 is the coordinate where $U = U_2 + 0.95\Delta U$ and y_2 is the coordinate where

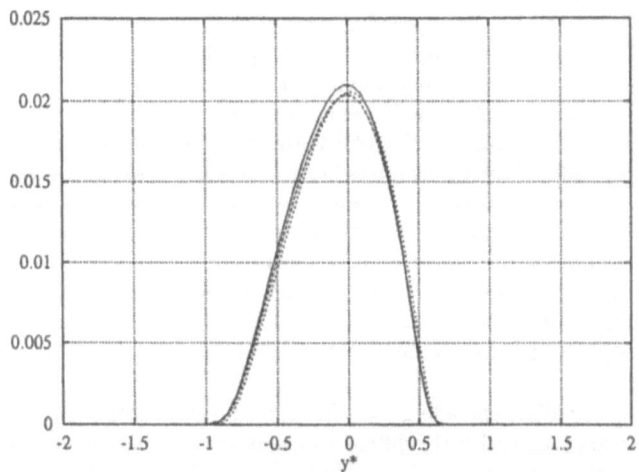

Figure 2: Turbulent kinetic energy similarity profiles Mc=1

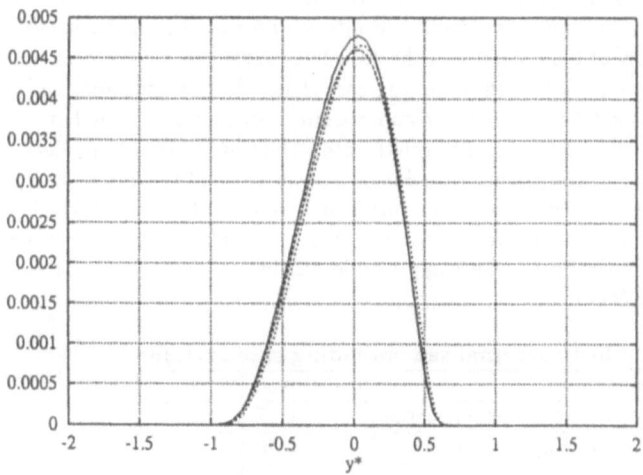

Figure 3: Dissipation similarity profiles Mc=1

$U = U_2 + 0.31\Delta U$. The spreading rate of the incompressible mixing layer with the freestream density ratio s=1 is equal to 0.0377 which is the value obtained at a convective Mach number of 0.45. However for the density ratio s=1.36, our numerical method does not converge for this incompressible case. At the convective Mach number of 0.45 the spreading rate of this mixing layer is equal to 0.034. As the compressibility effects are weak, the spreading rate should be almost the same as in the incompressible case. This value is then used to normalise the spreading rate at s=1.36. In the experiment of Goebel, the spreading rate is equal to 0.045, which is higher than the computed spreading rate. An empirical correlation is also given by Dussauge. The value of the spreading rate computed with this formula is equal to 0.0305 for s=1 and to 0.0312 for s=1.36 which is lower than the computed spreading rate.

The normalised spreading rate $F(Ma_c)$ in function of the convective Mach number computed with the different models and the experimental results of Papamoschou are presented in table 1a for s=1.36 and in table 1b for s=1. Results with the standard $k - \epsilon$ model, the Sarkar 1 model and the Sarkar 2 model are presented. Other models (Wilcox model, Zeman model) have also be tested in one case (s=1.36, Mc=1) but the results are worse than with the Sarkar 2 model (with the Wilcox model $F(Ma_c) = 0.73$ and with the Zeman model $F(Ma_c) = 0.79$). In the case of the mixing layer at a convective Mach number of 0.45, only the standard $k - \epsilon$ model is used because there is no compressibility effects. Results are presented for the convective Mach numbers of 0.45, 0.65, 1 in the case 1 (s=1.36) and for the convective Mach numbers of 0.45 and 1 for the case 2 (s=1). As expected, the standard $k - \epsilon$ model without the compressibility terms doesn't predict any effects of compressibility on the growth rate. The compressibility models predict a decrease of the growth rate due to a lower value of the kinetic energy, which induces a lower value of the turbulent viscosity. The model which gives the best results is the Sarkar 2 model. This model takes in account the two effects of compressibility (the compressible dissipation and the pressure-dilatation correlations).

Table 1a Normalised spreading rate s=1.36

M_c	0.45	0.65	1.
Papamoschou experiments	1.	0.77	0.5
Standard $k - \epsilon$ model	1.	0.96	1
Sarkar 1 model		0.91	0.79
Sarkar 2 model		0.85	0.70

Table 1b Normalised spreading rate s=1

M_c	0.45	1.
Standard $k - \epsilon$ model	1.	
Sarkar 1 model		0.76
Sarkar 2 model		0.70

Tables 2 and 3 give the maximum of k and of the Reynolds stress normalised by ΔU^2. The level of the maximum of k and of the turbulent Reynolds stress decreases with the growth rate and it is very important to adjust the inlet conditions for each case and each turbulent model. The level of the turbulent Reynolds stress in the mixing layer at a convective Mach number 0.45 is in good agreement with the experimental level and we can make the same remarks as for the spreading rate. The level decreases with the convective Mach number but not as strongly as in the experiment.

Table 2a Maximum of Reynolds stress s=1.36

M_c	0.45	0.65	1.
Experimental level	0.01	0.08	0.045
Standard $k - \epsilon$ model	0.0111	0.0108	0.0102
Sarkar 1 model		0.097	0.082
Sarkar 2 model		0.09	0.071

Table 2b Maximum of Reynolds stress s=1

M_c	0.45	1.
Standard $k - \epsilon$ model	0.0111	
Sarkar 1 model		0.084
Sarkar 2 model		0.075

Table 3a Maximum of the turbulent kinetic energy s=1.36

M_c	0.45	0.65	1.
Standard $k - \epsilon$ model	0.032	0.0308	0.029
Sarkar 1 model		0.0277	0.0226
Sarkar 2 model		0.0258	0.02

Table 3b Maximum of the turbulent kinetic energy s=1

M_c	0.45	1.
Standard $k - \epsilon$ model	0.0296	
Sarkar1 model		0.0244
Sarkar2 model		0.0215

Conclusion

Computations of mixing layers at different convective Mach numbers have been performed with the standard $k - \epsilon$ model and the $k - \epsilon$ model with compressibility terms. The standard $k - \epsilon$ model fails to predict important characteristics of the compressible mixing layer (the reduced spreading rate, the reduced level of the turbulent quantities). The model which takes in account compressibility effects (the compressible dissipation rate, the pressure dilatation correlation) improve the prediction of the spreading rate. However, the predicted decrease is not as important as in the experimental case. To improve the prediction, new turbulent models are needed based on now available direct simulation results.

References

[1] A.Dervieux "Steady Euler simulations using unstructured meshes",VKI,Lectures series 1884-04, (1985).

[2] Dussauge "Instructions for the computation of Test Case 1: Supersonic mixing layer".

[3] S.G. Goebel, J.C. Dutton, H.Krier, J.P. Renie "Mean and turbulent velocity measurements of supersonic mixing layers", Experiments in Fluids, vol. 8, pp. 263-272, (1990).

[4] "L. Hallo, JL. Munier, M. Buffat, G. Brun "iterative methods for solving implicit non-structured finite volume discretization of Euler equations" International Journal for Numerical Methods in Fluids, (1994).

[5] C. Le Ribault "Simulation dans écoulements turbulents compressibles par une methode mixte éléments finis /volumes finis", Thèse Ecole Centrale de Lyon (1991).

[6] D. Papamoschou, A. Roshko "The compressible turbulent sheart layer: an experimental study", J. Fluid. Mech., vol. 197, pp. 453-477, (1988).

[7] S. Sarkar, G. Erlabacher, M.Y. Hussaini "Compressible homogeneous shear: simulation and modeling", Eight Symposium on turbulent shear flows, Munich sept. 9-11 (1991).

[8] D.C. Wilcox "Turbulence modeling for CFD", Griffin Printing, Griffin Printing, Glendale, California, (1993).

[9] O. Zeman "Dilatational dissipation: the concept and application in modeling compressible mixing layers" Physics of Fluids A,2, 178-188 (1990).

SUPERSONIC MIXING LAYER

S. Tsangaris[1], A. Pentaris[1] and M. Thomadakis[2]

[1]National Technical University of Athens, Laboratory of Aerodynamics,

P.O. Box 64070, 15710 Athens

[2]Public Gas Corporation of Greece
DEPA SA

Mesogion 207, 11525 Athens

Summary

The computational results for an unsteady mixing layer flow are presented. An implicit numerical scheme for the solution of the unsteady NS equations has been used in junction with an algebraic turbulent model. Computations were performed using both first and second order time accuracy. The effect of the grid density on the quality of the solution is examined. The results obtained were in close agreement to the corresponding experimental data. The main difference between computational results and experimental data lies in the mixing layer growth at the downstream part of the domain.

1. Introduction

Shear layer mixing is an important fluid physics phenomenon in scramjet combustors. CFD analysts have to face two major difficulties in their effort to adequately predict compressible mixing layer flows. The first is the inherent unsteadiness of the flow structure, which requires efficient and robust numerical algorithms. The second is the turbulence model to be used when turbulent simulation is attempted.

In the present paper efforts have been focused on the flow solver. Our intention was to investigate the ability of our numerical tools in predicting such kind of flows. Standard turbulence models have been used without any modification for compressibility effects and unsteady phenomena.

2. Governing equations

The unsteady compressible Navier-Stokes equations written in the strong conservation form for a perfect gas, are the governing equations. A generalized coordinates' transformation from the physical (x,y,t) to the computational (ξ, η, τ) domain is performed:

$$\partial_\tau Q + \partial_\xi F + \partial_\eta G = \partial_\xi F_v + \partial_\eta G_v \tag{1}$$

where

$$Q = J^{-1}(\rho, \rho u, \rho v, \rho e)^T \tag{2}$$

$$F = J^{-1}(\rho U, \rho u U + \xi_x p, \rho v U + \xi_y p, (e+p)U - \xi_t p)^T$$

$$G = J^{-1}(\rho V, \rho u V + \eta_x p, \rho v V + \eta_y p, (e+p)V - \eta_t p)^T$$

$$F_v \quad , \quad G_v = (\text{Re}\, J)^{-1} \begin{pmatrix} 0 \\ a_x \tau_{xx} + a_y \tau_{xy} \\ a_x \tau_{xy} + a_y \tau_{yy} \\ a_x(u\tau_{xx} + v\tau_{xy}) + a_y(u\tau_{xy} + v\tau_{yy}) + \\ + \dfrac{1}{M_\infty^2(\gamma-1)}\left(\dfrac{\mu}{\text{Pr}} + \dfrac{\mu_t}{\text{Pr}_t}\right)(a_x T_x + a_y T_y) \end{pmatrix}$$

with $a=\xi$ for F_v and $a=\eta$ for G_v , J is the Jacobian of the transformation and τ_{xy} are the components of the shear stress tensor. Finally U and V are the contravariant velocities. For the simulation of turbulent flows, the algebraic eddy viscosity model of Baldwin and Lomax in its original form is implemented for the definition of the turbulent viscosity.

3. Numerical Scheme

For the solution of the system of equations (1) the method developed in [1] has been used. The method follows an implicit solution procedure [2] permitting the first or second order time accuracy of the computation. A generalized time differencing is applied for the approximation of the temporal derivative:

$$Q_\tau^n = \frac{1}{\Delta t}\frac{(1+\zeta)\Delta - \zeta\nabla}{1+\theta\Delta} Q^n + O\left[\left(\theta - \frac{1}{2} - \zeta\right)\Delta\tau + \Delta\tau^2\right] \tag{3}$$

where Δ and ∇ are the forward and backward difference operators. When $\theta=1$ and $\zeta=0$ the first order Euler scheme is obtained, while when $\theta=1$ and $\zeta=0.5$ the second order three - point - backward scheme is obtained. Second order central differences are used for the convective fluxes along with explicitly added blended second and fourth order dissipation terms. The viscous fluxes are discretized by central differences. They are temporally linearized in a manner which retains the second order time accuracy. This implementation has been shown

[3] to permit the use of longer values of the time step leading thus to the reduction of the CPU time necessary for the simulation of unsteady flows.

4. Results

The experimental setup and data are presented in [4]. Two grids were used for the numerical simulation of the flow. A coarse one with 45x43 points and a dense with 90x86 points, which is shown in figure 1. In figures 2a and 2b the mean velocity profiles at several streamwise positions are presented. Both the grids provide results which are in good agreement with the experimental data. Some differences in the growth of the mixing layer occur in the last three positions. In figure 3 the time histories of the velocity and pressure at various positions are presented. It is clear that the use of first or second order time accuracy leads to similar results. The periodicity of the flow is clear at every location of the flow. In addition the period of the phenomenon is constant. Finally, in figure 4 the power spectra density provided by the velocity and pressure time histories at several streamwise positions, are shown. The main frequencies are in good agreement to the experimental data and constant in the flow field.

REFERENCES

[1] Thomadakis M.P., PhD Dissertation, NTUA, Fluids Section, 1991.
[2] Beam R.M. and Warming R.F., AIAA J., Vol.16, No 4,1978, pp. 393-402.
[3] Thomadakis M.P. and Tsangaris S., 10th AIAA Applied Aerodynamics Conference,1992.
[4] Barre S., Quine C. and Dussauge J.P., Journal of Fluid Mechanics, Vol. 259, Jan 1994, pp. 47-78.
[5] Dupont P., Muscat P. and Dussauge J.P., ETMA, Deliverable WP1-T3D4,1994.

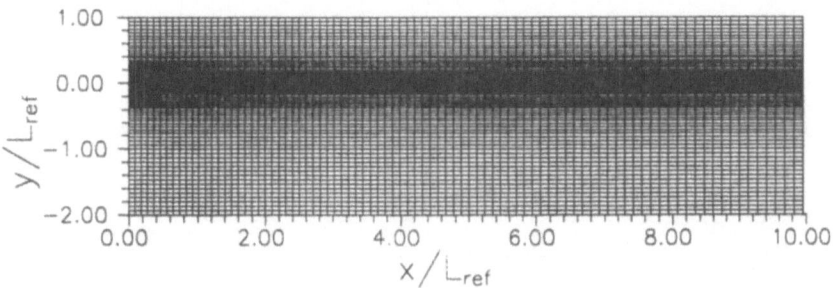

Figure 1. The 90x86 grid.

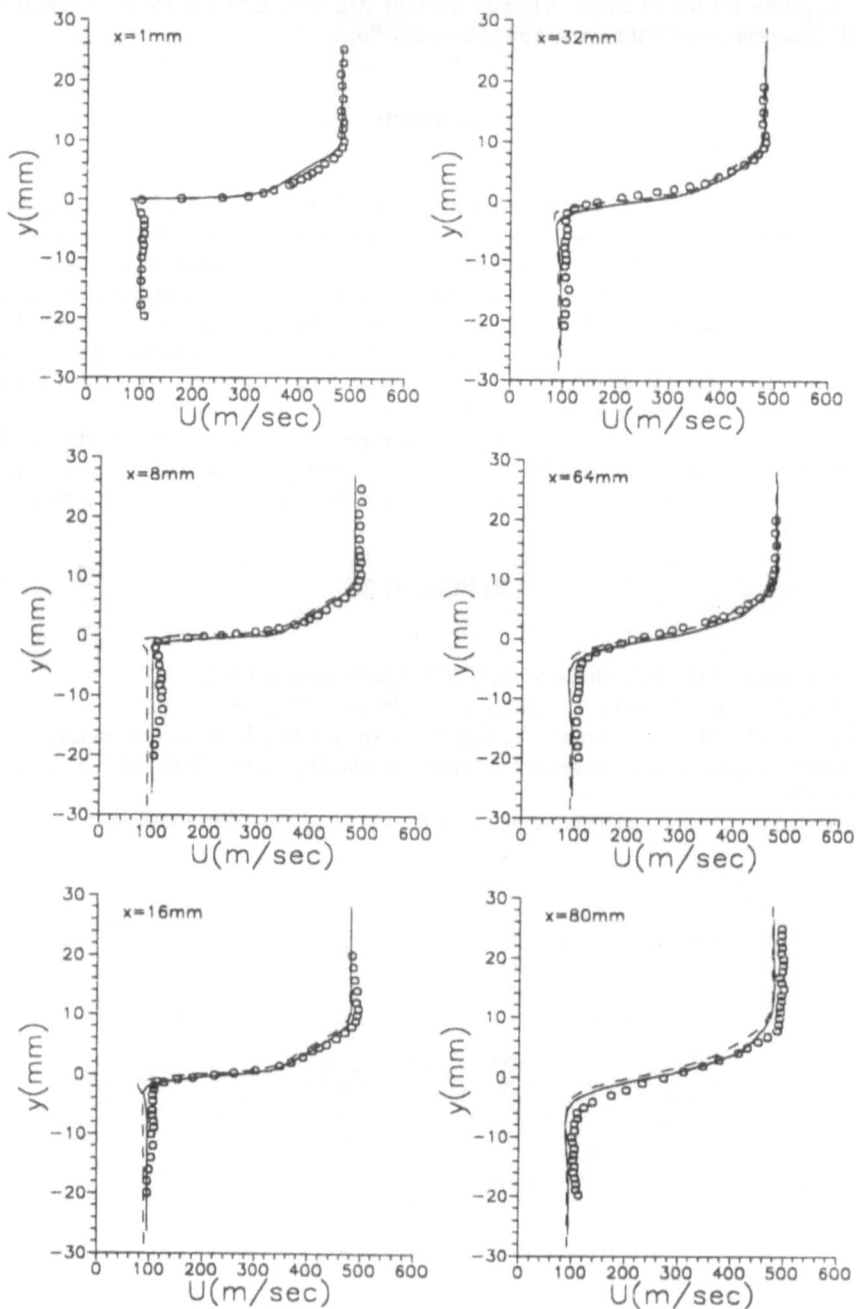

Figure 2a. Mean velocity profiles at various streamwise positions
oooo   experimental data   ———————   grid 45 x 43   - - - - - - - - -   grid 90 x 86

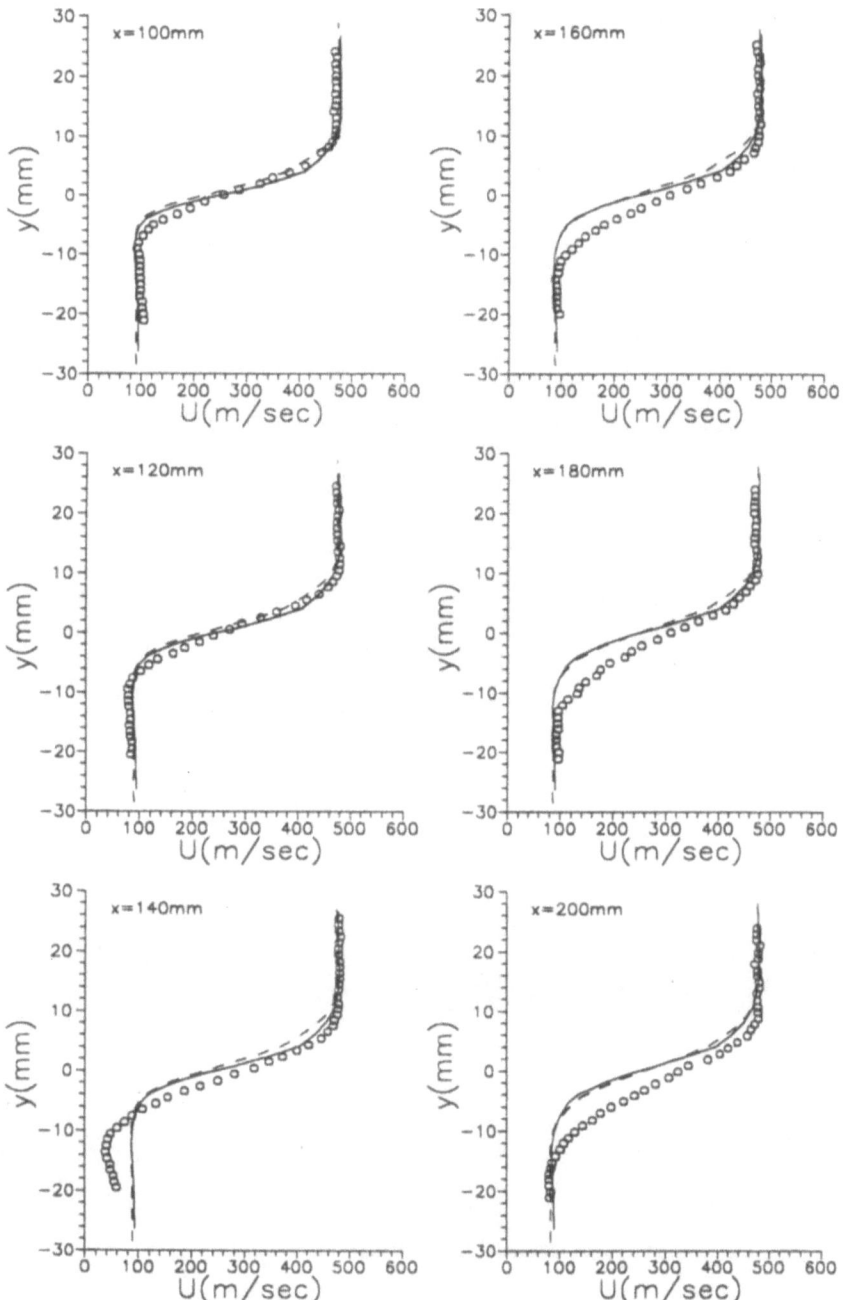

Figure 2b. Mean velocity profiles at various streamwise positions
oooo   experimental data   ———   grid 45 x 43   ---------   grid 90 x 86

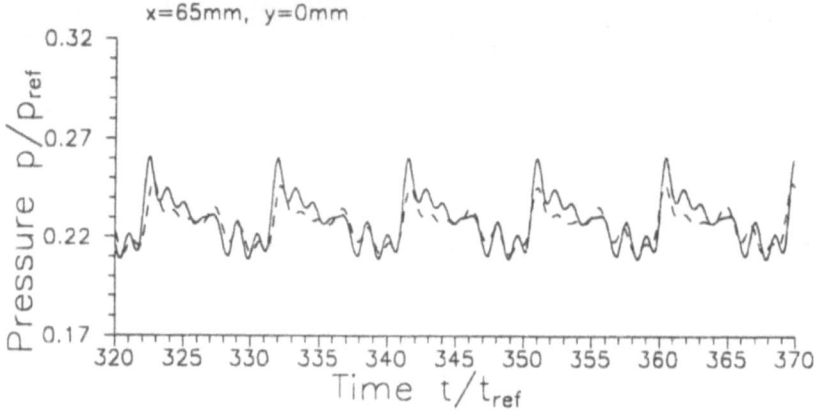

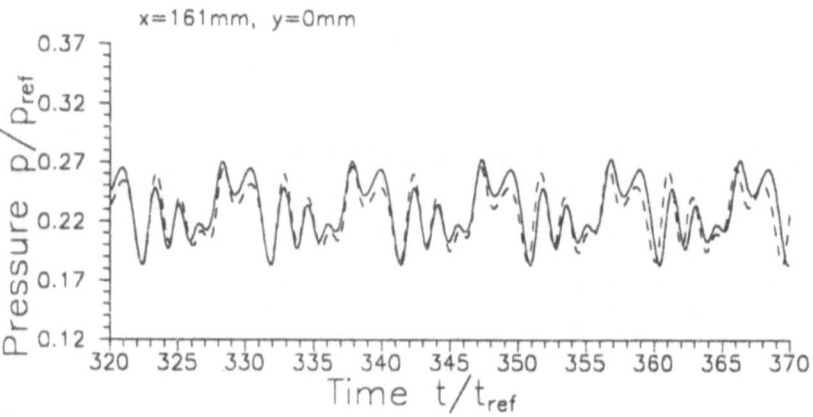

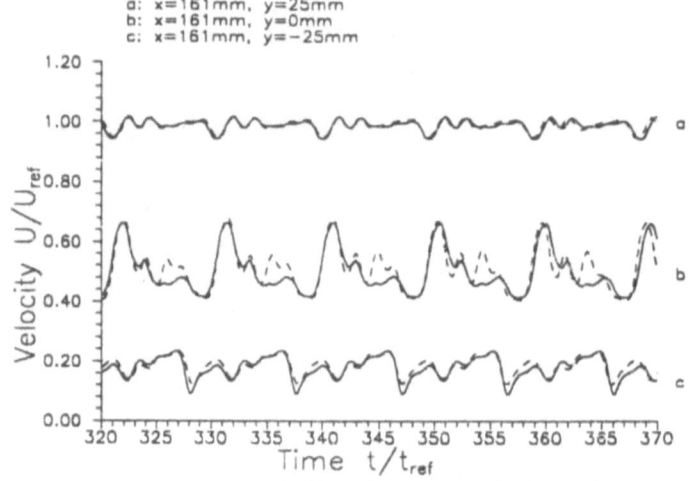

Figure 3. Pressure and velocity time histories.
——————— 1st order time accuracy
------------- 2nd order time accuracy

50

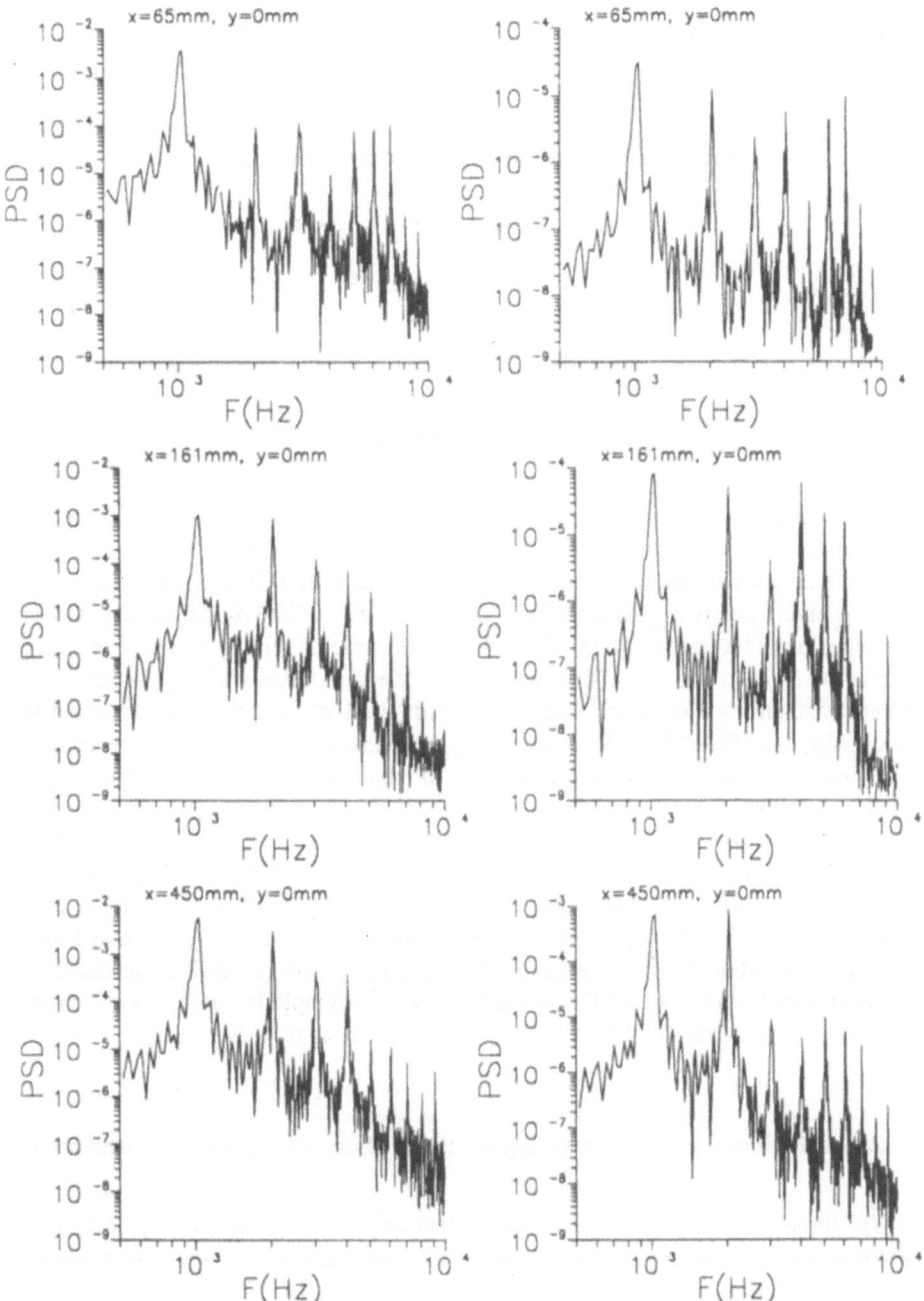

Figure 4. Power spectra for the velocity (left) and pressure (right)
at various streamwise positions (first order time accuracy)

Synthesis on compressible mixing layers

J-P. Dussauge
Institut de Recherche sur les Phénomènes Hors Equilibre
12, avenue Général Leclerc
13003 MARSEILLE, FRANCE

1. Introduction

In the workshop, five computations of compressible mixing layers have been presented; three of them (INRIA Sophia-Antipolis, University of Seville, ECL, Laboratoire de Mécanique des Fluides et d'Acoustique) were relative to TC100 (Steady supersonic mixing layer) and two of them (IMFT Toulouse and NTUA-Aerodynamics Athens) were performed to represent TC102 (unsteady supersonic mixing layers). For clarity, it may be recalled that TC 100 is dedicated to the computation of supersonic mixing layers, statistically steady and fully developed: similarity solutions are expected for flows with unity Prandtl numbers, and averages are made on all the scales of turbulence. The importance of having this test-case was to compare the results of computation on the same formulation for the normalized spreading rate vs. convective Mach number M_c (Langley curve). In TC 102, the large scale structures (or organized eddies) are computed explicitly, while the fine grained turbulence is modelled: there is no periodic forcing of the layer, only the natural modes of the layer are computed as unsteady perturbation. In TC100, asymptotic properties were computed (compilation of several experiments), while TC 102 was made for one particular flow in which the unsteadiness is documented , and in which the development of an asymptotic mixing layer from out of similarity conditions at the trailing edge of a separating plate is given.

2. Synthesis on TC 100 Compressible mixing layer (steady) by J.P. Dussauge:

The three computations by three contributors using the same type of method (mixed finite elements/finite volume methods on non structured meshes). Although the structure of the grid was not exactly the same (1800 nodes for ECL and 2469 nodes for Inria, and from 290 to 2280 nodes for Seville), it was checked that the solutions can be considered as mesh-independent. The models were different. INRIA and ECL used standard $k - \epsilon$ models with corrections for compressibility effects: the classical equations (Jones, Launder 1972, [12]) were used for k and ϵ, and compressibility corrections were introduced for the dilatation dissipation and for the pressure divergence terms. Four models were used: the models by Erlebacher, Hussaini, Kreiss, Sarkar (1990) [8] , by Zeman (1990) [19], and by Wilcox (1993) [18] for the dilatation dissipation and the model of Sarkar, Erlebacher, Hussaini (1992) [16] for the pressure divergence terms. These models are essentially functions of the turbulent Mach number. Seville used homogenization techniques [6] to build a $k - \epsilon - MPP$ model, and tested two sets of closure terms.

All the model computed successfully the incompressible mixing layer. The compressible cases consisted in three convective Mach numbers (0.45, 0.65 and 1), with one value of the density ratio, $s = 1.56$ which corresponds to a case of cross-gradients: the density and velocity gradient have opposite signs. The computations of Inria and ECL, using the same type of solver were in good agreement, since results made in comparable conditions agree within 5% accuracy. In Seville's results, testing a first set of closure terms resulted in numerical instabilities and too large values of the spreading rate. The computations with the other set were successful in subsonic cases but found no variation of the rate of spread with Mach number. The conclusion was that the homogenisation technique used together with compressible turbulence $k - \epsilon$ model requires further development to be fitted with compressibility effects.

All the computations agree to find small effects up to $M_c = 0.5$. It was checked that, for these models the convective Mach number appears as a similarity parameter, at least for the supersonic-supersonic combination, since the same value of the normalized spreading rate was obtained, for the same convective Mach number for a density ratio of 1.56 and for 1. It was also underlined that the results could be rather sensitive to the definition of the turbulent time scale in the equation for ϵ. The existing models gave a reasonable evolution of the Langley curve vs. M_c. The turbulent quantities k and $\overline{u'v'}$ were accordingly in agreement with the existing experimental data: it was known that these quantities varies approximately like the spreading rate. This behaviour is reproduced by the model. Moreover, recent work (Menaa 1996 [13]) confirms that for self-preservative solutions, the normalized turbulent fluxes decrease with M_c and do not depend much on the density ratio. However, the main conclusions that can be drawn is that none of the tested models can give a satisfactory evolution in the whole range of M_c: computations in good agreement for high values of M_c give poor results for low M_c, and conversely. More precisely, the computations have difficulties to find the right evolution of $\nu_t k^2 / \epsilon$ vs. M_c. As the predicted level of k is quite correct, this suggests that the level of ϵ vs. M_c is the problem to be addressed for such models. Similarly, as it is know that the tested compressible models are not efficient for supersonic turbulent boundary layers, it seems that a reassessment of the formulation of these models has to be made to improve their generality.

3. TC 102: Synthesis on compressible mixing layer (unsteady), by M. Braza.

3.1. Context of the study

The compressible mixing layer is a flow inherently unsteady, developing organised coherent structures. This is a physical situation occurring in many aerodynamic flows representing major industrial applications (e.g. the local dynamic characteristics downstream the trailing edge of the flow past wings, and past the separation point in a variety of flows around aerofoils, among other applications including internal aerodynamic flows and combustion, where the mixing layer dynamics governs the essentials of the flow process).

The methodology adopted is the Organised Eddy Simulation approach [3], [4], [5], [11], developed for flows characterised by pronounced periodicities. The principles of this approach are based on the physics of the flow and are briefly recalled in the present synthesis: a wide category of strongly sheared flows promotes the onset and the amplification of instabilities, owing to the inflexional shape of the streamwise velocity profiles and to a sufficiently high level of the Reynolds number, beyond a critical value. Owing to a high vorticity level, the amplification of instability waves lead to the formation of organized coherent structures, having distinctive characteristic length and time scales, comparing to the co-existing fine-scale turbulent motion. Consequently, the energy spectrum of these flows is not in statistical equilibrium according to the theory of Kolmogorov, because it presents predominant frequency peaks in the low and intermediate range, interspersed in continuous frequency windows and it has an overall slope which may be very different from the one of flows characterized by a statistical equilibrium. Despite the simultaneous development of the random turbulent motion, the persistence of the organised part on the apparently chaotic turbulence background constitutes a major physical characteristic which dictated the development of an adapted methodology. This methodology consists in solving the phase- averaged Navier-Stokes equations, according to the Organised Eddy Simulation approach, which adopts no more the Reynolds decomposition [14], but the phase-average one [5], for every unknown quantity. The derived phase-averaged Navier-Stokes equations are then solved by implementing one-point closure schemes, which have to be examined and reconsidered, in towards the existing closure schemes, which are mainly based on flows in statistical equilibrium. The ETMA research program allowed precisely the assessment of the efficiency of the OES approach in the context of free shear high-speed flows, by using, as a first step, simple algebraic turbulence models, modified though from the classical assumptions, in order to take into account the different energy exchange of the present unsteady flows.

3.2. Conclusions and future developments

The main conclusions of this collaborative study are summarised as follows:

54

- The OES approach is proven promising for the prediction of free shear high-speed flows with pronounced periodicities and this, by using two fundamentally different numerical codes by NTUA and IMFT respectively [17], [3]. This approach, which is not inherently three dimensional, allows the prediction of the organised part of the complex process of a turbulent high-speed flow and it offers generally a direct comparison of the phase-averaged quantities, because this averaging is a physically existing quantity and not a mathematical concept. These points constitute main differences from the LES approach.

- The predominant frequencies of the oscillations are predicted in good agreement between the two numerical studies and comparing with the physical experiment.

- The mean flow properties are provided by doing the statistics after computation over long physical time values. A good agreement of the mean flow parameters is achieved, especially for the near region. A less good agreement is obtained for the far region.

- Concerning the turbulence modelling implemented in the OES methodology, it is found that the modified closure [4] strengthens the amplitude of the predominant oscillations and provides a more clear organised pattern, comparing to the Baldwin-Lomax turbulence model. Moreover, a general conclusion from this study is that even by use of rather simple closure schemes, the principal flow characteristics are predictable because a major part of the complex physical processes is contained and resolved in the phase-averaging.

- The innovative character and expected achievements set by the ETMA program concerning the present category of flows have been realised through the present studies. Indeed, the successive steps to apply the OES approach on EXISTING numerical codes are clearly set through this project. The ability of even simple turbulence models the predict the frequency and amplitude of the inherently developed oscillations is also another important achievement of this project concerning the present category of shear flows and this fact interests particularly the industrial domain, where safe of CPU time constitutes a major challenge. Moreover, it is worthwhile to note that a major part of the compressibility effects is already contained in the resolved part of the turbulent motion through the OES approach and hence, there is less need to use compressibility corrections. However, for the sake of a complete study in a second phase, the implementation of higher order closures, including also compressibility corrections are worthwhile to be examined in the context of the OES approach in order to assess critically the overall behaviour, especially concerning the far field characteristics.

4. Conclusions and recommendations for future work

The comparisons made during the ETMA Workshop on TC100 and TC102 have shown very encouraging results. The one-point models indicate the right trend on the effect of compressibility on the rate of spread and on the turbulent quantities. They also show that in Organized Eddy Simulations the right frequencies are picked up by the simulations, even if the level seems more questionable: this is consistent with the assumption that in such free shear flows, the frequency depends mainly on the geometry and on external condition, but not critically on the damping due to viscosity (molecular or turbulent). However the workshop has underlined the limitations of the models in their present version. For the OES, it probably important to move to more general models. There are also hint for the improvement of the standard $k - \epsilon$ model. Huang, Bradshaw, Coakley (1994) [10] have underlined that the coefficients of this models had difficulties to reproduce the behaviour of simple flows with variable density. Working in this direction, together with some choices in the interpretation of the terms in the ϵ equation as suggested in El Baz and Launder (1994) [7] and perhaps also considering the influence of compressibility on turbulence anisotropy may be the first steps for building robust and accurate one-point models for compressible turbulence in shear flows.

Bibliography

[1] BALDWIN B. S., LOMAX H., "Thin layer approximation and algebraic model for separated turbulent flows." AIAA Paper, 78-257, January 1978.

[2] BARRE S., QUINE C., DUSSAUGE J.-P., "Compressibility effects on the structure of supersonic mixing layers: experimental results". J. Fluid Mech., vol. 259, pp.47-78, 1994.

[3] BRAZA M., HANINE H.,"Numerical simulation and modelling of an unsteady supersonic mixing-layer flow". Proceedings ETMA Workshop on Turbulence Modelling for Flows Arising in Aeronautics, UMIST, Manchester, Nov. 1994.

[4] BRAZA M., NOGUES P.,"Numerical simulation and modelling of the transition past a rectangular afterbody". Proceedings of the VIIIth Turbulent Shear Flow Conference, Munich, 9-11 Sept. 1991., pp. I-2-1, I-2-2.

[5] CANTWELL, B. J., & COLES, D. 1984 "An experimental study of entrainment and transport in the turbulent near wake of a circular cylinder". J. Fluid Mech. Vol. 136, 321-374.

[6] CHACON T. "Etude d'un modèle pour la convection des microstructures", Thèse de 3ème cycle, Univ. Paris VI, 1985

[7] EL BAZ A.M., LAUNDER B.E., "Second moment modelling of compressible mixing layer", Engineering Turbulence Modelling 1994and Experiments II, Ed. W. Rodi, M. Martelli, Firenze, Italy, 1993.

[8] ERLEBACHER G., HUSSAINI M.Y., KREISS H., SARKAR S., ICASE Report 90-15, 1990.

[9] HA MINH, H., VIEGAS , J. R., RUBESIN M. W., VANDROMME, D. D. & SPALART, P. 1989 Proceedings of the Seventh Symposium on Turbulent Shear Flows, Stanford, CA, USA, August 21-23.

[10] HUANG P.G., BRADSHAW P., COACKLEY T.J., "Turbulence models for compressible boundary layers", AIAA Jl., Vol. 32, N 4, 1994.

[11] JIN G.& BRAZA M. (1994) "A two-equation turbulence model for unsteady separated flows around airfoils". AIAA Journal , Vol. 32, N0 11, Nov. 94, pp. 2316, 2320.

[12] JONES W.P., LAUNDER B.E., "The prediction of laminarization with a two-equation model of turbulence", Int. Jl. Heat Mass Transfer, 15 :301-314, 1972.

[13] MENAA M., Thèse de Doctorat, Université de Provence, Marseille, France, in preparation,1995.

[14] REYNOLDS O. "An experimental investigation of the circumstances which determine whether the motion of water shall be direct of sinuous and the law of resistance in parallel channels. "Phys. Trans. Roy. Soc. London, 174, pp. 935-982, 1883.

[15] SARKAR S.,"The stability effect of compressibility in turbulent shear flow", J. Fluid Mech., 282, 163-186, 1995.

[16] SARKAR S., ERLEBACHER G., HUSSAINI M.Y., Icase Report 92-6, 1992.

[17] TSANGARIS S., PENTARIS A., THOMADAKIS M., "Supersonic mixing layer flow". Proceedings ETMA Workshop on Turbulence Modelling for Flows Arising in Aeronautics, UMIST, Manchester, Nov. 1994.

[18] WILCOX D.C., "Turbulence modeling for CFD", Griffin Printing, Glendale, California, 1993.

[19] ZEMAN O.,"Dilatation dissipation: the concept and application in modeling compressible mixing layers", Phys. Fluids A, Vol. 2, N 2, 1990, pp. 178-188.

CHAPTER 2 : COMPRESSIBLE BACK-STEP FLOW

The Supersonic Flow Over An Axisymmetric Rearward Facing Step. Synthesis of Results.

Kyriakos C. Giannakoglou
National Technical University of Athens,
Lab. of Thermal Turbomachines
P.O. Box 64069, 157 10 Athens, Greece
kgianna@central.ntua.gr

SUMMARY

This text concerns the numerical study of the the supersonic flow over an axisymmetric rearward facing step and is split into two parts. In the first part, specifications concerning data and the format of the requested results are provided. In the second part, numerical results obtained using two two-equation turbulence models are presented and compared with tha available experimental data.

INTRODUCTION

The prediction of the flow separation induced by a rearward facing step at supersonic flow conditions, remains a challenging problem. This case accumulates the difficulties of the numerical methods used in such configurations and the conditions at which classical turbulence models may provide a poor prediction. Turbulence is subjected to favourable and adverse pressure gradient and streamline curvature; furthermore compressible turbulence effects appear in the separated zone, while the accurate prediction of the reattachment length is of primary interest. The rearward facing step at supersonic flow conditions was examined experimentally by A. Roshko and G.J. Thomke [1] The three cases examined herein concern three step heights.

Numerical results obtained using two Navier-Stokes solvers, where closure is effected by two two-equation models, are presented and compared with experiments. Contributors , representing the Rockwell Science Center, Thousand Oaks, CA, USA [2] and the National Technical University of Athens, Greece [3] , describe their own methods in different articles in the same volume. The available experimental data are limited to the pressure coefficient distribution along the solid wall and the estimation of the reattachment length.

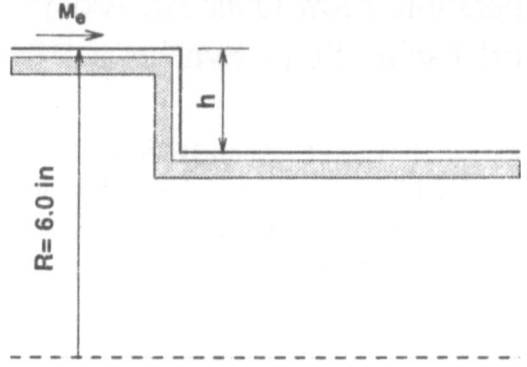

Figure 1: The axisymmetric body

TEST CASE SPECIFICATION

The flow is taking place around an axisymmetric body, which is sketched on figure 1. Upstream of the step, the forebody has a constant diameter equal to 12 in. The exact cylindrical part has a finite length of 1 in. only, but during the numerical experiments, the whole length of the inner surface, upstream of the step location, was modeled as an exact cylindrical surface. . The exact geometrical data and the different flow conditions, for which measurement are available, are listed in the aforementioned publication. The cases measured by Roshko and Thomke cover the range of Mach numbers from 2 to 4 approximately, while three different step heights (h=0.250 in. h=1.020 in. and h=1.675 in.) have been worked out. Two mandatory test cases have been proposed, in the ETMA Workshop, under the code number TC2, corresponding to extreme step height values. In both cases, the same Mach number M_s=2.09 is used, where M_s stands for the Mach number of the external flow just upstream of the separation point and h is the step height.

For the rearward facing step case, no recommended grids have been provided. The reason was that the geometry under consideration could be meshed using either a single- or a multi-block grid system. Thus, contributors were free to test their own grids.

REPORTING REQUIREMENTS

In order to keep a common style in the presentation of results and to facilitate their comparison, contributors were requested to follow the guidelines provided below:

62

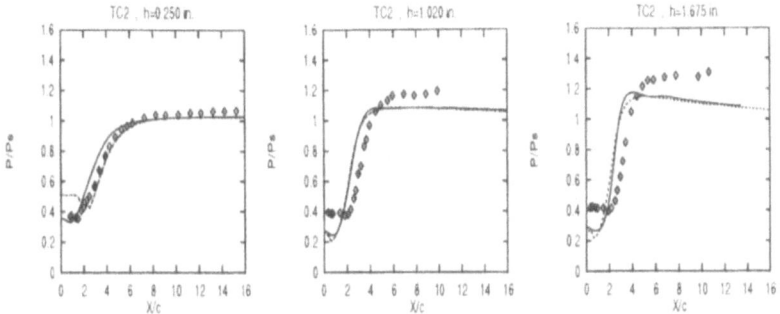

Figure 2: The wall pressure distribution

1. Plot the wall pressure normalized by the surface pressure just upstream of the separation point (P_s) versus the X-coordinate,normalized by the step height (abscissa). The X-coordinate is defined in the downstream direction of the step, its origin being the step location.

2. Plot the wall friction coefficient C_f along the X-coordinate normalized on the step height (abscissa).

3. Plot the Mach number distribution versus the Y-coordinate divided by the step height, at five successive stations, namely X = h,2h,3h,4h and 5h .

4. The reattachment lengths should be provided. In any case, these can be estimated by the C_f distribution plot.

COMPARISON OF THE NUMERICAL PREDICTIONS

The first numerical prediction presented herein, [3] , is based on an explicit fractional-step Navier-Stokes code using the low-Reynolds $k - \epsilon$ model. This will be refered to as *CODE1*. The second computational tool is based on a third-order accurate TVD formulation, solving for the turbulent kinetic energy and the undamped eddy viscosity (k^2/ϵ).This will be refered to as *CODE2*. Neither of the models used included curvature corrections.

In the comparisons which will be presented hereafter, a unique notation is followed. So, a continuous line corresponds to numerical predictions of [3] , while a dashed line corresponds to the numerical predictions of [2]. In case of available experimental data, these are given as marks in the figures. Each figure contains three graphs corresponding to the three step-heights examined, h=0.250 in. , h=1.020 in. and h=1.675 in.

The wall pressure distribution is shown in figure 2. For the low step height, the comparison of *CODE1* and *CODE2* with the pressure measurements is very satisfactory.

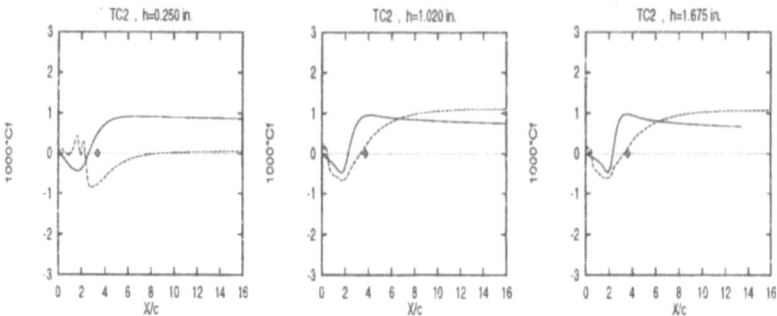

Figure 3: The friction coefficient distribution

Nevertheless, local examination of the pressure distribution within the recirculation zone reveals that the two codes behave in a different way. *CODE1* predicts a better pressure distribution very close to the step height, while *CODE2* better captures the pressure recovery. On the contrary, for the h=1.020 in. and h=1.675 cases, there are important discrepancies in the location of the pressure recovery, as well as in the pressure level along the part of the wall after the flow reattachement. Both codes underestimate the pressure level, by predicting exactly the same pressure level.

The friction coefficient distribution is presented in figure 3 ; the mark which appears in this figure demonstrates the location of flow reattachment, measured in [2], since no experimental C_f distribution is available. For the lower step height, the C_f distribution obtained using *CODE2* must be not taken into account, since much more reasonable results have been shown during the Workshop; as yet, these updated results are not available to the case responsible.

The standard low-Reynolds $k-\epsilon$ model used in *CODE1* was expected to underpredict the reattachment length, as it is usually shown , by numerous other users of this model, when applied in step-like configurations, regardless of the level of Mach number. It must kept in mind that, apart from differences related to the solution variables (*CODE1* solves the turbulence equations in k and ϵ, while *CODE2* solves in k and k^2/ϵ, see comments in [2]), differences also exist in the low Reynolds terms, implemented in the two models.

The Mach number profiles at five locations downstream of the step position, are plotted in figures 4, 5, 6, 7 and 8. These profiles reflect two situations, namely (a) how accurately the reattachment of the flow was predicted and (b) the external flow characteristics. The ordinate in this plots has been set to twice the step height, which is marginal for the estimation of (b) in the case of the very low height. In the other two cases, the agreement of the results of the two codes in the outer region is excellent. Of course, close to the solid wall of the inner cylinder, the two flow profiles remain separated up to a different percentage of the step height, since different reattachment lengths have been predicted. The oblique shock induced compression, in the vicinity of the reattachment point, can be seen in figures 7 and 8, at distances greater (approximately)

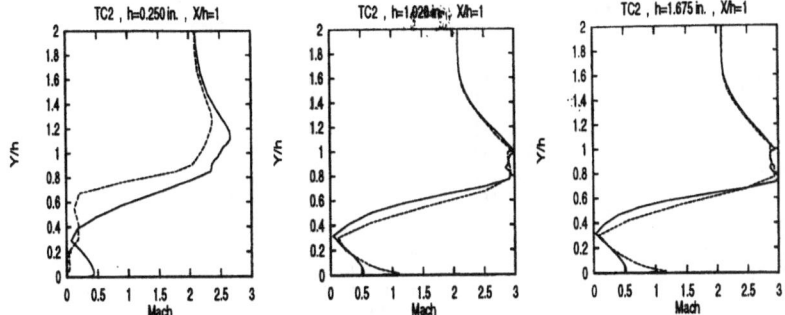

Figure 4: The Mach number profile at X/h=1

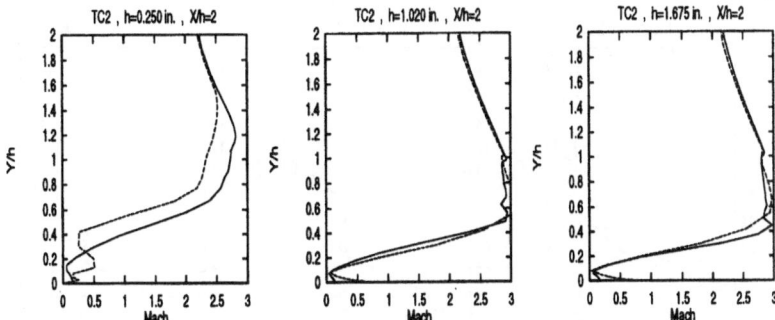

Figure 5: The Mach number profile at X/h=2

than four times the step height. In 8, the compression predicted by *CODE1* is found above the location of the corresponding compression as predicted via *CODE2* and this is reasonable since the reattachment occurs slightly upstream.

For the sake of completeness, some additional comments are listed below:

1. In both codes, two-grid topologies, have been used. The two-grid topologies used had considerable differences with respect to the size of the computational domain upstream and downstream of the step, the number of grid nodes used, the distance of the first grid line off the solid wall and the communication at the two grid interfaces. It is to be pointed out that, in [3], two grids overlap.

2. Different inlet profiles have been used in the two runs. They are described, in full detail, in the cited references, [2] and [3].

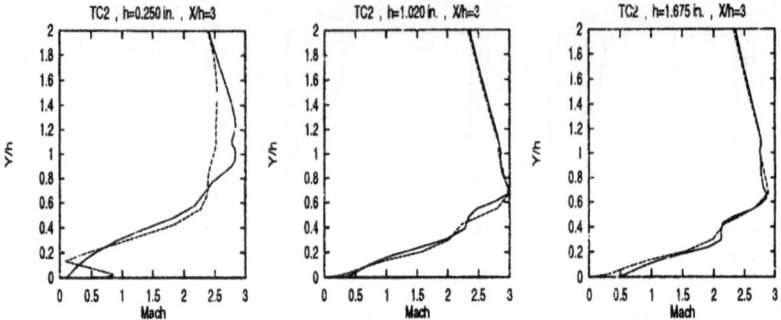

Figure 6: The Mach number profile at X/h=3

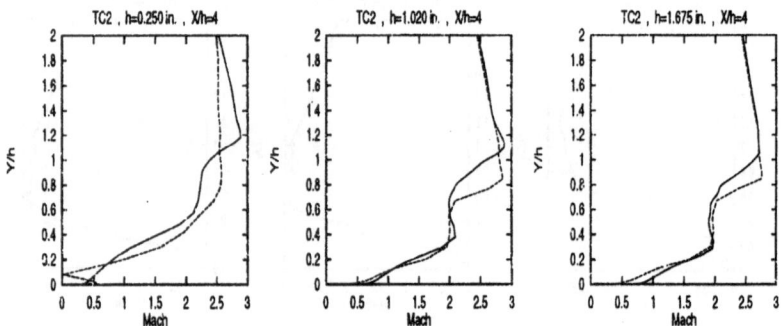

Figure 7: The Mach number profile at X/h=4

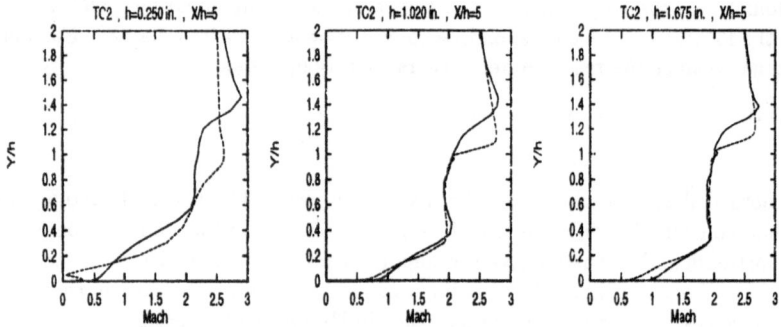

Figure 8: The Mach number profile at X/h=5

REFERENCES

[1] *Roshko, G. J.*, and *Thomke, J. E.* , "Observations. of Turbulent Reattachment behind an Axisymmetric Downstream-Facing Step in Supersonic Flow" , AIAA Journal **4**, No 6, (1966).

[2] *Goldberg, U. C.*, "ETMA Workshop. TC2: Supersonic Back Step " , ETMA Workshop, Manchester, U.K. (1994).

[3] *Vassilopoulos, C., Giannakoglou, K. C.* , and *Papailiou K. D.*, "Supersonic Rearward-Facing Step Calculations Using an Explicit Fractional-Step Method and a Two-Equation Turbulence Model" , ETMA Workshop, Manchester, U.K. (1994).

COMPUTATION OF AN AXISYMMETRIC SUPERSONIC BACK-STEP FLOW USING A POINTWISE $k - k^2/\epsilon$ TURBULENCE MODEL

Uriel C. Goldberg

CFD Department, Rockwell Science Center

Thousand Oaks, California 91360, USA

SUMMARY

A recently introduced $k - k^2/\epsilon$ turbulence model is used to compute supersonic flow over axisymmetric backward-facing steps with several step heights. experimental data are by Roshko and Thomke (1966). This case involves a separated flow region downstream of the step as well as an oblique shock interacting with the boundary layer in the reattachment region. It is, therefore, considered a challenging test for turbulence models. The model, in its present form, predicts the extent of the flow separation rather well, however, except for the case with the smallest step height, its prediction of the wall pressure distribution is fair at best.

INTRODUCTION

The standard high Reynolds number $k - \epsilon$ model is a good performer in free shear flows but its extension to low Reynolds number wall-bounded flows proved over the years a rather difficult task with only limited success.

One approach follows the method of Jones-Launder [1] and Launder-Sharma [2] in which the dissipation rate of turbulence kinetic energy, ϵ, is replaced by $\tilde{\epsilon} = \epsilon - 2\nu(\partial\sqrt{k}/\partial y)^2$ which, unlike ϵ itself, vanishes at solid surfaces. This necessitates adding the term $S = 2\nu\nu_t(\partial^2 U/\partial y^2)^2$ to the $\tilde{\epsilon}$ transport equation. The presence of the normal-to-wall first and second derivatives in $\tilde{\epsilon}$ and in S, respectively, introduces two disadvantages; a numerical and a conceptual one: these derivatives create sometimes severe numerical "stiffness", limiting the usable size of the time step and increasing the chance of numerical transients severe enough to terminate a computation; they also prevent the model from being pointwise (local) because of the need to find normal-to-wall unit vectors.

Another approach is to use the ϵ transport equation unaltered, such as in the Lam-Bremhorst [3] model, but then the presence of the unnatural wall boundary condition $\epsilon = \nu\partial^2 k/\partial y^2$ becomes the source of numerical difficulties and inability to render the model pointwise.

In both the above approaches there is a need for two or three near-wall functions to convey the effect of viscous damping and to ensure proper asymptotic behaviour in the viscous sublayer.

This state of affairs regarding low Reynolds number extensions to the standard $k - \epsilon$ model has been a constant source of difficulties and frustration to CFD users attempting to predict real life flow problems.

Since the source of trouble is evidently the dissipation rate equation, some researchers replaced it with alternative transport equations to determine the length scale. A significant example is the Wilcox $k-\omega$ model [4] in which ϵ is replaced by $\omega = \epsilon/(C_\mu k)$. The model performs better that the $k - \epsilon$ model in adverse pressure gradient flows and is compatible with near-wall flows without the need to use damping functions; it is, however, a poor performer in free shear flows because of its sensitivity to the freestream condition of ω. Another source of difficulty is the singularity of ω at solid surfaces. This requires setting it to its value at the first mesh point off walls, $\omega_1 = C(\nu/y^2)_1$ which is, again, a non-pointwise attribute and also introduces an extra measure of sensitivity to normal-to-wall mesh distribution.

Baldwin and Barth [5] and later Goldberg [6] introduced two equation models in which the ϵ equation was replaced by one for $R \equiv k^2/\epsilon$. With both k and R vanishing at walls the latter model is completely pointwise and involves only a single near-wall damping function. The model proved to be a good performer for wall-bounded flows including those involving separated flow regions. It does, however, exhibit some sensitivity to the freestream value of R which impairs its performance in predicting free shear flows. The reason for this behaviour is the action of the extra difusion term present in that model while it enhances turbulence propagation from solid surfaces into the flow, it inhibits such propagation in defect portions of boundary layers and in free shear flows. The next section explores this aspect of the model in detail and suggests a possible remedy to be explored in future work.

MODEL FORMULATION

We start with the standard high Reynolds number $k - \epsilon$ model.

$$\frac{Dk}{Dt} = \nabla \cdot \left(\frac{\nu_t}{\sigma_k} \nabla k \right) + P - \epsilon , \tag{1}$$

$$\frac{D\epsilon}{Dt} = \nabla \cdot \left(\frac{\nu_t}{\sigma_\epsilon} \nabla \epsilon \right) + C_{\epsilon 1} \frac{\epsilon}{k} P - C_{\epsilon 2} \frac{\epsilon^2}{k} . \tag{2}$$

The turbulence production term is given in terms of the Boussinesq concept

$$P = \left[\nu_t \left(\frac{\partial U_i}{\partial x_j} + \frac{\partial U_j}{\partial x_i} - \frac{2}{3} \frac{\partial U_k}{\partial x_k} \delta_{ij} \right) - \frac{2}{3} k \delta_{ij} \right] \frac{\partial U_i}{\partial x_j} . \tag{3}$$

Here k is the turbulence kinetic energy; ϵ is its dissipation rate; U_i are the Cartesian mean velocity components; x_i are the corresponding coordinates; ν and ν_t are the molecular and eddy kinematic viscosities, respectively; and D/Dt is the material derivative. The constants appearing in these transport equations will be discussed later.

A transport equation for the undamped eddy viscosity, $R \equiv k^2/\epsilon$, may be derived from these equations. For example, $\nabla \epsilon = (2k/R)\nabla k - (k/R)^2 \nabla R$. Carrying out the algebra and assuming that $\sigma_k = \sigma_\epsilon = \sigma$ results in the following equation, written in conservation form:

$$\begin{aligned}
\frac{D\rho R}{Dt} = \nabla \cdot \left(\frac{\mu_t}{\sigma} \nabla R \right) &+ (2 - C_{\epsilon 1}) \frac{R}{k} \rho P - (2 - C_{\epsilon 2}) \rho k \\
&- \frac{2}{\sigma} C_\mu \rho \nabla R \cdot \nabla R \quad (I) \\
&- \frac{2}{\sigma} \frac{\mu_t R}{k^2} \nabla k \cdot \nabla k \quad (II) \\
&+ \frac{4}{\sigma} \frac{\mu_t}{k} \nabla k \cdot \nabla R \quad (III) .
\end{aligned} \tag{4}$$

Here ρ is the density, and $\mu_t = \rho \nu_t = C_\mu \rho R$.

In the process of deriving their one-equation model, Baldwin and Barth [5] neglected the last two diffusion terms, retaining only the first in the slightly different form $-2/\sigma \rho \nabla \nu_t \cdot \nabla R$. Goldberg [6] proposed a pointwise version of the Baldwin-Barth model, still using this term only. It is of interest, therefore, to reexamine the effect of the three terms on the behaviour of the model. Experience indicates that the original models, using term (I) only, enable good prediction of near-wall flows but may fail in the defect layer and in free shear flows. The reason and possible remedy may be understood from evaluating the normal-to-wall entrainment velocity. By Eq.(4) this velocity is given as

$$\tilde{v} = v + 2\frac{C_\mu}{\sigma}\frac{\partial R}{\partial y} + v_{II} - \frac{4}{\sigma}\frac{\nu_t}{k}\frac{\partial k}{\partial y} \ , \tag{5}$$

where v_{II} is the contribution of term (II) which is not readily translatable into a normal-to-wall velocity but behaves similar to v_I. Eq.(5) provides a rather clear picture: in the near-wall region $\partial R/\partial y$ and $\partial k/\partial y$ are both positive, so that v_I enhances while v_{III} reduces diffusion away from the wall, making the former desirable and the latter undesirable; in defect layers the situation is reversed: now both $\partial R/\partial y$ and $\partial k/\partial y$ are negative, hence v_{III} becomes instrumental in diffusing the turbulence into the non-turbulent zone. A possible criterion for incorporating term (III) is to invoke it only if

$$[(\mathbf{V}\cdot\nabla)R](\nabla k \cdot \nabla R) < 0 \tag{6}$$

where $\mathbf{V}$ is the velocity vector. This constitutes a pointwise generalisation of the one-dimensional criterion $v(\partial k/\partial y) < 0$.

A simple yet useful initial test case is fully developed flow in a channel, since it includes both near-wall and outer flow regimes. Using Wilcox's [7] PIPE flow solver the two velocity terms v_I and v_{III} are plotted against wall distance in Fig. (1). The regions of positive and negative entrainment velocities from the two diffusion terms, (I) and (III), are clearly seen, in correspondence with Eq.(5). In particular, the negative portion of v_{III} adjacent to the walls is observed; this is the portion that must be avoided to circumvent the corresponding reduction in entrainment velocity.

The proposed transport equation for R, written in low Reynolds number form, thus reads

$$\frac{D\rho R}{Dt} = \nabla\cdot[(\mu + \mu_t/\sigma)\nabla R] + (2 - C_{\epsilon 1})\frac{R}{k}\rho P - (2 - C_{\epsilon 2})\rho k$$
$$- 2\frac{C_\mu}{\sigma}\rho\nabla R\cdot\nabla R + \mathcal{D} \tag{7a}$$

where the cross-diffusion term is given by

$$\mathcal{D} = \frac{4}{\sigma}\frac{\mu_t}{k}\nabla k\cdot\nabla R \tag{7b}$$

if the criterion in Eq.(6) is met, otherwise $\mathcal{D} = 0$.

Evaluating Eq.(7) at the logarithmic overlap, the following relation results:

$$\sigma = \frac{\kappa^2}{\sqrt{C_\mu}(C_{\epsilon 2} - C_{\epsilon 1})} \tag{8}$$

where $\kappa = 0.41$ and $C_\mu = 0.09$. The other model constants are chosen to be those suggested by Myong and Kasagi [8], namely, $C_{\epsilon 1}=1.42$, $C_{\epsilon 2}=1.83$, and from Eq.(8) it follows that $\sigma=1.367$.

The eddy viscosity is now given by

$$\mu_t = C_\mu f_\mu \rho R \tag{9}$$

where the near-wall damping function is chosen as

$$f_\mu = \frac{1 - e^{-A_\mu R_T}}{1 - e^{-A_\epsilon R_T}} \tag{10}$$

using the turbulence Reynolds nubmer $R_T \equiv R/\nu$. The constants here are determined by optimisation for flat plate near-wall flow, resulting in $A_\mu = 0.017$ with $A_\epsilon = C_\mu^{3/4}/2\kappa = 0.2$.

The two transport equations, Eqs.(1) and (7) (the former in low Reynolds number conservation form with ϵ replaced by its equivalent, k^2/R), are subject to the following boundary conditions:

(i) Solid Walls

$$k = 0 , \ R = 0 , \tag{11}$$

(ii) freestream (and initial) conditions

$$k/U_{\text{ref}}^2 \ll 1 , \ R/\nu_{\text{ref}} < 1 . \tag{12}$$

In general k_∞ is determined by the level of freestream turbulence, otherwise a value of $k/U_{\text{ref}}^2 = 10^{-6}$ is used. R_∞ is kept below the level of the molecular viscosity.

The $k - R$ turbulence model, with $\mathcal{D} \equiv 0$ in Eq.(7), was included in the USA Reynolds-averaged Navier-Stokes multi-block structured grid flow solver, which is an up to 3rd order accurate solver based on an upwind TVD scheme for the convection terms within a finite volume framework [9].

Supersonic Axisymmetric Back-Step Flow (ETMA TC2)

A Mach 2.1 flow calculation over an axisymmetric backward-facing step geometry is compared with experimental data by Roshko and Thomke [10]. The expanding flow detaches from the step corner, forming a free shear layer, then reattaches onto the lower cylindrical surface through an oblique shock- induced compression. A recirculation bubble is formed between the step corner and the reattachment point. The diameter of the cylinder downstream of the step is variable, so that flows with various step heights can be produced. Figure 2 shows a sketch of the geometry including some flow features.

The computations were done on a two zone grid topology, with 10×55 points in the zone upstream of the step, starting 1" upstream of the corner, and 100×90 points in the downstream zone, extending fifty step heights from the step. The upper boundary was located 25 step heights from the step. The minimum grid spacing off the cylindrical surfaces was $7.5 \times 10^{-5}h$ and that off the vertical step was $2.0 \times 10^{-2}h$, where h is the step height. Figure 3 shows a typical grid topology.

The flow was initialized with freestream values of velocity, pressure, and density. The inflow velocity profile followed a 1/7th power law using the experimentally observed boundary layer thicknesses: δ_S/h=0.548, 0.1343, 0.0818, respectively, for the step heights h=0.25, 1.02, and 1.675". Other inflow conditions were: M_S=2.09, $Re_S = 10^6$/inch, P_S/P_∞=0.81, and the stagnation temperature was assumed to be $T_{0,\infty}$ =300 K, which was also the wall temperature. The eddy viscosity profile was set to $\mu_t = A(y/\delta_S)\exp\{-B(y/\delta_S)\}$ where the constants A and B were such that the peak value, $\mu_t \approx 10$, occurred at $y/\delta_S \approx 0.2$. A similar treatment was employed for the inflow turbulence kinetic energy profile: $k = 10^{-3}(y/\delta_S)\exp\{-B(y/\delta_S)\}$.

Figs. 4(a) and (b) are wall pressure and skin friction distributions for the h=0.25" case. Figs. 5(a,b) and 6(a,b) are the corresponding ones for the h=1.02 and 1.675" cases, respectively. The reattachment locations are rather well predicted for all step heights. The pressure distribution, however, is well predicted only for the h=0.25" case; For the larger steps, pressure is only fairly predicted: the plateau in the separated flow region is not found, the recompression shock is predicted too early, and the pressure level downstream of reattachment is underpredicted, suggesting a weaker shock than the one observed experimentally. Fig. 7 is an isoMach contour plot showing the expansion fan at the step shoulder and the oblique shock interacting with the boundary layer at the end of the separation bubble.

CONCLUSIONS

This paper analysed the effect of the extra diffusion terms which arise from deriving a transport equation for the undamped eddy viscosity, k^2/ϵ, from the standard $k - \epsilon$ turbulence model. Based on evaluation of the diffusion velocities it was shown that a selective utilisation of the cross-diffusion term, in addition to the diffusion term already used in the original model, may improve predictive capability of boundary layer outer regions and free shear flows. While this potential improvement remains subject to future investigation, the original model was used here to compute supersonic flow over axisymmetric backward-facing steps using three step heights. These involve a separated flow region downstream of the step, and an oblique shock/boundary layer interaction in the flow reattachment zone. The model's performance was good in predicting the extent of flow separation, but the pressure distribution was only fairly predicted except in the case of the smallest step height.

Since the free shear layer at the outer edge of the separation bubble is of major importance in these flows, the effect of the cross-diffusion term, described in this paper, on the performance of this model for such flows must be investigated.

REFERENCES

[1] Jones, W. P., Launder, B. E.: "The Prediction of Laminarization with a Two-Equation Model of Turbulence", *International Journal of Heat and Mass Transfer*, **15** (1972), pp. 301–314.

[2] Launder, B. E., Sharma, B. I.: "Application of the Energy-Dissipation Model of Turbulence to the Calculation of Flow Near a Spinning Disc", *letters in Heat and Mass Transfer*, **1** (1974), pp. 131–138.

[3] Lam, C. K. G., Bremhorst, K. A.: "Modified Form of $k - \epsilon$ Model for Predicting Wall Turbulence", *ASME Journal of Fluids Engineering*, **103** (1981), pp. 456–460.

[4] Wilcox, D. C.: "Reassessment of the Scale Determining Equation for Advanced Turbulence Models", *AIAA Journal*, **26** (1988), pp. 1299–1310.

[5] Baldwin, B. S., Barth, T. J.: "A One-Equation Turbulence Transport Model for High Reynolds Number Wall-Bounded Flows", NASA TM 102847 (1990).

[6] Goldberg, U. C.: "Toward a Pointwise Turbulence Model for Wall-Bounded and Free Shear Flows", *ASME J. Fluids Eng.*, **116** (1994), pp. 72–76.

[7] Wilcox, D. C.: *Turbulence Modeling for CFD*, DCW Industries, Inc., La Cañada, California (1993).

[8] Myong, H. K., Kasagi, N.: "Prediction of Anisotropy of the Near-Wall Turbulence With an Anisotropic Low-Reynolds-Number $k - \epsilon$ Turbulence Model", *ASME Journal of Fluids Engineering*, **112** (1990), pp. 521–524.

[9] Chakravarthy, S. R., Szema, K.-Y., Haney, J. W.: "Unified 'Nose-to-Tail' Computational Method for Hypersonic Vehicle Applications", AIAA Paper 88-2564 (1988).

[10] Roshko, A., Thomke, G. J.: "Observations of Turbulent Reattachment behind an Axisymmetric Downstream-Facing Step in Supersonic Flow", *AIAA Journal*, **4**, No. 6 (1966), pp. 975–980.

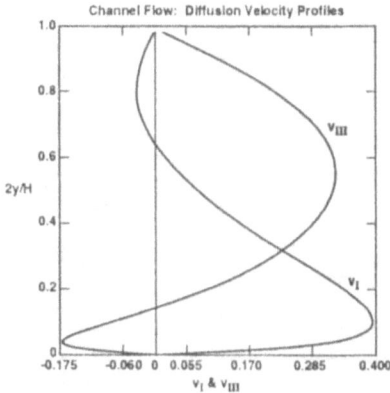

Fig. 1. Channel flow: diffusion velocity profiles

SC.1435E.011095

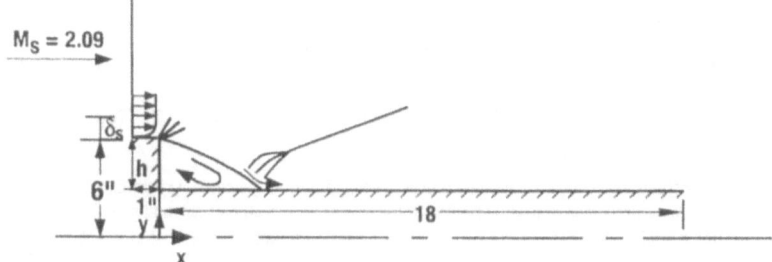

Fig. 2. Supersonic axisymmetric back-step flow: geometry and flow features

SC.1362E.112294

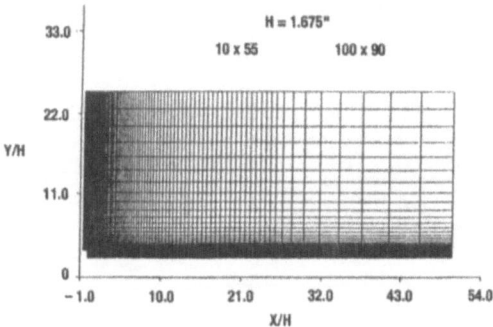

Fig. 3. Supersonic axisymmetric back-step flow: typical grid topology

73

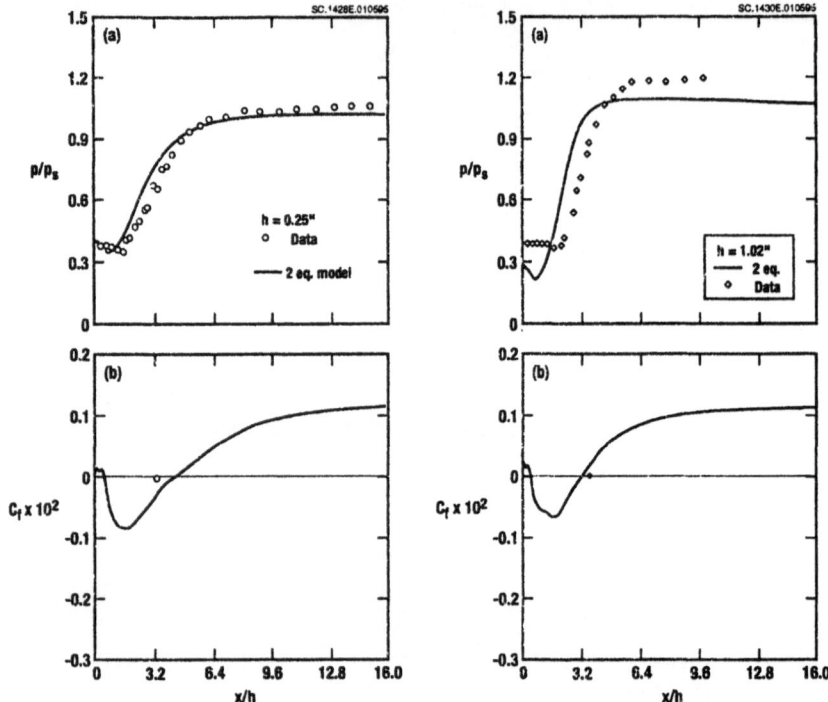

Fig. 4(a,b). Supersonic Axisymmetric back-step flow: wall pressure and skin friction profiles for the $h = 0.25"$ case

Fig. 5(a,b). Supersonic Axisymmetric back-step flow: wall pressure and skin friction profiles for the $h = 1.02"$ case

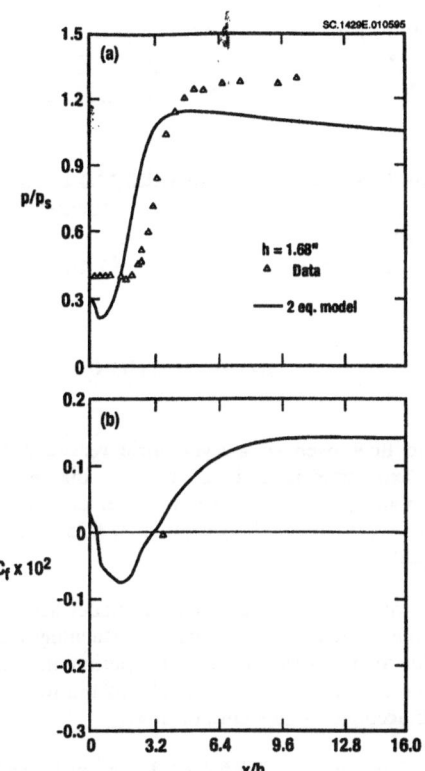

Fig. 6(a,b). Supersonic axisymmetric back-step flow:
wall pressure and skin friction profiles for
the $h = 1.675"$ case

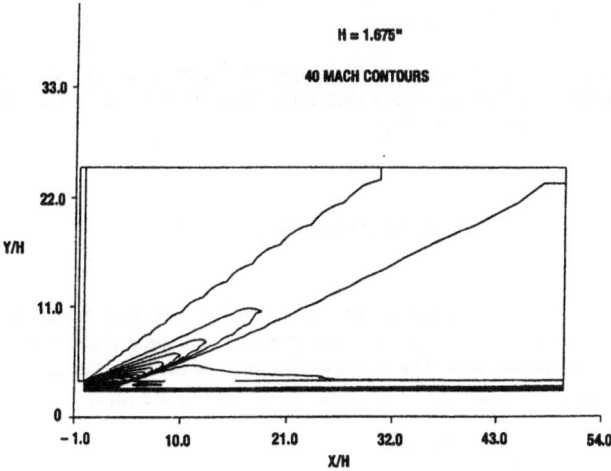

Fig. 7. Supersonic axisymmetric back-step flow:
isoMach contours

SUPERSONIC REARWARD-FACING STEP CALCULATIONS USING AN EXPLICIT FRACTIONAL-STEP METHOD AND A TWO-EQUATION TURBULENCE MODEL

C. Vassilopoulos, K.C. Giannakoglou and K.D. Papailiou
Lab. of Thermal Turbomachines
National Technical University of Athens
P.O. Box 64069, 157 10 Athens, Greece.

INTRODUCTION

The supersonic flow over an axisymmetric rearward facing step, placed on the circumference of an axisymmetric body of a 6 in. diameter, has been experimentally investigated by Roshko and Thomke, [1]. Among other flow conditions, they have published measurements for three different step heights, namely 0.25, 1.02 and 1.675 in., at an external flow Mach number equal to 2.09.

In the present work, the low-Reynolds k-ε model has been used for the numerical modelling of the aforementioned flow problems. Unfortunately, a complete set of comparisons cannot be presented, due to the restricted available experimental data. Nevertheless, the present study must be considered as a demonstration of capabilities as well as weaknesses of the standard k-ε model, to account for this kind of flows.

A modified version of code ATHENA (A Turbulent Hyperbolic Explicit Navier-Stokes Algorithm) developed in the Lab. of Thermal Turbomachines of NTUA, [2], [3], constitutes the basic numerical kernel for the present study. This is an explicit, time-marching fractional-step method which solves the unsteady Navier-Stokes equations by means of a successive application of one-dimensional operators. The code, was initially programmed with the intention to handle planar flows. For the purpose of the present investigation, it has been appropriately modified to account for axisymmetric flows as well.

It would go beyond the scope of this paper to discuss in detail those numerical aspects which are related to the fractional step algorithm. This information is included in two recent publications by the authors, [4] and [5].

GOVERNING EQUATIONS

Let t, ϱ, U_z, U_r, E_t, p, k and ε denote time, density, velocity components in the axial and radial directions, total energy per unit volume, pressure, turbulent kinetic energy and turbulent energy dissipation, respectively. The conservative form of the two-dimensional axisymmetric, Favre-averaged Navier-Stokes equations, may be written in the form

$$\frac{\partial Q}{\partial t} + \frac{\partial}{\partial z}(F^H - F^P) + \frac{\partial}{\partial r}(G^H - G^P) = S . \tag{1}$$

When the k-ε model is used to effect closure, the unknown variable vector Q reads

$$Q= [r\rho, r\rho U_z, r\rho U_r, rE_t, r\rho k, r\rho e]^T.$$

Equation (1) includes inviscid (denoted by the superscript H) and viscous (superscripted by P) contributions. The inviscid-hyperbolic flux vector F^H is given by

$$F^H= \left[r\rho U_z, r\rho U_z^2+r p_{eff}, r\rho U_z U_r, r(E_t+p_{eff}) U_z, r\rho U_z k, r\rho U_z e\right]^T$$

while the viscous-parabolic one F^P is

$$F^P= \left[0, r\tau_{zz}, r\tau_{zr}, r(\tau_{zz}U_z+\tau_{zr}U_r)+\gamma\left(\frac{\mu}{Pr}+\frac{\mu_t}{Pr_t}\right)r\frac{\partial e}{\partial z}, r\mu_k\frac{\partial k}{\partial z}, r\mu_e\frac{\partial e}{\partial z}\right]^T.$$

Similar expressions hold for the flux vectors G^H and G^P. The effective pressure is defined as

$$p_{eff}=p+\frac{2}{3}\rho k$$

where the static pressure p is deduced from the equation of state. The total energy per unit volume E_t is defined by

$$E_t=\rho e+\frac{1}{2}\rho(U_z^2+U_r^2)+\rho k, \quad e=\frac{p}{(\gamma-1)\rho}.$$

Because of the axial symmetry of the flow, a non-zero right-hand term appears in the radial momentum equation which, along with the source terms in the k-ε equations, constitute the non-zero entries in S. Thus, S is given by

$$S=[0, 0, p_{eff}-\tau_{nn}, 0, S_k, S_e]^T. \tag{2}$$

The four non-zero stress tensor components are

$$\tau_{zz}= \mu_{eff}\left[2\frac{\partial U_z}{\partial z}-\frac{2}{3}\left(\frac{\partial U_z}{\partial z}+\frac{\partial U_r}{\partial r}+\frac{U_r}{r}\right)\right]$$

$$\tau_{zr}= \mu_{eff}\left(\frac{\partial U_z}{\partial r}+\frac{\partial U_r}{\partial z}\right)$$

$$\tau_{rr}= \mu_{eff}\left[2\frac{\partial U_r}{\partial r}-\frac{2}{3}\left(\frac{\partial U_z}{\partial z}+\frac{\partial U_r}{\partial r}+\frac{U_r}{r}\right)\right]$$

$$\tau_{nn}= \mu_{eff}\left[2\frac{U_r}{r}-\frac{2}{3}\left(\frac{\partial U_z}{\partial z}+\frac{\partial U_r}{\partial r}+\frac{U_r}{r}\right)\right].$$

The turbulent kinetic energy k does not appear in the above expressions, since it is has been previously included in the definition of the effective pressure.

The form of the low-Reynolds turbulence model used herein, is based on the standard k-ε model, introduced by Jones and Launder [6]. The involved source terms have the following form

$$S_k = rP - r\rho\varepsilon - 2\mu r\left(\frac{\partial\sqrt{k}}{\partial n}\right)^2$$

$$S_\varepsilon = c_{\varepsilon_1} rP\frac{\varepsilon}{k} - c_{\varepsilon_2} f_2 r\rho\frac{\varepsilon^2}{k} + 2\frac{\mu\mu_t}{\rho} r\left(\frac{\partial^2 V_t}{\partial n^2}\right)^2$$

(3)

where V_t denotes the velocity component tangential to solid walls and n is the normal to the wall direction. The production term P, is given by

$$P = \mu_t\left(\frac{\partial U_z}{\partial r} + \frac{\partial U_r}{\partial z}\right)^2 + 2\mu_t\left[\left(\frac{\partial U_z}{\partial z}\right)^2 + \left(\frac{\partial U_r}{\partial r}\right)^2 + \left(\frac{U_r}{r}\right)^2\right]$$

$$- \frac{2}{3}\mu_t\left(\frac{\partial U_z}{\partial z} + \frac{\partial U_r}{\partial r} + \frac{U_r}{r}\right)^2 - \frac{2}{3}\rho k\left(\frac{\partial U_z}{\partial z} + \frac{\partial U_r}{\partial r} + \frac{U_r}{r}\right) .$$

The eddy viscosity concept is taken into account by assuming the following effective viscosity coefficients

$$\mu_{eff} = \mu + \mu_t \quad , \quad \mu_k = \mu + \frac{\mu_t}{Pr_k} \quad , \quad \mu_\varepsilon = \mu + \frac{\mu_t}{Pr_\varepsilon}$$

with the Prandtl numbers Pr_k and Pr_ε equal to 1.0 and 1.3 respectively. The laminar viscosity μ is calculated through the Sutherland's law while the turbulent viscosity μ_t is obtained from the Prandtl-Kolmogorov relation,

$$\mu_t = c_\mu f_\mu \frac{\rho k^2}{\varepsilon} .$$

The constants C_μ, $C_{\varepsilon 1}$ and $C_{\varepsilon 2}$ are set equal to 0.09, 1.44 and 1.92 respectively and the functions f_μ and f_2 are given by

$$f_2 = 1 - 0.3\exp(-Re_t^2) \quad , \quad f_\mu = \exp\left[\frac{-3.4}{\left(1 + \dfrac{Re_t}{50}\right)^2}\right] \quad , \quad Re_t = \frac{\rho k^2}{\mu\varepsilon} .$$

NUMERICAL ASPECTS

According to the fractional step algorithm, the dependent variable array Q is solved through the successive application of one-dimensional operators. The fractional-step procedure can be cast in the symbolic form

$$Q^{n+2} = L_\xi^H L_\eta^H L_\xi^P L_\eta^P L^{ST} L^{ST} L_\eta^P L_\xi^P L_\eta^H L_\xi^H Q^n .$$

(4)

Since all calculations are carried out in the computational space (ξ,η), the expression (4) has been adapted to the transformed form of the governing equations in the (ξ,η) space. Each one-dimensional operator L has been subscribed by ξ or η, to indicate whether sweeps are performed along the η or ξ = constant grid lines. Three different superscripts, namely H, P and ST have been introduced, in order to distinguish the nature of the part of the governing equation which has been assigned to each operator Thus, the L^H operator corresponds to the inviscid part of the equations (where H stands for Hyperbolic), the L^P operator solves for the viscous part (P = Parabolic), while the L^{ST} operator handles the source terms that appear in the radial momentum and the turbulence equations.

In expression (4), the double and inverse sequence of the one-dimensional operators ensures second order accuracy in time. The second order accuracy in space is obtained using the predictor-corrector MacCormack scheme for the hyperbolic and the parabolic operators.

The source terms which are numerically assigned to operator L^{ST} may cause numerical instabilities, during the early stages of the computation. This problem could be efficiently overcome, if a semi-implicit treatment is used when solving the L^{ST} operator, [4], [5]. This operator is split in a non-positive and a non-negative part and the last one is further Newton-linearized to enhance the diagonal dominality.

All dependent variables are stored at the same nodes which are defined by the intersecting grid lines. Governing equations are discretized using centered finite-difference schemes on a collocated grid which normally leads to odd-even decoupling effects and oscillations close to shock waves. This is circumvented by explicitly adding an extra dissipation term. According to the widely used Jameson scheme [7], it has the form of a blend of second- and fourth-order derivatives, with appropriate control parameters which depend upon pressure differences. Herein, the standard Jameson scheme is further modified, by creating a new control parameter which depends on the spatial variations of the turbulent energy dissipation and adjusts the amount of extra dissipation added to the turbulence equations.

MULTI-DOMAIN APPROACH

The numerical modelling of geometries involving backward-facing steps require a particular effort in constructing efficient computational grids. The easiest way to solve the problem is to incorporate a single H-type orthogonal grid which fills only the after-step part of the domain. Such an approach results to a solution field which depends strongly on the inlet velocity profile at the edge of the step. It is also evident that an a priori specification of the flow profile in this position is not an easy task. On the other hand, the generation of a single H-type curvilinear grid, covering both parts of the domain, before and after the step, is not convenient under the particular stretching requirements dictated by Navier-Stokes codes.

For these reasons, in the present study, the multi-domain technique was employed, combining two overlapping grids. The first one starts in a certain distance upstream of the step corner and ends slightly downstream of the step. The second one fills the region at the right of the step corner. In the overlapping part of the domain, there is a non-identical nodal distribution in both the radial and the axial directions.

The communication procedure at the inlet of the second domain is straightforward (a one-dimensional interpolation is required), since it coincides with a grid line of the first domain. A two-dimensional interpolation is required at the exit and the last lower part of the first domain. By scrutinizing the obtained results, one may observe that the solution field remains smooth across the interfaces. In the two-dimensional solution fields which will be presented in the results section, slight inconsistencies in the overlaid parts of the domain could be observed; these are attributed to the 'linear' graphics software and are of minor importance.

RESULTS AND DISCUSSION

For all three examined cases, corresponding to the three different step heights, the generated computational grids extended between the same starting and ending x-coordinates, in the Cartesian frame. This means that in all cases, the length of the second domain was the same in absolute coordinates, but different if expressed in terms of step height (h) units. In the Cartesian frame, with its origin located at the step position, the calculation domain starts at x=-0.04m and ends at x=0.572m. Consequently, for the higher step-height which is the most severe case, the calculation starts one step height before the step and uses a length of approximately 14h in order to allow a proper reattachment and development. Since, in this case, the reattachment takes place at approximately four step-heights, it seems that this domain is quite appropriate for this calculation, while having reasonable memory requirements.

In the figures illustrating iso-Mach contours, it could be seen that the second domain is a non-orthogonal one, with straight grid lines. Moving in the streamwise direction, the total height of the second domain is increased, aiming at a better handling of the Mach line patterns emanating from the step; these lines had to be directed towards the exit, rather than the upper boundary of the domain.

Concerning boundary conditions, apart from the infinite Mach number (M=2.09), the non-dimensional velocity profile at the step corner was available from reference [1]. The boundary layer thickness just ahead of the step was also available; for the examined cases, this was equal to

$\delta/h = 0.5480$, for h = 0.250 in.
$\delta/h = 0.1343$, for h = 1.020 in.
$\delta/h = 0.0818$, for h = 1.675 in.

Using the Kuhn-Nielsen expression for turbulent velocity profiles [8] and the van Driest transformation which accounts for the compressibility effects, inlet flow profiles have been built at the inlet of the first domain. By means of a trial and error procedure, we have finally succeeded in reproducing the aforementioned velocity profile at the step corner. Further information about how turbulence quantities have been obtained at the inlet, could be found in [9]. Any other boundary layer data provided in [1] have not been directly used during the above procedure. Nevertheless, the corresponding calculated quantities do not generally decline from the measured ones.

The grids had 31x51 and 87x79 nodes in the first and the second domain, respectively. They were appropriately stretched close to vertical and horizontal solid walls, in order to ensure a maximum non-dimensional distance from the wall less than to 1.5.

Figure (1) shows the calculated iso-Mach contours for the three step-heights. For the two greater step-heights, the whole calculation domain is presented while in the smallest step-height case, a blow-up near the step corner is shown. It could be easily seen that the iso-Mach curves remain continue and smooth across the domains interfaces.

Figure (2) shows the non-dimensional pressure distribution along the lower wall. These are compared with the available measured distributions. In all cases the position where the shock reflects over the lower wall is slightly underestimated. Figure (3) illustrates the friction coefficient distribution along the same wall. No C_f measurements were available and the only comparison could be made in terms of the reattachment length. The measured reattachment lengths are included in the same figure in the form of a single mark. It is quite obvious that

80

the reattachment length is underestimated. The overprediction of the reattachment length using the standard k-ε model was expected, and this is a common feature of this model that could be found in similar incompressible and compressible studies.

Finally, figure (4) shows the velocity profiles at several different locations along the lower wall (x/h = 1,2,3 and 4). These profiles are presented in non-dimensional form, in the range from y/h = 0 up to y/h = 1. At this height, the non-dimensional velocities are much greater than 1 and three or four step-heights are required to establish the infinite velocity.

CONCLUSIONS

The axisymmetric rearward facing step geometry, at three different step-heights and at supersonic flow conditions has been numerically analyzed, using the standard k-ε turbulence model and an explicit numerical kernel. In general, the calculated results are in good agreement with the available measurements. The standard k-ε model reproduces the experimental data with satisfactory accuracy but, as expected, underpredicts the reattachment length.

The multi-domain technique was used to mesh the computational flow domain. From a numerical point of view, the communication of the two overlaid domains, taking place in a region where complex physical phenomena occur, operates in a very satisfactory way for both mean-flow and turbulence quantities. No problems are to be reported concerning the explicit scheme stability, when combined with the two-equation turbulence model.

REFERENCES

[1]   Roshko, A., Thomke, G.J.: "Observations of turbulent reattachment behind an axisymmetric downstream-facing step in supersonic flow", AIAA Journal, 4, No. 6, June (1966), pp. 975-980.

[2]   Simandirakis, G.: "Numerical solutions of Navier-Stokes equations for transonic flows inside turbine bladings", Ph.D. Thesis, Athens, February 1992.

[3]   Giannakoglou, K., Simandirakis, G., Papailiou, K.D.: "Turbine cascade calculated through a fractional step Navier-Stokes algorithm', ASME Paper 91-GT-55 (1991).

[4]   Dejean, F., Vassilopoulos, C., Simandirakis, G., Giannakoglou, K., Papailiou, K.D.: "Analysis of transonic turbomachinery flows using a 2-D explicit low-Reynolds k-ε Navier-Stokes Solver", ASME Paper 94-GT-63 (1994).

[5]   Simandirakis, G., Vassilopoulos, C., Giannakoglou, K., Papailiou, K.D.: "Steady and unsteady two-dimensional flow calculations using an explicit fractional step algorithm', Proceedings of the Second European Computational Fluid Dynamics Conference, Stuttgart, September, 5-8 (1994).

[6]   Jones, W.P., Launder, B.E.: "The Prediction of laminarization with a two-equation model of turbulence', Int. Journal Heat Mass Transfer, 15 (1972), pp. 301-314.

[7]   Jameson, A., Schmidt, W., Turkel, E.: "Numerical solutions of the Euler equations by finite volume methods using Runge-Kutta time stepping schemes", AIAA paper (1981), pp. 81-1259.

[8]   Kuhn, G.D., Nielsen, J.N.: "Prediction of turbulent separated boundary layers", AIAA paper 73-663 (1973).

[9]   Vassilopoulos, C., Giannakoglou, K., Papailiou, K.D.: "Study of the supersonic compression ramp flow using an explicit fractional-step technique and the k-ε turbulence model", ETMA Workshop On Turbulence Modeling for Compressible Flow Arising in Aeronautics, Manchester, November, 14-17 (1994).

Acknowledgements

Part of the work concerning turbulence modelling was funded by the ETMA (Efficient Turbulence Models for Aeronautics) Project (BRITE-EURAM-2076/2032). The authors would like to acknowledge Dr. G. Simandirakis' contribution to the development of the fractional step numerical algorithm.

h = 0.250 in.

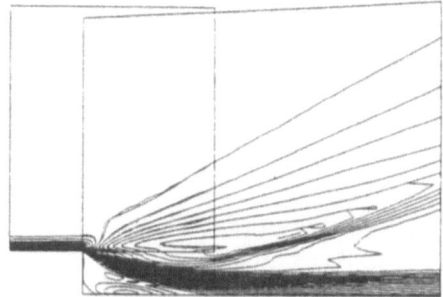

h = 1.020 in.

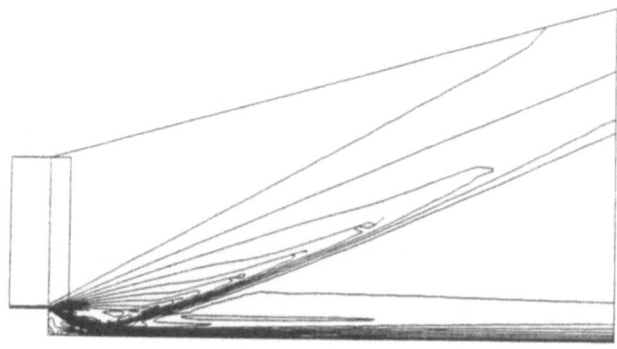

h = 1.675 in.

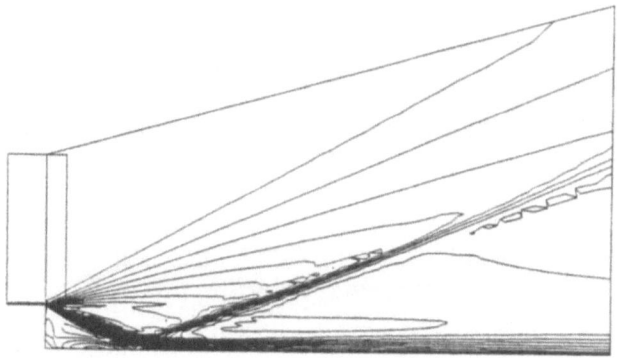

Fig. 1   Calculated iso-Mach contours for the three step-heights (increment 0.1).

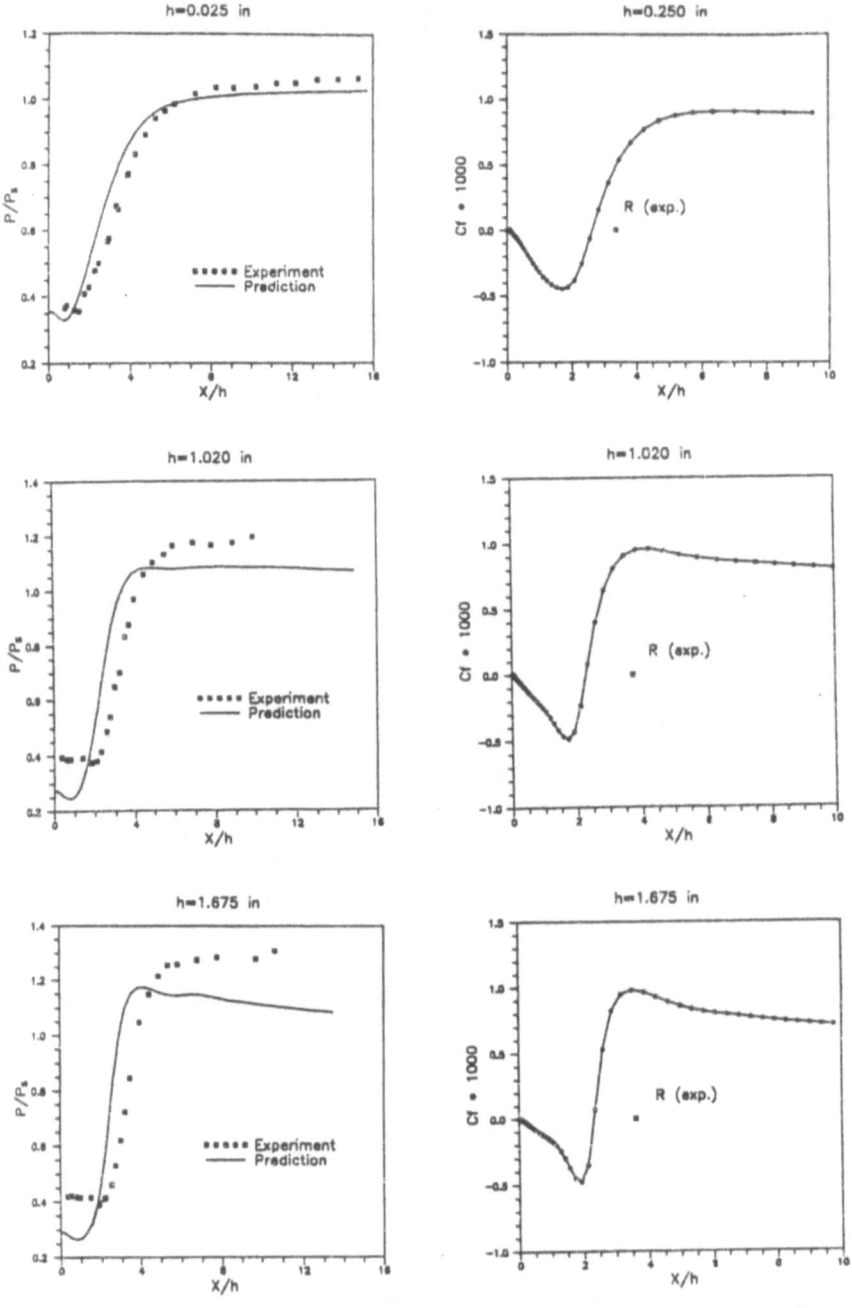

Fig. 2   Non-dimensional pressure distribution along the lower wall.

Fig. 3   Friction coefficient distribution along the lower wall.

h = 0.250 in.

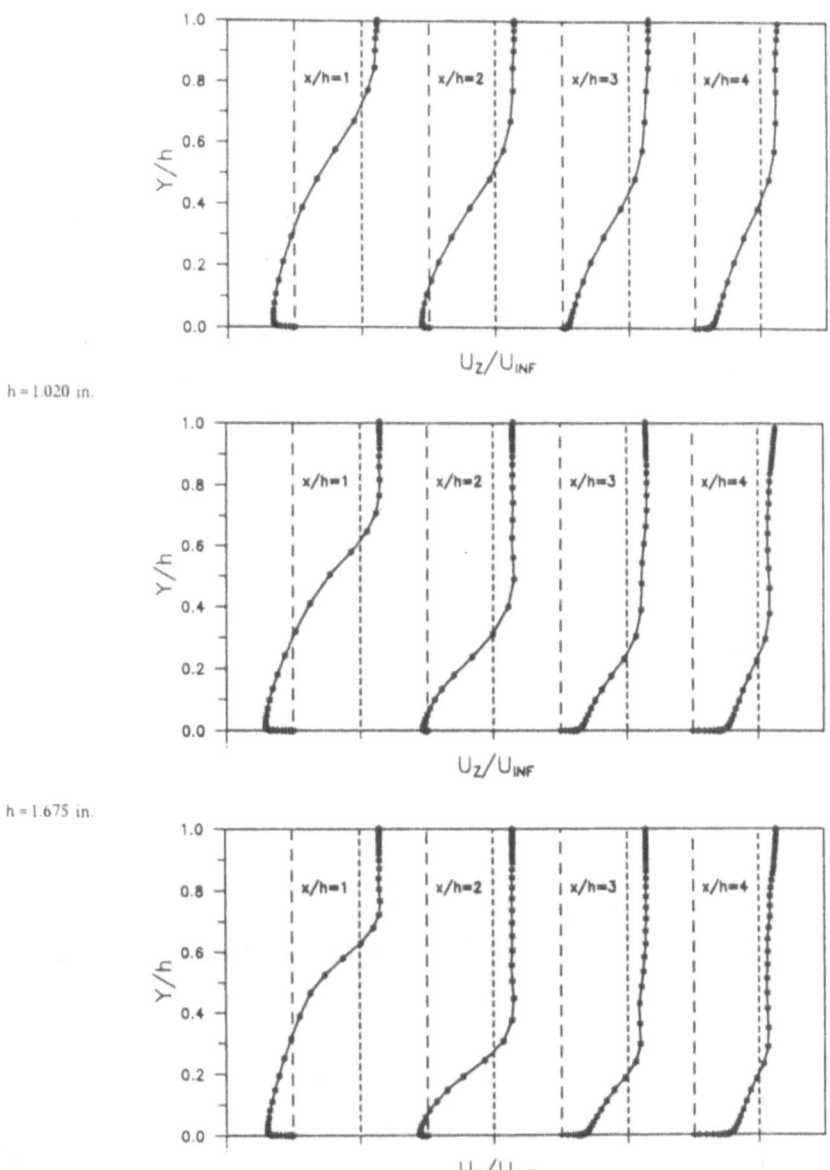

Fig. 4   Calculated non-dimensional axial velocity profiles at four streamwise positions.

85

CHAPTER 3 : INCOMPRESSIBLE
WALL FLOWS WITH SEPARATION

PRESENTATION OF TEST CASES TC-2A, TC-2B, TC-2C, TC-2D TWODIMENSIONAL, INCOMPRESSIBLE, WALL FLOWS WITH SEPARATION

P. S. Larsen

Department of Fluid Mechanics, Technical University of Denmark
Building 404, DK-2800 Lyngby, Denmark

The four test cases comprise the backfacing step at high Re-number (TC-2A) and low Re-number (TC-2B), a low Re-number boundary layer flow past a thin obstacle, fence-on-wall (TC-2C), and a high Re-number developed channel flow past a square obstacle (TC-2D). Geometry, test conditions and available data are described. The flow conditions constitute basic tests for the ability of turbulence models to handle complex, turbulent, near-wall flows with separation and recirculation at low and high values of the Reynolds number.

INTRODUCTION

Among the many twodimensional test cases in the literature involving a turbulent flow that separates from a backward-facing step the four cases listed in Table 1 have been selected because detailed data are available for these cases. Figure 1 shows schematically the geometries, and Table 1 lists details on geometry, flow and data.

The first two cases are standard, involving boundary layer flows in a plane channel with a sudden expansion that is symmetrically double-sided (TC-2A) and one-sided (TC-2B), respectively. The latter two cases involve obstacles in plane channel flows: a boundary layer flow, with free-stream turbulence, past a small fence of

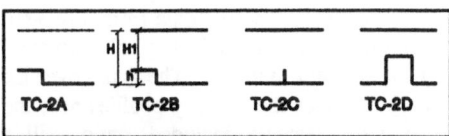

Fig.1 Test cases (schematic)

height about twice that of the boundary layer thickness but only one-seventh of the channel width (TC-2C), and a fully developed flow past a square obstacle blocking half the width of the channel (TC-2D). Here, TC-2C has the character of a free flow, where the pressure distribution over the separation zone is largely governed by the nearly uniform freestream, while TC-2D has the character of a confined flow, where the proximity of the opposing wall strongly affects the pressure distribution over the separation zone.

For all test cases the objectives of predicting the flows are focussed on several features: the length of the separation zone, X_r/h, profiles of mean velocity, the turbulence level, $k = \frac{1}{2}(\langle u^2 \rangle + \langle v^2 \rangle + \langle w^2 \rangle)$, particularly of the shear layer bounding the recirculation region, the recovery of the wall layer downstream, and coefficients of pressure and skin friction along the wall downstream of the step.

TC-2A, BACKWARD-FACING STEP, Re = 5,000

Experiments by Jović and Driver [1] were carried out in a wind tunnel facility, having a rectangular inlet section of height 96.5 mm followed by a 9.65 mm double-sided rearward-facing step, giving an expansion ratio of 1.2. At atmospheric conditions, a freestream velocity of U_{ref} = 7.72 m/s upstream corresponds to Re = 5000 based on step height. The upstream boundary layer was hightly turbulent, Re_θ = 610 at x/h = -3, x being measured from the step.

LDA-data provided 6 streamwise profiles of mean fields U, V, and Reynolds stresses $\langle u^2 \rangle$, $\langle uv \rangle$, $\langle v^2 \rangle$. The data cover the range x/h = -3 to 19, where the downstream station is interpreted as the recovery length. Other measurements included static wall pressure and the coefficient of skin friction along the lower wall downstream of the step. The recommended computational domain extends from x/h = -10 to 20, where adequate inlet conditions should be generated from performing a separate, constant-pressure boundary layer calculation to reach Re_θ = 610 at x/h = -3.

TC-2B, BACKWARD-FACING STEP, Re = 37,000

Experiments by Driver and Seegmiller [2] were carried out in a low-speed wind tunnel facility, comprising a 1.0 m long x 0.151 m wide x 0.101 m high rectangular duct followed by a 12.7 mm rearward-facing step in the floor. The opposing upper wall was hinged at a location 50 mm upstream of the step, permitting a pressure gradient to be imposed on the freestream over the recirculation zone. Two angles (0° and 6°) of the hinged wall were considered. At atmospheric conditions, a freestream velocity of U_{ref} = 44.2 m/s upstream corresponds to a Mach number of 0.128 and Re = 37000 based on step height. The upstream tripped boundary layer on the step-side wall was hightly turbulent, Re_θ = 5000 at x/h = -4, x being measured from the step.

Counter-processed data from a two-component LDA-system provided 22 streamwise profiles of mean fields U, V, Reynolds stresses $\langle u^2 \rangle$, $\langle uv \rangle$, $\langle v^2 \rangle$, and tripple-product correlations $\langle u^3 \rangle$, $\langle u^2 v \rangle$, $\langle uv^2 \rangle$, and $\langle v^3 \rangle$. The data cover the range x/h = -4 to 32. Only far downstream (x/h > 16) does the velocity profiles near the wall show any evidence of a log-law region. Other measurements included static wall pressure and, by use of an oil-flow laser interferometer, the coefficient of skin friction along the lower wall downstream of the step. Tunnel wall divergence reduces the level of skin friction in the attached flow and increases the reattachment length considerably, as expected.

TC-2C, FENCE-ON-WALL, Re = 3,000

Experiments by Larsen et al [3] were carried out in a low speed wind tunnel fitted with a turbulence generating grid (M = 39 mm) after the contraction, at the inlet to the 0.3 x 0.6 m² test section. The 40 x 10 x 600 mm fence was mounted on one side-wall, 760 mm downstream of the inlet. At 22°C a freestream velocity of 1.17 m/s corresponds to Re ≈ 3000 based on fence height. The transparent glass bottom gave access for a two-component, back-scatter LDA-system. Counter-processed data provided mean fields U, V, Reynolds stresses $\langle u^2 \rangle$, $\langle uv \rangle$, $\langle v^2 \rangle$, as well as third and fourth moments of the two velocity components. The data cover a rectangular region in the x,y-midplane of the wind tunnel, reaching 8h

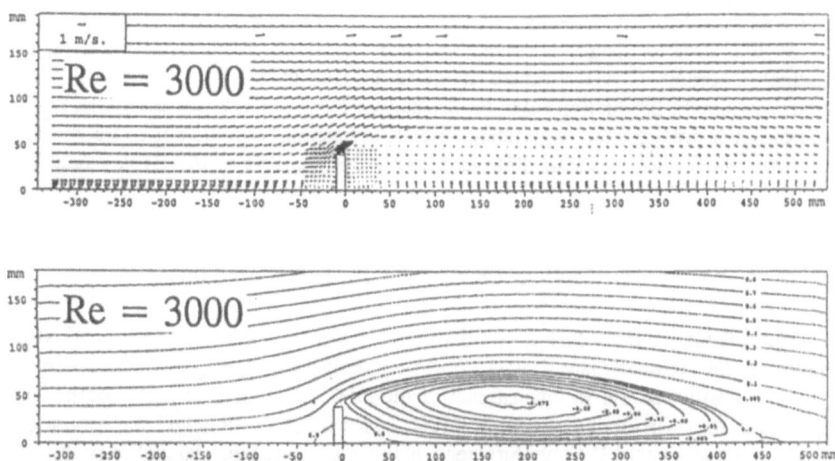

Fig.2 Velocity vector plot (top) and streamline plot computed from smoothed LDA-data.

upstream, $13h$ downstream and $4.5h$ out from the wall. The data point spacing is 10 mm in x and y, except in the near field and along walls where it is refined to 5 mm and 1 mm, respectively. For an overall view, figure 2 shows a velocity vector plot of the raw data, as well as a computed, smoothed, divergence-free streamline plot [5]. The small upstream separation bubble starts at $x = -34$ mm (x being measured from the downstream face of the fence), and the downstream reattachment point is located at $x = 470$ mm ($X_r/h = 11.7$). The coefficient of skin friction, $C_f = 2\tau_w/(\rho U_{ref}^2) = 2\nu(\partial u/\partial y)_w/U_{ref}^2$, at selected points along the wall with the fence has been estimated from the limiting value of velocity gradient at the wall.

To further characterize the turbulent boundary layer flow, detailed measurements of the velocity profile (from about $y^+ = 0.3$ to 1000, figure 3), at the position where the fence would be mounted, showed that the grid turbulence assured the development of a reasonable turbulent boundary layer with a logarithmic range. Here, the friction velocity was estimated from a linear fit to the data in the sublayer. The coefficient of skin friction is $C_f \approx 0.0067$, and integration of the profile gives a Reynolds number based on the momentum thickness, $Re_\Theta = \Theta U_{ref}/\nu \approx 88$.

The proposed computational domain covers the width of the channel, beginning 330 mm upstream of the fence ($x/h = -8.25$) and extending 1300 mm downstream ($x/h = 32.5$). At the inlet, profiles of U and V are measured directly, the turbulent kinetic energy is estimated from $k = 0.75(\langle u^2 \rangle + \langle v^2 \rangle)$, and the

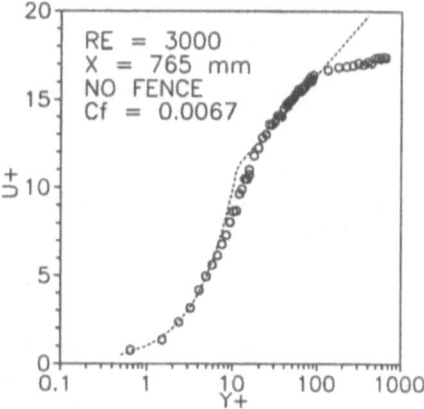

Fig.3 $u+$ versus $y+$ of turbulent boundary layer at location of fence, $Re = 3000$.

91

dissipation of turbulent kinetic energy was estimated from $\epsilon \approx k^{3/2}/\ell$ in the freestream, where ℓ denotes the integral length scale estimated from autocorrelations of time series of longitudinal u-velocity. This estimate is close to that made from measured decay of turbulence, assuming production equal to dissipation. Near walls, ϵ was estimated from wall laws. These inlet profiles, the experimenteal data and estimated C_f-values at Re = 3000, as well as raw data for Re = 1500, 2000 and 2500, are available in the ETMA data base at INRIA [7].

TC-2D, OBSTACLE-IN-CHANNEL, Re = 42,000

Experiments by Dimaczek et al [4] were carried out in a 3 m long by 0.05×0.55 m rectangular water channel. The $25 \times 25 \times 550$ mm obstacle, blocking half the width of the channel, was positioned 1.4 m downstream of the inlet. At 25.5°C a mean (bulk) velocity of $U_{ref} = 1.64$ m/s corresponds to Re = 42000 based on obstacle height h. Windows gave access for one-component LDA-measurements in the midplane of the channel, providing mean fields U, V, Reynolds stresses $\langle u^2 \rangle$, $\langle uv \rangle$, $\langle v^2 \rangle$ and, at certain locations, W and $\langle w^2 \rangle$. The data cover the width of the channel, reaching $36h$ upstream (x = -900 mm) and $50h$ downstream (x = 1250 mm), where x is measured from the upstream face of the obstacle. Figure 4 shows samples of axial velocity profiles downstream of the obstacle, indicating the recirculation

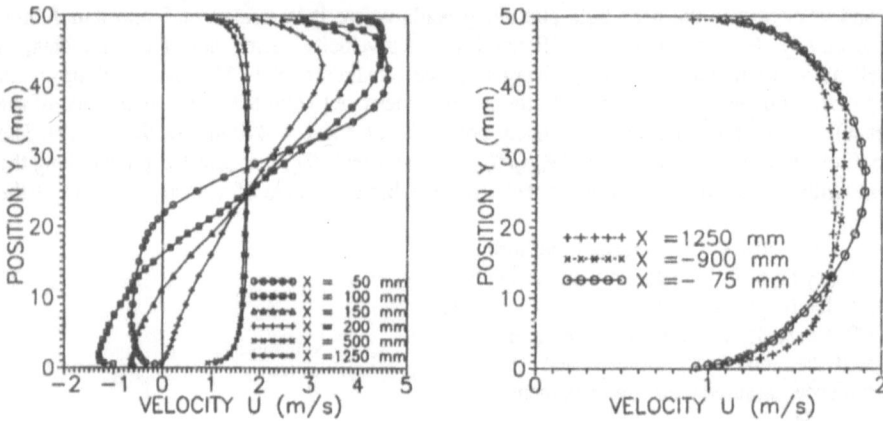

Fig.4 Profiles of axial velocity component at selected locations downstream of obstacle (left) and upstream to show asymmetry (right).

region and subsequent recovery. Also shown are U-profiles at x = -900, -75 and 1200 mm to indicate the prevailing upstream asymmetry. The small upstream separation bubble starts at x = -12.5 mm, and the downstream reattachment point is located at x = 228.4 mm (X_r/h = 8.13). The coefficient of skin friction, $C_f = 2(U_\tau/U_{ref})^2$, at selected points along the wall with the obstacle, has been estimated by fitting appropriate parts of the velocity profiles to the log-law. As an illustration, figure 5 shows the profile at x = 1000 mm. Although this approach is of questionable accuracy it was considered to be important to have some additional measure to check predictions. At far upstream and downstream stations (x = -625 mm and x = 1250 mm, respectively) the aboveapproach gave $C_f = 0.00385$ and 0.00452,respectively. For comparison, the data of Beavers et al [8] for fully developed turbulent flow in a plane channel, gives $C_f = 0.00344$, while the Blasius formula gives 0.00391. These results confirm that estimates in the database are of the right order of magnitude, but they also suggest the range of uncertainty.

The proposed computational domain covers the width of the channel, beginning 75 mm upstream of the obstacle ($x/h = -3$) and extending 1250 mm downstream ($x/h = 30$). The inlet conditions at $x = -75$ mm are given from experimental data for U, V and $k = \frac{1}{2}(\langle u^2 \rangle + \langle v^2 \rangle + \langle w^2 \rangle)$, while the dissipation was estimated from $\epsilon = k^{**}(3/2)/(0.3^*D_h)$, where the hydraulic diameter is $D_h = 0.1$ m. These inlet profiles, the experimental data and estimated C_f-values are available in the ETMA data base at INRIA [8].

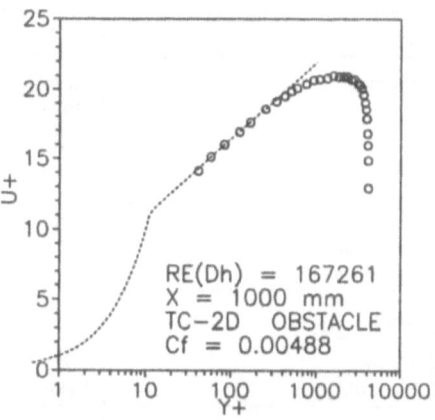

Fig.5 TC-2D. Data fit to log-law, $x = 1000$ mm.

REFERENCES

[1] Jović,S, and Driver,D.M., 1993. Unpublished data.

[2] Driver,D.M., and Seegmiller,H.Lee, 1985. "Features of a Reattaching Turbulent Shear Layer in Divergent Channel Flow". AIAA,J., 23, 163-171.

[3] Larsen,P.S., Marxen,U. and Jørgensen,O.A., 1994. "Fence-on-wall experiment" (in preparation, see also [5] and [6]).

[4] Dimaczek,G., Kessler,R., Martinuzzi,R. and Tropea,C. 1989. "The flow over two-dimensional, surface-mounted obstacles at high Reynolds numbers". Proc. 7th Symposium on Turbulent Shear Flows, Standford, CA.

[5] Marxen,U. and Jørgensen,O.A. 1993. Master's Thesis, Dept.of Fluid Mech., Techn. Univ.of Denmark, Report AFM-EP 93-06.

[6] Larsen,P.S., Westergaard,C.H., Koch,C.W. and Schmidt,J.J., 1993. "12-month intermediate report", Report ETMA/AFM/93-02, (see also [5]).

[7] Larsen,P.S. 1995, "Database on TC-2C and TC-2D, Report ETMA/AFM/95-01, Feb.1995.

[8] Beavers,G.S., Sparrow,E.M. and Lloyd,J.R. 1971, J.Basic Eng., 93, 296.

Table 1 Summary of test case parameters

Test case	TC-2A	TC-2B	TC-2C	TC-2D
Reference	[1] Jović & Driver (1993)	[2] Driver & Seegmiller (1985)	[3] Larsen et al (1994)	[4] Dimaczek et al (1989)
Geometry	back-facing step	back-facing step	fence-on-wall	obstacle-in-channel
Heights, h (mm)	9.65	12.7	40	25
$H1/H$ (mm), see Fig.1	96.5/115.9	101/113.7	260/300	25/50
Expansion ratio, ER $= H/H1$	1.2	1.126	1.154	2.0
Fluid	air	air	air	water
$\nu \times 10^6$ (m²/s)	≈ 15	≈ 15	15.6	0.9805
U_{ref} (m/s)	7.72	44.2	1.17	1.64
Re $= hU_{ref}/\nu$	5000	37000	3000	42000
Experiment, Method		2-d LDA,	2-d LDA	1-d LDA
Coordinate system $(x,y)=(0,0)$	downstream wall corner	downstream wall corner	downstream wall corner	upstream wall corner
Data domain, x (mm)/y (mm)			850/180	2150/50
Upstream x/h	-3.0	-1.05	-8.25	-36.0
Downstream x/h	19	-	13.0	50.0
Experimental data				
reattachment length, X_r/h	6.0	6.1	11.7	8.13
recovery length X_{rec}/h	19.0	51	-	-
Available data:	C_p, C_f $U,V,\langle u^2\rangle,\langle uv\rangle$ $\langle v^2\rangle$	C_p, C_f $U,V,\langle u^2\rangle,\langle uv\rangle$ $\langle v^2\rangle$	C_f-estimate $U,V,\langle u^2\rangle,\langle uv\rangle$ $\langle v^2\rangle$, S, F	C_f-estimate $U,V,\langle u^2\rangle,$ $\langle uv\rangle,\langle v^2\rangle,$ and $W,\langle w^2\rangle$ at some stations
profile locations: x/h	3,4,6,10,15, 19	-1.05, 0.21 ..., 6.58		
k-budget	-	yes	-	-
Computational domain, x/h	-10 to +20	-10 to +60	-8.25 to 32	-3 to 30
y/h	wall to center	wall to wall	0 to 7.5	0 to 2
inlet profiles	(see data)	(see data)	U,V,k,ϵ	U,V,k,ϵ

INCOMPRESSIBLE RECIRCULATING FLOWS

TC2-A   LOW-RE BACKWARD-FACING STEP
TC2-B   HIGH-RE BACKWARD-FACING STEP

Robert D. Harper & Michael M. Gibson
Mechanical Engineering Department, Imperial College, London SW7 2BX

SUMMARY

A new low-Reynolds number, two-equation turbulence model based on the variables $q \equiv \sqrt{k}$ and $\zeta \equiv \tilde{\varepsilon}/2q$ is described, this model together with the low-Re k-ε model of Launder & Sharma and the 'standard' k-ε model with wall functions are used to compute two backward-facing step flows. The first computation is for a low-Re flow for which experimental data, [5] and DNS data, [6] are available. The second computation, for high Re flow, is based on reference [7]. The results highlight that the low-Re models resolve more of the flow detail than the wall function approach and that the q-ζ model performs slightly better then the low-Re k-ε model. Some savings in computing time for the q-ζ model have been observed due to coarser grids being used in the near-wall regions.

INTRODUCTION

The k-ε model of turbulence suffers from the grave defect that its constituent equations are ill-adapted to integration to the wall. Consequently its application is effectively restricted to high Reynolds number turbulence outside the viscous layers, which have to be bridged by wall functions. The lack of a natural wall boundary condition for the ε-equation presents problems which may only be surmounted by the introduction of terms which require an excessively fine grid for their resolution. These difficulties are avoided by changing to variables which are zero at the wall and vary linearly with distance close to it. The q-ζ model is a k-ε model replacement designed primarily to improve computational efficiency. The improvement in predictive accuracy shown in the present results is an unexpected but welcome bonus.

This report documents the results of 2-dimensional separated flow calculations using the new q-ζ turbulence model. Comparisons are made with results from the low and high Reynolds number versions of the k-ε model and also with experimental and DNS data where applicable. Calculations were performed for the low-Re backward-facing step, [5, 6], TC2-A and also for TC2-B the high-Re backward-facing step, [7]. Geometry and flow conditions are given in the test case specification.

A full derivation of the q-ζ model is supplied in [1, 2]. A companion paper within these proceedings, [3] also includes this derivation, together with a critical comparison of the k-ε and q-ζ models. Only an outline of the model will be given here.

The variable q is defined by $q \equiv \sqrt{k}$, while ζ is its dissipation rate. Transport equations for q and ζ are obtained by transforming the 'standard' k and ε equations and can be written as shown below, (note that pressure diffusion in the q-equation is assumed negligible).

$$U_i \frac{\partial q}{\partial x_i} = \frac{\partial}{\partial x_j}\left\{\left(v + \frac{v_t}{\sigma_q}\right)\frac{\partial q}{\partial x_j}\right\} + Q - \zeta$$

$$U_i \frac{\partial \zeta}{\partial x_i} = \frac{\partial}{\partial x_j}\left\{\left(v + \frac{v_t}{\sigma_\zeta}\right)\frac{\partial \zeta}{\partial x_j}\right\} + \frac{\zeta}{q}\left(C_{\zeta 1}f_{\zeta 1}Q - C_{\zeta 2}f_{\zeta 2}\zeta\right) + \psi.$$

Q, the production rate of q, and ψ, the secondary mean-flow production of ζ, are given by:

$$Q \equiv \gamma_t \frac{\partial U_i}{\partial x_j}\left(\frac{\partial U_i}{\partial x_j} + \frac{\partial U_j}{\partial x_i}\right), \qquad \psi = 2v\gamma_t\left(\frac{\partial^2 U_i}{\partial x_j \partial x_k}\right)\left(\frac{\partial^2 U_i}{\partial x_j \partial x_k}\right).$$

Other quantities, including eddy viscosity, v_t, are defined as;

$$\gamma_t \equiv \frac{C_\mu f_\mu}{4}\frac{q^2}{\zeta}, \qquad v_t \equiv C_\mu f_\mu \frac{q^3}{2\zeta}.$$

While the viscous-layer damping functions are taken from Launder and Sharma [4]:

$$f_\mu = \exp\left\{\frac{-A_\mu}{(1 + Re_t/50)^2}\right\}, \qquad C_\zeta f_\zeta \equiv 2C_\varepsilon f_\varepsilon - 1, \qquad f_{\varepsilon 2} = 1 - 0.3\exp\left(-Re_t^2\right).$$

Re_t is the turbulence Reynolds number $q^3/2v\zeta$, and

$$f_\varepsilon = 1.0, \quad C_{\varepsilon 1} = 1.88, \quad C_{\varepsilon 2} = 2.84, \quad A_\mu = 6.0, \quad \sigma_q = 1.0, \quad \sigma_\zeta = 1.3.$$

The wall boundary conditions are $q = 0$, $\zeta = 0$.

The model constants and damping functions used here need not necessarily be those of Launder and Sharma, these are used because Re_t is a better parameter for general flow calculations than the distance from the wall and the Launder-Sharma model is sufficiently well-established to form a suitable benchmark for comparison. A limited number of calculations with alternative constants and functions have been performed in order to calibrate the model [1,2] these results will not be presented here.

Predictions for all flows were obtained using a two-dimensional elliptic code with co-ordinate based grids, staggered for velocity and pressure, with hybrid differencing and pressure correction using the SIMPLE algorithm. The number of grid nodes required to achieve grid independent solutions obviously depends both on the geometry of the flow and the model considered, however a typical grid for a low-Re turbulence model calculation would consist of approximately 180×300 nodes in the cross-stream and streamwise directions respectively. For all low-Re model calculations the first grid node was at no more than $y^+ = 1.0$.

The test case geometries and inlet conditions were all prescribed in the test case specification. Standard conditions were applied at the remaining boundaries i.e., no slip conditions at a wall and Neumann conditions at the exit plane.

RESULTS

TC2-A: Low Re Backward-Facing Step

Calculations have been performed using the standard k-ε model with wall functions, and the q-ζ and k-ε low-Reynolds-number models with Launder-Sharma functions. Results for the low-Re case are shown in figures 1-3. Comparison are made with the measurements of Jovic and Driver [5] and the direct simulations of Le et al [6]. Note that the vertical (y) scale is magnified. Examination of the skin friction along the step wall shows that the least satisfactory results are given by the wall-function method. Just downstream of the step, the experimentally observed secondary recirculation region approximately 2-2.5 step heights in length is scarcely noticed, the low-Re model calculations, however, quite accurately predict the size of this recirculation zone, with the q-ζ model doing better than the k-ε model. Comparison with the DNS data, however, show that the distribution of c_f within this region is poorly predicted by both low-Re models, especially within 1 step height of the step where a peak in c_f should be observed. Further downstream it can be seen that both low-Re models predict similar reattachment lengths for the main recirculation zone at 6.5 step heights while the wall function calculation predicts reattachment at 5.5 step heights, this result compares with 6.1 step heights for experiment and 6.0 step heights for the DNS data. In the region around reattachment the distribution of c_f given by the low-Re k-ε model shows c_f to be over-predicted both upstream and downstream of the reattachment point, but not by the q-ζ model. Figure 2 shows measured and predicted velocity distributions at different streamwise distances. Good agreement is registered for all models in the region upstream of the step, however, within the recirculation region, at $x/h = 4$, it can be seen that the wall-function calculation gives too small a recirculation zone. Both the low-Re models exaggerate the depth of the recirculation region, while the magnitude of the maximum velocity is well predicted. Further downstream, at $x/h = 6$, it is seen that the low-Re models overpredict the length of the recirculation region, where the wall-function calculation indicates reattachment. As the boundary layer develops in the reattached flow, all the calculations show slow recovery compared to the data. The wall-function calculations show the most rapid recovery, partly due early reattachment and a shallower recirculation zone. Profiles of the shear stress at various distances along the wall are shown in figure 3. In general these suggest that the low-Re models predict the magnitude and location of peak values better than the wall-function approach at $x/h > 6$, but that they also pick up more of the near-wall flow details such as negative shear stress near the wall at $x/h = 4$. However in the far downstream location, at $x/h = 19$ in particular, the predicted shear stress falls to zero very rapidly for both the low-Re models, most noticeably in the q-ζ case. The shear layer edge at $x/h = 19$ is a region where the flow field depends to some degree upon convection from the inlet plane. Calculations, which are not reproduced here, have shown that by improving the profile of ε near the inlet boundary layer edge gives a smoother transition to zero for $-\overline{uv}$, the results away from this region are unaffected. The second reason for the behaviour of $-\overline{uv}$ is due to Re_t, which falls towards the outer region of the shear layer, this reduces value of f_μ and hence artificially suppresses $-\overline{uv}$ in this region, the effect is enhanced in the q-ζ model by the change in A_μ from 3.4 to 6.0.

TC-2B: High Re Backward-Facing Step

Results for the high-Reynolds-number backstep flow of Driver and Seegmiller [7] are shown in figures 4-7. The skin-friction distribution shows many of the features evident in the low-Re flow calculations of the previous section. The wall-function calculation gives a reattachment length approximately 1 step height shorter than the data indicate and fails to recover the secondary recirculation region. The low-Re model calculations show the existence of such a zone, but of only half the length. Perhaps the most striking feature of the results is the very marked under and over-shoot of cf obtained with the low-Re k-ε model, much less evident in the q-ζ results, and entirely absent in the high-Re wall-function calculations. The cf gradients at the reattachment point given by both low-Re models are excessive. The streamwise velocity profiles of figure 5 are similar to those of the low-Re flow. At the upstream end of the recirculation zone, $x/h < 1.0$, the velocity distribution is fairly accurately predicted by the two low-Re models but the accuracy deteriorates further away from the step. As has often been remarked on previous calculations, the performance of the wall-function method throughout most of the recirculating flow and in the vicinity of the reattachment point is surprisingly good, in view of the restrictions implied by the use of the wall law. Downstream of the reattachment point the calculated flows all recover too slowly as before. The profiles for the turbulence kinetic energy and Reynolds shear stress shown in figures 6 and 7 generally exhibit similar features to those of the low-Re flow discussed previously.

CONCLUSIONS

These calculations demonstrate the feasibility of general turbulent-flow calculations with a new two-equation model related to, but not the same as, the k-ε model of turbulence. The chief merit of the new model is that, unlike the equations for k and ε, the q and ζ equations are well conditioned near the wall, where the boundary conditions are specifically defined, and the dependent variables vary linearly with distance from the wall when the wall is close. The equations prove to be less stiff numerically, while linear variations in the sublayer permit the use of a coarser grid with consequent savings in computer time, a reduction of approximately 20-25% was observed for TC2-A, the low-Re backstep.

The k-ε and q-ζ equations do not give exactly the same results when integrated through the sublayers to the wall because the ε and ζ equations are not exact transforms of each other.

Gradient diffusion of ζ is not exactly the same process as gradient diffusion of ε. Thus the results from the new form are not the same as those of the k-ε model from which it was derived.

The q-ζ model allows one of the principal problems of current CFD research to be approached from a new angle, the potential of the established models being almost exhausted. It is regarded as a useful vehicle for further development, not least in the possibility of using the length-scale-determining ζ-equation in higher order closures. But it is likely that the model is some way from the optimum form. The constants and damping functions chosen for these demonstration calculations have known limitations in k-ε modelling, and the search for improvements has become a major research activity, thus far largely fruitless. This is a research area demanding further intensive study, extended well beyond the scope of fitting curves to the results of channel flow simulations. In continuing work on the model we plan to explore the possibility of inertial damping in the outer sublayers $y^+ > 5$ say, and that of recasting the source terms of the ζ-equation in more physically realistic and robust form.

REFERENCES

1. Gibson, M.M. & Dafa'Alla, A.A. A two-equation model for turbulent wall flow, AIAA J (to appear).
2. Gibson, M.M. & Dafa'Alla, A.A. 1994 The q-ζ model for turbulent wall flow, Fluid Dynamics Division, American Physical Society, 47th Annual Meeting, Atlanta, Georgia.
3. Dafa'Alla, A.A, Harper, R.D. & Gibson, M.M. 1995 ETMA Workshop TC3: Flat Plate Boundary Layers.
4. Launder, B.E. & Sharma, B.I. 1974 Application of the energy-dissipation model of turbulence to the calculation of the flow near a spinning disk, Letters in Heat and Mass Transfer 1, 131-138.
5. Jovic, S. & Driver, D.M. 1993 Unpublished data cited in [6].
6. Le, H., Moin, P. & Kim, J. 1993 Direct numerical simulation of turbulent flow over a backward-facing step. 9th Symposium on Turbulent Shear Flows, Kyoto, Japan, 13-2.
7. Driver, D.M. & Seegmiller, H.L. 1985 Features of a reattaching shear layer in divergent channel flow, AIAA J. 23, 163-171.

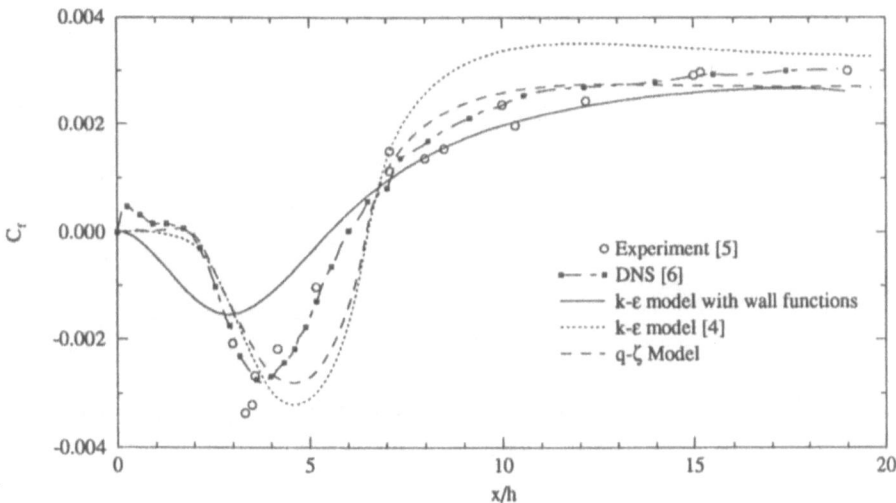

Figure 1. Skin friction coefficient for low-Re flow over a backward-facing step.

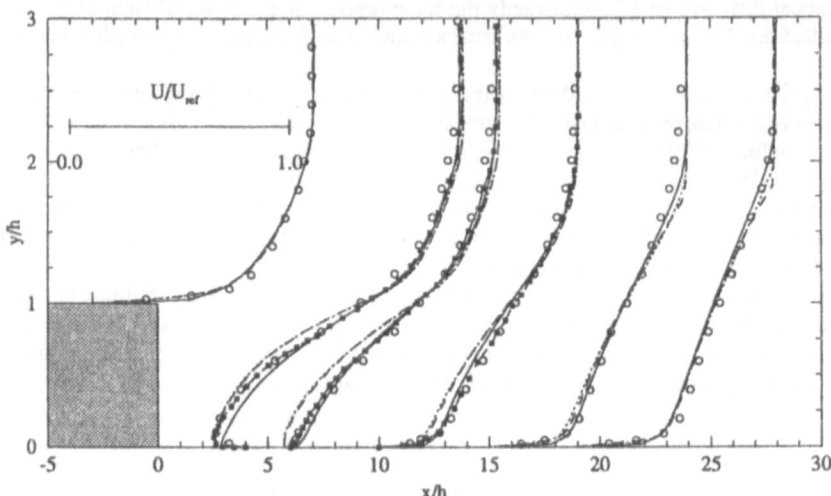

Figure 2. Mean streamwise velocity for low-Re flow over a backward-facing step, see figure 1 for legend.

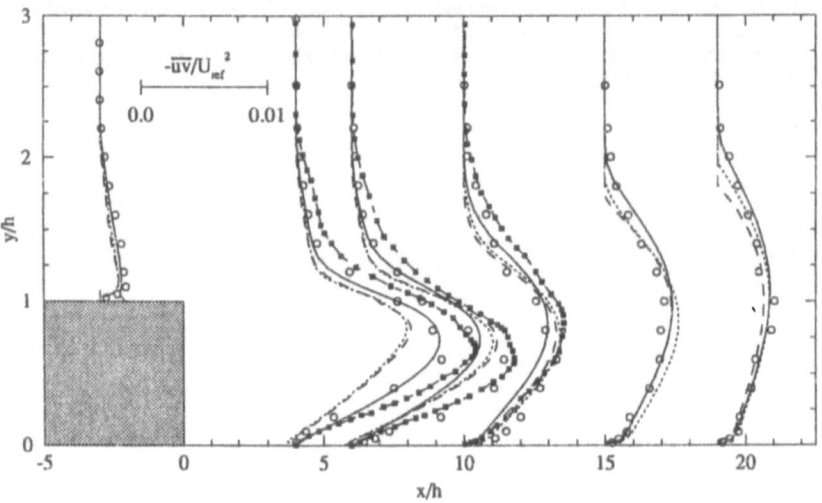

Figure 3. Turbulence shear stress for low-Re flow over a backward-facing step, see figure 1 for legend

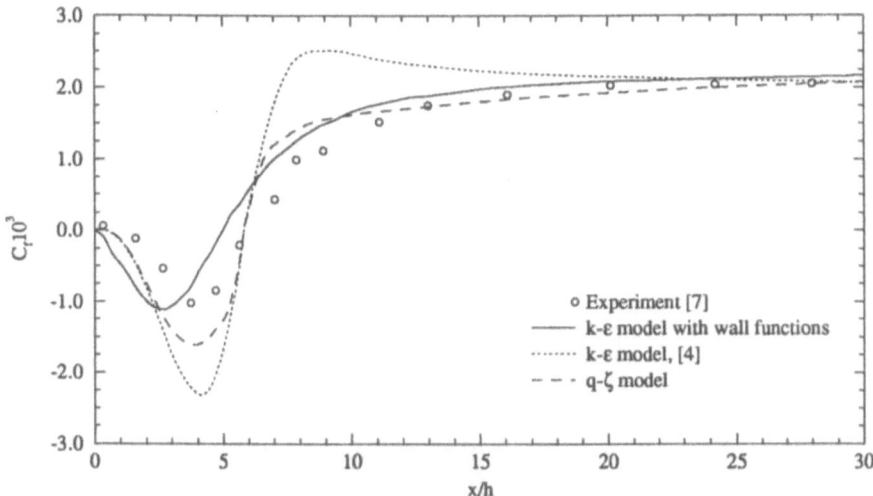

Figure 4. Skin friction coefficient for high-Re flow over a backward facing step.

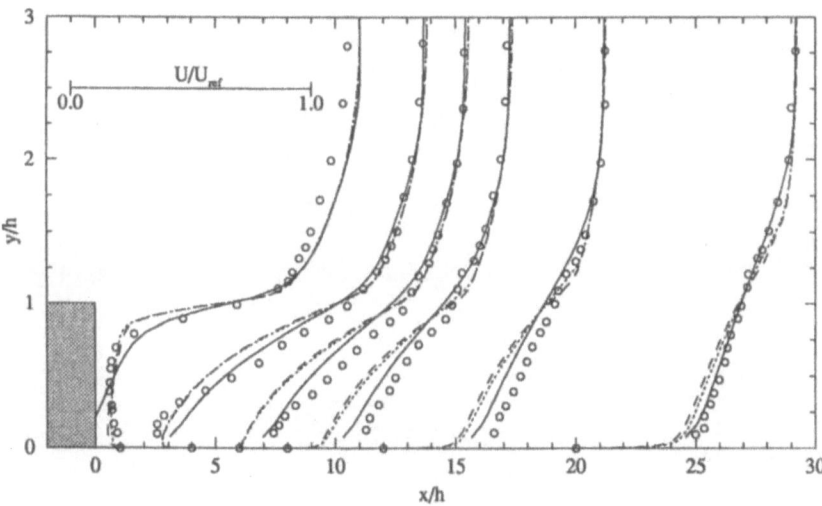

Figure 5. Mean streamwise velocity for high-Re flow over a backward facing step, see figure 4 for legend.

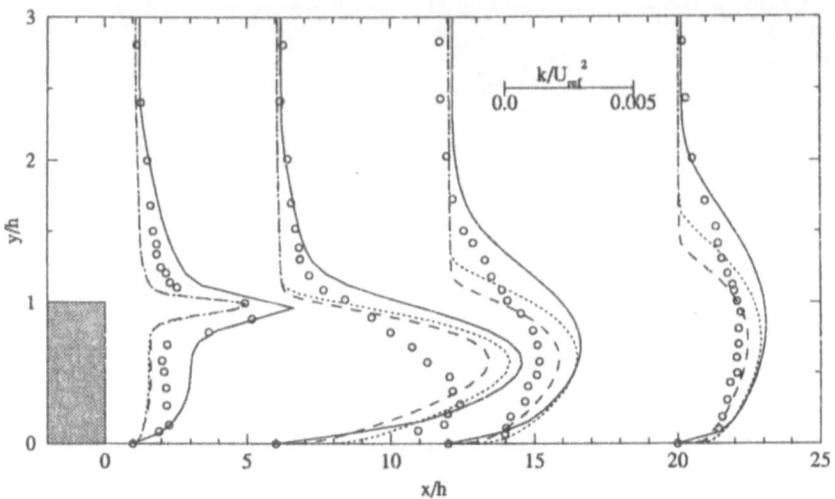

Figure 6. Turbulence kinetic energy for high-Re flow over a backward-facing step, see figure 4 for legend.

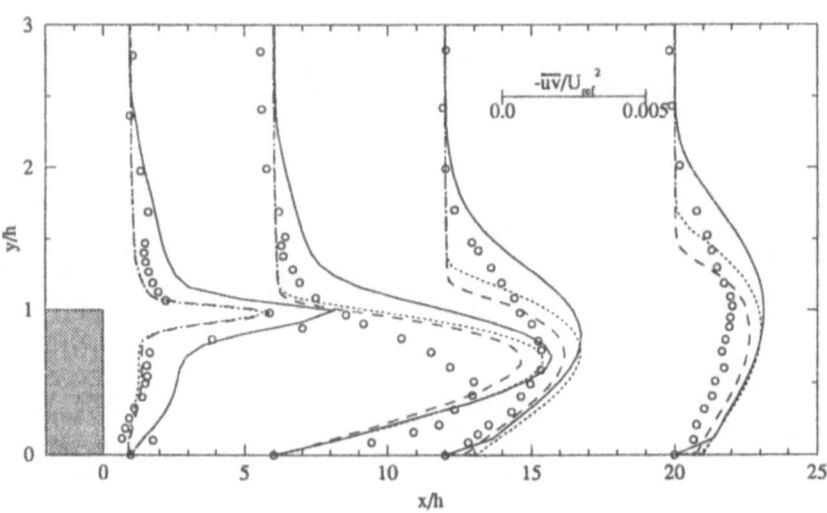

Figure 7. Turbulence shear stress for high-Re flow over a backward-facing step, see figure 4 for legend.

COMPUTATIONAL RESULTS ON TEST CASES TC-2C AND TC2-D TWO DIMENSIONAL, INCOMPRESSIBLE FLOWS WITH RECIRCULATION

J.J.Schmidt and P.S.Larsen

Department of Fluid Mechanics, Building 404

Technical University of Denmark,DK-2800 Lyngby, Denmark

SUMMARY

Computational results of mean velocity, Reynolds stress, turbulent kinetic energy and wall stress for 4 turbulence models (standard high-Re k-ε, low-Re k-ε, low-Re k-τ, and the standard high-Re Reynolds stress model, RSM, for reference) are compared to available LDA-data for 2 test cases, a boundary layer flow (TC-2C, fence-on-wall) and a confined flow (TC-2D, obstacle-in-duct), both involving complex low-Reynolds number near-wall flows with recirculation, reattachment and recovery. The high-Re k,ε model appears to perform best, although all two-equation models have severe deficiencies, such as the underprediction of the turbulence level as well as the reattachment length. Only for the confined flow and then only the RSM can predict the extent of the separation bubble. For both flows, both the RSM and the low-Re k-ε model can predict the overshoot and recovery of skin friction downstream of the separation bubble.

BACKGROUND

Incompressible two-dimensional flow over a backward-facing step can be roughly divided into two classes. One class deals with free (boundary layer) flows past a wall-mounted obstacle, such as TC-2C of the fence-on-wall test case [5]. The other class deals with confined flows, such as TC-2D [2]. The primary difference between these classes of flow is the role played by the pressure distribution through the effect of the adjacent wall.

The ability of turbulence models to predict turbulent backstep flows is often focussed on their ability to predict the length to reattachment, x_r, but other features, such as the turbulence level and anisotropy, are equally important. Generally, when the step height, h, is large, of the order of 50% of the channel height downstream, H (as for TC-2D), predictions of x_r tend to be good. However when h/H is small, models usually underpredict x_r. In most cases, models predict too low turbulence levels in the shear layer. The greater h/H, the more decisive is the pressure distribution for the resulting flow distribution, hence the less sensitive is the flow to the turbulence model. Therefore, TC-2C may be a a good test for turbulence models.

For the usual backstep flow, although involving a simple geometry, the flow is of considerable fundamental interest since it involves separation, a stagnation flow region at reattachment, a reattaching shear layer, and a recirculating zone. Also, turbulence levels are high. Among the numerous references, the experimental study of Tropea & Gackstatter [8] shows the influence of step height, and that of Dimaczek et al, [2] shows characteristics of flow past various two-dimensional obstacles.

103

TURBULENCE MODELS

The standard high-Re and the low-Re k,ε-models are described by Jones & Launder [3] and Launder & Sharma [4]. The k,τ-model is described by Speziale et al. [7] and the high-Re Reynolds Stress Model is described by Clarke and Wilkes [1].

The boundary conditions at a solid wall for the turbulent quantities are $k = 0$ and $\tilde{\varepsilon} = 0$ for the low-Re k,ε-model and $k = 0$ and $\tau = 0$ for the k,τ-model. The boundary conditions for the high-Re k,ε-model and the high-Re RSM are that $\partial k/\partial y = 0$ at a solid wall, that ε in the cell adjacent to the wall is a function of k in that cell, and that u^+ in the cell adjacent to the wall is given by wall functions with y^+ based on k and a switch between viscous sub-layer and the logarithmic region at $y^+ = 11.6$.

The transport equation for $\tau = k/\varepsilon$, proposed by Speziale et al. [7], is given by:

$$\frac{D\tau}{Dt} = (1 - c_1)\frac{\tau}{k}(-\overline{u_i u_j})\frac{\partial U_i}{\partial x_j} + (c_2 f_2 - 1) + \frac{\partial}{\partial x_j}\left[\left(\nu + \frac{\nu_t}{\sigma_{\tau 1}}\right)\frac{\partial \tau}{\partial x_j}\right]$$

$$+\frac{2}{k}\left(\nu + \frac{\nu_t}{\sigma_{\tau 2}}\right)\frac{\partial k}{\partial x_i}\frac{\partial \tau}{\partial x_i} - \frac{2}{\tau}\left(\nu + \frac{\nu_t}{\sigma_{\tau 3}}\right)\frac{\partial \tau}{\partial x_i}\frac{\partial \tau}{\partial x_i} \quad ,$$

where constants are $c_1 = 1.44$ and $\sigma_{\tau 1} = \sigma_{\tau 2} = \sigma_{\tau 3} = 1.36$. The damping functions c_2, f_2 and f_μ are given as:

$$c_2 = 1.83\left[1 - \frac{2}{9}\exp\left(-\frac{Re_t^2}{36}\right)\right] \quad ,$$

$$f_2 = \left[1 - \exp\left(-\frac{y^+}{4.9}\right)\right]^2 \quad ,$$

$$f_\mu = \left(1 + \frac{3.45}{\sqrt{Re_t}}\right)\tanh\left(\frac{y^+}{70}\right) \quad ,$$

where the turbulent Reynolds number is defined as $Re_t = k\tau/\nu$. The value of y^+ for the damping functions is not determined from U_τ based on the velocity gradient at the wall since the flow is deattached in regions. Instead, the wall friction velocity from $U_\tau = C_\mu^{1/4}k^{1/2}$ is evaluated at a distance from the wall where $y^+ = 11.6$.

NUMERICAL METHODS

The computations are performed with the commercial elliptic code Flow3D ver.3.2.1 from Harwell, which employs a cell centered finite volume method. The differencing schemes are *Quick* for velocities and *Hybrid* for turbulent quantities. The pressure correction algorithm applied is SIMPLEC.

The high-Re k-ε model, low-Re k-ε model and high-Re Reynolds Stress Model are originally implemented in the code, whereas the k-τ model has been implemented by the authors.

The computations on TC-2C employ a 200 x 125 mesh for low-Re models and a 200 x 75 mesh for computations with high-Re models. The computations on TC-2D employ a 200 x 140 mesh for low-Re models and a 200 x 80 mesh for computations with high-Re models. The values of y^+ in the cell adjacent to the wall are below 0.6 for the low-Re meshes and around 6 for the high-Re meshes.

Comparisons between the present calculations and calculations performed with the same code on the recommended refined meshes in the ETMA database show good agreement, confirming that mesh-independence had been reached.

Results include selected profiles of mean velocity U, turbulent kinetic energy k, and Reynolds stress $\overline{uv}$ (Fig.1-4), as well as skin friction C_f (Fig.5 and 6). Reattachment lengths are given in Table 1.

TC-2C: FENCE-ON-WALL. The high-Re models succeeds in predicting the C_f values (fig. 6) upstream of the fence, while the C_f values downstream of the fence are highly over- or underpredicted by all models. The reattachment length is underpredicted by all models, with the best results obtained by the RSM and the low-Re k-ε model.

While the U-profiles (fig. 4) are nearly identical and all models correctly predict the recirculation on top of the fence, the k-profiles show significant differences, especially in the recirculation bubble at the station 200 mm downstream of the fence.

TC-2D: OBSTACLE-IN-DUCT. All C_f values (fig. 5) deviates considerably from the experimental values, except far downstream. Especially the k-τ model highly overpredicts the values of C_f.

In summary all two-equation models have deficiencies, e.g. in underestimating the turbulence level in the shear layer over and downstream of the recirculation zone and the reattachment length. The high-Re k,ε-model performs quite well for both cases. Experience shows that the k-τ model is computationally ill-conditioned, hence costly.

For reference, the RSM gives generally satisfactory results, being able to predict skin friction overshoot and recovery, as well as reattachment length for the confined flow. However, it also has deficiencies in underestimating the turbulence level and, for the unconfined flow, the reattachment length.

Table 1 Reattachment length x_r downstream of obstacle

Test case	Re	Exp.	Hi-Re k,ε	RSM	Low-Re k,ε	k,τ
Fence	3000	468 mm	357 mm	434 mm	421 mm	383 mm
Obstacle	42000	203 mm	166 mm	202 mm	164 mm	180 mm

ACKNOWLEDGEMENTS

This research was partially supported by CEC contract AERO-CT92-0051 (ETMA project) and by the Danish Technical Research Council under grant STVF 5.26.16.31.

REFERENCES

[1] Clarke and Wilkes, N.S. 1989. "The calculation of turbulent flows in complex geometries using a Differential Stress Model". AERE-R 13428.

[2] Dimaczek,G., Kessler,R., Martinuzzi,R. and Tropea,C. 1989. "The flow over two-dimensional, surface-mounted obstacles at high Reynolds numbers". Proc. 7th Symposium on Turbulent Shear Flows, Standford, CA.

[3] Jones,W.P. and Launder,B.E. 1972. "Some properties of sink-flow turbulent boundary layers", J.Fluid Mech., 56, 337-351.

[4] Launder,B.E. and Sharma,B.I. 1974. "Application of the energy-dissipation model of turbulence to the calculation of flow near a spinning disc. Lett. Heat Mass Transfer, 1, 131-138.

[5] Marxen,U. and Jørgensen,O.A. 1993. Master's Thesis, Dept. of Fluid Mech., Techn. Univ.of Denmark, Report AFM-EP 93-06.

[6] Patel,V.C., Rodi,W. and Scheuerer,G. 1985. "Turbulence models for near-wall and low Reynolds number flows: A review". AIAA J., 23, 1308-1319.

[7] Speziale,C.G., Abid,R. and Anderson,E.C. 1992. "Critical evaluation of two-equation models for near-wall turbulence". AIAA J., 30, 324-331.

[8] Tropea,C.D. and Gackstatter,R. 1985. "The flow over two-dimensional surface-mounted obstacles at low Reynolds numbers". J.Fluids Engrg., 107, 489-494.

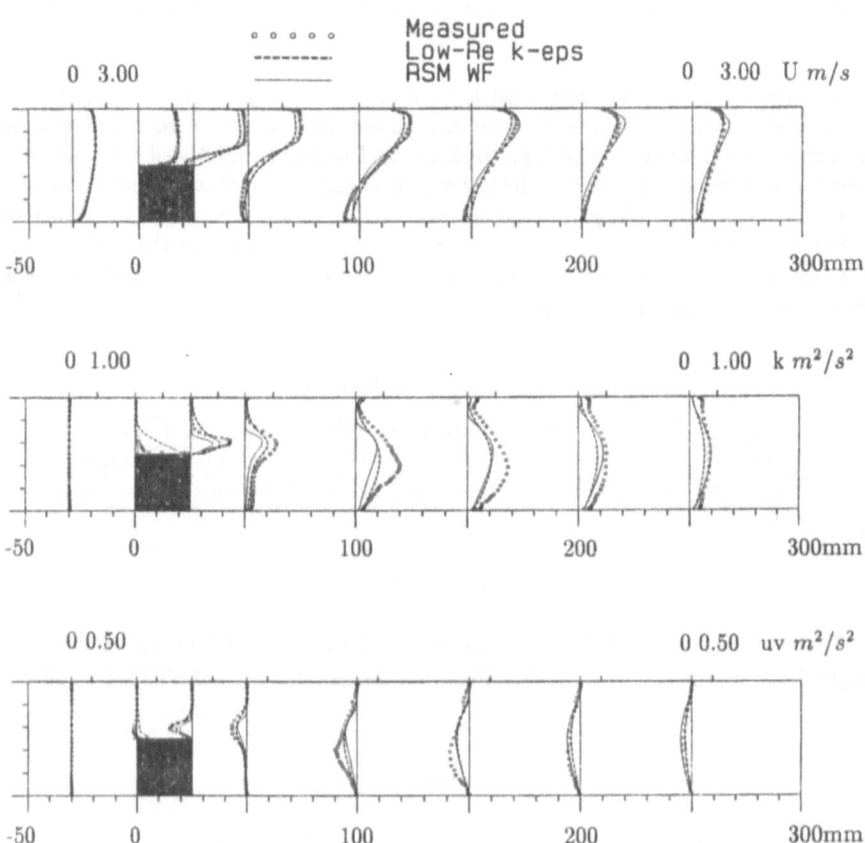

Figure 1: Obstacle. U, k and uv profiles with experimental and numerical data for low Reynolds number k-epsilon model and Reynolds Stress model.

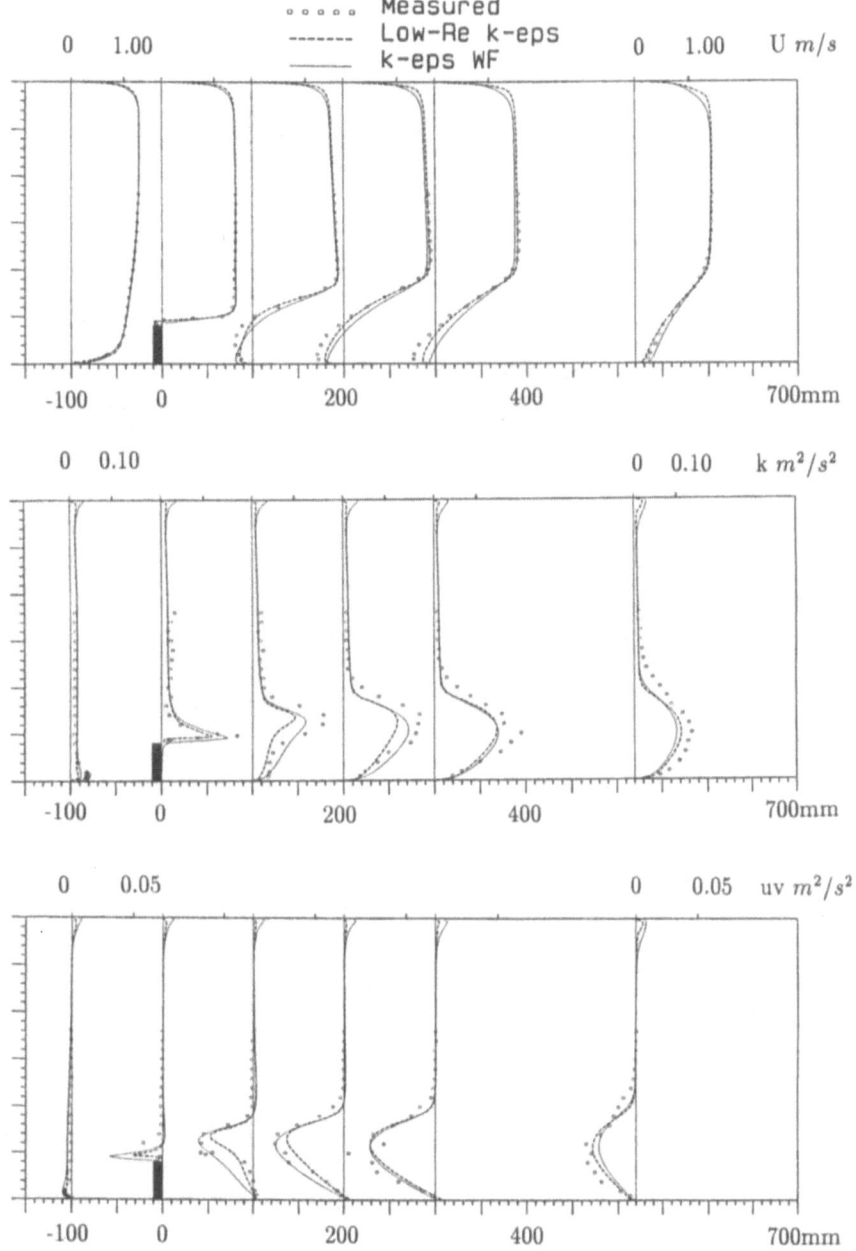

Figure 2: Fence-on-wall. U, k and uv profiles with experimental and numerical data for high and low Reynolds number k-epsilon models.

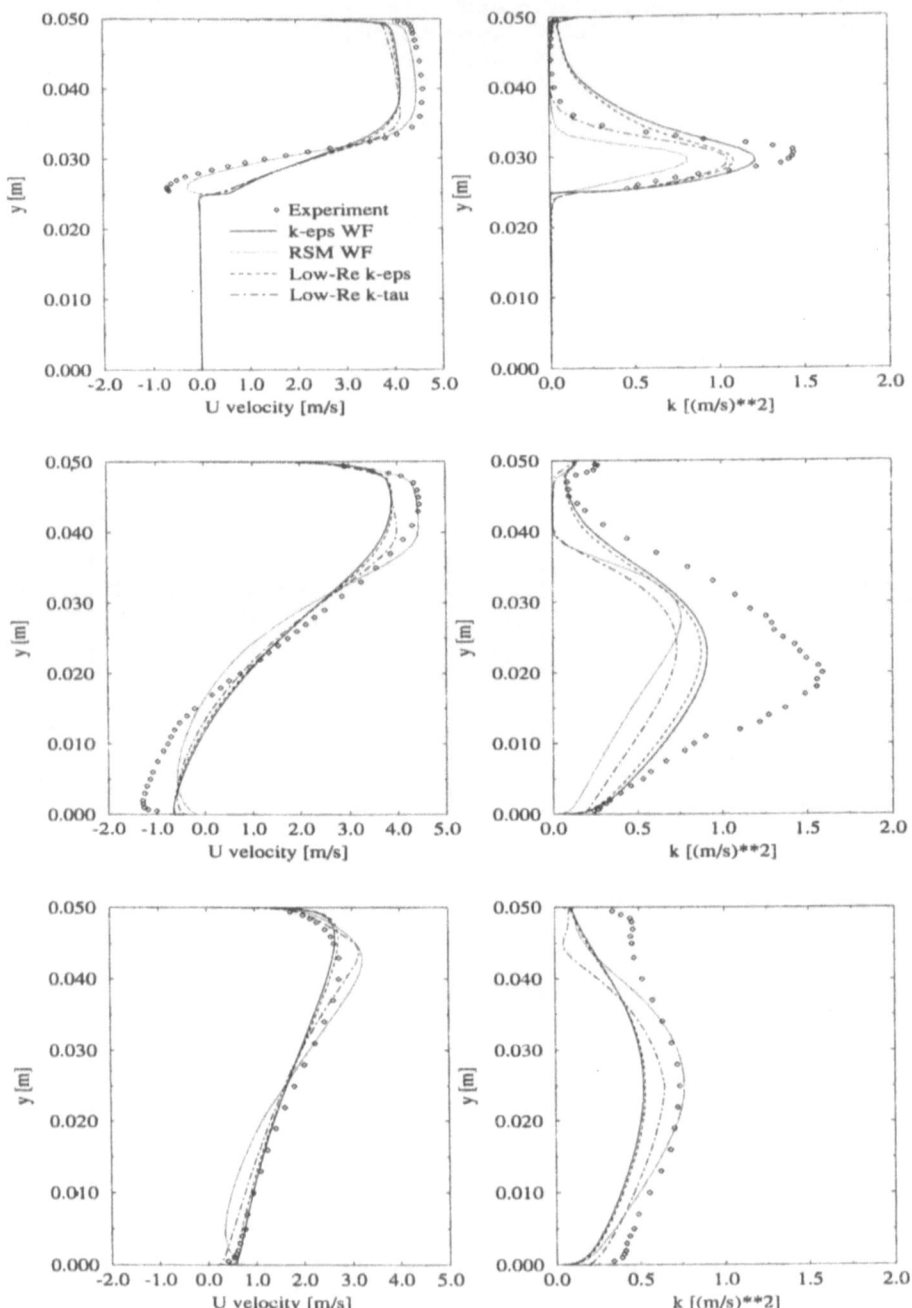

Figure 3: Obstacle test case with experimental and numerical results for U and k profiles for four turbulence models at (top to bottom) x=25, 100 and 250 mm.

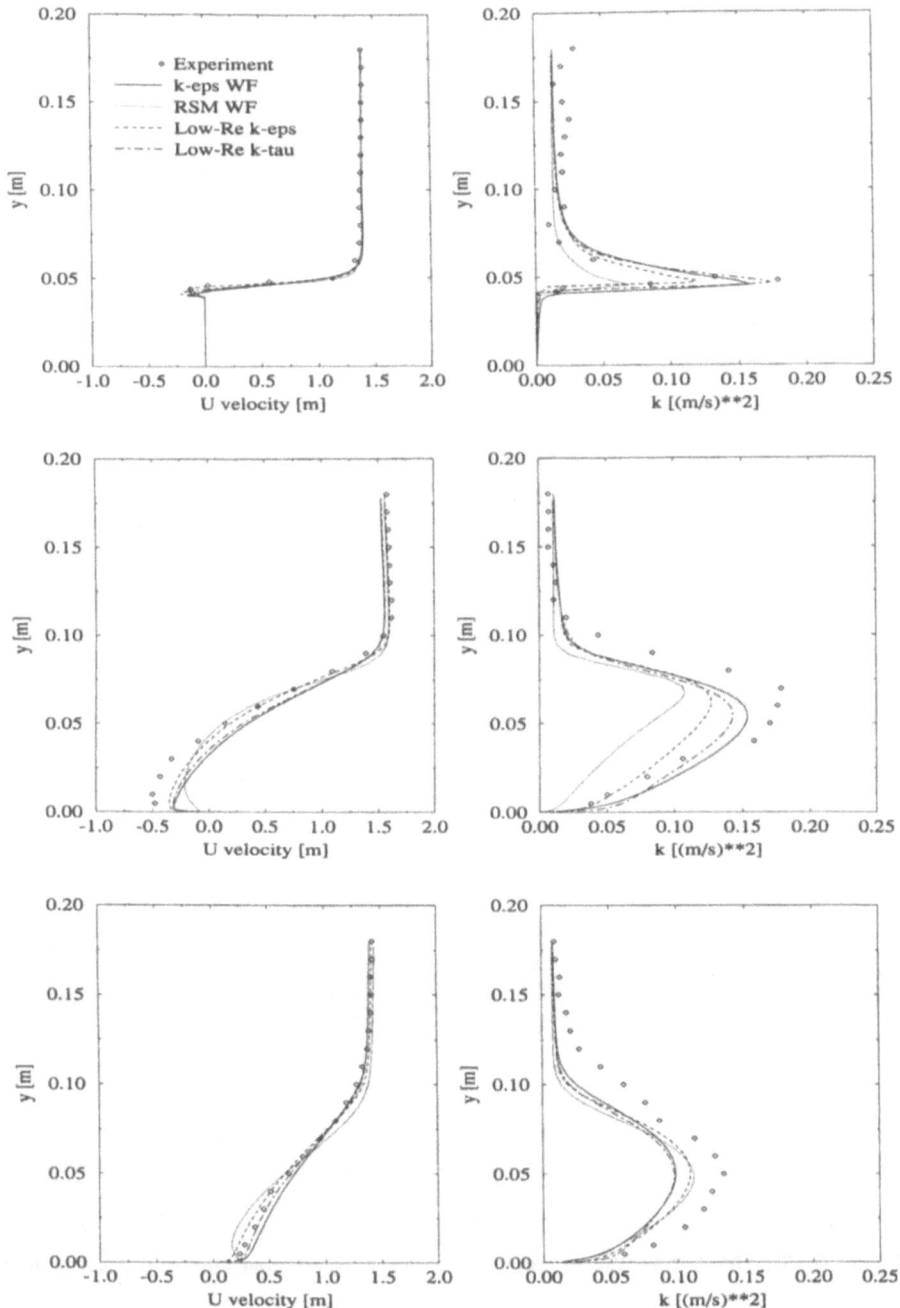

Figure 4: Fence-on-wall TC-2C with experimental and numerical results for U and k profiles for four turbulence models at (top to bottom) x=0, 200 and 520 mm.

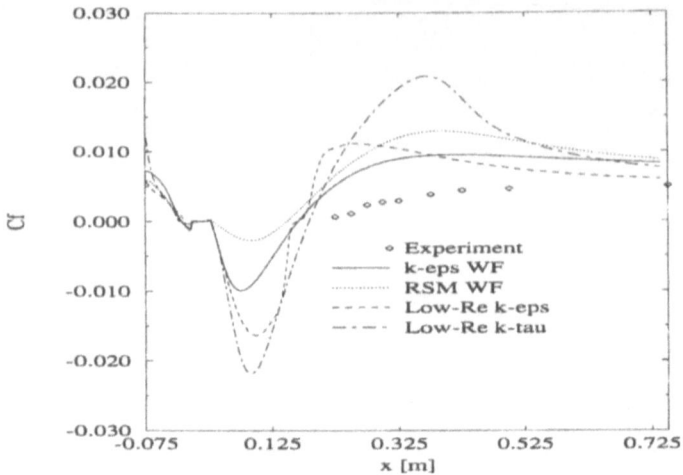

Figure 5: Skin friction for obstacle TC-2D, numerical data for four turbulence models.

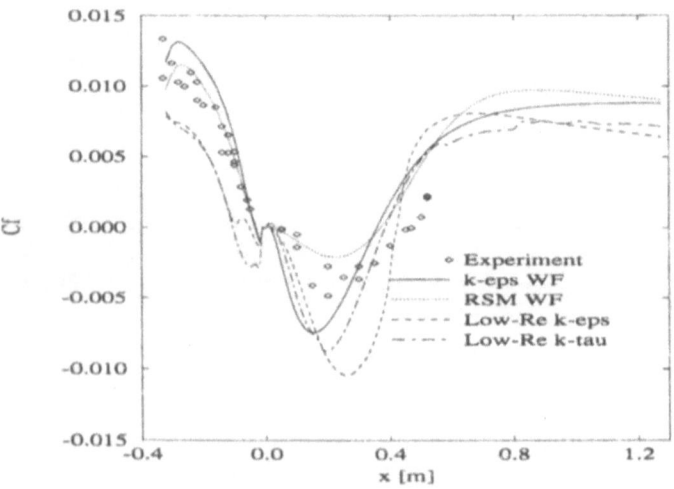

Figure 6: Skin friction for Fence-on-wall TC-2C, experimental and numerical data for four turbulence models.

INCOMPRESSIBLE RECIRCULATING FLOWS

TC2-C FENCE-ON-A-WALL
TC2-D OBSTACLE-IN-CHANNEL

Robert D. Harper & Michael M. Gibson
Mechanical Engineering Department, Imperial College, London SW7 2BX

SUMMARY

A new low-Reynolds number, two-equation turbulence model based on the variables $q \equiv \sqrt{k}$ and $\zeta \equiv \tilde{\varepsilon}/2q$ is presented, this model together with the low-Re k-ε model of Launder & Sharma and the 'standard' k-ε model with wall functions are used to compute two flows involving separation over wall mounted obstacles. The first computation is for flow over a fence-on-a-wall [5] while the second is for the flow over an obstacle-in-a-duct [6]. The results shown highlight that the low-Re models appear to resolve more of the flow details than the wall function calculations and the q-ζ model is seen to perform slightly better then the low-Re k-ε model.

INTRODUCTION

The k-ε model of turbulence suffers from the grave defect that its constituent equations are ill-adapted to integration to the wall. Consequently its application is effectively restricted to high Reynolds number turbulence outside the viscous layers, which have to be bridged by wall functions. The lack of a natural wall boundary condition for the ε-equation presents problems which may only be surmounted by the introduction of terms which require an excessively fine grid for their resolution. These difficulties are avoided by changing to variables which are zero at the wall and vary linearly with distance close to it. The q-ζ model is a k-ε model replacement designed primarily to improve computational efficiency. The improvement in predictive accuracy shown in the present results is an unexpected but welcome bonus.

This report documents the results of 2-dimensional separated flow calculations using the new q-ζ turbulence model and compares the results with predictions from the low and high Reynolds number versions of the k-ε model and also with experimental data. Calculations were performed for TC2-C, fence-on-a-wall, and for TC2-D, obstacle-in-channel. The geometrical and flow details for these cases are given in the test case specification.

THE q-ζ TURBULENCE MODEL

A full derivation of the q-ζ model is supplied in [1, 2]. A companion paper within these proceedings, [3] also includes this derivation, together with a critical comparison of the k-ε and q-ζ models. Only an outline of the model will be given here.

The variable q is defined by $q \equiv \sqrt{k}$, while ζ is its dissipation rate. Transport equations for q and ζ are obtained by transforming the 'standard' k and ε equations and can be written as shown below, (note that pressure diffusion in the q-equation is assumed negligible).

$$U_i \frac{\partial q}{\partial x_i} = \frac{\partial}{\partial x_j}\left\{\left(\nu + \frac{\nu_t}{\sigma_q}\right)\frac{\partial q}{\partial x_j}\right\} + Q - \zeta$$

$$U_i \frac{\partial \zeta}{\partial x_i} = \frac{\partial}{\partial x_j}\left\{\left(\nu + \frac{\nu_t}{\sigma_\zeta}\right)\frac{\partial \zeta}{\partial x_j}\right\} + \frac{\zeta}{q}\left(C_{\zeta 1}f_{\zeta 1}Q - C_{\zeta 2}f_{\zeta 2}\zeta\right) + \psi \; .$$

Q, the production rate of q, and ψ, the secondary mean-flow production of ζ, are given by:

$$Q \equiv \gamma_t \frac{\partial U_i}{\partial x_j}\left(\frac{\partial U_i}{\partial x_j} + \frac{\partial U_j}{\partial x_i}\right), \qquad \psi = 2\nu\gamma_t\left(\frac{\partial^2 U_i}{\partial x_j \partial x_k}\right)\left(\frac{\partial^2 U_i}{\partial x_j \partial x_k}\right),$$

Other quantities, including eddy viscosity, ν_t, are defined as;

$$\gamma_t \equiv \frac{C_\mu f_\mu}{4}\frac{q^2}{\zeta}, \qquad \nu_t \equiv C_\mu f_\mu \frac{q^3}{2\zeta} \; .$$

While the viscous-layer damping functions are taken from Launder and Sharma [4]:

$$f_\mu = exp\left\{\frac{-A_\mu}{(1 + Re_t/50)^2}\right\}, \qquad C_\zeta f_\zeta \equiv 2C_\epsilon f_\epsilon - 1, \qquad f_{\epsilon 2} = 1 - 0.3 exp\left(-Re_t^2\right).$$

Re_t is the turbulence Reynolds number $q^3/2\nu\zeta$, and

$$f_\epsilon = 1.0, \quad C_{\epsilon 1} = 1.88, \quad C_{\epsilon 2} = 2.84, \quad A_\mu = 6.0, \quad \sigma_q = 1.0, \quad \sigma_\zeta = 1.3.$$

The wall boundary conditions are $q = 0$, $\zeta = 0$.

The model constants and damping functions used here need not necessarily be those of Launder and Sharma, these are used because Re_t is a better parameter for general flow calculations than the distance from the wall and the Launder-Sharma model is sufficiently well-established to form a suitable benchmark for comparison. A limited number of calculations with alternative constants and functions have been performed in order to calibrate the model [1,2] these results will not be presented here.

COMPUTATIONAL DETAILS

Predictions for all flows were obtained using a two-dimensional elliptic code with co-ordinate based grids, staggered for velocity and pressure, with hybrid differencing and pressure correction using the SIMPLE algorithm. The number of grid nodes required to achieve grid independent solutions obviously depends both on the geometry of the flow and the model considered, however a typical grid for a low-Re turbulence model calculation would consist of approximately 180×300 nodes in the cross-stream and streamwise directions respectively. For all low-Re model calculations the first grid node was at no more than $y^+ = 1.0$.

The test case geometries and inlet conditions were all prescribed in the test case specifictaion. Standard conditions were applied at the remaining boundaries i.e., no slip conditions at a wall and Neumann conditions at the exit plane.

RESULTS

TC-2C: Fence-on-a-Wall
The results for this test case [5] are shown in figures 1-4. The skin-friction distributions plotted in figure 1 display the under and overshoot characteristic of the low-Re modelling methods, but on this occasion. Overall, the wall-function calculation gives surprisingly good results, though it fails to reproduce the recirculation zones upstream and on top of the fence (identified here by negative skin friction). Nor does this method account for the secondary recirculation. But the gross features of the flow appear to be adequately predicted. By contrast, the two low-Re methods, though picking up such small-scale details as the minor recirculations, underpredict the length of the main recirculation region by 10% or so. Nor, in this flow, do the predictions from the two methods differ significantly. The streamwise velocity profiles shown in figure 2 are broadly similar for all calculations. The profiles of k and $\overline{uv}$ are shown in figures 3 and 4.

TC-2D: Obstacle-in-a-Channel
The water flow over a rib, or thick fence, is described by Dimaczek et al [6]. Comparative calculations for this final test case are shown in figures 5-8. Skin-friction data have not been derived from the measurements but the observed extent of the recirculation zones are indicated in figure 5. The streamwise velocity profiles highlight some interesting features of the flow over the upper surface of the rib. All three computations predict reattachment on the upper surface of the rib while the experiments show reverse flow along the whole of this region. This result could in turn be responsible for the discrepancies between all the model predictions and the data in the region immediately downstream of the rib, the agreement improves as the flow recovers. The profiles of k and $\overline{uv}$ are shown in figures 7 and 8.

CONCLUSIONS

These calculations demonstrate the feasibility of general turbulent-flow calculations with a new two-equation model related to, but not the same as, the k-ε model of turbulence. The chief merit of the new model is that, unlike the equations for k and ε, the q and ζ equations are well conditioned near the wall, where the boundary conditions are specifically defined, and the dependent variables vary linearly with distance from the wall when the wall is close. The equations prove to be less stiff numerically, while linear variations in the sublayer permit the use of a coarser grid with consequent savings in computer time.

The k-ε and q-ζ equations do not give exactly the same results when integrated through the sublayers to the wall because the ε and ζ equations are not exact transforms of each other. Gradient diffusion of ζ is not exactly the same process as gradient diffusion of ε. Thus the results from the new form are not the same as those of the k-ε model from which it was derived.

The q-ζ model allows one of the principal problems of current CFD research to be approached from a new angle, the potential of the established models being almost exhausted. It is regarded as a useful vehicle for further development, not least in the possibility of using the length-scale-determining ζ-equation in higher-order closures. But it is likely that the model is some way from the optimum form. The constants and damping functions chosen for these demonstration calculations have known limitations in k-ε modelling, and the search for improvements has become a major research activity, thus far largely fruitless. This is a research area demanding further intensive study, extended well beyond the scope of fitting curves to the results of channel flow simulations. In continuing work on the model we plan to explore the possibility of inertial damping in the outer sublayers $y^+ > 5$ say, and that of recasting the source terms of the ζ-equation in more physically realistic and robust form.

REFERENCES

1. Gibson, M.M. & Dafa'Alla, A.A. A two-equation model for turbulent wall flow, AIAA J (to appear).
2. Gibson, M.M. & Dafa'Alla, A.A. 1994 The q-ζ model for turbulent wall flow, Fluid Dynamics Division, American Physical Society, 47th Annual Meeting, Atlanta, Georgia.
3. Dafa'Alla, A.A, Harper, R.D. & Gibson, M.M. 1995 ETMA Workshop TC3: Flat Plate Boundary Layers.
4. Launder, B.E. & Sharma, B.I. 1974 Application of the energy-dissipation model of turbulence to the calculation of the flow near a spinning disk, Letters in Heat and Mass Transfer 1, 131-138.
5. Westergaard, C.H., Kock, C.W., Larsen, P.S., Buchave, P. 1993 Application of PIV to turbulence studies - an exploratory study, Dept. Fluid Mech. Rep. AFM-93-07, Technical University of Denmark.
6. Dimaczek, G., Kessler, R., Martinuzzi, R. & Tropea, C. 1989 The flow over two-dimensional, surface mounted obstacles at high Reynolds numbers, 7th Turbulent Shear Flows Symposium, Stanford, 10.1.

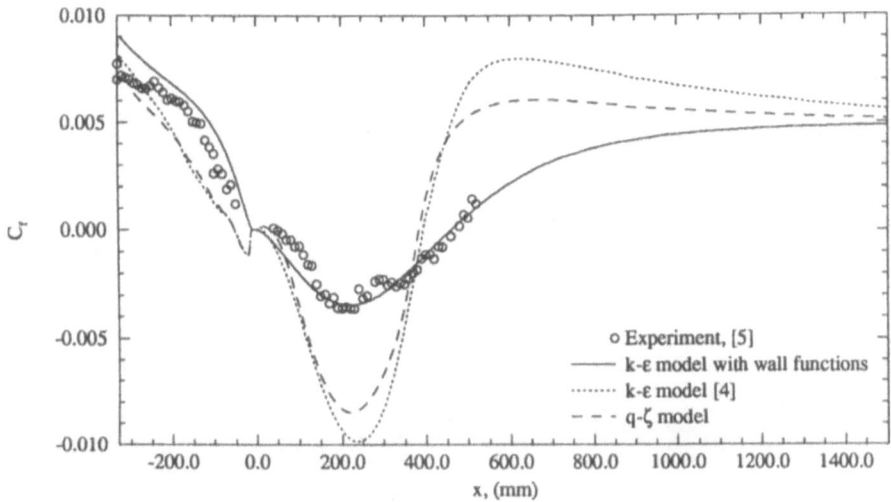

Figure 1. Skin friction coefficient, C_f for flow over a fence.

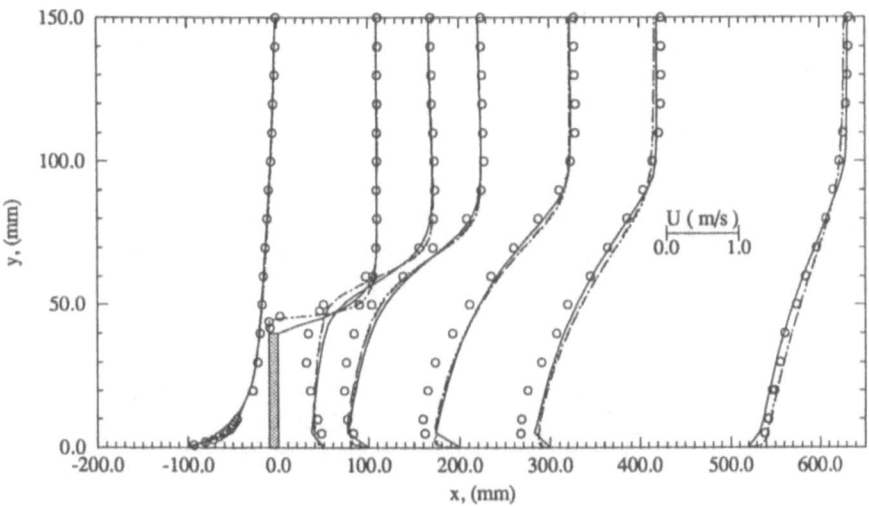

Figure 2. Mean streamwise velocity, U, for flow over a fence, see figure 1 for legend.

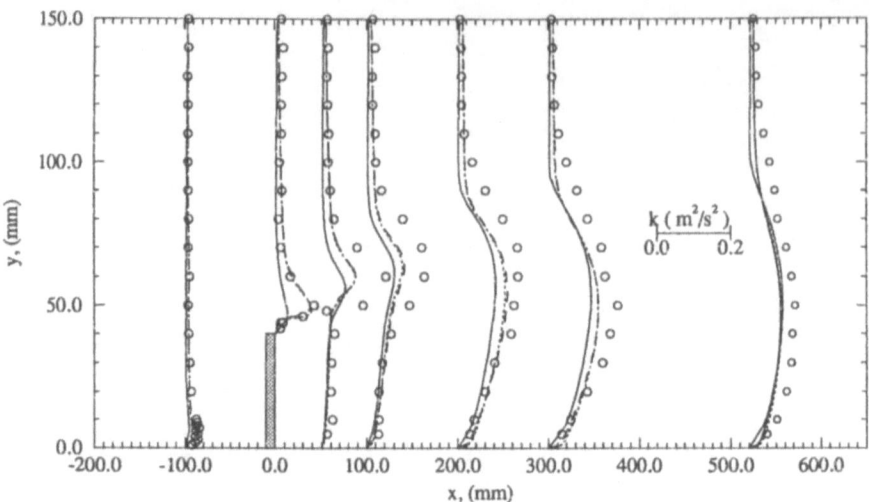

Figure 3. Turbulence kinetic energy, k, for flow over a fence, see figure 1 for legend.

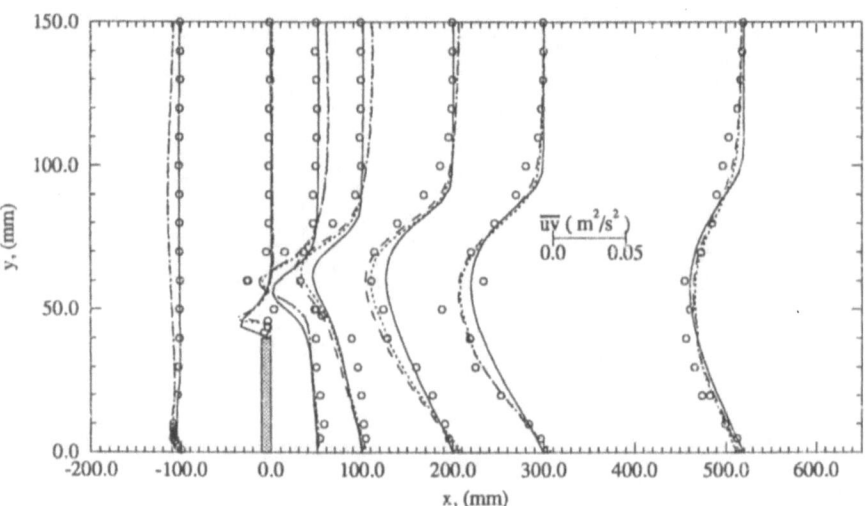

Figure 4. Turbulence shear stress, uv, for flow over a fence, see figure 1 for legend.

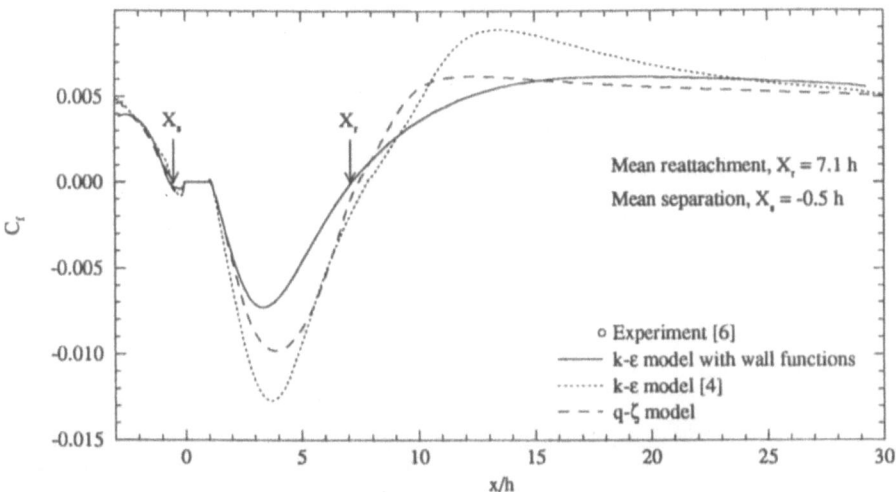

Figure 5. Skin friction coefficient, C_f, for flow over a rib.

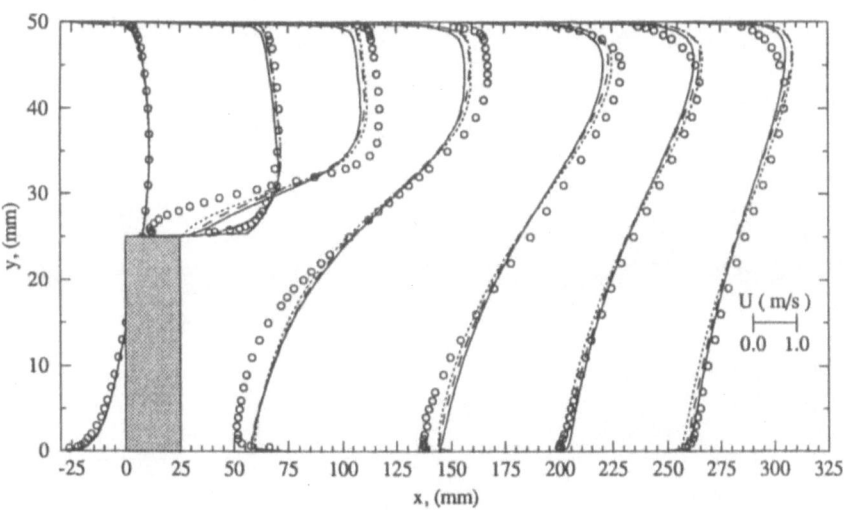

Figure 6. Mean streamwise velocity for flow over a rib, see figure 5 for legend

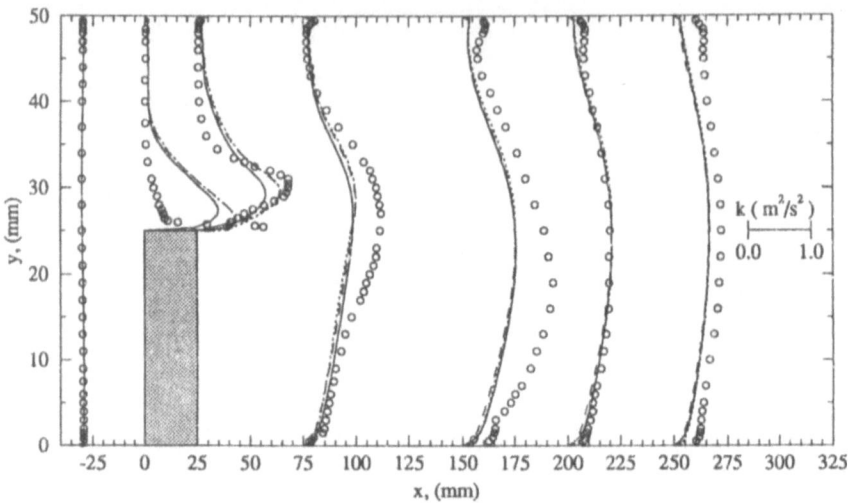

Figure 7. Turbulence kinetic energy, k, for flow over a rib, see figure 5 for legend

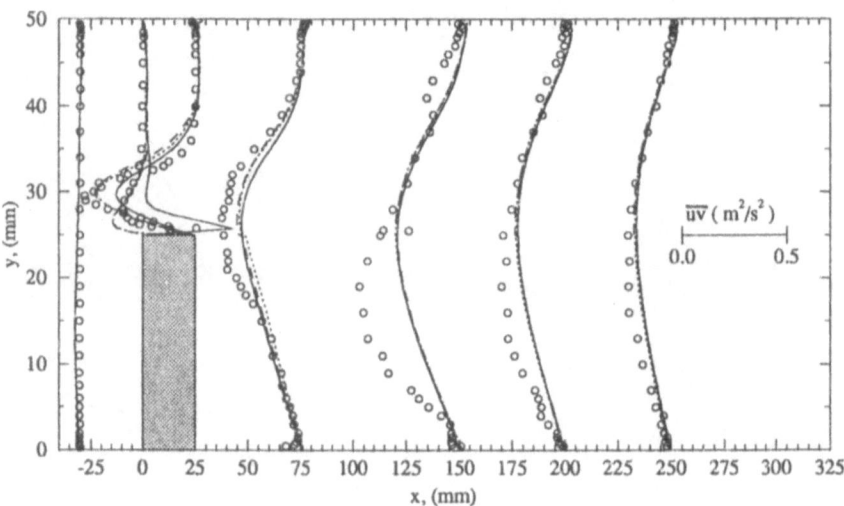

Figure 8. Turbulence shear stress, uv, for flow over a rib, see figure 5 for legend.

118

NUMERICAL SOLUTION OF THE TURBULENT FLOW OVER A FENCE USING TWO EQUATION MODELS

J.C.F. Pereira, M.H. Kobayashi and N.P.C. Marques
Instituto Superior Técnico/Technical University of Lisbon
Mechanical Engineering Department
Av. Rovisco Pais
1096 Lisboa Codex - Portugal

SUMMARY

The comprehended test case is a mounted fence-on-wall (TC-2C). Two two-equation models are studied. The accuracy and applicability of this turbulence models to incompressible, turbulent flows at low Reynolds numbers on regions where large flow separation occurs, is assessed by comparing the computed results with reported experimental data.

1. INTRODUCTION

Two turbulence models (two equation ones) are tested against available LDA data for a fence on wall test case. It is therefore possible to evaluate their capability to predict the Mean Velocity, Reynolds Stress, Turbulent Kinetic Energy and wall stress along the wall, before and after the mounted Fence. The general goal is an evaluation of the prediction capability of recently developed as well as a commonly used turbulence model with low Reynolds number corrections to compute incompressible, separated flows. The turbulence models are the Standard k-ε model (with wall functions) and a low Reynolds model by Huang and Coakley. The Standard k-ε model provides better results for almost all of the variables and sections. The Huang and Coakley model has extended the Standard k-ε model to low Reynolds number flows to describe the near wall region. The model incorporates a wall damping effect by means of turbulent Reynolds number dependent coefficients.

2. MATHEMATICAL AND NUMERICAL MODEL

2.1 *Governing Equations*

For general steady incompressible viscous flows the governing equations for mass and momentum are, in a Cartesian system using eddy viscosity/diffusivity assumptions, as following:

$$\frac{\partial}{\partial x_i}(u_i) = 0 \tag{1}$$

$$\frac{\partial}{\partial x_j}\left[\left(\rho\, u_i\, u_j + \widehat{T}_{ij}\right)\right] = S_i \tag{2}$$

$$\frac{\partial}{\partial x_j}\left(\rho\, u_j\, H + q_j\right) = \frac{\partial}{\partial x_j}\left[u_i\, \widehat{T}_{ij} + \left(m + \frac{\mu_t}{\sigma_1}\right)\frac{\partial k}{\partial x_j}\right] \tag{3}$$

$$\frac{\partial}{\partial x_i}\left[\rho\, u_i\, \phi + q_i^\bullet\right] = S_\phi \tag{4}$$

$$\frac{\partial}{\partial x_j}[\rho u_j s_i + r_{ij}] = R_i \tag{5}$$

$$r_{ij} = -\left(\mu - \frac{\mu_t}{\sigma_i}\right)\frac{\partial s_i}{\partial x_j} \tag{6}$$

where s_1 ($\equiv k$) and s_2 ($\equiv \varepsilon$) are additional field variables so that μ_t is a function of s_1 and s_2. The modelling constants, σ_i, are turbulent Prandtl numbers and R_1, R_2 stand for the production and dissipation of s_1 and s_2, respectively. The total viscous stress tensor is given by

$$\widehat{T}_{ij} = -2\mu\, S_{ij} + T_{ij} . \tag{7}$$

The Boussinesq approximation is evoked, yielding a Reynolds stress tensor which is proportional to the mean strain-rate tensor, in symbols:

119

$$T_{ij} = -2\mu_i \, S_{ij} + \frac{2}{3} \rho \, k \, \delta_{ij} \, . \tag{8}$$

Finally, the heat flux vector q_i and the scalar flux vector q_j^ϕ are approximated as:

$$q_i = -\left(\frac{\mu}{Pr} + \frac{\mu_t}{Pr_t}\right)\frac{\partial h}{\partial x_j} \tag{9}$$

$$q_j^\phi = -\left(\Gamma_\phi + \Gamma_t\right)\frac{\partial \phi}{\partial x_j} \, . \tag{10}$$

2.2 The Tested Turbulence Models

Two turbulence models are compared, the standard high Reynolds number k-ε model with wall functions and k-ε model of Coakley and Huang [1], denoted by HC. This latest model is implemented on it's low Reynolds version according to

$$\mu_t = C_\mu \, f \, \rho \, \frac{k^2}{\varepsilon} \tag{11}$$

where C_μ is a modeling constant and f a damping function which contemplates the low Reynolds effects. On the Standard k-ε model, f is taken equal to 1. The turbulent source terms can be written on a compact form as:

$$R_i = \left[C_{i1} \, C_\mu \, f_i \left(\frac{S}{\omega}\right)^2 - C_{i2}\right]\rho\omega s_i \tag{12}$$

where s=k, C_{i1}, C_{i2}, and f_i are model dependent parameters, ω the specific dissipation rate or frequency scale (ε/k) and S^2 is defined as:

$$S^2 = \left(\nabla v \, . \, \nabla v^T\right) : \nabla v \, . \tag{13}$$

The models defining equations and related parameters are presented in the following table:

Table 1 — Wall Damping Functions and basic Modeling Constants

Model	Damping Functions	Model Constants
HC	$f_\mu = \tanh(y^+/43)^2 g$ $g = \max(1, 10.1/R_t)$ $y^+ = R^+(1 + (1 + 2(15/R^+)^2)^{1/2})^{1/2}$ $R^+ = (C_\mu/4)^{1/4} R_q.$ $R_q = k^{1/2} y/\nu$ $R_t = k^2/\nu\varepsilon.$	$C_{21} = 1.5; \; C_{22} = 1.8$ $\sigma_k = 1.3; \; \sigma_\varepsilon = 1.87$ $C_\mu = 0.09$
Std. k-ε	$f = 1.$	$C_{21} = 1.44; \; C_{22} = 1.92$ $\sigma_k = 1.0; \; \sigma_\varepsilon = 1.3$ $C_\mu = 0.09$

About the boundary conditions relevant to each of the models, Table 2 summarizes the ones used.

Table 2 — Solid Wall Boundary Conditions

Model	Conditions
k-ε HC	$k = 0, \; \varepsilon_1 = 2\nu_1 k_1 / y_1^2$
Std. k-ε	wall function

2.3 Numerical Schemes

The Finite-Volume method based on a non-staggered (or collocated) grid system is used. Discretization procedure follows Peric [2] and Kobayashi and Pereira [3], yielding for a cell:

$$J_e - J_w + J_n - J_s = \int_{dVp} S \, dV \tag{14}$$

120

where the surface integrals have been denoted J_e, J_w, J_n and J_s. Each of these terms is the result of two contributing physical processes: diffusion and convection. As an example, on the east face, we have $J_e = J_e^D + J_e^C$ where:

$$J_e^D = - \int_{A_e} \left[\frac{(\Gamma_\phi + \Gamma_t)}{J} \left(\frac{\partial \phi}{\partial x_1} B_1^i + \frac{\partial \phi}{\partial x_2} B_2^i \right) \right] dx^2 \approx J_e^{DN} + J_e^{DC} \tag{15a}$$

$$= - \left(\frac{(\Gamma_\phi + \Gamma_t)}{\delta vol} \right)_e D_{1e}^1 (\phi_E - \phi_P) - \left(\frac{(\Gamma_\phi + \Gamma_t)}{\delta vol} \right)_e D_2^1 (\phi_{ne} - \phi_{se}) \tag{15b}$$

$$J^C = F_{ic} \phi_e J^C = F_{ic} \phi_e \tag{16}$$

$$F_{ic} = \rho_e (u_1 b_1^i + u_2 b_2^i)_e \tag{17}$$

and ϕ_e is obtained using the convection discretization Minmod scheme, see Kobayashi and Pereira [4]. The following compact equation results for the "e" face of a non-uniform mesh:

$$R(x,\phi) = \overline{\phi}_P + \frac{sp}{h} (x - x_P) \qquad x (x_w, x_e) \quad \text{and} \quad h = x_e - x_w \tag{18}$$

$$sp = \text{minmod} (s_e, s_w) \quad \text{with} \quad s_e = \left(\overline{\phi}_E - \overline{\phi}_P \right) \quad \text{and} \quad s_w = \left(\overline{\phi}_P - \overline{\phi}_w \right) \tag{19}$$

where

$$\phi_e = a^+ \phi_{he}^+ + a^- \phi_{he}^- \text{ with } \phi_{he}^+ = R(x_e^+, \phi) \quad \text{and} \quad \phi_{he}^- = R(x_e^-, \phi) \tag{20}$$

and

$$a^+ = \frac{u_{1e} + |u_{1e}|}{2u_{1e}} \quad \text{and} \quad a^- = \frac{u_{1e} - |u_{1e}|}{2u_{1e}}. \tag{21}$$

Writing (14) for every control volume in the grid, we obtain, after assemblage of a matrix and corresponding forcing vector, the system:

$$[A] \{\phi\} = \{b\}. \tag{17}$$

This system is solved with the implementation of the "Strongly Implicit Procedure", according to Stone [5] and Azevedo et al. [6].

After having solved the previous system, the pressure and velocity field are to be corrected by a SIMPLE like method. Altogether, we can summarize the following algorithm for the solution of viscous incompressible flows as follows:

1. Initialize all variables.
2. Calculate the coefficients of the momentum equation and solve it to obtain a new velocity field.
3. Calculate new values of mass fluxes trough PWIM interpolation.
4. Calculate the coefficients of the pressure-correction equation and solve it to obtain the pressure-correction field.
5. Use this pressure-correction to update the velocity, pressure and mass fluxes fields.
6. Solve for other scalar (turbulent energy, turbulent energy dissipation, etc.).
7. Return to step 2 and repeat until convergence.

2.4 *Boundary Conditions*

Besides the specific boundary conditions related to the turbulence models used and which were already presented, the following list presents the implemented boundary conditions for the rest of the variables:

1. Experimental values at inlet.
2. No slip condition for solid and north faces.
3. No gradient at outlet.

3. RESULTS

The results of the present study are summarized in the figures below.

As expected, and observing the combined C_f plot, if the HC model over-predicts the reattachment length, while the Standard k-ϵ model under-predicts it. This model also allows for a better prediction on C_f evolution over x, as can also be seen from the figure. Also in the velocity profiles, it can clearly be seen that better results are attained trough the use of the Standard k-ϵ model with wall functions. This is specially true for the third station. Due to the slower recovery in the reattachment process in the HC model (see figures), the turbulent and mean quantities are all badly off the experimental results. A possible reason to this behavior can be attributed to the values of the constants of the model (see table 1). In fact, if we recall that in the decay of homogeneous turbulence, the power law of the turbulence decay, directly correlates a decrease of C_{22} to a decrease in the value of μ_t, we see that the lower value of C_{22} in the HC model gives rise to a decrease in the turbulent energy and a corresponding decrease in the eddy viscosity. Also, from a local equilibrium shear-layer, it is possible to compute the "effective" C_μ values of the model in this condition: $C_{\mu(eff.)} = 0.0729$. Both factors can contribute to a lower value of the eddy viscosity and a slow down of the process of momentum transport from the outer (high velocity) region to the bubble. The prediction of $\overline{uv}$ for the HC model only reflects this fact.

4. CONCLUSIONS

In conclusion, and somewhat unexpectedly, the Standard k-ε model with wall functions, yielded the best results for the mean quantities as well as turbulent quantities in this particular incompressible, low Reynolds number, recirculating flow.

REFERENCES

1. Coakley, T.J. and Huang, P.G. , "Turbulence Modelling for High-Speed Flows", AIAA paper 92-0436, Reno NV, 1992;

2. M. Peric. "A Finite Volume Method for the Prediction of Three Dimensional Fluid Flows in Complex Ducts", PhD Thesis, University of London, 1985;

3. M.H. Kobayashi and J.C.F. Pereira, "Numerical Comparison of Momentum Interpolation Methods and Pressure-Velocity Algorithms Using Non-Staggered Grids", *Communications in Applied Numerical Methods*, vol. 7, pp. 173-186, 1991;

4. M.H. Kobayashi and J.C.F. Pereira, "Predictions of Compressible Viscous Flows at All Mach Number Using Pressure Correction, Collocated Primitive Variables and Non-Orthogonal Meshes", AIAA paper 92-0426, Reno, NV, 1992;

5. H.L. Stone, "Iterative Solution of Implicit Approximations of Multi-Dimensional Partial Differential Equations", *SIAM J. Num. Analysis*, vol. 5, No.3, pp. 530-573, 1968;

6. J.L.T. Azevedo, F. Durst and J.C.F. Pereira, "Comparison of Strongly Implicit Procedures for the Solution of the Fluid Flow Equations in Finite Difference Form", *App. Math. Modelling*, vol. 12, pp. 51-62, 1988.

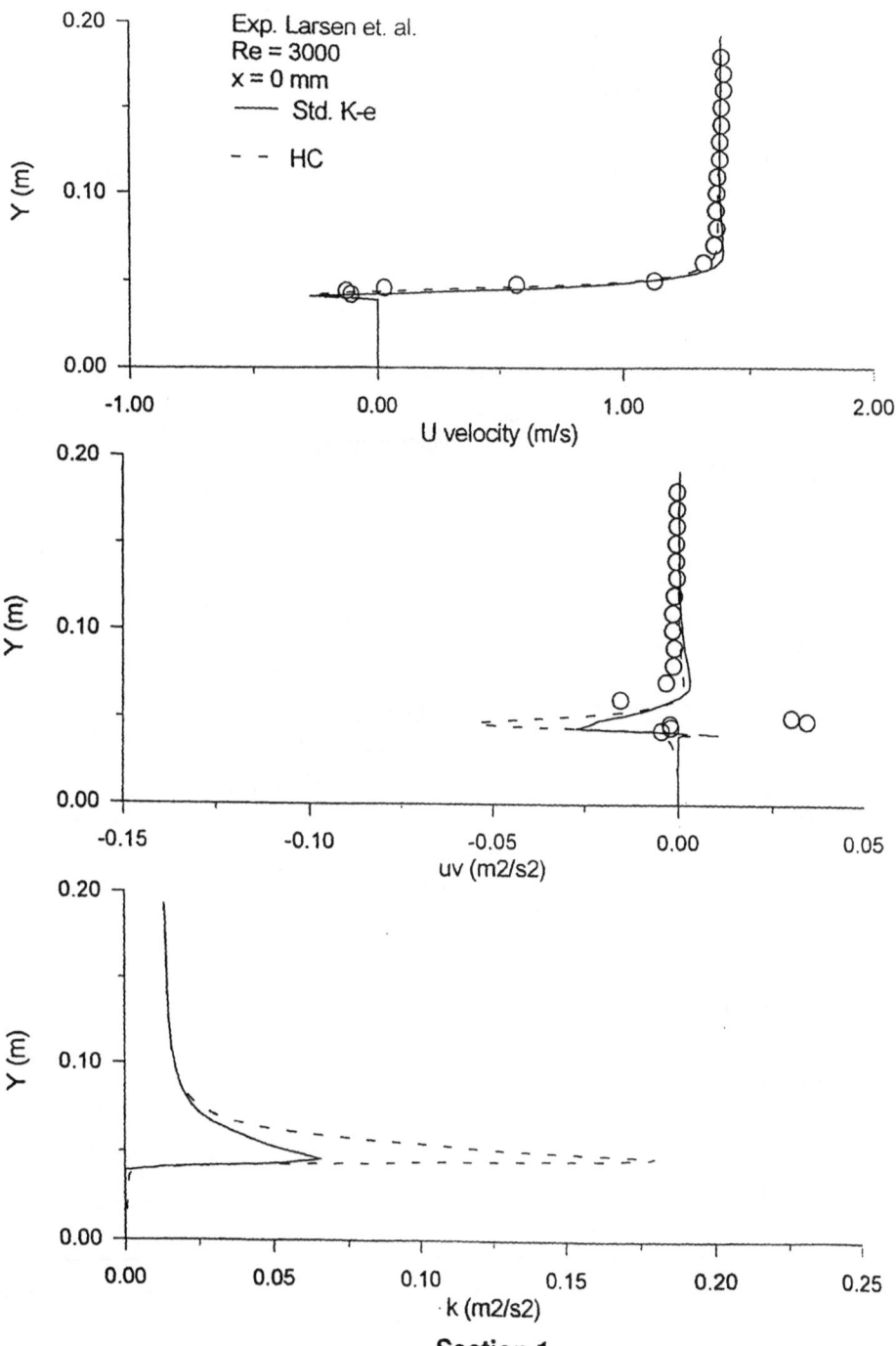

Section 1

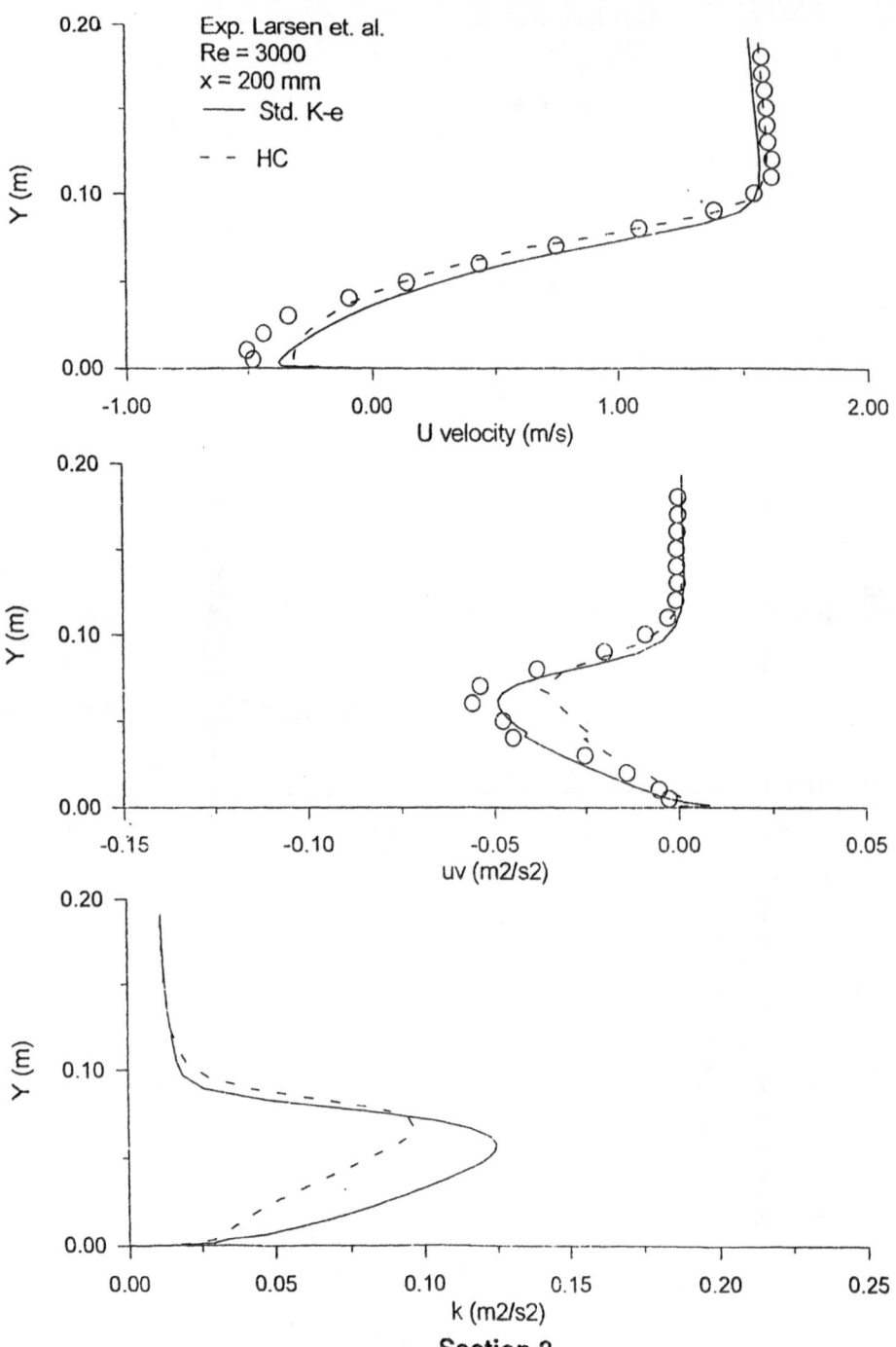

Section 2

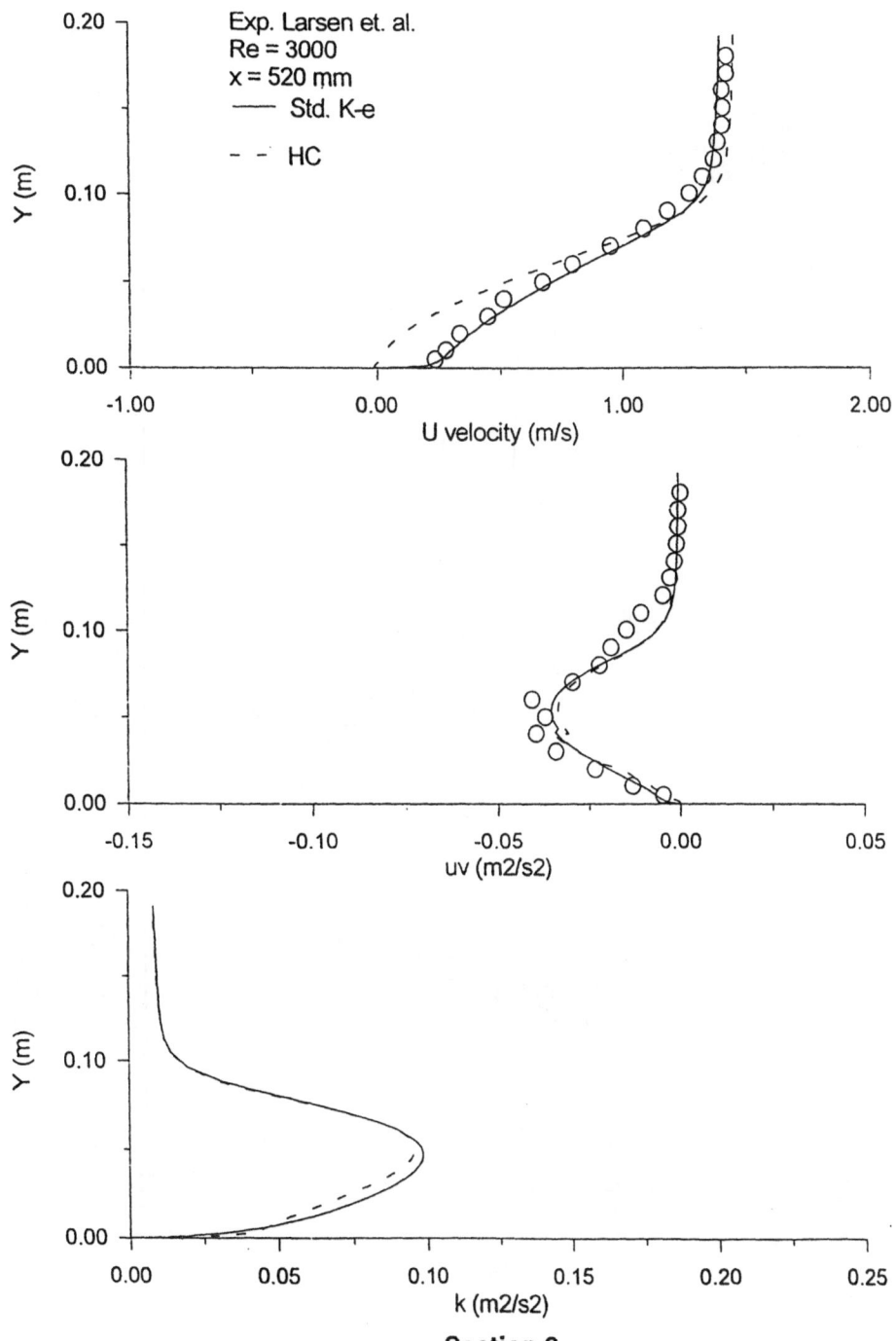

Section 3

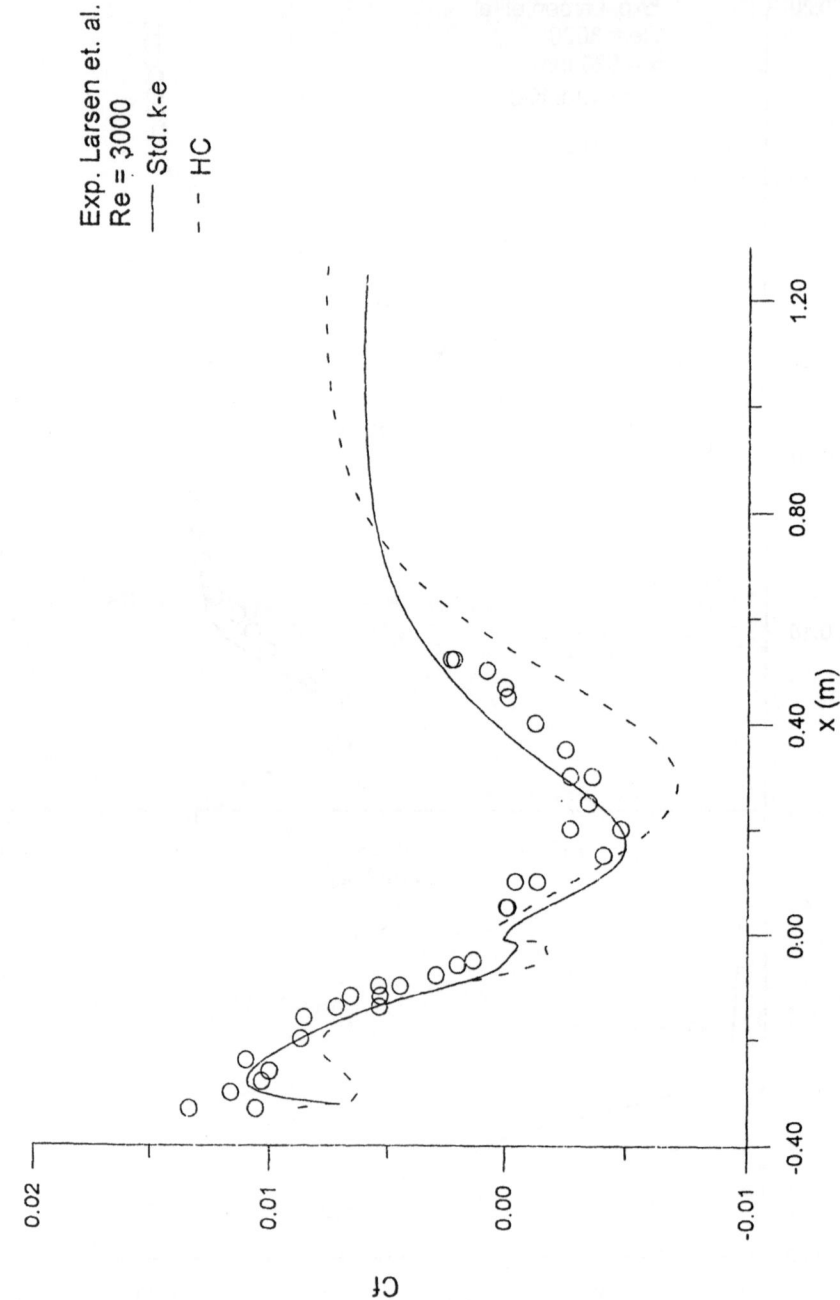

126

COMPUTATIONS OF SEPARATING AND REATTACHING FLOWS WITH HIGH- AND LOW-Re-NUMBER SECOND-MOMENT CLOSURES

S. Jakirlić[1], K. Hanjalić[2] and I. Hadžić[1]

[1]*LSTM, University of Erlangen, Cauerstr. 4, 91058 Erlangen, Germany*

[2] *Delft University of Technology, Lorentzweg 1, 2628 CJ Delft, The Netherlands*

SUMMARY

This work outlines the method and presents some results of the computation of four two-dimensional separating and reattaching flows, which served as test cases in the ETMA project[1]. The first two cases are the turbulent flow behind a backward-facing step, at low and high Re-numbers (test cases TC2A and TC2B, experiments of Jović & Driver (1993) and Driver & Seegmiller (1985), respectively). The third case is the turbulent flow over a fence mounted on the wall (the test case TC2C, experiments of Larsen et al., 1994) and the fourth is the flow over an obstacle in the channel (test TC2D, experiments of Dimaczek et al., 1989). The solutions were obtained with several turbulence models of varying complexity, with the emphasis on a second-moment (Reynolds stress) closure model with new modifications for viscosity and wall proximity effects, which allow the integration up to the wall. For the backward-facing step cases, the results with the latter model show substantial improvements in the reproduction of the experimental results in all important zones of the flow: shear layer, recirculating zone, in the corner, around the reattachment point and in the recovery region downstream from the reattachment. However, for the flows over the fence and obstacle, the results are only marginally better than those obtained by standard $k - \varepsilon$ and high-Re-number stress model.

INTRODUCTION

Flow separation and impingement to the solid surface are probably the two most challenging flow phenomena in aeronautics. Each of them is associated with abrupt changes of the flow properties, which impose strong effects on the bounding walls (steep variation of pressure, friction, heat and mass transport) which is difficult to predict by any existing analytical or computational method. Other specific flow features, inevitably associated with the former, are the flow recirculation, curved shear layer which separates the recirculating region from the external fluid, its bifurcation at the reattachment line or point, as well as downstream recovery of the boundary layer. Each of the locality mentioned is characterized by different turbulence dynamics, usually far from energy equilibrium pertaining to attached wall-parallel flows. Some interactions dominate in some flow regions, but play only marginal role in others (i.e. turbulence production in the shear layer, or turbulent transport and redistribution around

---

[1]Efficient Turbulence Models for Aeronautics, the EC research project within the BRITE-EURAM programme, 1992-95

the reattachment), though multiple overlapping of regions with different dynamics are also present. Hence, the prediction of a separating flow requires a comprehensive model of turbulence which is able to adequately account for all important processes in the evolution of turbulence within a relatively narrow flow regions.

Of all aspects, perhaps the most challenging is the prediction of location of the separation and reattachment points or lines in complex three-dimensional flows over curved surfaces. The reason for this is a high sensitivity of the location of the streamline singularities to the upstream and external flow conditions and the local orientation of the solid surface to the incoming fluid flow, which is often characterized by inherent unsteadiness and periodicity, producing in essence multiple solution to the problem. In order to study at least some aspects of separating flows, unburdened by uncertainty and ambiguity related to the separation location, most studies in the past have been confined to recirculating flows with a fixed separation point or line at a sharp trailing edge at the surface discontinuity of the bounding wall, such as a flow behind a backward facing step or a sudden duct or pipe expansion. Yet, even this relatively simple class of flows proved to be still a challenge to most existing turbulence models, particularly if a broader range of step heights, relative to flow depth or width, and of Reynolds numbers are considered. Accurate measurements, particularly of turbulence properties, in separated region are not easy to perform so that the validation of performances of turbulence models in separating flows, brings in additional degree of uncertainty [1].

The present paper discusses the performances of several existing and some new versions of turbulence models in three types of two-dimensional turbulent separating flows, which have been investigated experimentally within the framework of the ETMA project, aimed at providing further insight into the flow and turbulence structure, as well as supplying reliable data for validation of turbulence models for aeronautical flows. In addition to the classic backward facing step, which was considered at two very different Reynolds numbers, other two geometries include a flow over a fence and over a two-dimensional rectangular obstacle, both mounted on the wall of a channel. The considered three geometries have much in common: the fixed flow separation at the sharp trailing edge of the step, fence or obstacle, recirculating bubble, flow reattachment and boundary layer recovery downstream of it. Yet, the different upstream conditions, with different turbulence structure and the incoming streamline curvature, showed a substantial influence upon the flow downstream of the separation, which apparently none of the tested model was able to reproduce in full. Hence the paper reports only partial success, mainly in improving the predictions of the flow behind a step, whereas the disagreement of the predicted flow properties (mainly the turbulence energy and shear stress) with measurements in other two flows remains to be clarified.

TURBULENCE MODELS

The flow cases considered have been computed with four turbulent models:
- Standard high-Re-number k-ε model, (Std k-ε),
- Low-Re-number k-ε model of Jones, Launder and Sharma (1972/74),(LS low k-ε), [6],
- Basic Re-stress model (IP) of Launder, Reece, Rodi and Gibson (1975/78), (LRRG), [7,8],

- Hanjalić and Jakirlić low-Re-number full stress model, (HJ low RSM), [1,11].

The latter model consists of the following set of modelled equations:

$$\frac{D\overline{u_i u_j}}{Dt} = \frac{\partial}{\partial x_k}\left[\left(\nu + C_s \frac{k}{\varepsilon}\overline{u_k u_l}\right)\frac{\partial \overline{u_i u_j}}{\partial x_l}\right] - \left(\overline{u_i u_k}\frac{\partial U_j}{\partial x_k} + \overline{u_j u_k}\frac{\partial U_i}{\partial x_k}\right) + \Phi_{ij} - \varepsilon_{ij} \quad (1)$$

$$\Phi_{ij,1} = -C_1\varepsilon a_{ij} \qquad \Phi_{ij,2} = -C_2\left(P_{ij} - \frac{2}{3}P_k\delta_{ij}\right) \quad (2)$$

$$\Phi_{ij,1}^w = C_1^w f_w \frac{\varepsilon}{k}\left(\overline{u_k u_m}n_k n_m \delta_{ij} - \frac{3}{2}\overline{u_i u_k}n_k n_j - \frac{3}{2}\overline{u_k u_j}n_k n_i\right) \quad (3)$$

$$\Phi_{ij,2}^w = C_2^w f_w\left(\Phi_{km,2}n_k n_m \delta_{ij} - \frac{3}{2}\Phi_{ik,2}n_k n_j - \frac{3}{2}\Phi_{kj,2}n_k n_i\right) \quad (4)$$

$$C_1 = C + \sqrt{A}E^2 \quad C = 2.5AF^{1/4}f \quad F = \min\{0.6; A_2\} \quad (5)$$

$$f = \min\left\{\left(\frac{Re_t}{150}\right)^{3/2}; 1\right\} \qquad f_w = \min\left[\frac{k^{3/2}}{2.5\varepsilon x_n}; 1.4\right] \quad (6)$$

$$C_2 = 0.8A^{1/2} \quad C_1^w = \max(1 - 0.7C; 0.3) \quad C_2^w = \min(A; 0.3) \quad (7)$$

$$A = 1 - \frac{9}{8}(A_2 - A_3) \quad A_2 = a_{ij}a_{ji} \quad A_3 = a_{ij}a_{jk}a_{ki} \quad (8)$$

$$E = 1 - \frac{9}{8}(E_2 - E_3) \quad E_2 = e_{ij}e_{ji} \quad E_3 = e_{ij}e_{jk}e_{ki} \quad (9)$$

$$a_{ij} = \frac{\overline{u_i u_j}}{k} - \frac{2}{3}\delta_{ij} \quad e_{ij} = \frac{\varepsilon_{ij}}{\varepsilon} - \frac{2}{3}\delta_{ij} \quad Re_t = \frac{k^2}{\nu\varepsilon} \quad (10)$$

$$\varepsilon_{ij} = f_s\varepsilon_{ij}^* + (1 - f_s)\frac{2}{3}\delta_{ij}\varepsilon \quad f_s = 1 - \sqrt{A}E^2 \quad f_d = (1 + 0.1Re_t)^{-1} \quad (11)$$

$$\varepsilon_{ij}^* = \frac{\varepsilon\left[\overline{u_i u_j} + (\overline{u_i u_k}n_j n_k + \overline{u_j u_k}n_i n_k + \overline{u_k u_l}n_k n_l n_i n_j)f_d\right]}{1 + \frac{3}{2}\frac{\overline{u_p u_q}}{k}n_p n_q f_d} \quad (12)$$

$$\frac{D\varepsilon}{Dt} = \frac{\partial}{\partial x_k}\left[\left(\nu + C_\varepsilon\frac{k}{\varepsilon}\overline{u_k u_l}\right)\frac{\partial\varepsilon}{\partial x_l}\right] - C_{\varepsilon_1}\frac{\varepsilon}{k}\overline{u_i u_j}\frac{\partial U_i}{\partial x_j} - C_{\varepsilon_2}f_\varepsilon\frac{\varepsilon\tilde{\varepsilon}}{k}$$
$$+ C_{\varepsilon_3}\nu\frac{k}{\varepsilon}\overline{u_j u_k}\frac{\partial^2 U_i}{\partial x_j\partial x_l}\cdot\frac{\partial^2 U_i}{\partial x_k\partial x_l} + S_l \quad (13)$$

$$f_\varepsilon = 1 - \frac{C_{\varepsilon_2} - 1.4}{C_{\varepsilon_2}}\exp\left[-\left(\frac{Re_t}{6}\right)^2\right] \quad (14)$$

$$S_l = \max\left\{\left[\left(\frac{1}{C_l}\frac{\partial l}{\partial x_n}\right)^2 - 1\right]\left(\frac{1}{C_l}\frac{\partial l}{\partial x_n}\right)^2; 0\right\}\frac{\tilde{\varepsilon}\varepsilon}{k}A \quad l = \frac{k^{3/2}}{\varepsilon} \quad (15)$$

$$C_s = 0.22 \quad C_\varepsilon = 0.18 \quad C_{\varepsilon_1} = 1.44 \quad C_{\varepsilon_2} = 1.92 \quad C_{\varepsilon_3} = 0.25 \quad C_l = 2.5.$$

The new model version is based on the standard second-moment closure with linear, first order models of the pressure strain term (Rotta, Naot) for high Re-number flows, with Gibson and Launder (1978), [8], wall reflections, which serves as the high-Re-number asymptote. The major novelty is a new extension of the model to account for the viscosity and wall-proximity effects. These modifications involve several empirical functions, but all of them are defined in terms of scalar parameters, invariant to the wall topography: turbulence Reynolds number $Re_t = k^2/\varepsilon\nu$ and the invariants of the stress tensor anisotropy A_2, A_3, and of the dissipation-rate tensor anisotropy, E_2 and E_3. The other novelty, introduced specifically to deal with the separating and reattaching flows, is the additional source term in the dissipation equation - S_l based on the idea of Yap (1987), [9], which suppresses the excessive growth of the length scale in non-equilibrium flows with strong adverse pressure gradient and, particularly, in the reattachment region. While following the same arguments, we have replaced the ratio of the turbulence length scale to the local distance, $k^{3/2}/(C_l y \varepsilon)$, used by Yap, by the invariant derivative of the length scale, not related directly to the local wall distance. It should be noted that the application of the original Yap expression performed visibly inferior as compared with the new term.

THE NUMERICAL METHOD

The solution of the modelled set of transport equations was performed by a finite-volume Navier-Stokes numerical solver for two-dimensional flows in an orthogonal coordinate system with colocated variable arrangement, (Obi, Perić and Scheuerer, 1991, [10]). Care was taken to eliminate any numerical ambiguity so that each case was computed with at least two different grids. Also different differencing schemes for discretizing the convective terms have been tested. UDS, LUDS and CDS have been applied for the computations with high-Re-number turbulence models (wall functions approach) and for the computations with LS low $k - \varepsilon$ model and the blended scheme (UDS+0.5*CDS) for the computations with HJ low-Re-number Reynolds stress model. For the high-Re-number models with wall functions the final mesh had 90x46 control volumes (CV) for TC2A, 140x110 CV for TC2B. 200x60 CV for TC2C and 200x80 CV for TC2D, all clustered in the regions close to the walls and in the shear layer, as shown in Fig 8. In the case of the low-Re-number models, the final mesh was 180x100 CV for TC2A, 175x145 CV for TC2B and 210x110 CV for TC2C, with the first $y^+ \leq 0.5$ along the walls for all cases.

RESULTS AND DISCUSSION

We discuss first the flow behind a backward facing step. Two cases considered (test cases TC2A and TC2B) correspond to the relative step sizes ('expansion ratio') $ER = (H + h)/h = 1.2$ and 1.125 (where H is the step size and h is the flow depth prior to the expansion). The first case, investigated experimentally by Jović and Driver (1993), had a relatively low Reynolds number, $Re_H = 5000$, whereas in the second case, investigated by Driver and Seegmiller (1985), the Reynolds number was much higher, $Re_H = 37500$.

130

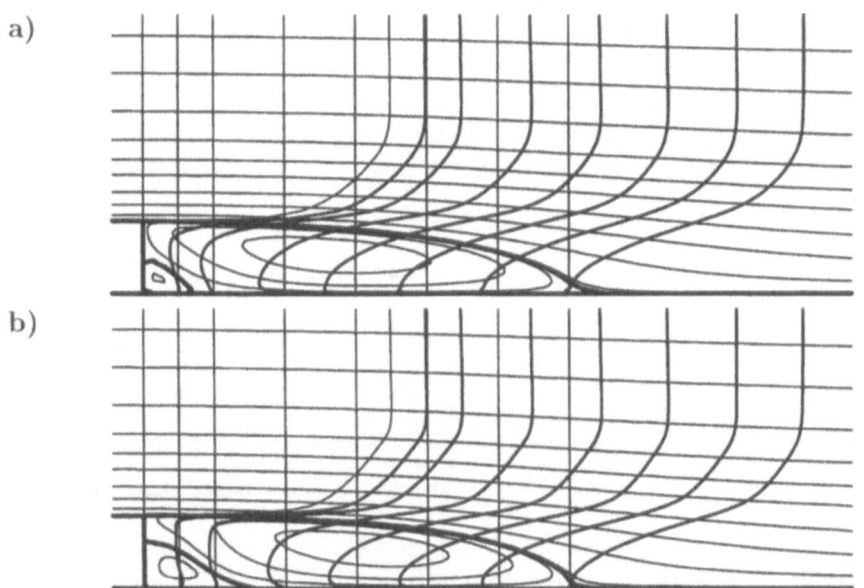

Fig 1 Computed streamlines and velocity profiles for the low-Re-number back step flow -TC2A
($Re_H = 5000, ER = 1.2$), a) with $LRRG + S_l + WF$, b) with $HJlowRSM$

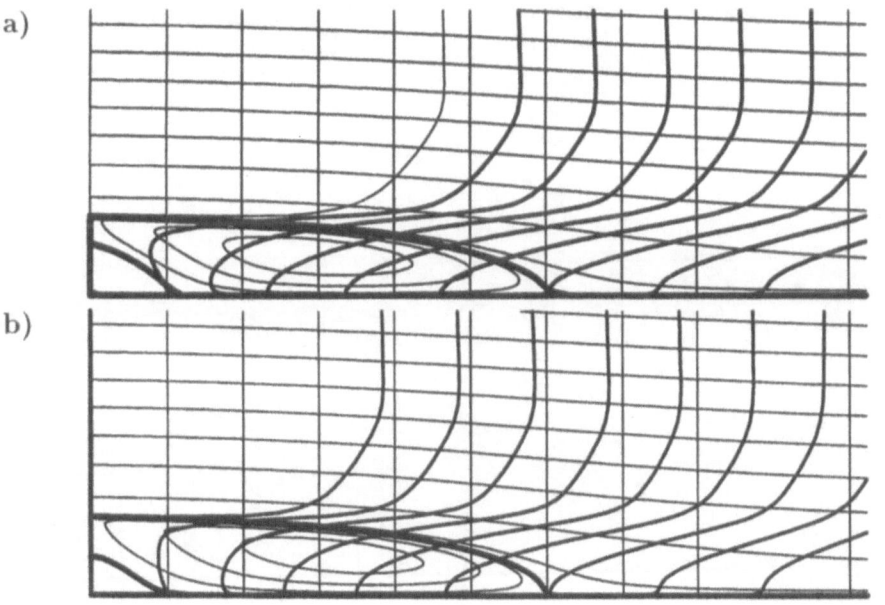

Fig 2 Computed streamlines and velocity profiles for the high-Re-number back step flow -TC2B
($Re_H = 37500, ER = 1.125$), a) with $LRRG + S_l + WF$, b) with $HJlowRSM$

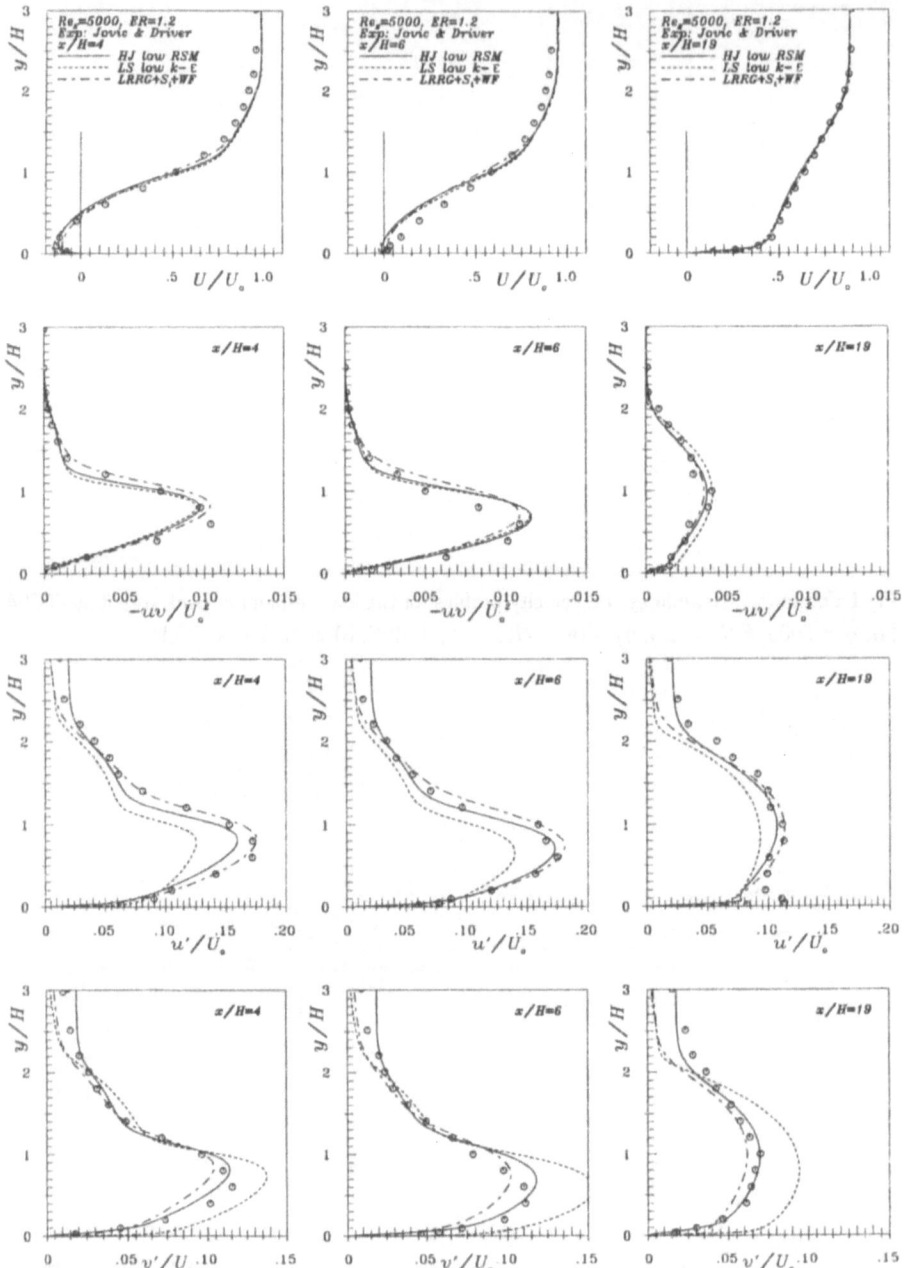

Fig 3 Mean velocity, turbulent shear and normal stress profiles at selected stations in the low-Re-number backward facing step flow - TC2A

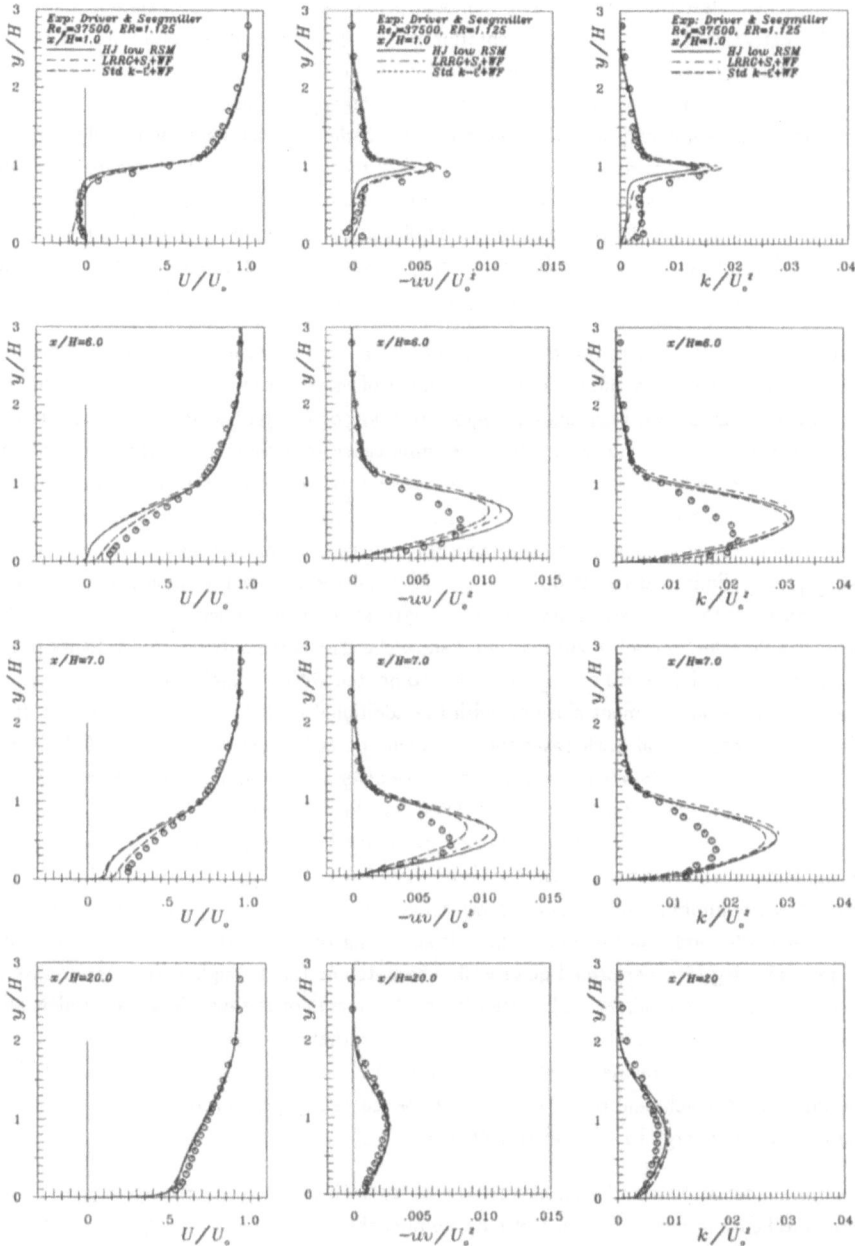

Fig 4 Mean velocity, turbulent shear stress and kinetic energy profiles at selected stations in the high-Re-number backward facing step flow - TC2B

133

It is well known that the standard $k - \varepsilon$ model produces too short recirculating zone and too early reattachment, but also poor reproduction of the flow in other zones (no secondary recirculation in the corner, too mild and gradual variation of the friction coefficient and a too slow recovery of the flow downstream of the reattachment). The use of the wall functions is highly questionable not only in the corner and around the reattachment, as usually implied, but also much further downstream, because of a slow recovery of the wall boundary layer and the equilibrium conditions pertinent to the logarithmic velocity law. The application of low-Re-number $k - \varepsilon$ models (in the present case we employed the Jones and Launder model (1972), [6], with modifications of Launder Sharma, 1974) obviates the wall functions and associated deficiency, but did not show much of an improvement outside the near wall region.

The second moment closures, both with wall functions and with the low-Re-number modifications, produced generally the desired improvement of most of the flow features, such as the shape and the size of the recirculation region, position of the reattachment, as well as the recovery process downstream, illustrating the importance of a more exact treatment of the dynamics of turbulent stress tensor and its anisotropy. However, the second-moment model produced a curious anomaly close to the reattachment, where the dividing streamline bends backwards and shows a kink in the velocity profile. This anomaly, discussed in more details in ref. [1], was eliminated by introducing an additional term S_l in the dissipation equation, which suppresses the excessive growth of the length scale, which is usually produced by the uncorrected Re-stress model. Figures 1 and 2 show the streamline patterns and velocity profiles for two cases of backward facing step flow, computed with the high- and low-Re-number versions of the second moment closures, with the additional term S_l in the $\varepsilon-$ equation. It is interesting to note that in both cases the streamline patterns look very similar. Slight differences are noted in the detailed plot of the mean velocity and components of turbulent stress tensor at several stations, Fig. 3 for $Re = 5000$ and Fig. 4 for $Re = 37500$. For the lower Re number both second-moment closures perform much better than the $k - \varepsilon$ model and in close agreement with the experiments of Jović and Driver (1993), [2]. The complete model with the low-Re-number modifications shows some improvements as compared with the high-Re-number model and wall functions, but it is surprising that even the latter model, in spite of non-existent log-law, performed quite well. The outcome for the higher Re-number case is quite different: here the difference between the performances of the two Re-stress models and of the standard $k - \varepsilon$ model with wall functions are rather small and all models considered show some serious disagreements with the experimental data of Driver and Seegmiller (1985), [3], around the reattachment point. Particularly noticeable are the overpredictions of the shear stress and kinetic energy by all models considered.

The variation of the friction factor along the wall for both cases considered is compared with experimental data in Fig. 5. For the lower Re number the new low-Re-number second-moment closure (HJ) shows the best agreement, though the extreme peak prior to the reattachment, has not been reproduced. The low-Re-number $k - \varepsilon$ model was the only among the models considered, which yielded the extreme C_f in the recirculating region, close in the value to the measured one, but at a wrong position. The overall bahaviour of the C_f obtained by the latter model is very unrealistic, and in view of the experimental uncertainties associated with the two points of extreme wall shear, which were not reproduced in recent DNS by Le, Moin and Kim (1993), the performance of the low-Re-number $k - \varepsilon$ model can be viewed as very unsatisfactory. Other three models reproduced the behaviour of C_f better, with a slight superiority of the new general second-moment model. For the higher Re-number flow, three models considered show similar performances, all reasonably close to the experimental results.

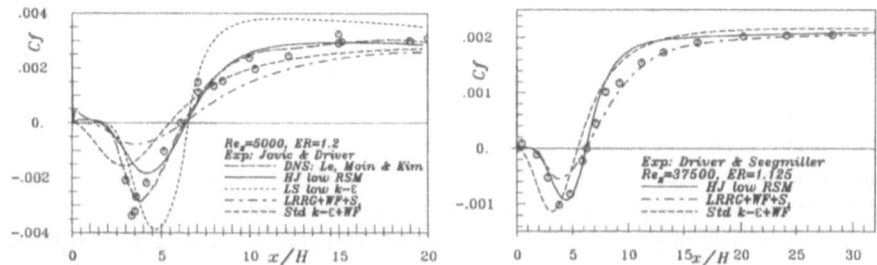

Fig 5 Friction coefficients in the low- (TC2A) and high-Re-number backward facing step flow (TC2B), respectively

Next two figures illustrate further the performances of the new model. Fig. 6 shows first the budget of the kinetic energy very close to the reattachment point, in good agreement with the DNS results of Le, Moin and Kim, except very close to the wall where the model failed to reproduce sharp peaks of the dissipation rate and balancing turbulent transport. Fig. 7 compares the semi-log plot of the mean velocity profiles at three positions within the recovery zone downstream of the reattachment. Excellent agreement is achieved for all three cases, each showing a substantial departure from the logarithmic law.

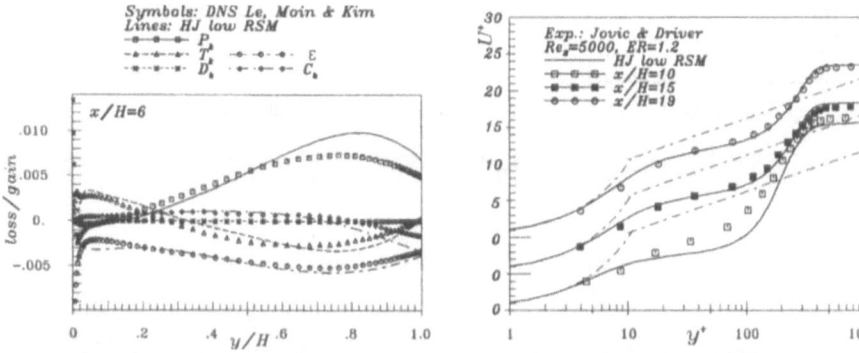

Fig 6 Budget of kinetic energy in the low-Re-number backward facing step flow - TC2A

Fig 7 Mean velocity profiles at selected locations in recovery region of the low-Re-number backward facing step flow - TC2A

The next two cases considered are the flows over a fence and over an obstacle both mounted on the wall of a plane channel. In the fence flow (test case TC2C), investigated experimentally by Larsen et al. (1994), [4], the flow blockage was mild with the fence relative height $H/h = 1/7.5$ (ER=1.154) and the flow Reynolds number was relatively small, $Re_H = 3000$. In the other case (Dimaczek, 1989), [5], the two-dimensional square-sectioned obstacle extending half width of the channel (ER=2), caused a substantial flow blockage. This case also had a much higher Reynolds number, $Re_H = 42000$. Finite length of the obstacle as compared with the thin fence is the major difference in the flow geometry, though, probably, more influential is the difference in the flow blockage. In the first case the small blockage diminishes the effect of pressure and emphasizes the role of turbulent stresses on the mean velocity field and importance of their accurate modelling, as compared with the second flow dominated by pressure field. There are also difference in the inflow conditions, but these may not play particularly important role, considering a strong flow skewing when encountering the obstacle. Figures 9 and 12 show the computed streamline patterns and the mean velocity profiles for the two cases, each computed by two models. The first case was computed by the high- and low-Re-number second moment closures, each containing the additional S_l term in the ε equation. Both patterns look very similar, characterized by a sharp bending of streamlines in front of the fence, with a small embedded recirculation zone, and highly curved streamlines separated from the fence edge. A close look at the friction coefficient, Fig. 11, shows some differences, with the low-Re-number model reproducing somewhat better the overall variation along the flow, particularly before the fence, though both models produced too long recirculating bubble. The standard second moment closure with wall functions (LRRG+WF) reproduced best the friction coefficient over the whole domain, except at the flow inlet, though the scatter of the experimental points in the near-zone behind the fence makes it difficult to reach unambiguous judgment on the models performances. It is interesting to note that the standard $k - \varepsilon$ model performed also rather well. However, unlike in the backstep flow, the addition of the term S_l overextended the length of the recirculation zone. Fig. 10 shows the mean velocity, shear stress and kinetic energy profiles at selected position downstream. The agreement with experimental data is considerably less satisfactory, than in the case of the backward facing step flow, with standard $k - \varepsilon$ model producing results closer to the experiments than any of the tested second-moment closures. This unexpected outcome remains still to be clarified since the new second-moment model gave substantial improvements in reproducing the backward facing step flow, as well as several classes of non-equilibrium wall flows [11].

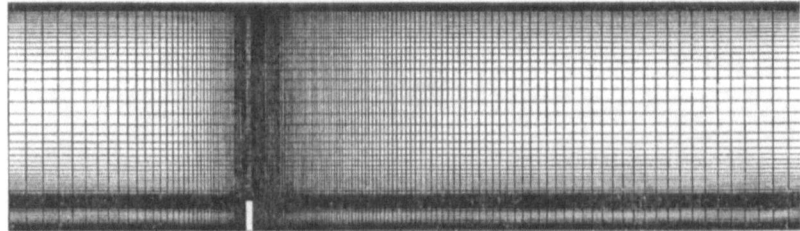

Fig 8 Numerical grid with 210x110 CV for the computations of the flow over a fence mounted on the wall with low-Re-number turbulence models

We return now to the flow over the obstacle. Because of a high Re number, the solution with the low-Re-number second-moment model is very tedious and requires a very fine grid clustered in the wall region and a large number of iterations. We expect, however, that the low-Re-number modification to the second-moment closure will have similar effects as in the case of the backward facing step flow. Fig. 12 compares the two solutions obtained by the same high-Re-number second-moment models, first without, and in the second case with the additional term S_l. Although the streamline patterns are very similar, the recirculating region in the second figure is longer, than found by experiment, whereas the standard model without S_l produces fairly accurate reattachment position. However, the standard model exhibits again the anomaly in the streamline behaviour around the reattachment, leading to the conclusion that the S_l terms cures the anomaly and extends the recirculation, which in the present and previous cases becomes too long. This is also visible in Fig. 11b where the standard LRRG model reproduces best the friction factor (at least in the region downstream of the reattachment, where the experimental data exist). Like in the fence flow, the shape of the friction factor curve obtained by the modified LRRG model with S_l term is similar as in the experiment, except for a delay due to elongation of the recirculation bubble. Finally, Fig. 13 compares the velocity, shear stress and kinetic energy profiles at several stations. To our surprise, here the basic $k-\varepsilon$ model with wall functions shows best agreement with experiments at some stations, though around the reattachment all models produce too low turbulence levels as compared with experiments. In light of numerous examples of separating flows where the second moment closures yielded superior predictions, this result is unexpected and can not be explained at present. It is symptomatic, however, that a major disagreement between predictions and experiments appears in regions of very high peaks of turbulence energy and shear stress at high Re number flows, where the measurements, particularly with the hot wire anemometers, tend to be often in error due to high relative turbulence intensity. To this conclusion leads also a relative closeness of the predicted turbulence properties by all models considered, in spite of large difference in their formulations.

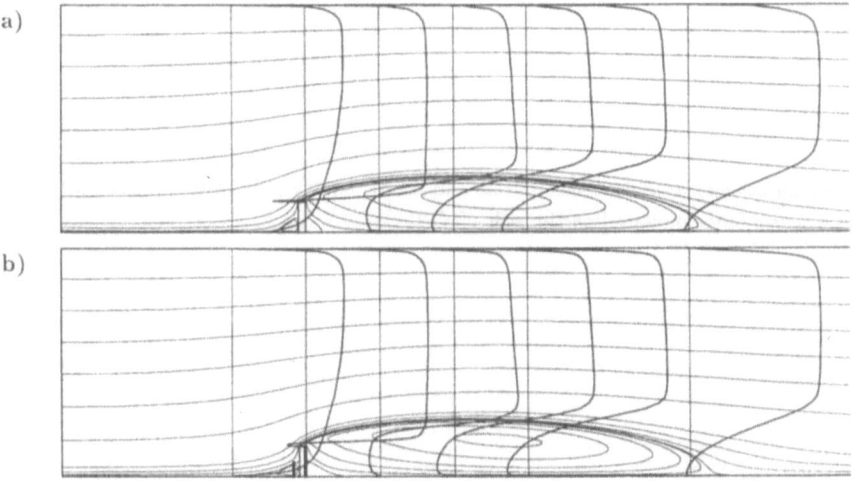

Fig 9 Computed streamlines and velocity profiles for the flow over a fence mounted on the wall - TC2C ($Re_H = 3000, ER = 1.154$), a) with $LRRG + S_l + WF$, b) with $HJlowRSM$

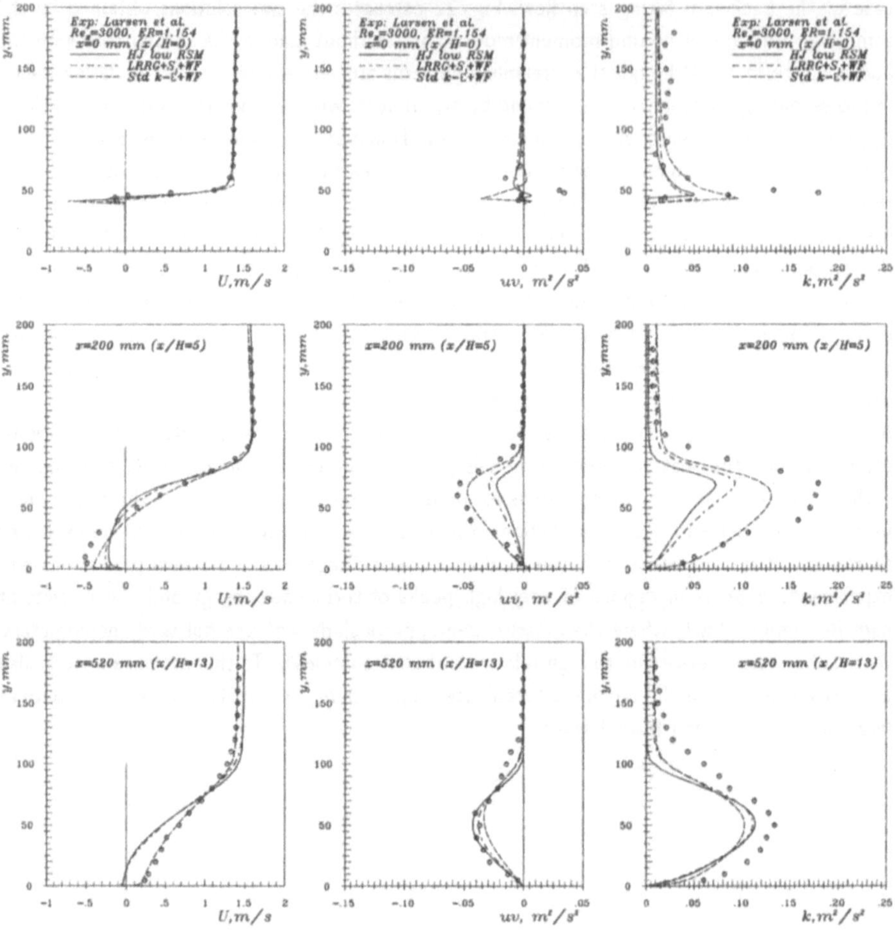

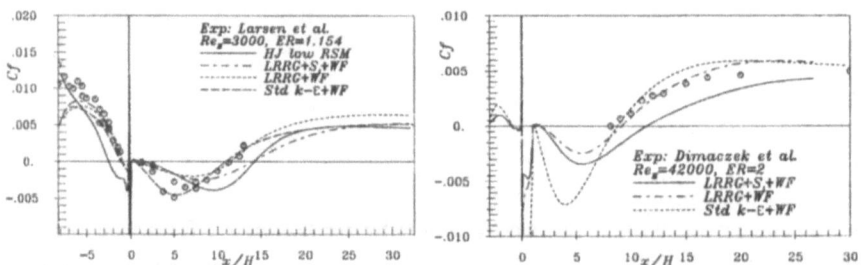

Fig 10 Mean velocity, turbulent shear stress and kinetic energy profiles at selected stations in the flow over a fence mounted on the wall - TC2C

Fig 11 Friction coefficients in the flow over a fence mounted on the wall (TC2C) and over an obstacle in the channel (TC2D), respectively

138

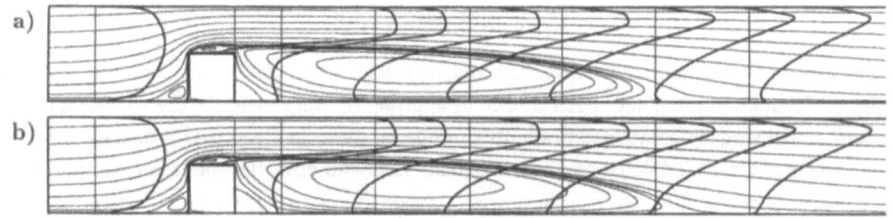

Fig 12 Computed streamlines and velocity profiles for the flow over an obstacle in the channel - TC2D ($Re_H = 42000, ER = 2$), a) with $LRRG + WF$, b) with $LRRG + S_l + WF$

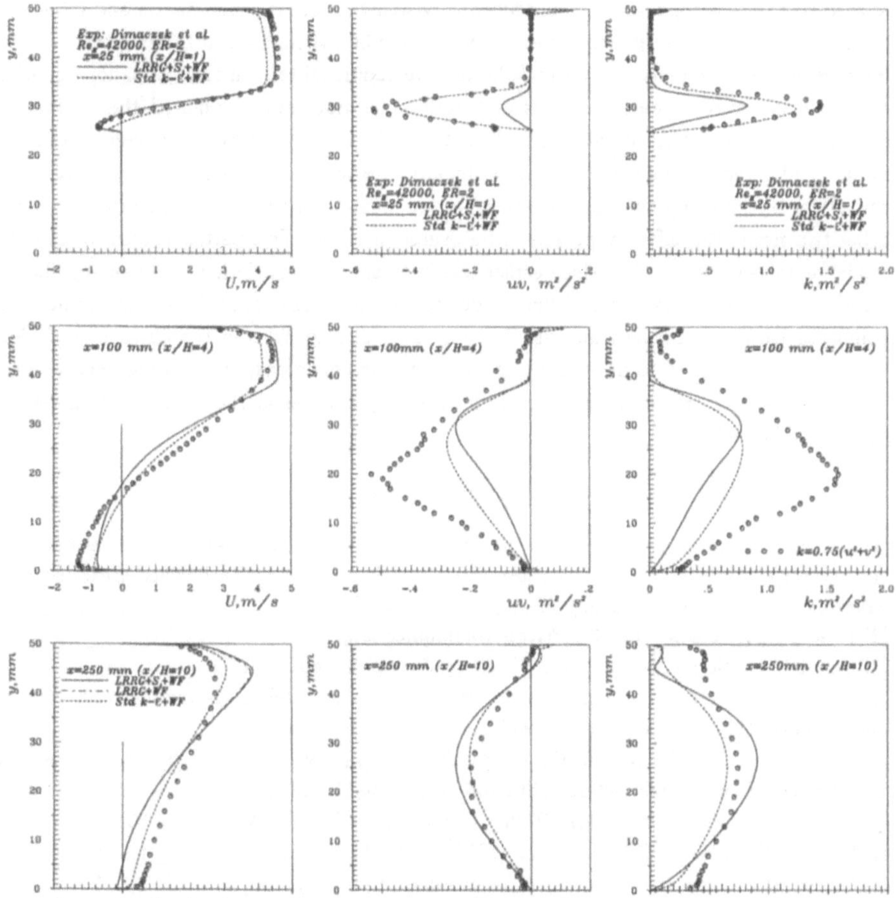

Fig 13 Mean velocity, turbulent shear stress and kinetic energy profiles at selected stations in the flow over an obstacle in the channel- TC2D

CONCLUSIONS

Four cases of two-dimensional separating and reattaching turbulent flows, covering a broad range of Reynolds number have been studied by applying several variants of the second-moment closure model with different modifications. Considered were the flows over a backward facing step, over a sharp-edged fence and over a square-sectioned obstacle, the latter two mounted on a wall of a plane channel. The flows were solved with the standard high-Re-number- and with a new low-Re-number versions of the second-moment closure, both with additional term in the dissipation equation. For comparison, all flows have been solved also with the standard high- and low-Re-number $k - \varepsilon$ models. The new low-Re-number second moment closure, with modifications for viscous and wall proximity effects formulated in invariant form not directly related to the wall topography, reproduced very well the recent experimental and DNS data for a flow over a backward facing step at low Re number, though failed to bring the expected improvements in the fence and obstacle flows. The additional term in the dissipation equation, expressed in invariant form in terms of length scale gradients, eliminated the anomaly in streamline pattern close to the reattachment, which is produced by the standard Re-stress model, and extended the length of the recirculation region. In the backward facing step flow, this effect brought improvement in predicting the location of the reattachment point, but in other two flows, the effect appeared to be overemphasized. This partial success of the new model remains to be clarified, since earlier test in a variety of equilibrium and non-equilibrium, steady and unsteady wall parallel flow, produced in most cases substantial improvements as compared with the standard models. Perhaps some discrepancies in the obstacle and fence flows can be attributed to uncertainty of the upstream conditions, and to experimental error, particularly in the measurements of the turbulence properties in the region of high relative turbulence intensity.

REFERENCES

[1] Jakirlić S., Hanjalić K., (1994.): *On the Performance of the Second-Moment High- and Low-Re-Number Closures in Reattaching Flows*, Proc. Int. Symposium on Turbulence, Heat and Mass Transfer, Paper No. 3.5, Lisbon, 9-12 August,

[2] Jović S., Driver D.M., (1993): *Private communication*,

[3] Driver D.M., Seegmiller H.L., (1985), AIAA J., Vol. 23, pp 163-171,

[4] Larsen P.S., Westergaard C.H., Koch C.W.,Schmidt J.J., (1993): *12-month intermediate report*, Report ETMA/AFM/93-02, Technical University of Denmark,

[5] Dimaczek G., Kessler R., Martinuzzi R., Tropea C., (1989), Proc. 7th Symposium on Turbulent Shear Flows, Paper No. 10.1, Stanford University, 21-23 August,

[6] Jones W.P., Launder B.E., (1972), Int. J. Heat Mass Transfer, Vol. 15, pp 301-314,

[7] Launder B.E., Reece G.J., Rodi W., (1975), J. of Fluid Mech., Vol. 68, pp 537-566,

[8] Gibson M.M., Launder B.E., (1978), J. of Fluid Mech., Vol. 86, pp 491-511,

[9] Yap C., (1987): *Turbulent heat and momentum transfer in recirculating and impinging flows*, Ph.D. thesis, Faculty of Technology, University of Manchester,

[10] Obi S., Perić M., Scheuerer G., (1991): *Second-moment calculation procedure for turbulent flows with colocated variable arrangement*, AIAA J. Vol. 29, pp 585-590.

[11] Jakirlić S., Hanjalić K., Hadžić I., (1995.): *Computation of Non-Equilibrium Wall Boundary Layers with a Low-Re-Number Second Moment Closure*, ETMA Workshop.

INCOMPRESSIBLE RECIRCULATING FLOWS

A Critical Comparison of Computations for Low- and High-Reynolds Number Flow over a Backward-Facing Step

Robert D. Harper* & Suad Jakirlić[†]

*Mechanical Engineering Department, Imperial College, London SW7 2BX, UK
[†]LSTM, University of Erlangen, Cauerstr. 4, 91058 Erlangen, Germany

SUMMARY

The computations from two contributors for high- and low-Reynolds number flow over a backward-facing step are compared and discussed. A total of six turbulence models have been used, including wall-function and low-Reynolds-number formulations of two-equation and Reynolds stress models. Baseline computations using the same models, but from different contributors, have shown excellent agreement allowing valuable conclusions to be drawn from the remaining calculations. The reattachment length can only be predicted well by some models, although all models show poor predictions of the redeveloping boundary layer. Overall comparison shows Reynolds stress models provide superior results to two-equation models with low-Reynolds number formulations performing better than the wall function based models.

INTRODUCTION

The aim of this report is to compare and comment upon the results for the contributions to TC2A and TC2B which are concerned with the flow over a backward facing step at both low and high step-height based Reynolds numbers. The low Re case of Jovic and Driver, TC2A, is described experimentally in [1] and has also been directly simulated in [2], the high Re flow of Driver and Seegmiller, TC2B, is described in [3]. The geometrical details and inlet conditions provided to each participant, together with the motivation behind these test cases is provided in the test case specification, [4].

The computational results were compared with each other and with all available experimental data. For both cases comparisons were made with mean velocity, skin friction co-efficient, reattachment length, turbulence kinetic energy and Reynolds shear stress, for TC2A comparisons were also made with the normal stresses, $\overline{u^2}$ and $\overline{v^2}$. Because of the similarities between TC2A and TC2B the results will usually be considered together.

Participants in the test cases were Imperial College (IC), producing computations based on low-Re two-equation models, [5], while the second contribution, from LSTM, concentrated on Reynolds stress modelling (RSM) with one computation for a low-Re k-ε model, [6], both contributors produced calculations using the standard k-ε model with wall functions (WF) for both cases.

A total of 6 different turbulence models were used, the standard k-ε model with wall functions, the low-Re k-ε model of Launder and Sharma, [7], the low-Re q-ζ model of Gibson & Dafa'Alla [8], the basic Reynolds stress model of Launder, Reece & Rodi, [9] with wall functions employing the pressure-strain correlation of Gibson and Launder [10] (LRRG + WF), the full low-Re RSM of Hanjalic and Jakirlic, [11] (HJ low RSM) and the LRRG + WF model with the additional source term S_1 from [11]. A summary of which models were used for each test case is given in table 1.

Table 1: Turbulence models, mesh size and differencing scheme used by each contributor.

	IC		LSTM	
	TC2A	TC2B	TC2A	TC2B
'Standard' k-ε model + WF	90×48 Hybrid	180×93 Hybrid	90×46 Blended	140×110 Blended
Low Re k-ε model, [7]	280×180 Hybrid	320×150 Hybrid	180×100 Blended	-
Low-Re q-ζ model [8]	280×180 Hybrid	320×150 Hybrid	-	-
LRRG + WF, [9, 10]	-	-	90×46 LUDS	100×70 UDS
LRRG + WF + S_1	-	-	90×46 LUDS	140×110 LUDS
HJ low RSM, [11]	-	-	180×100 Blended	175×145 Blended

RESULTS

Before any useful conclusions can be drawn from the results presented here it is necessary to deduce whether the results from the different contributors are consistent with one another. The codes used by IC and LSTM were based on slightly different computational methodologies and also utilised different methods for the implementation of the wall functions, with the results being achieved on differently sized grids. Both methods rely upon co-ordinate based meshes, although the IC calculations were produced using hybrid differencing on a staggered mesh with the SIMPLE pressure correction algorithm while the LSTM calculations were performed using either upwind (UDS), linear upwind (LUDS) or blended differencing on a co-located mesh, again details are given in table 1. Consistency between the codes was assessed by computing both TC2A and TC2B with the 'standard' k-ε model with wall functions and also by calculating TC2A using the low-Re k-ε model.

Figures 1, 2 and 5 show the comparative results for TC2A from both contributors, figure 1 shows the distribution of skin friction coefficient for the 'standard' and low-Re variations of the k-ε model, figure 2 shows the distribution of pressure coefficient and figure 5 shows profiles of U, k and $-\overline{uv}$ at various downstream locations for the low-Re k-ε model only. The agreement achieved between the same models in the different codes is excellent despite differences in the differencing schemes and meshes.

The comparison between the computed skin friction distributions for TC2B is shown in figure 8. Agreement between these results is not as good as for TC2A, this is because a coarser mesh has been used in the IC calculation which was necessary in order to ensure that the first grid point from the wall was at $y^+ > 15$ for the whole domain. The LSTM calculation of TC2B and both comparison calculations of TC2A only applied the restriction on y^+ outside the recirculation zone only thus allowing a finer computational mesh to be used. The consistency in the results for TC2A and explanations for the differences in the wall function calculations of TC2B are such that valid comparisons between the results from the remaining models can be made irrespective of the code used to produce them.

Table 2: Calculated and experimental values of mean reattachment length, X_r/h.

	IC		LSTM	
	TC2A	TC2B	TC2A	TC2B
'Standard' k-ε + WF	5.48	4.9	5.47	5.4
Low Re k-ε, [7]	6.53	5.89	6.52	-
q-ζ model [8]	6.49	5.92	-	-
LRRG + WF, [9, 10]	-	-	5.71	5.34
LRRG + WF + S_1	-	-	6.26	6.21
HJ low RSM, [11]	-	-	6.38	6.20
Experiment: TC2A[1], TC2B[3]	6.1	6.26	6.1	6.26
Direct Simulation: TC2A[2]	6.28	-	6.28	-

The distributions of skin friction coefficient, C_f vs. x/h are shown in figures 1 and 3 for TC2A and in figures 8 and 9 for TC2B, table 2 lists the length of the recirculation zones, X_r for both cases. From these it has been observed that the 'standard' k-ε model with wall functions has failed to predict any secondary recirculation zones and has also under predicted the main reattachment length, X_r, inspite of this, the distributions of C_f exhibit the trends and gradients of the data reasonably well. The low-Re two-equation model calculations show improvements over the wall function based k-ε model, a secondary recirculation zone is present, although its length is too small, and the main reattachment lengths are closer to the data, for TC2A however, X_r suffers a slight overprediction whereas in TC2B, a slight underprediction in X_r is evident.

The C_f distributions for these models are unfortunately a little less encouraging. For both test cases the low-Re k-ε model shows an overshoot in both the negative and positive peaks of C_f around reattachment, this is most evident in TC2B, figure 9. The q-ζ model has removed these errors to some degree but the gradient of C_f with x around reattachment is still too steep. The RSM computations show differing degrees of under- and overprediction of X_r. The LRRG + WF computations consistently under estimate X_r, and also exhibit an underprediction in the negative peak of C_f upstream of reattachment. The low-Re full stress model, HJ low RSM, shows a very good distribution of C_f with x, although the negative peak in C_f upstream of reattachment in TC2A is slightly too small. The success of the S_l term in this model, as detailed in [11], is evident from figure 4, many researches have noticed that RSM calculations of backward-facing steps exhibit a so called 'back bending' of the dividing stream line at reattachment, the S_l term removes this anomaly by suppressing the growth of the turbulence length scale within the reattachment region. The addition of this term to the LRRG model led to substantial improvement of the reattachment length predictions for both cases (table 2).

The mean velocity profiles can be seen in figures 5, 6 and 7 for TC2A and figure 10 for TC2B. At all locations for TC2A and for all models the results are remarkably similar. At x/h =10   all models exhibit slow recovery especially in the region of   y/h ≤ 0.5 . Further deviations from the data and hence modelling insight could possibly be gained by making further comparisons closer to the step in the recirculation region, but unfortunately these have not been made due to a lack of experimental data and unavailability of the DNS data. The TC2B results show a little more scatter than the TC2A computations although the conclusions are broadly similar. At   x/h =1   , the location of reattachment for the secondary recirculation region, the low-Re   k-ε   and   q-ζ models and the RSM predictions are superior to the   k-ε model with wall functions. At   x/h =6   , close to the main reattachment point, the   k-ε model with wall functions appears to predict the data well, this can however be explained, to some degree, by its early reattachment. The redeveloping boundary layer is again poorly predicted, the RSM predictions show a higher degree of recovery but this is too slow when compared to the data. Further explanation for these disagreements could be attributed to the fact that the experimental results are not absolutely reliable: the reattachment length deduced from the velocity profiles at   x/h =5.5   (not shown here) and   x/h = 6   (x_r/h < 5.5) is not consistent with the value obtained from the friction coefficient redistribution (x_r /h = 6.26).

The remaining parameters to be compared are the turbulence quantities. Profiles of k and $-\overline{uv}$ are again shown in figures 5, 6 and 7 for TC2A and figures 11 and 12 respectively for TC2B. The conclusions that can be drawn from these parameters are not consistent between test cases. The results from TC2A imply that all models perform reasonably well at all locations with accurate predictions of the magnitude and trends of both k and $-\overline{uv}$ across the whole of the shear layer, although the predictions of k at x/h = 4 do show some discrepancies against the data. The results for TC2B, shown in figures 11 and 12, highlight that all computations show similar trends but for this case the magnitudes of k and $-\overline{uv}$ are constantly overpredicted and only downstream of reattachment is the correct position of their maximum captured.

Comparisons for the normal stresses were only made for TC2A and are shown in figures 6 and 7 in the form y/h vs. u' or v', the importance of these stresses in the overall solution is far less than for the shear stress, their effects are moderated as only their gradients appear in the equations of motion. The figures show that the full RSM predictions are superior to those of the q-ζ model which consistently underpredicts u' and overpredicts v', this however is hardly surprising as the Bousinesq approximation is used to extract these data. The predictions of the HJ low-Re RSM appear better than those of the LRRG + WF model for these quantities.

144

CONCLUSIONS

A declaration of the 'best' model for these test cases is very difficult because the performance of each model can be seen to differ between the parameters compared, the locations considered and the test case chosen. In general, particularly when friction coefficients and reattachment lengths are compared, superior results have been achieved by employing Reynolds stress models rather than two-equation models with the low-Re formulations of both model types achieving better results than their wall function based counterparts. This however leaves somewhat of a 'grey area' when choosing between low-Re two-equation models and Reynolds stress models employing wall functions which appear to produce results of comparable accuracy.

It cannot be denied that the k-ε model with wall functions can give a rough idea of the flow field within reasonable computational time and provides what could be described as an 'engineering approximation', however, the theory on which wall functions are based is simply not valid for flows of this type. Integration of the governing equations to the wall is therefore required in order to allow the application of exact wall boundary conditions and thus give greater accuracy when more complex flow configurations or scalar transport problems are considered.

One important point that can be identified from this exercise is that by using the low-Re two-equation models presented here it is possible to achieve a reasonably accurate value for the reattachment length although the overprediction of the length scale in the reattachment needs addressing, a possible solution to this is proposed with the S_1 function applied to the low-Re RSM model. The performance of all models in the recovery region downstream of reattachment also needs to be addressed. The re-developing boundary layer recovers too slowly, especially in the region immediately after reattachment and particularly for the low-Re two-equation models in TC2B. This indicates that the modelling of the physics within the reattachment region and the redeveloping boundary layer should be an area of possible future investigation.

REFERENCES

1. Jovic, S. & Driver, D.M. 1993 Unpublished data cited in [2].
2. Le, H., Moin, P. & Kim, J. 1993 Direct numerical simulation of turbulent flow over a backward-facing step. 9th Symposium on Turbulent Shear Flows, Kyoto, Japan, 13-2.
3. Driver, D.M. & Seegmiller, H.L. 1985 Features of a reattaching shear layer in divergent channel flow, AIAA J. 23, 163-171.
4. Larsen, P.S. 1995 Presentation of Test cases TC-2A, TC-2B, TC-2C, TC-2D; Two-Dimensional, incompressible wall flows with separation, ETMA Workshop.
5. Harper, R.D. & Gibson, M.M., 1995 ETMA Workshop TC2: Incompressible Recirculating Flows; TC2-A Low-Re Backward-Facing Step, TC2-B High-Re Backward-Facing Step.
6. Jakirlic, S., Hadzic, I. & Hanjalic, F., 1995 TC2-A, TC2-B, ETMA Workshop.
7. Launder, B.E. & Sharma, B.I. 1974 Application of the energy-dissipation model of turbulence to the calculation of the flow near a spinning disk, Letters in Heat and Mass Transfer 1, 131-138.
8. Gibson, M.M. & Dafa'Alla, A.A. A two-equation model for turbulent wall flow, AIAA J (to appear).
9. Launder, B.E., Reece G.J. & Rodi, W. 1975 Progress in the development of a Reynolds-stress turbulence closure, J. Fluid Mech. 68, 537.
10. Gibson M.M. & Launder, B.E. 1978 Ground effects on pressure fluctuations in the atmospheric boundary layer. J. Fluid Mech. 86, 491.

11. Jakirlic S., Hanjalic K., 1994 On the performance of the second-moment high- and low-Re-number closures in reattaching flows, Proc. Int. Symposium on Turbulence, Heat and Mass Transfer, Lisbon, Portugal, August 9-12, Paper No 3.5 .

* * *

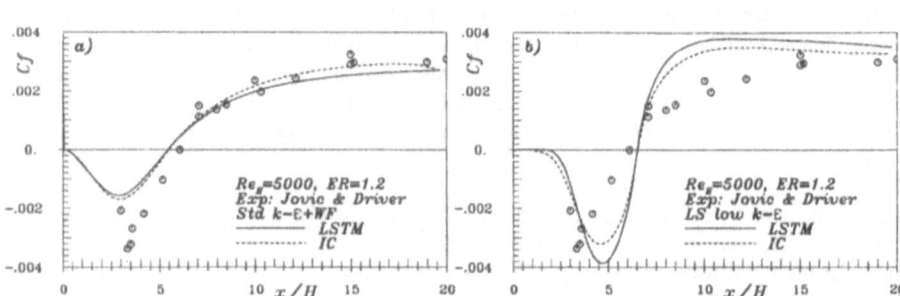

Fig 1 Friction coefficients in the low-Re-number backward facing step flow (TC2A) obtained by standard $k - \varepsilon$ models with wall functions approach (a) and integration up to the wall (b)

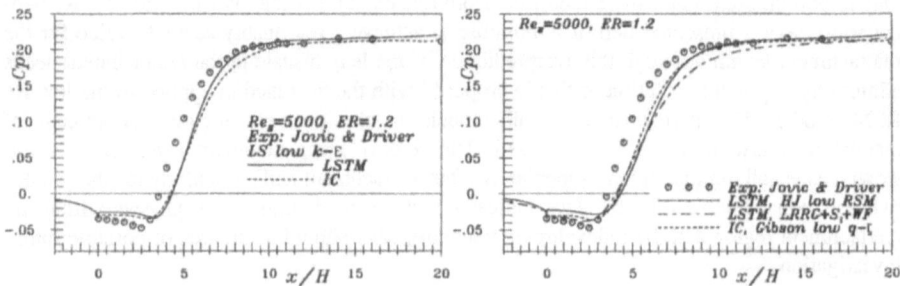

Fig 2 Pressure coefficients in the low-Re-number backward facing step flow (TC2A) obtained by different low-Re-number models, LS-Launder, Sharma and HJ- Hanjalic, Jakirlic

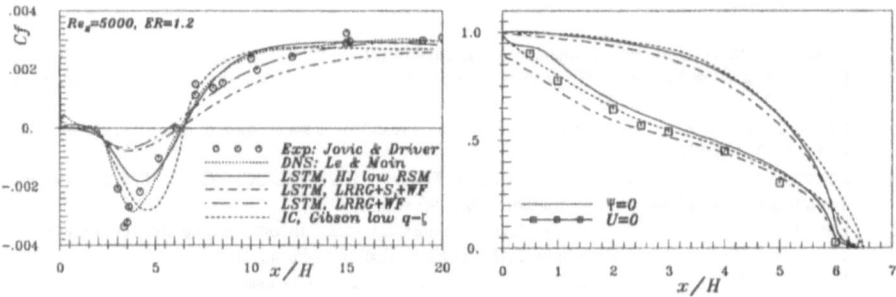

Fig 3 Friction coefficients in the low-Re-number backward facing step flow (TC2A) obtained by different models, key as Fig 2

Fig 4 Dividing streamlines and zero velocity lines, key as Fig 2

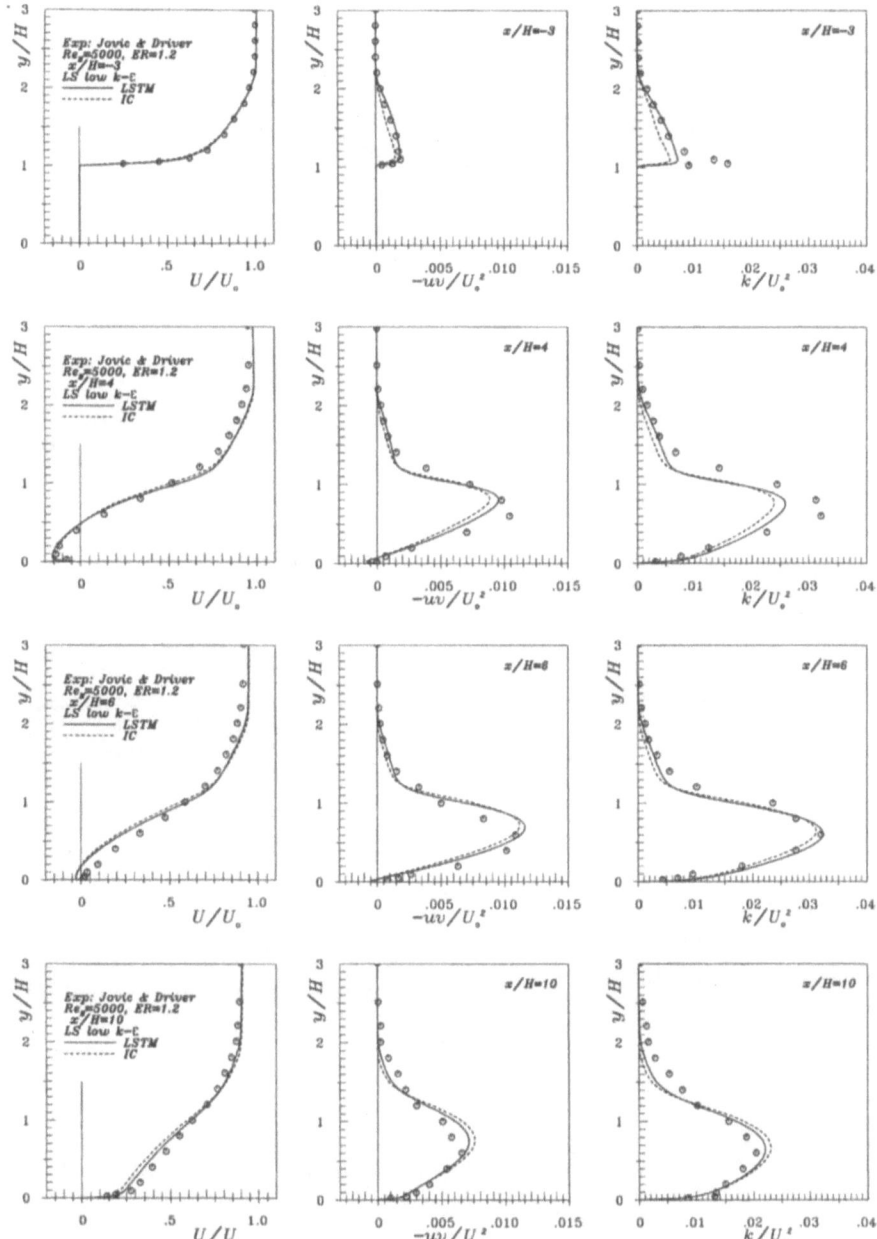

Fig 5 Mean velocity, turbulent shear stress and kinetic energy profiles at selected stations in the low-Re-number backward facing step flow (TC2A) obtained by Launder, Sharma low-Re-number $k - \varepsilon$ model

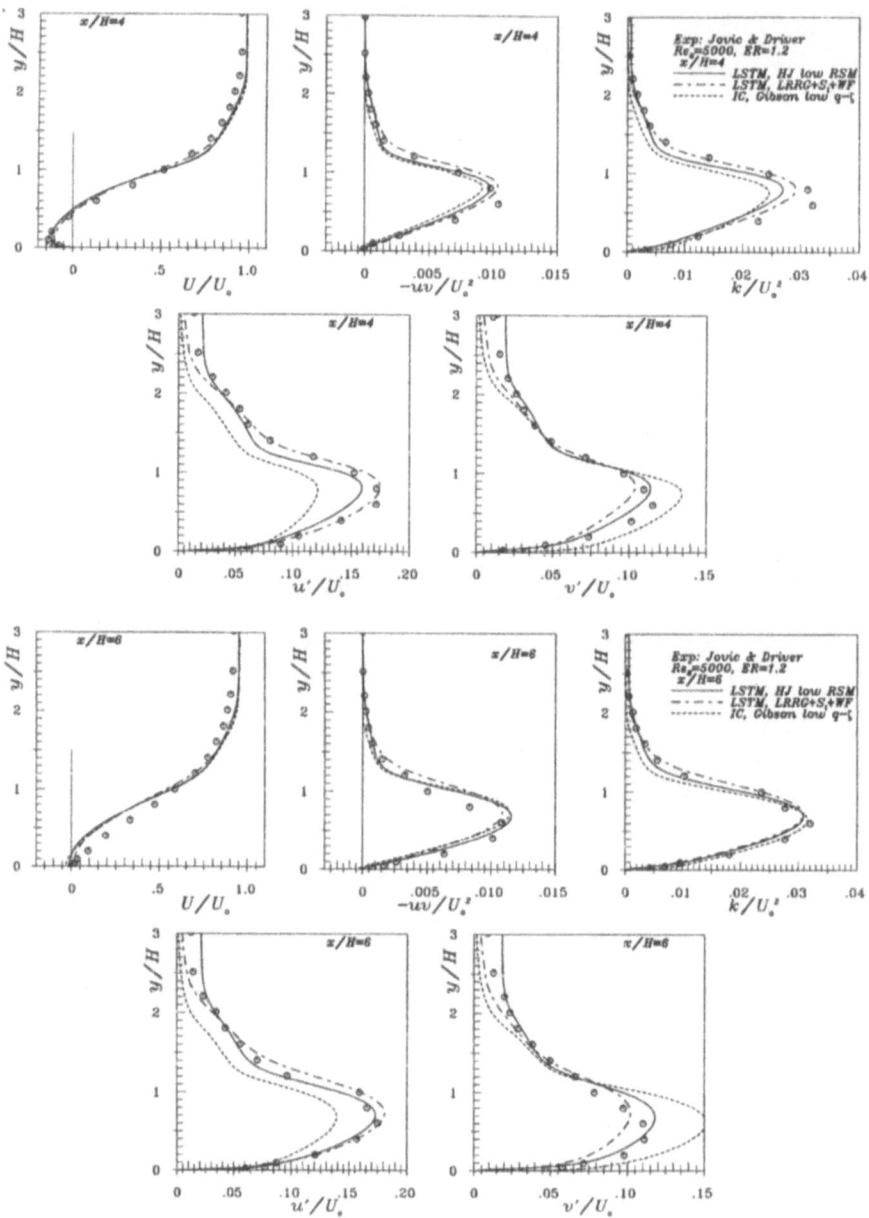

Fig 6 Mean velocity, turbulent shear stress, kinetic energy and turbulent normal stress profiles at the stations x/H=4.0 and 6.0 in the low-Re-number backward facing step flow (TC2A) obtained by different models of turbulence, LRRG-Launder, Reece, Rodi, Gibson model, WF-wall functions

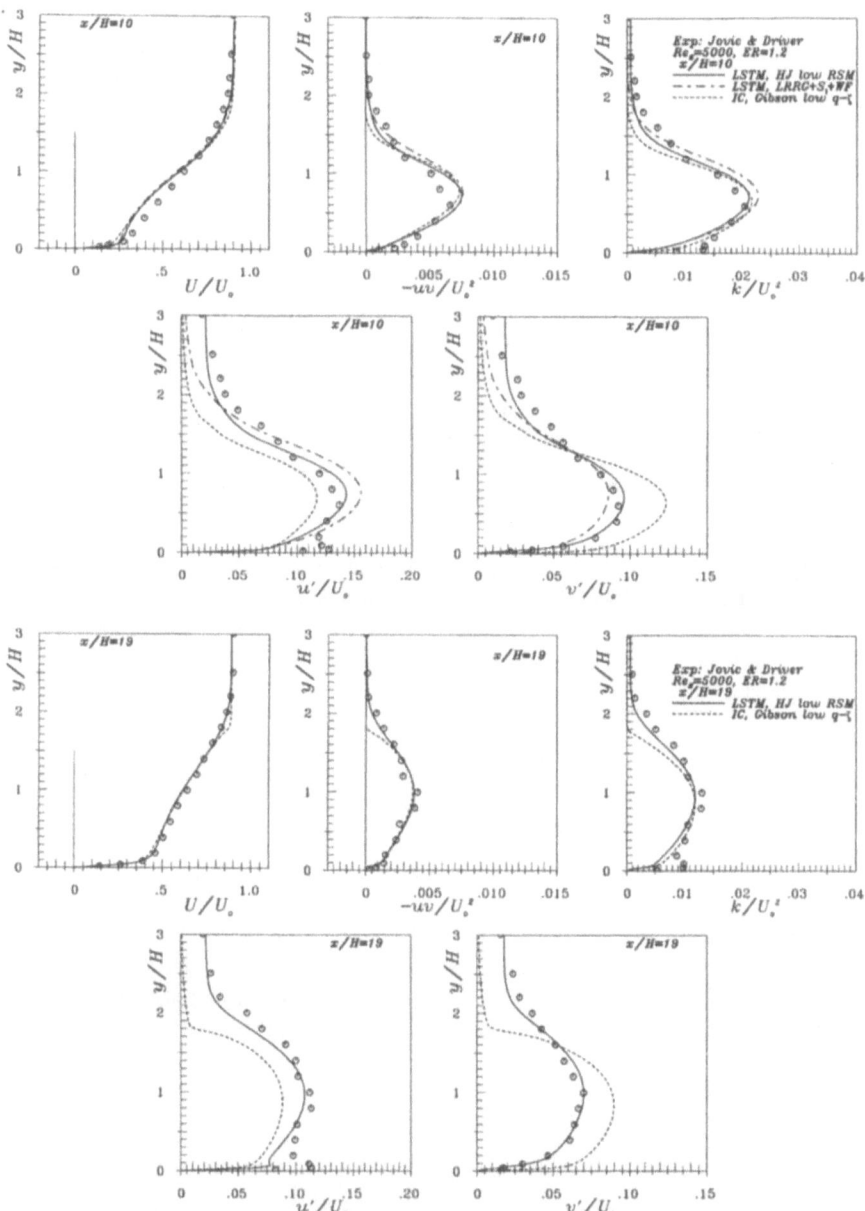

Fig 7 Mean velocity, turbulent shear stress, kinetic energy and turbulent normal stress profiles at the stations x/H=10.0 and 19.0 in the low-Re-number backward facing step flow (TC2A) obtained by different models of turbulence, key as Fig 6

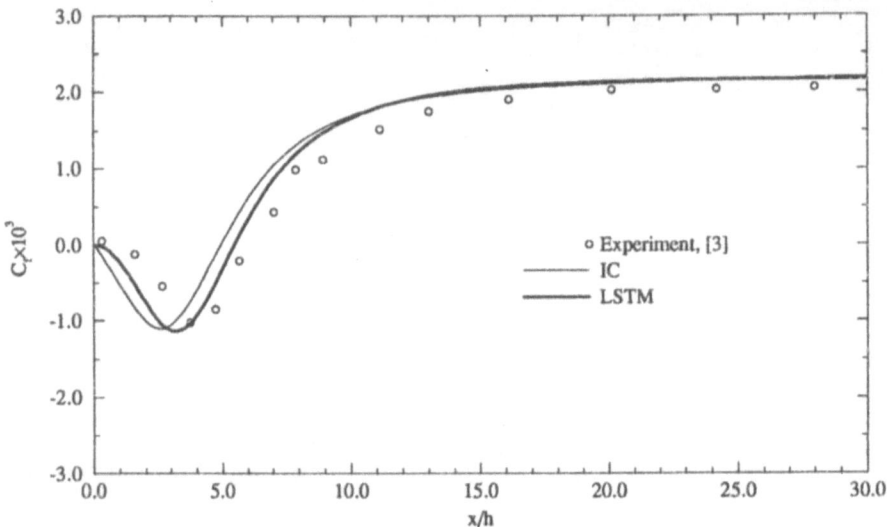

Figure 8: Skin friction coefficient for high-Re flow over a backward-facing step. Comparison of k-ε model with wall function calculations from IC and LSTM.

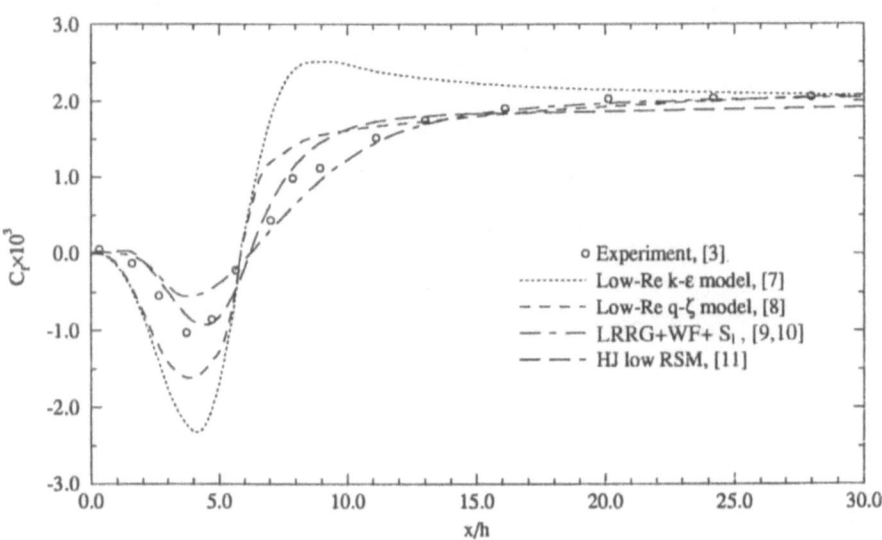

Figure 9: Skin friction coefficient for flow high-Re flow over a backward-facing step. Comparison of results from various models of turbulence.

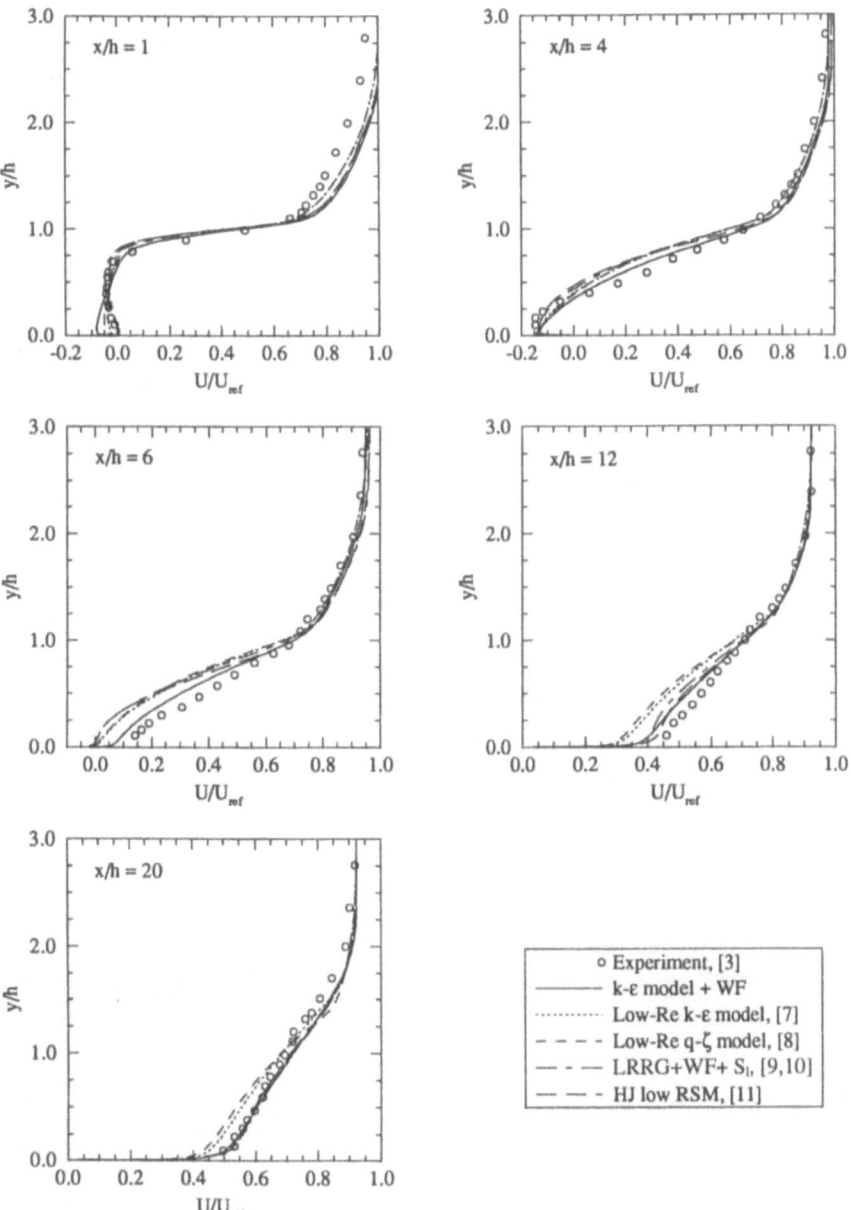

Figure 10: Mean streamwise velocity, U, for high-Re flow over a backward-facing step.

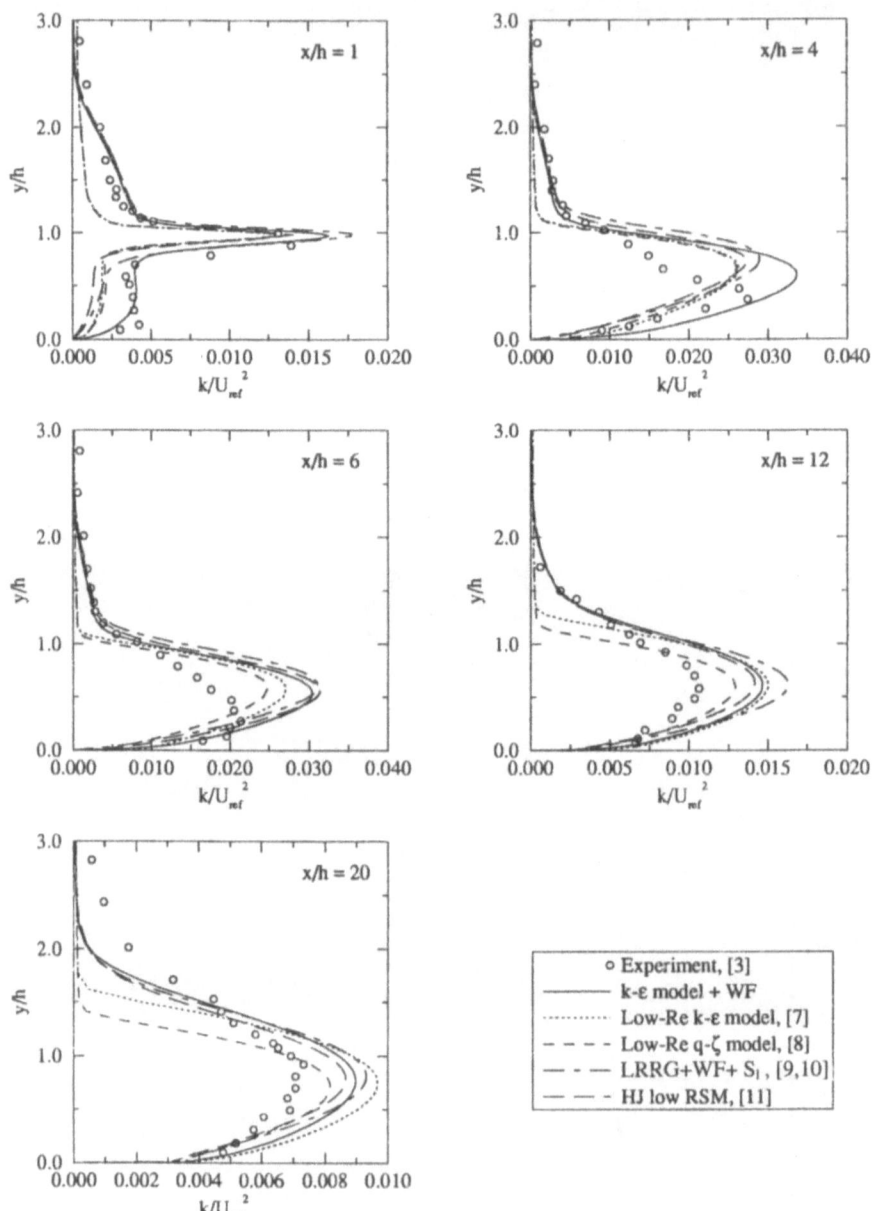

Figure 11: Turbulence kinetic energy, k, for high-Re flow over a backward-facing step.

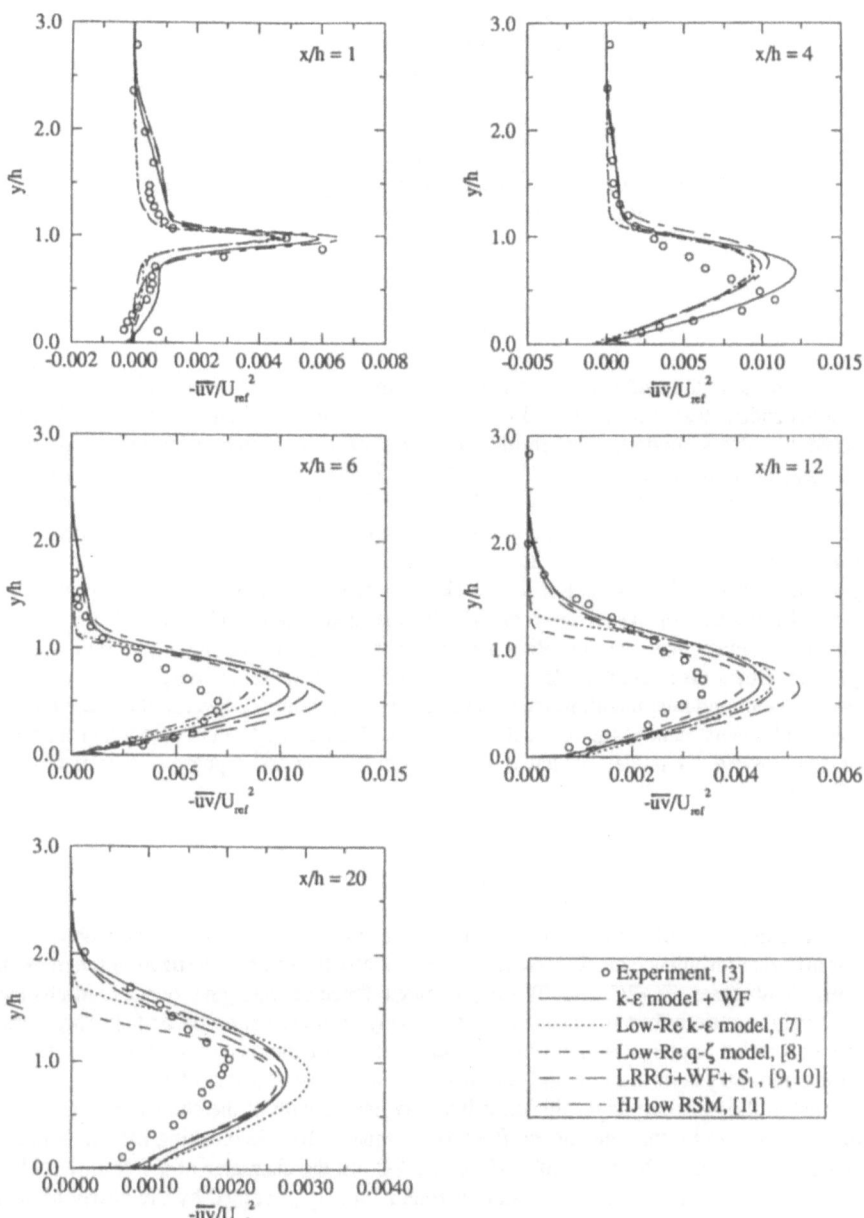

Figure 12: Turbulence shear stress, $-\overline{uv}$, for high-Re flow over a backward-facing step.

SYNTHESIS OF TEST CASES TC-2C AND TC-2D TWODIMENSIONAL, INCOMPRESSIBLE FLOW PAST WALL-MOUNTED OBSTACLES

P. S. Larsen

Department of Fluid Mechanics, Technical University of Denmark
Building 404, DK-2800 Lyngby, Denmark

SUMMARY

A summary and evaluation is given of the computational predictions made by 3 independent teams using a total of 4 different two-equation models and 3 different Reynolds stress models. The results, in terms of mean velocity, turbulent kinetic energy, Reynolds stress and coefficient of skin friction, are compared mutually and to available experimental results obtained by laser Doppler anemometry. Differences in results obtained with the same turbulence model are ascribed to differences in implementation and meshes, while differences between models clearly reveal strengths and deficiencies of models. Results on skin friction, in particular, show surprisingly large differences for the different models. The standard k,ϵ-model with wall functions performs better than the low-Re k,ϵ-model throughout the recirculation region. In general, Reynolds stress models, particularly those for low-Re number, are superior to two-equation models, but all models underpredict the turbulence level in the shear layer over and downstream of the recirculation region. The new q,ζ-model performs roughly as the low-Re k,ϵ-model, which it replaces, but it has an improved computational efficiency.

INTRODUCTION

The objective of the present computational exercise is to compare results obtained by use of different turbulence models, employing essentially the same discretization schemes and meshes. Test cases TC-2C and TC-2D, denoted fence-on-wall and obstacle-in-channel, respectively, are described in detail in an accompanying presentation paper [1]. These cases involve obstacles in plane channel flows at two values of Reynolds number Re, based on step height: a boundary layer flow, with free-stream turbulence, past a small fence of height about twice that of the boundary layer thickness but only one-seventh of the channel width (TC-2C at Re = 3000), and a fully developed flow past a square obstacle blocking half the width of the channel (TC-2D at Re = 42,000). Here, TC-2C has the character of a free flow, where the pressure distribution over the separation zone is largely governed by the nearly uniform freestream, while TC-2D has the character of a confined flow, where the proximity of the opposing wall affects the pressure distribution over the separation zone.

The flow predictions are focussed on several features: the length of the separation zone, X_r/h, profiles of mean velocity, the turbulence level, k, and shear stress, $-uv$, (particularly in the shear layer bounding the recirculation region), the recovery of the wall layer downstream, and the coefficient of skin friction along the wall with obstacle.

NUMERICAL PROCEDURE AND TURBULENCE MODELS

Three teams have contributed with computations, referred to as IMP [2], ERL [3], and AFM [4]. Common to all contributions is the use of finite volume discretization of the full elliptical Navier-Stokes equations on a structured Cartesian mesh. Contributions furthermore use the same computational domains with same specified upstream inflow conditions [1]. In addition, a number of standard meshes supplied by ERL, and now available in the ETMA database [5], were recommended, see Table 1. Here, MESH-1 ensures $y_P^+ > 11.63$ for both test cases, and is the recommended mesh for reference calculations using the standard high-Re k,ϵ-model with wall functions. MESH-2 has twice the mesh points while MESH-3 is adapted to have more mesh points in the wall vicinity and in the shear layer. MESH-4 for TC-2C is the recommended mesh for the low-Re k,ϵ-model. The 7 turbulence models employed by the three teams, as well as final meshes and upwinding schemes, are shown in Table 2. Here, 'blended' signifies an upwind difference scheme (uds) plus 50% central difference scheme (cds). The turbulence models, according to numbers, are:
1. Standard high-Re k,ϵ-model with wall functions; 2. Jones-Launder-Sharma low-Re k,ϵ-model [6],[7]; 3. Low-Re k,τ-model of Speziale [8]; 4. New low-Re q,ζ-model [9]; 5. Basic high-Re Reynolds stress model (RSM) with wall functions and wall reflection of Launder-Reece-Rodi-Gibson [10],[11]; 6. High-Re Reynolds stress model of Hanjalić-Jakirlić with wall functions and a Yap correction S_l in the ϵ-equation [3]; 7. Low-Re RSM of Hanjalić-Jakirlić with S_l [3]. For further details, see [2]-[4].

Table 1 Recommended meshes for k,ϵ-model calculations

MESH-no. Turbulence model	1 Hi-Re k,ϵ	2 Hi-Re k,ϵ	3 Hi-Re k,ϵ	4 Lo-Re k,ϵ
TC-2C	101 x 42	202 x 84	202 x 60	210 x 110
TC-2D	204 x 80			

Table 2 Mesh and upwinding scheme (* included in present comparison)

Team Test case	IMP [2] TC-2C TC-2D	ERL [3] TC-2C TC-2D	AFM [4] TC-2C TC-2D
1. Hi-Re k,ϵ + WF	200x80* 200x80* hybrid hybrid	200x60* 200x60* cds cds	200x80* 200x80* hybrid hybrid
2. Lo-Re k,ϵ (LS)	260x130* 300x200* hybrid hybrid	-- -	200x80* 200x80* hybrid hybrid
3. Lo-Re k,τ (Speziale)	-- -	-- -	200x80 200x80 uds uds
4. Lo-Re q,ζ (Gibson)	260x130 300x200 hybrid hybrid	-- -	-- -
5. Hi-Re RSM+WF (LRRG)	-- -	200x60 200x80 uds uds	200x80* 200x80* hybrid hybrid
6. Hi-Re RSM-HJ+S_l+WF	-- -	200x60* 200x80* blended blended	-- -
7. Lo-Re RSM-HJ+S_l	-- -	210x110* blended	-- -

COMPARISON OF RESULTS

For a summary of results, Table 3 shows the computed reattachment lengths X_r/h obtained with different turbulence models for the two test cases. The consistently lower values obtained by AFM for the standard high-Re k,ϵ-model must be ascribed to a deficiency in the implementation of the wall functions. This trend is seen to be more pronounced for TC-2D (Re = 42,000) than for TC-2C (Re = 3000). As a whole, the results of Table 3 are not encouraging, indicating that low-Re models are not better than high-Re models with wall functions, a fact that has been known for some time.

Table 3 Computed values of reattachment length X_r/h

| Team | IMP [2] | | ERL [3] | | AFM [4] | |
Test case	TC-2C	TC-2D	TC-2C	TC-2D	TC-2C	TC-2D
1. Hi-Re k,ϵ + WF	10.43	7.15	10.89	8.75	8.93	6.64
2. Lo-Re k,ϵ	9.86	7.92			10.53	5.56
3. Lo-Re k,τ					9.58	7.20
4. Lo-Re q,ζ	9.61	7.43				
5. Hi-Re RSM + WF			11.43	9.12	10.85	8.08
6. Hi-Re RSM-HJ + S_i + WF			14.00	11.44		
7. Lo-Re RSM-HJ + S_i			14.48			
Experimental values	11.7	8.13	11.7	8.13	11.7	8.13

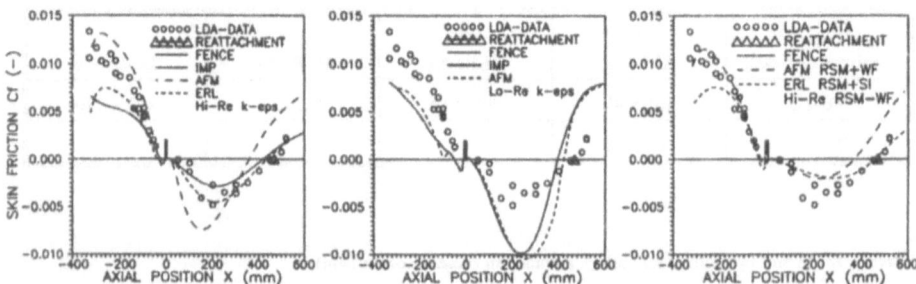

Fig.1 TC-2C. Coefficient of skin friction; standard high-Re k,ϵ-model with wall functions (left); low-Re k,ϵ-model (middle); RSM-models (right)

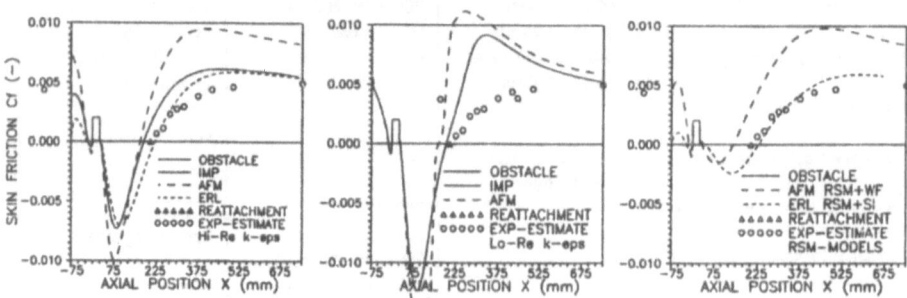

Fig.2 TC-2D. Coefficient of skin friction; standard high-Re k,ϵ-model with wall functions (left); low-Re k,ϵ-model (middle); RSM-models (right)

156

Results on the coefficient of skin friction along the wall with obstacle are shown in Figures 1 and 2 for TC-2C and TC-2D, respectively. Results are grouped according to turbulence model in three graphs: high-Re k,ϵ; low-Re k,ϵ; Reynolds stress models (high- and low-Re), respectively. Agreement with data is surprisingly good for the standard high-Re k,ϵ-model with wall functions, consistently poor for the low-Re k,ϵ-model, while only the low-Re RSM with modifications of ERL [3] performs satisfactorily for the two test cases. However, the scatter of experimental data should be noted. It reflects the approximate approach used in estimating C_f from the LDA-data on velocity profiles [5].

It should be noted that, starting from from the specified inlet locations ($x = -330$ mm and $x = -75$ mm for TC-2C and TC-2D, respectively), C_f of nearly all predictions show first an increase and then a decrease. This anormaly is probably due to the insufficient accuracy, consistency and completeness of the specified inlet profiles. The effect relaxes after a short distance in the parabolically dominated upstream region, and it is believed that it does not affect the results further downstream. However, it might have been desirable, as in TC-2A, to start the computations still further upstream and then adjust the assumptions there to match the specified inlet profiles.

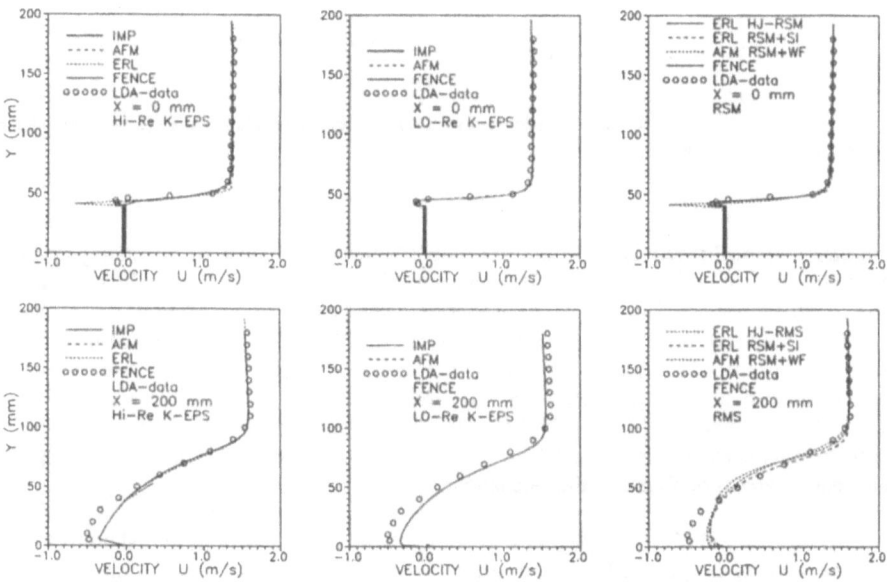

Fig.3 TC-2C. Velocity profiles $U(y)$ at $x = 0$ mm (top row) and at $x = 200$ mm (lower row).

Further details emerge when profiles of mean velocity, turbulent kinetic energy and the Reynolds stress of shear are considered. Consider first the low-Re number flow past the fence, TC-2C. Here Figure 3 (with results grouped according to turbulence model as in Figures 1 and 2) shows profiles of mean velocity $U(y)$ at the location of the backface of the fence, $x = 0$ mm, and in the recirculation region, $x = 200$ mm, respectively. On the top of the fence, data clearly indicate moderate backflow, which is absent in one prediction by IMP and overpredicted in two cases by ERL. At $x = 200$ mm, there is generally good agreement

157

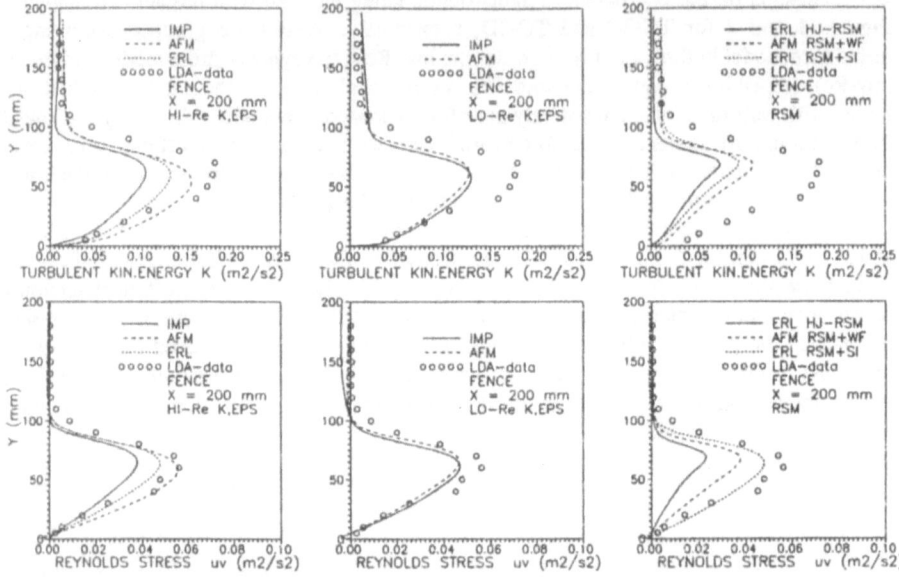

Fig.4 TC-2C. Profiles of turbulent kinetic energy k (top row) and Reynolds stress uv (lower row) at $x = 200$ mm (in recirculation region).

between numerical results. Predictions of the distribution of turbulent kinetic energy $k(y)$ and Reynolds stress in shear uv at $x = 200$ mm are shown in Figure 4. Generally, all models underpredict the level of turbulent kinetic energy in the shear layer bounding the recirculation region, particularly the RSM of second order closure. The high-Re k,ϵ-model predictions show a disturbing disagreement which must be ascribed to differences in meshes and in implementation of the wall functions. For the low-Re k,ϵ-model, IMP and AFM get essentially the same results. For the Reynolds stress, on the other hand, levels are almost correctly predicted by several models. These results on k and uv indirectly suggest that deficiencies are to be found in the ϵ-equation.

Next, consider the high-Re flow past the obstacle, TC-2D. Figure 5 shows profiles of mean velocity $U(y)$ at the location of the backface of the obstacle, $x = 25$ mm, and in the recirculation region, $x = 100$ mm, respectively. On the top of the obstacle, data clearly indicate significant backflow, which is absent in low- and high-Re k,ϵ-models of AFM and IMP, but is well modelled by the second order RSM. At $x = 100$ mm, each group of predictions show quite good agreement, while deviations from the experimental results are due to the inability of the models to describe the extent of the separation region.

Predictions of the distribution of turbulent kinetic energy $k(y)$ and Reynolds stress in shear uv at $x = 100$ mm are shown in Figure 6. Again, all models underpredict the level of k in the shear layer bounding the recirculation region. For each group of models (except the ERL results for the case of the high-Re k,ϵ-model), the mutual agreement of predictions is good, clearly better at this higher Re number (42,000) than at the lower Re number (3000) for the fence, although the underprediction of the level of k is more pronounced at the higher Re number (see Fig.4). For the Reynolds stress, the mutual agreement in each group of

158

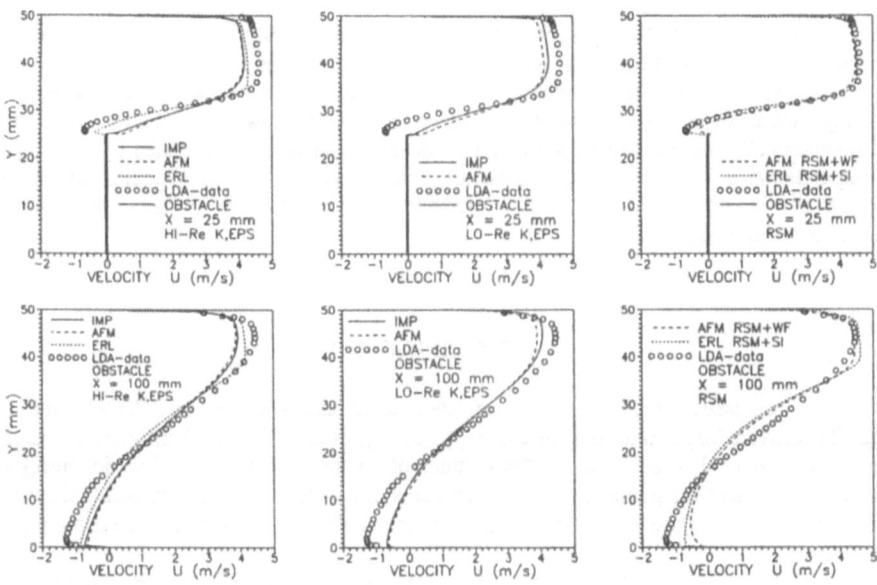

Fig.5 TC-2D. Velocity profiles $U(y)$ at $x = 0$ mm (top row) and at $x = 100$ mm (lower row).

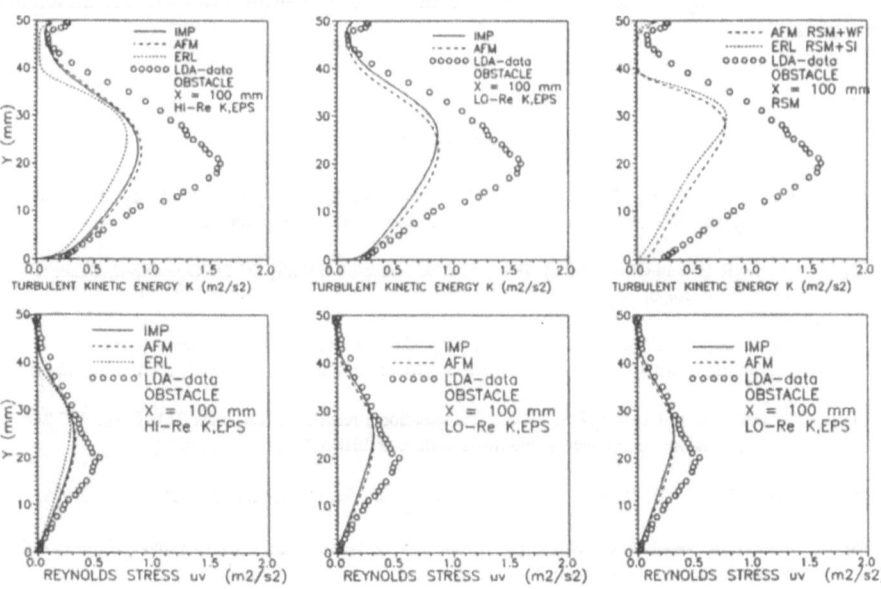

Fig.6 TC-2D. Profiles of turbulent kinetic energy k (top row) and Reynolds stress uv (lower row) at $x = 100$ mm (in recirculation region).

159

models is quite good. The agreement with data is also generally good, except at the peak value of stress in the shear layer above the recirculation region.

To limit the number of curves, two model calculations have been left out. That of the new q,ζ-model of IMP gives results very close to those of the low-Re k,ϵ-model, while being computationally more efficient, see [2]. That of the k,τ-model, implemented by AFM, gave computational difficulties and unsatisfactory results, see [4]. For the models considered, results at other axial sections of the flows show tendencies that are similar to those shown, hence they are not included.

CONCLUSIONS

A comparison has been made of computational results for test cases TC-2C and TC-2D [1], provided by 3 independent teams [2],[3],[4], using a total of 7 different turbulence models. In general, predictions of distributions of velocity and turbulent kinetic energy are in good agreement for similar two-equation models and second-order closure RSM. However, predictions of the more sensitive parameter of the coefficient of skin friction show greater differences.

The predictions have also been compared to experimental LDA-measurements [5]. While agreement on velocity and skin friction is generally fair, the predictions of levels of turbulent kinetic energy are consistently lower than those measured, while predictions of the Reynolds shear stress is in fair agreement with data. These findings indicate some principal shortcomings of all of the models for these test cases. It also confirms the view that there is a need for further improvements of turbulence models for near-wall flows with separation and recirculation.

REFERENCES

[1] Larsen,P.S. 1995, "Presentation of test cases TC-2A, TC-2B, TC-2C, TC-2D; Two-dimensional, incompressible wall flows with separation", *these proceedings*.

[2] Harper,R.D. and Gibson,M.M. 1995, "TC-2C Fence-on-a-Wall; TC-2D Obstacle-in-Channel", *these proceedings*.

[3] Jakirlić,S., Hadžić,I. and Hanjalić,K. 1995, "Computations of turbulent separating and reattaching flows with high- and low-Re-number second-moment closures", *these proceedings*.

[4] Schmidt,J.J. and Larsen,P.S. 1995, "Computational results on test cases TC-2C and TC-2D; Two-dimensional, incomnpressible flows with recirculation", *these proceedings*.

[5] Larsen,P.S. 1995, "Database on TC-2C and TC-2D, Report ETMA/AFM/95-01, Feb.1995.

[6] Jones,W.P. and Launder,B.E. 1972, "Prediction of laminarization with a two-equation model of turbulence, Int.J.Heat Transfer, *15*, 301-314.

[7] Launder,B.E. and Sharma,B.I. 1974, "Application of the energy-dissipation model of turbulence to the calculation of flow near a spinning disc". Lett.Heat Mass Transfer, *1*, 131-138.

[8] Speziale,C.G., Abid,R. and Anderson,E.C. 1992, "Critical evaluation of two-equation models for near-wall turbulence", AIAA J., *30*, 324-331.

[9]   Gibson,M.M. and Dafa'Alla,A.A. 1994, "The $q\zeta$-model for turbulent wall flow", Fluid Dynamics Division, Amer.Phys.Soc., 47th Annual Meeting, Atlanta, Georgia.

[10]   Launder,B.E., Reece,G.J. and Rodi,W. 1975, "Progress in the development of a Reynolds-stress turbulence closure", J.Fluid Mech., 68, 537-566.

[11]   Gibson,M.M. and Launder,B.E. 1978, "Ground effects on pressure fluctuations in the atmospheric boundary layer", J.Fluid Mech., 86, 491-511.

161

CHAPTER 4 : FLOWS PAST
A FLATE PLATE

SPECIFICATION OF TEST CASE TC3

(FLAT PLATE BOUNDARY LAYERS)

Ch. Hirsch and Erbing Shang

Dept. of Fluid Mechanics, Vrije Universiteit Brussel
Pleinlaan 2, 1050 Brussels, Belgium

SUMMARY

The ETMA workshop test case, TC3, is the flat plate boundary layer flow. It consists of four sub-test cases:

TC3-1. Klebanoff zero pressure gradient boundary layer;

TC3-2. Samuel & Joubert adverse pressure gradient flow;

TC3-3. Spalart sink flow ;

TC3-4. Mabey supersonic boundary layer on adiabatic wall.

TC3–1. KLEBANOFF ZERO PRESSURE GRADIENT FLOW

DESCRIPTION OF THE EXPERIMENT

Klebanoff's test case is a zero pressure gradient incompressible flow over a flat plate [1]. The incoming boundary layer is artificially thickened. The free stream velocity is

$$u_e = 15.24 \ (m/s) \ .$$

At the measuring station, the boundary layer thickness is $\delta = 76.2$ mm. The wall friction parameter [2] is

$$\sigma = u_\tau/u_e = 0.037,$$

The Reynolds number is

$$\text{Re} = u_e X_0/\nu = 4.2 \times 10^6$$

based on a turbulent boundary layer development length of $X_0 = 4.32816$ m.

COMPUTATIONAL DOMAIN

The numerical simulation model length L is suggested as 0.4 m. The computational domain is 0.4m × 0.05m. The turbulent boundary layer development length involved in the Reynolds number, $\text{Re} = X_0 u_e/\nu = 4.2 \times 10^6$, is scaled from $X_0 = 4.32816$ m to $X_0 = 0.35$ m.

MESH

The mesh contains 41×81 grid points in the streamwise and normal directions, respectively, as shown in Fig.1. The mesh spacing is constant in the streamwise direction and determined by a constant stretching parameter of 1.1 in the normal direction. For the first grid points, $y^+ = 0.1 \sim 1$. Within the boundary layer, there are $50 \sim 65$ grid points.

BOUNDARY CONDITIONS

Since the free stream temperature and static pressure are not mentioned in the Klebanoff's paper [1], standard conditions are assumed for the temperature and static pressure.

Inlet

Assuming oncoming flow is uniform.

Freestream velocity	$u_e = 15.24$	(m/s)
Static pressure	$P = 108309$	(Pa)
Temperature	$T = 293$	(K) .

Flat Plate

An adiabatic wall boundary condition is applied.

Freestream and Outlet

Static pressure	$P = 108309$	(Pa).

OUTPUT FORMATS

Output Quantities

1. Velocity distribution (at x = 0.35m) compared with Klebanoff data

$\log_{10}(y^+)$, u^+.

The quantities will be plotted in the range: $-1 \leq \log_{10}(y^+) \leq 5$, $\quad 0 \leq u^+ \leq 30$.

2. Wall friction coefficient distribution compared with Nikuradse correlation [3]

   Rex, C_f.

   The quantities will be plotted in the range:

   $0 \leq \text{Rex} \leq 4.8 \times 10^6$, $\quad 2 \times 10^{-3} \leq C_f \leq 6.5 \times 10^{-3}$.

Format for Paper Output

The distribution outputs are to be plotted in figures of a standard size $10 \times 10 \text{cm}^2$.

NOMENCLATURE

C_f	—	wall friction coefficient $= 2\sigma^2 = 2\tau_w/(\rho u_e^2)$
L	—	length of flat plate
P	—	static pressure
Re	—	Reynolds number based on X_0, $\text{Re} = X_0\, u_e/\nu$
Rex	—	Reynolds number based on x, $\text{Rex} = x\, u_e/\nu$
u	—	velocity component in x direction
u_e	—	free stream velocity
u_τ	—	wall friction velocity, $u_\tau = \sqrt{\tau_w/\rho}$
u^+	—	non-dimensional u, $u^+ = u/u_\tau$
T	—	temperature
x	—	streamwise coordinate
X_0	—	turbulent boundary layer development length
y	—	normal distance to the wall
y^+	—	non-dimensional y, $y^+ = yu_\tau/\nu$
δ	—	boundary layer thickness
σ	—	wall friction parameter, $\sigma = u_\tau/u_e$.

REFERENCES

[1] Klebanoff., P. S., 1954, "Characteristics of Turbulence in a boundary layer with zero pressure gradient", NACA Technical Note 3178.

[2] Hinze, J. O., 1959, "Turbulence", McGraw-Hill Book Company, pp.493.

[3] Schlichting, H., 1979, 'Boundary Layer Theory', 7th edition. McGraw-Hill.

TC3–2. SAMUEL & JOUBERT ADVERSE PRESSURE GRADIENT FLOW

DESCRIPTION OF THE EXPERIMENT

This experiment is performed in a long region of a flexible-roof tunnel with a finely controlled streamwise pressure distribution. Mean flow and fluctuating quantities are obtained in an incompressible turbulent boundary layer developing on a smooth wall in a pressure domain where both dp/dx and d^2p/dx^2 are positive (increasingly adverse pressure gradient). The boundary layer is tripped by pins center spacing at a streamwise position of $x = 0.161$m. This flow has a strong enough pressure gradient that it appears to be approaching separation at the end of the working region, but no separation is present. The centerline velocities are obtained within the working region 0.8m $\leq x \leq 3.5$m. For the details one may refer to [1] and the STANFORD 80-81 database (Case f0141a) [2].

COMPUTATIONAL DOMAIN

Since the smoothed velocity profile is provided at $x_0 = 1.04$m in the STANFORD 80-81 database (Case f0141a), and the centerline velocity distribution is given in the range 0.8m $\leq x \leq 3.5$m, the length of the flat plate is chosen as $L = 3.5 - 1.04 = 2.46$ m.

The tunnel centerline, where u_e is measured, is located approximately at $h \approx 0.2$m. Hence the size of the computational domain is 2.46x0.2 m.

MESH

The mesh contains 41×81 grid points in the streamwise and normal directions, respectively, as shown in Fig.2. The mesh spacing is constant in the streamwise direction and determined by a constant stretching parameter of 1.1 in the normal direction. For the first grid points, $y^+ = 0.1 \sim 1$. Within the boundary layer, there are $50 \sim 65$ grid points.

REFERENCE VALUES

P_{ref}	$= 100641$	(Pa)
T_{ref}	$= 296$	(K)
u_{ref}	$= 26.15$	(m/s)
ρ_{ref}	$= 1.185$	(Kg/m^3)
Re/L	$= u_{ref}/\nu_{ref} = 1.71 \times 10^6$.	

BOUNDARY CONDITIONS

Inlet

$$P_1 \qquad = 100649 \qquad (Pa)$$
$$T_1 \qquad = 296 \qquad (K)$$
$$u_{e1} \qquad = 26.15 \qquad (m/s) .$$

The velocity profile is imposed as

$$u = u(y), v = 0.$$

The data of $u = u(y)$ are provided in [2] (Case f0141a, $x_0 = 1.04m$).

Flat Plate

An adiabatic wall boundary condition is applied.

Freestream

The static pressure distribution is imposed from the distribution of the given centerline velocity $u_e(x')$ as follows by assuming no changes of total pressure along the centerline:

$$M(x') = \frac{u_e(x')}{\sqrt{\gamma R T_{ref}}} \qquad \text{and} \qquad P(x') = P_1 \left[\frac{1 + \frac{\gamma - 1}{2} M_1^2}{1 + \frac{\gamma - 1}{2} M(x')^2} \right]^{\gamma/(\gamma-1)} .$$

For the data of the centerline velocity distribution $u_e(x')$, one may refer to [2].

Outlet

$$P_2 \qquad = 100847 \qquad (Pa) .$$

OUTPUT FORMATS

Output Quantities

1. Velocity distribution compared with the experimental data,

   $u/u_e, \quad y$

   at $x' = 0.12, 0.72, 1.22, 1.52, 1.83$ and 2.00 m (coinciding with the experimental measuring positions $x = 1.16, 1.76, 2.26, 2.56, 2.87$ and $3.04m$).

   The quantities will be plotted in the range: $0 \le u/u_e \le 1.2, \quad 0 \le y \le 0.12m$.

2. Reynolds stress compared with the experimental data,

   $y, \quad -\overline{u'v'}/u_e^2$

at x' = 0.4, 0.75, 1.34, 1.85 and 2.35 m (coinciding with the experimental measuring positions x = 1.44, 1.79, 2.38, 2.89, and 3.39 m).

The quantities will be plotted in the range: $0 \le y \le 0.012$m, $\quad 0 \le -\overline{u'v'}/u_e^2 \le 3\times10^{-3}$.

3. Wall friction coefficient compared with the experimental data,

$\quad$ x', $\quad C_f$.

The quantities will be plotted in the range: $0 \le x' \le 3$m, $\quad 0 \le C_f \le 3\times10^{-3}$.

Format for Paper Output

The distribution outputs are to be plotted in figures of a standard size 10x10cm^2.

NOMENCLATURE

C_f — wall friction coefficient $C_f = 2\tau_w/(\rho_{ref}u_{ref}^2)$

C_p — static pressure coefficient $C_p = 2(P - P_{ref})/(\rho_{ref}u_{ref}^2)$

L — length of flat plate

P — static pressure

R — gas constant, R = 287 (m^2/s^2 K)

Re — Reynolds number based on L, $Re = L\upsilon_{ref}/\nu_{ref}$

T — temperature

u, v — velocity components in x and y direction

u_e — streamwise velocity external to the boundary layer

u_τ — wall friction velocity, $u_\tau = \sqrt{\tau_w/\rho}$

$-\overline{u'v'}$ — Reynolds stress

x — experimental streamwise coordinate

x' — streamwise coordinate for numerical calculation, $x' = x - x_0$

y — normal distance to the wall

γ — specific heat ratio, $\gamma = 1.4$

ν — laminar kinetic viscosity

ρ — density .

Subscripts

1 — inlet value

2 — outlet value

ref — reference value .

REFERENCES

[1] Samuel, A.E. and Joubert, P.N., 1974, "A Boundary Layer Development in an Increasingly Adverse Pressure Gradient", Journal of Fluid Mech., Vol.66, pp.481-505.

[2] STANFORD 80-81 database (Case f0141a).

TC3–3. SPALART SINK FLOW

DESCRIPTION OF THE EXPERIMENT

The incompressible flow in a two dimensional convergent channel is called sinkflow. In the sinkflow the boundary layer edge velocity is inversely proportional to distance from sink.

Let the flow move in the positive x-direction toward $x = x_0$, where the sink itself is. Let X denote $(x_0 - x)$, 2β be the angle the walls and $2\beta Q$ be the total mass flux. The velocity on the boundary layer edge is

$$U_e = \frac{Q}{X} \cdot$$

The statistical quantities of the turbulent sinkflow are self-similar. The experiments show that the self-similarity is to be reached asymptotically as one approaches the sink [1]. In such a solution all the non-dimensional quantities like the acceleration parameter $K = \nu/U_e^2 (dU_e/dx)$, wall friction coefficient C_f, shape factor H, and thickness Reynolds number $Re_{\delta*}$ and Re_θ are independent of the streamwise coordinate x. The entrainment is zero, in the sense that the edge of the boundary layer is also a streamline. The sinkflow boundary layer is the purest example of an 'equilibrium' turbulent boundary layer, a boundary layer with a shape that is invariant in the streamwise direction [2]. The flow is defined by the acceleration parameter

$$K = \frac{\nu}{U_e^2} \frac{dU_e}{dx} = \frac{\nu}{U_e X} \cdot$$

The Reynolds number is defined by K as

$$Re = \frac{U_e X}{\nu} = \frac{1}{K} \cdot$$

The Spalart DNS test case with $K = 1.5 \times 10^{-6}$ is used for the numerical simulation. In the self-similar region,

$$C_f = 4.99 \times 10^{-3}, \qquad Re_\theta = 690 \cdot$$

For the details, one may refer to the file 'Sink_K.150' in the STANFORD 80-81 database [3] and [2].

COMPUTATIONAL DOMAIN

The length of the flat plate is chosen as $L = 1.8$ m. The sink position is assumed to be located at $x_0 = 2.2$m. The channel half opening angle is $\beta = 28°$. The inlet and outlet are circular arcs of radii $X_1 = 2.2$ m and $X_2 = 0.4$ m, with their centers located at the sink position.

MESH

The mesh contains 41×81 grid points in the streamwise and normal directions, respectively. It is shown in Fig.3. The mesh spacing is constant in the streamwise direction and determined by a constant stretching parameter of 1.1 along the circular arc. For the first grid points, $y^+ = 0.1 \sim 1$. Within the boundary layer, there are $50 \sim 65$ grid points.

BOUNDARY CONDITIONS

Inlet

$$
\begin{aligned}
P_1 \quad &= 101330 \quad \text{(Pa)} \\
T_1 \quad &= 293 \quad \text{(K)} \\
U_{el} \quad &= 22.4545 \quad \text{(m/s)} .
\end{aligned}
$$

The velocity profile is imposed as

$$u = u(y)$$
$$v = 0.$$

The inlet velocity profile is defined by the Spalart nondimensional DNS data, and is derived as follows

$$u = u^+ \left(\frac{u_\tau}{U_{el}} \right) U_{el} = u^+ \left(C_f/2 \right)^{1/2} U_{el}$$

$$y = y^+ \left(\frac{\nu}{U_{el} X_1} \frac{U_{el}}{u_\tau} X_1 \right) = y^+ K \left(C_f/2 \right)^{-1/2} X_1 .$$

For the Spalart DNS data, one may refer to the file 'Sink_K.150' in the STANFORD 80-81 database [3].

Flat Plate

An adiabatic wall boundary condition is applied.

Freestream

A symmetry boundary condition is applied.

Outlet

$$P_2 \quad = 94293 \quad \text{(Pa)} \,.$$

OUTPUT FORMATS

Output Quantities

1. Streamwise velocity on the edge of the boundary layer compared with the theoretical formula $U_e = Q/X$,

   $x, \quad U_e$.

   The quantities will be plotted in the range: $0 \leq x \leq 2.4\text{m}, \quad 20 \leq U_e \leq 140$ m/s.

2. Non-dimensional velocity distribution (at $x = 1.305\text{m}$) compared with the DNS data,

   $\log_{10}(y^+), \quad u^+$.

   The quantities will be plotted in the range: $-1 \leq \log_{10}(y^+) \leq 5, \quad 0 \leq u^+ \leq 24$.

3. Non-dimensional Reynolds stress distribution (at $x = 1.305\text{m}$) compared with the DNS data,

   $y^+, \quad -\overline{u'v'}/u_\tau^2$.

   The quantities will be plotted in the range: $0 \leq y^+ \leq 300, \quad 0 \leq -\overline{u'v'}/u_\tau^2 \leq 0.9$.

Format for Paper Output

The distribution outputs are to be plotted in figures of a standard size $10\text{x}10\text{cm}^2$.

NOMENCLATURE

C_f — wall friction coefficient $C_f = 2\tau_w/\left(\rho_e U_e^2\right)$

K — acceleration parameter $K = \nu/U_e^2 \left(dU_e/dx\right)$

L — length of flat plate

P — static pressure

Re — Reynolds number based on X, $Re = XU_e/\nu$

$Re_{\delta*}$ — Reynolds number based on displacement thickness, $Re_{\delta*} = \delta^* U_e/\nu$

Re_θ — Reynolds number based on momentum thickness, $Re_\theta = \theta U_e/\nu$

T	—	temperature
u, v	—	velocity components in x and y direction
U_e	—	streamwise velocity external to the boundary layer
u_τ	—	wall friction velocity, $u_\tau = \sqrt{\tau_w/\rho}$
u^+	—	nondimensional u, $u^+ = u/u_\tau$
$-\overline{u'v'}$	—	Reynolds stress
Q	—	mass flux per unit angle
x	—	streamwise coordinate
x_0	—	sink position
X	—	distance to the sink position, $X = x_0 - x$
y	—	coordinate along arclength
y^+	—	nondimensional y, $y^+ = y\, u_\tau /\nu$
β	—	channel half opening angle .

Subscripts

1	—	inlet value
2	—	outlet value .

REFERENCES

[1] Jones, W.P. and Launder, B.E., 1972, 'Some Properties of Sinkflow Turbulent Boundary Layers', J. of Fluid Mech., Vol.56, pp337-351.

[2] Spalart, P. R., 1986, 'Numerical Study of Sinkflow Boundary Layers', J. of Fluid mech., Vol.172, pp.307-328.

[3] STANFORD 80-81 database (Sink_K.150).

TC3–4. MABEY SUPERSONIC BOUNDARY LAYER ON ADIABATIC WALL

DESCRIPTION OF THE EXPERIMENT

Mabey's test case considers the development of the turbulent boundary layer on an adiabatic flat plate with zero pressure gradient at a freestream Mach number of 4.5.

The experiment is performed on a flat plate with a length of 1.65m. Surface roughness is within 0.64 μm except for the transition strip from x = 2.584 to 5.08 mm where small glass spheres of 0.28 mm diameter are distributed sparsely.

The profiles are measured with a combined Pitot and total temperature probe at streamwise positions x = 0.368, 0.623, 0.876, 1.130 and 1.384m. The good agreement of experimental data with both the logarithmic law and the outer law proves that the boundary layer is fully developed in this experiment [1].

The working fluid is treated as a perfect gas with constant specific heats. The perfect gas properties of air assumed are

Gas constant R = 287.1387 (m^2/s^2 K),   Specific heat ratio γ = 1.40.

The flow Reynolds number of unit length is

Re/L = 2.82×10^7.

For the details, one may refer to the test case ID No.7402 in [2].

COMPUTATIONAL DOMAIN

The computational domain is from the first measuring position x= 0.368m (ID No. 74021801) to the flat plate trailing edge x = 1.65m. So the length of flat plate L = 1.282m. The height of the numerical domain is 0.2 m.

MESH

The mesh proposed contains 113×81 grid points in the streamwise and normal directions, respectively, as shown in Fig.4. The mesh spacing is determined by a constant stretching parameter of 1.11 in the normal direction. For the first grid points from the wall, y^+ = 0.1 ~ 1. In the streamwise direction, a parabolic distribution is applied with clustering at the leading edge.

BOUNDARY CONDITIONS

Inlet

The streamwise velocity and temperature profiles measured at x= 0.368m are imposed.

Static pressure	P_1	= 3119.30	(Pa)
Normal velocity	v_1	= 0	

The experimental data of velocity and temperature are provided in [2].

Outlet and Upper Boundary

Static pressure	P_{fre}	= 3119.30	(Pa)
Temperature	T_{fre}	= 61.805	(K)
Streamwise velocity	u_{fre}	= 711.97	(m/s)

Normal velocity $\qquad v_{fre} \quad = 0$.

Flat Plate

An adiabatic wall boundary condition is applied.

OUTPUT FORMATS

For comparison, the data are nondimensionlized with the corresponding values on the boundary layer edge, which is defined at the position where local total pressure reaches 99 percent of the freestream total pressure, i.e., $P_t = 0.99\ P_{te}$ [2].

Output Quantities

1. Velocity distribution compared with the experimental data

u/u_e, y.

The quantities will be plotted in the range: $0 \leq u/u_e \leq 1.2$, $0 \leq y \leq 0.03$ at the streamwise positions: x= 0.623, 0.876, 1.130 and 1.384m.

2. x-momentum distribution compared with the experimental data

$\rho u/\rho u_e$, y.

The quantities will be plotted in the range: $0 \leq \rho u/\rho u_e \leq 1.2$, $0 \leq y \leq 0.03$ at the streamwise positions: x= 0.623, 0.876, 1.130 and 1.384m.

3. Temperature distribution compared with the experimental data

T/T_e, y.

The quantities will be plotted in the range: $0 \leq T/T_e \leq 1.2$, $0 \leq y \leq 0.03$ at the streamwise positions: x= 0.623, 0.876, 1.130 and 1.384m.

4. Mach number distribution compared with the experimental data

M/M_e, y.

The quantities will be plotted in the range: $0 \leq M/M_e \leq 1.2$, $0 \leq y \leq 0.03$ at the streamwise positions: x= 0.623, 0.876, 1.130 and 1.384m.

5. Total temperature distribution compared with the experimental data

T_t/T_{te}, y.

The quantities will be plotted in the range: $0.90 \leq T_t/T_{te} \leq 1.02$, $0 \leq y \leq 0.03$ at the streamwise positions: x= 0.623, 0.876, 1.130 and 1.384m.

6. Pitot total pressure distribution compared with the experimental data

P_{tt}/P, y.

The quantities will be plotted in the range: $0 \leq P_{tt}/P \leq 30$, $0 \leq y \leq 0.03$ at the streamwise positions: x= 0.623, 0.876, 1.130 and 1.384m.

7. Wall friction coefficient distribution compared with the experimental data

x, C_f.

The quantities will be plotted in the range: $0 \leq x \leq 1.8$m, $6 \times 10^{-4} \leq C_f \leq 4.5 \times 10^{-3}$

Format for Paper Output

The distribution outputs are to be plotted in figures of a standard size $10 \times 10 \text{cm}^2$.

NOMENCLATURE

C_f	—— wall friction coefficient, $C_f = 2\tau_w/\left(\rho_{fre}u_{fre}^2\right)$
L	—— length of flat plate
M	—— Mach number
P	—— static pressure
P_{tt}	—— Pitot total pressure
Re	—— Reynolds number based on L, Re= $\rho_{fre}u_{fre}L/\mu_{fre}$
T	—— temperature
u, v	—— velocity components in x and y direction
x	—— streamwise coordinate
y	—— normal distance to the wall
ρ	—— density .

Subscripts

e	—— value on boundary layer edge
fre	—— freestream value
t	—— total quantity
w	—— value on flat plate .

REFERENCES

[1] Fernholz, H.H. and Finley, P.J. 1980, "A Critical Commentary on Mean Flow Data for Two-dimensional Compressible Turbulent Boundary Layers", AGARDograph 253.

[2] Fernholz, H.H. and Finley, P.J. 1977, "A Critical Compilation of Compressible Turbulent Boundary Layer Data", AGARDograph 223.

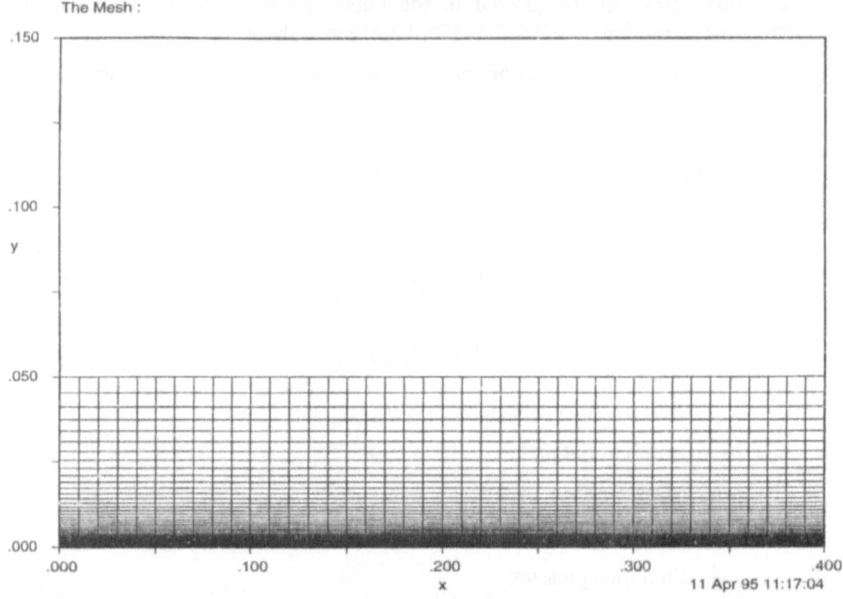

Figure 1: Mesh proposed for Klebanoff zero pressure gradient flow (TC3-1)

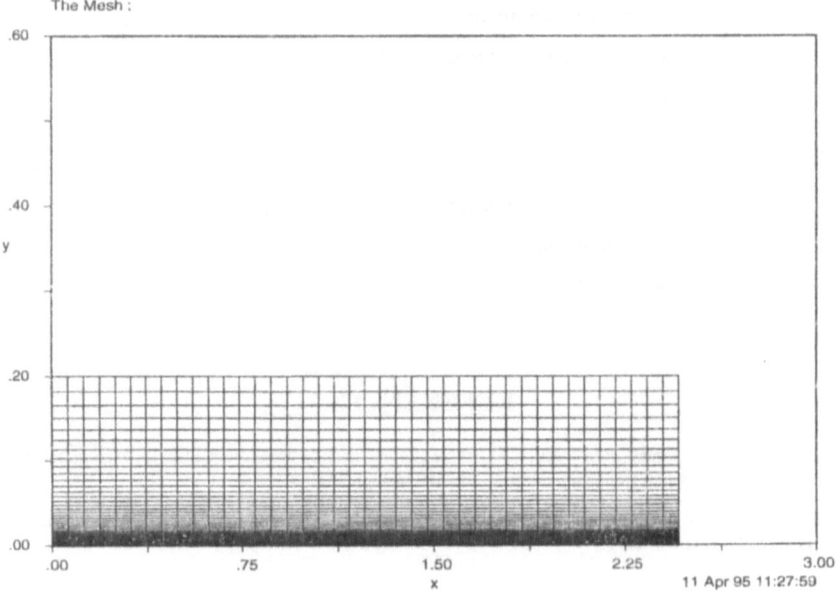

Figure 2: Mesh proposed for Samuel and Joubert adverse pressure gradient flow (TC3-2)

178

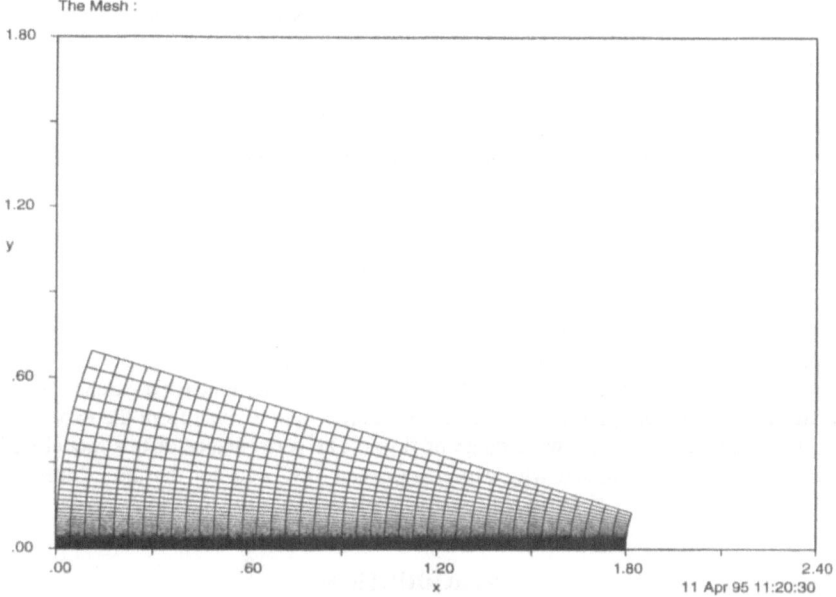

Figure 3: Mesh proposed for Spalart sink flow (TC3-3)

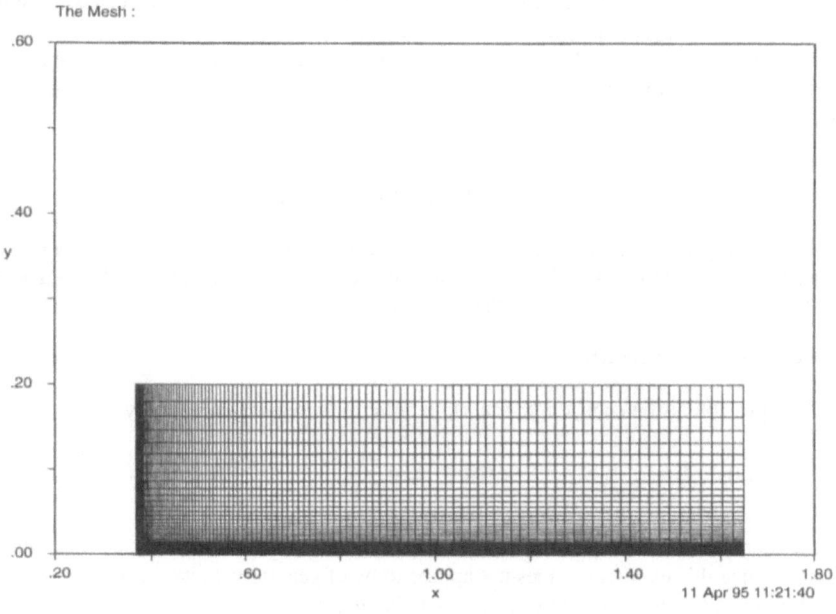

Figure 4: Mesh proposed for Mabey supersonic flow over adiabatic wall (TC3-4)

A NUMERICAL EVALUATION OF A NEW
ALGEBRAIC TURBULENCE MODEL

Ch. Hirsch and Erbing Shang

Dept. of Fluid Mechanics, Vrije Uiversiteit Brussel, Pleinlaan 2, 1050 BRUSSEL

SUMMARY

In this paper, a new algebraic closure model formulation is presented, which is independent of wall distance. Based on comparisons with experiments of zero pressure gradient flow, all the constants involved in the model are determined. To extend the model into flows with pressure gradient, two extra correlations are derived from the experimental data processed which cover a wide range of flows. The performance of this model will be illustrated through comparison with experiments and DNS data for the three cases of the ETMA project.

INTRODUCTION

In order to solve turbulent flows, there is a need for simple turbulent models. The algebraic eddy viscosity turbulence models are still widely used for their simplicity and engineering accuracy. The most successful algebraic turbulence models are Cebeci-Smith [1] and Baldwin-Lomax models [2]. However, these models take the normal distance to the wall as the turbulence length scale, like y^+, the determination of which is not straightforward in complex geometries. Within the framework of Navier-Stokes computations, it would be more appropriate to base the formulation of turbulence models on local properties of the mean flow.

CLOSURE MODEL FORMULATION

For the algebraic model, the choice for the length scale is more difficult than one and two equation models because k and ε can not be used. The freedom is limited to the combination of velocity and their gradients.

A nondimensional parameter

The possible turbulence length scales, in order of increasing complexity, are

$$\textbf{(a)} \quad L = \frac{u}{|du/dy|} \qquad \textbf{(b)} \quad L = \frac{|du/dy|}{d^2u/dy^2} \; .$$

If expression **(b)** is introduced, the von Karman formula for turbulent eddy viscosity can be deduced. Since this expression does not appear to be of general validity, only expression **(a)** is used for the later analysis. The viscous length is selected as

$$L_v = \nu/|u| \tag{1}$$

180

where |u| is the amplitude of the relative mean velocity in a frame fixed to the solid boundary. A nondimensional parameter R_L, based on L is constructed as

$$R_L = \frac{L}{L_v} = \frac{u^2}{v\,|du/dy|}\,.$$

(2)

R_L represents the ratio of the turbulent length scale L to the viscous length scale L_v. The quantity $X = R_L^{1/2}$ is chosen as the new nondimensional parameter for simulating turbulent eddy viscosity. In general, it can be expressed as

$$X = R_L^{1/2} = \frac{|\vec{u}|}{\left[v\,|\vec{\omega}|\right]^{1/2}} = o(y)\,.$$

(3)

Here $|\vec{u}|$ and $|\vec{\omega}|$ are the magnitude of velocity and vorticity, respectively. One can see that $R_L^{1/2}$ is a unique function of y^+, as shown in Fig.1. It may be conveniently used to replace y^+.

Observations based on experiment and *DNS* data

From the experimental data of turbulent eddy viscosity [3,4,5], the following conditions have to be satisfied:

i). On the wall, the turbulent eddy viscosity is zero;

$$y = 0, \qquad v_t = 0\,.$$

(4)

ii). In the logarithmic region, the turbulent eddy viscosity increases almost linearly;

$$v_t \approx \alpha\, y^+\,, \qquad \text{for small } y^+ \text{ values}\,.$$

(5)

iii). At a definite distance from the wall, the eddy viscosity reaches its maximum value;

$$y = y_m, \qquad v_t = v_{tm}\,.$$

(6)

iv). The turbulent eddy viscosity reduces to zero at the edge of the boundary layer;

$$y \rightarrow \delta, \qquad v_t \rightarrow 0\,.$$

(7)

v). Smooth transition of the turbulent eddy viscosity from the inner region to the outer region;

vi). **DNS** data confirm the near wall region behaviour:

$$-\overline{u'v'}^+ \approx a_{uv} y^{+3} + \cdots$$

(8)

with

$$a_{uv0} = (7.2 \sim 13.) \times 10^{-4} \qquad \text{(for flat plate flow [6])}$$

(9-1)

$$a_{uv0} = (7.0 \sim 9.5) \times 10^{-4} \qquad \text{(for channel flow [7])}\,,$$

(9-2)

Model formulation

In the model development, three components are considered: a near wall damping function $f_\mu(R_L^{1/2})$, a main part $(R_L^{1/2})^{C_2}$, and an outer region intermittency function $F_{int}(R_L^{1/2})$. If the wall damping function takes the following form

$$f_\mu(X) = 1 - \exp\left[-X^{C_3}/C_\mu\right] \tag{10}$$

and the outer region intermittency function is selected as

$$F_{int} = \frac{1}{1 + C_4 (X/X_m)^{C_5}} \tag{11}$$

the model can be expressed as:

$$\frac{v_t}{v} = C_1 \frac{\left[1 - \exp\left(-X^{C_3}/C_\mu\right)\right] X^{C_2}}{1 + C_4 (X/X_m)^{C_5}} . \tag{12}$$

Considering the conditions mentioned above, this leads to

$$C_1 = \frac{L_m}{X_m^{C_2}} \frac{C_5}{(C_5 - C_2)} = a_{uv} C_\mu \quad, \qquad C_2 + C_3 = 3, \qquad C_4 = \frac{C_2}{C_5 - C_2} . \tag{13}$$

L_m and X_m are chosen as

$$L_m = 190, \qquad X_m = 340 \tag{14}$$

based on the experimental data of flat plate turbulent boundary layer [3], as shown in Fig.2. After numerous tests, it is found that the best options for the constants are

$$C_2 = 1.65, \ C_3 = 1.35, \ C_5 = 5, \ C_\mu = 16 \text{ (corresponding to } a_{uv0} = 11.8 \times 10^{-4}), \tag{15}$$

Therefore the model is established as

$$\frac{v_t}{v} = a_{uv} C_\mu \frac{X^{C_2}\left[1 - \exp\left(-X^{C_3}/C_\mu\right)\right]}{1 - C_2/(C_5 - C_2)(X/X_m)^{C_5}} . \tag{16}$$

For zero pressure gradient flat plate flow, L_m and X_m are correlated as

$$L_m = L_{m0} = (C_{v3} \, C_f)^{-C_{v4}}, \qquad C_{v3} = 134, \, C_{v4} = 5.1 \tag{17}$$

$$X_m = X_{m0} = (C_{m1} \, C_f)^{-C_{m2}}, \qquad C_{m1} = 55, \, C_{m2} = 3.08 . \tag{18}$$

If the near wall behaviour (Eq.8) is assumed as universal, eq.(13) provids a relation between L_m and X_m as

$$X_m \sim L_m^{1/C_2} . \tag{19}$$

Applying this expression to L_{m0}, one has

$$X_m = X_{m0} = \left(C_{m1} \, C_f\right)^{-C_{m2}}, \qquad C_{m1} = 56, \quad C_{m2} = 3.09 \tag{20}$$

in excellent agreement with Eq.(18).

Compared with Townsend, Klebanoff and Schubauer data [3] of turbulent eddy viscosity and Reynolds stress, the proposed model gives a good prediction, see in Fig.4 and 5.

Extension to flows with pressure gradients

To apply the model into the flows with pressure gradients, it is necessary to investigate the pressure gradient effect on the two most important parameters, a_{uv} and X_m. Totally, five sets of experimental data *(Samuel and Joubert [8], Simpson [9], Van Den Berg [10], Patel [11], Pozzorini [10])*, one set of DNS data *(Spalart [12])*, two sets of numerical results (Deutsche Aerospace's results for test cases:*Wieghardt, Ludwieg and Tillmann [13]*) are processed. The data evaluated cover the favourable and (moderate, mild) adverse pressure gradient approaching to separation, thin and thick boundary layer, internal and external subsonic flow cases.

The data are plotted against several pressure gradient parameters: non-dimensional pressure gradient parameter $[p^+ = \left(v/\rho u_*^3\right)(dp/dx) \,]$, Clauser pressure gradient parameter $[\beta = \delta^*/\tau_w \cdot (dp/dx)]$ and Coles wake parameter (Π). The experimental data show that for the fixed pressure gradient parameter $(p^+, \beta$ and $\Pi)$, with wall friction coefficient increasing, L_m and X_m decrease. For fixed wall friction coefficient, with increasing pressure gradient parameter, L_m and X_m decrease also.

To remove the wall friction effect on L_m and X_m, the derived data L_m and X_m are divided by the zero pressure gradient expression L_{m0} (17) and X_{m0} (18). Since only p^+ can be used into the model reformulation, the data are plotted against p^+, as shown in Fig.3 and 6. The data are falling together, and show that with pressure gradient level increasing, L_m/L_{m0} and X_m/X_{m0} decreases for adverse pressure gradient and increases for favourable pressure gradient.

The data of X_m/X_{m0} can be fitted by

$$X_m/X_{m0} = \left\{ \begin{array}{ll} \left[1 + \left(200 \, p^+\right)^{2.5}\right]^{-1} & \text{for } p^+ \geq 0 \\[2mm] \exp\left[-\left(66p^+ + 0.924\right)^2 + 0.85\right] & \text{for } p^+ < 0 \,. \end{array} \right. \tag{21}$$

The value of a_{uv} can be deduced if (13) is considered as a relationship among three parameters a_{uv}, L_m and X_m. The data show that a_{uv} is a function of pressure gradient parameter, see Fig.7. This hints that the near wall region behaviour (8) is not valid anymore for the flows with pressure gradients. Considering (13), one can have

$$a_{uv}/a_{uv0} = \frac{C_5}{a_{uv0}\left(C_5 - C_2\right) C_\mu} \frac{L_m}{X_m^{C_2}} \,. \tag{22}$$

Applying the experimental data of L_m and X_m into the above expression, (22) is plotted vs. p^+ in Fig.7. The data of a_{uv}/a_{uv0} is correlated by

$$a_{uv}/a_{uv0} = \left(24 \, p^+\right)^2 + \left(40 \, p^+\right) + 1 \tag{23}$$

by weighting all the experimental data. With the correlation (21) and (23), the model can be applied into a wide range of flows with pressure gradients.

CLOSURE MODEL PERFORMANCE

The model just described has been incorporated into the EURANUS code. The general features of EURANUS are described in Hirsch [14]. The model is tested on three test cases of flat plate for ETMA projec. For the description, mesh and boundary conditions of each test case, one can refer to Hirsch and Shang [15].

Klebanoff zero pressure gradient flow (TC3-1)

For zero pressure gradient flow, the model performance is very good. The numerical results agree quite well with Klebanoff's experimental data [16] for velocity distribution, see Fig.8. And the wall friction coefficient distribution matches with Nikuradse correlation result, as shown in Fig.9. This also proves the correct selection on the constants involved in the model.

Samuel and Joubert adverse pressure gradient flow (TC3-2)

Samuel and Joubert flow is a moderate adverse pressure gradient flow. Included in Fig.10 is the skin friction distribution. Except the leading edge of the plate, the good agreement between the model result and the experimental data could be achieved for the majority part of the plate. During the model development mentioned above, one may notice that the near wall region parameter a_{uv} plays an important role. It determines the effects of turbulence on the near wall region flow. The correlation for the near wall region parameter a_{uv} is proposed by weighting all the experimental data, and is very close to Samuel & Joubert data, see Fig.7. The accurate prediction on a_{uv} gives the correct values on the wall friction coefficient.

Fig. 11 to 16 present the velocity distribution and the Reynolds stress distribution are shown in Fig.17 to 21. At the computational positions x' = 0.12 and 0.72m, where the pressure gradients are so small, the numerical results agree quite well with the experimental data, plotted in Fig.11 and 12. The Reynolds stresses in this region, see Fig.17 and 18 for x' = 0.4m and x' = 0.75m, respectively, are correctly predicted. Starting from x' = 1.22m, where the pressure gradient is strong, the small underprediction in the outer region appears as shown in Fig.13. In the even stronger pressure gradient region, e.g., at the positions x' = 1.52, 1.72 and 2.0m, the model still shows reasonable results, the slight discrepancy between the numerical results and the experiment still exists in the outer region as demonstrated in Fig.14 to 16. This difference is due to the overprediction on X_m. This in turn results in a somewhat higher value of the turbulent eddy viscosity, or the larger value of the Reynolds stress in the outer region, see Fig.19 to 21.

Spalart sinkflow (TC3-3)

Saplart sinkflow is a very strong favourable pressure gradient flow. Since the flow is self similar, the comparison is being made at the middle of the flat plate. The velocity profile is plotted in Fig.22. The agreement between the model result and the DNS data is very good. For the Reynolds stress, the new model demonstrate a nice performance for both the inner region and outer regions, see Fig.23. This is due to the correlation for the two parameters X_m and a_{uv} matching well with DNS data of Spalart sinkflow.

CONCLUSIONS

A new algebraic turbulence model, which is independent of y^+, is proposed. This model takes into account the correct near wall behaviour of turbulent shear flows. It is a one formula model covering both inner and outer regions. For the theoretical prediction of turbulent eddy viscosity and Reynolds stress for zero pressure gradient, the new model agrees with Townsend, Klebanoff and Schubauer experimental data better than the Cebeci-Smith and Baldwin-Lomax models. The evaluation on the experimental data and DNS data shows a_{uv}/a_{uv0} and X_m/X_{m0} are unique functions of local pressure gradient parameter p^+, i.e. with increasing of p^+, a_{uv} increase and X_m/X_{m0} decrease. Based on these properties, they are correlated with p^+. With these correlation, the model can reasonably predict the turbulent eddy viscosity for the flows with favourable, zero and moderate adverse pressure gradients. To make the model work more generally, further model validation tests are still needed on more complex turbulent flows.

REFERENCES

[1] Cebeci,T., and Smith, A.M.O., 1974, "Analysis of Turbulent Boundary Layers," Academic Press.

[2] Baldwin, B.S. and Lomax, H., 1978, "Thin Layer Approximation and Algebraic Model for Separated Turbulent Flows," AIAA-Paper 78-257.

[3] Hinze, J.O., 1959, "Turbulence", McGraw-Hill Book Company.

[4] Abid, R., 1988, "Extension of the Johnson-King Turbulence Model to the 3-D Flows", AIAA paper, 88-0223.

[5] Kavsaoglu, M.S., et al, 1991, "Three-Dimensional Application of The Jonson-King Turbulence model for a Boundary-Layer Direct Method", Computers & Fluids, Vol.19, No.3/4, pp363-376.

[6] So., R. M. C., Zhang, H. S., and Speziale, C.G., 1991, "Near Wall Modelling of the Dissipation Rate Equation", AIAA Journal, Dec., pp2069-2076.

[7] Antonia, Bisset and Kim, 1991, "An Eddy Viscosity Calculation Method for a Turbulent Duct Flow", J. of Fluid Engineering, Vol.113, pp616-619.

[8] Samuel, A.E. and Joubert, P.N., 1974, "A Boundary Layer Development in an Increasing Adverse Pressure Gradient", J. of Fluid Mechanics, Vol.66, pp481-505.

[9] Simpson, R. L., Chew, Y.T, Shivaprasad, B.G., 1981, "The Structure of a Separating Turbulent Boundary Layer", J. of Fluid Mechanics, Vol.113, pp23-51.

[10] Stanford Database 80-81.

[11] Patel, V.C., 1974, "Measurements in the Thick Axisymmetric Turbulent Boundary Layer Near the tail of Body of Revolution", J. of Fluid Mechanics, Vol.63, pp345-367.

[12] Spalart, P.R., 1986, "Numerical Study of Sinkflow Boundary Layer", J. of Fluid Mechanics, Vol.172, pp307-328.

[13] Haase, W., et al, 1992, "EUROVAL– A European Initiative on Validation of CFD Codes", Notes on Numerical Fluid Mechanics, Vol. 42, Vieweg, April.

[14] Hirsch, Ch. and Lacor, C., Dener, C. and Vucinic, D., 1992, "An Integrated CFD System for 3D Turbomachinery Applications", AGARD-CP-510.

[15] Hirsch, Ch. and Shang Erbing, 1994, "Specification of Test Case TC3", ETMA Report VUB-7, ETMA database.

[16] Klebanoff., P.S., 1954, "Characteristics of Turbulence in a boundary Layer with Zero Pressure Gradient", NACA Technical Note 3178.

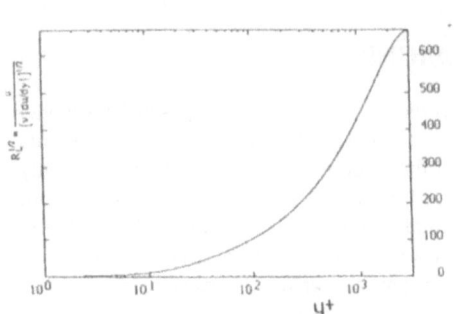

Figure 1: Nondimensional parameter $Rl^{**}(1/2)$ v.s. $y+$

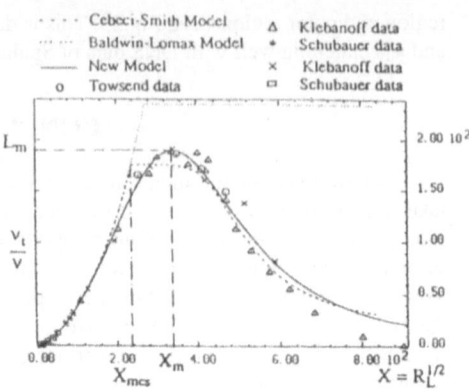

Figure 4: Nondimensional turbulent eddy viscosity distribution

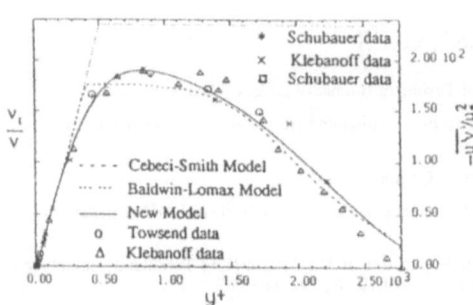

Figure 2: Definition for maximum value of eddy viscosity and its position

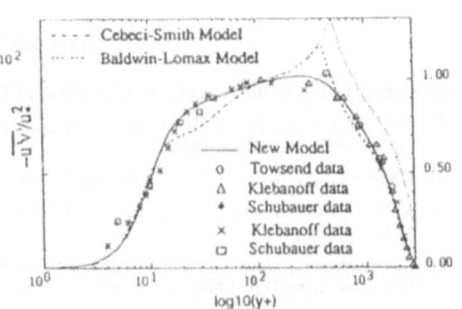

Figure 5: Nondimensional Reynolds stress distribution in log-scale

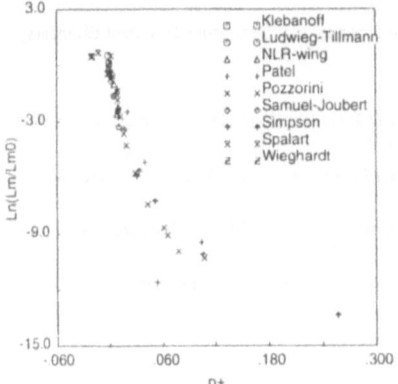

Figure 3: Ratio of maximum value of eddy viscosity $Lm/Lm0$ v.s. $p+$

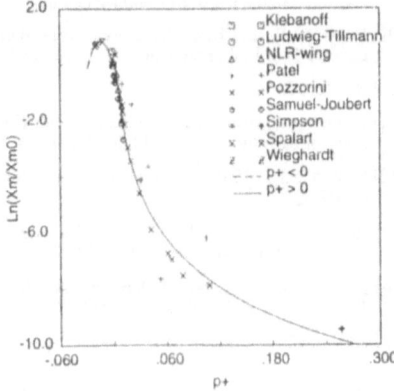

Figure 6: Ratio of maximum eddy viscosity position $Xm/Xm0$ v.s. $p+$

186

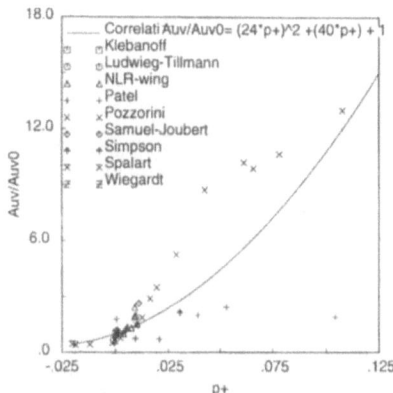

Figure 7: Ratio of near wall turbulence parameter v.s. p+

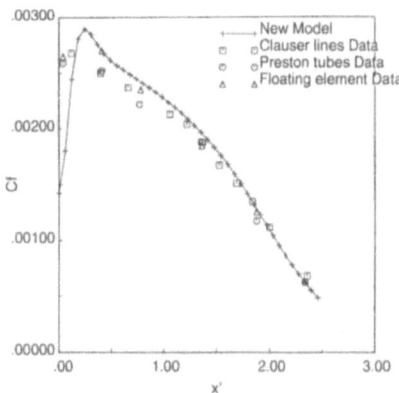

Figure 10: Wall friction coefficient distribution for Samuel-Joubert flow (TC3-2)

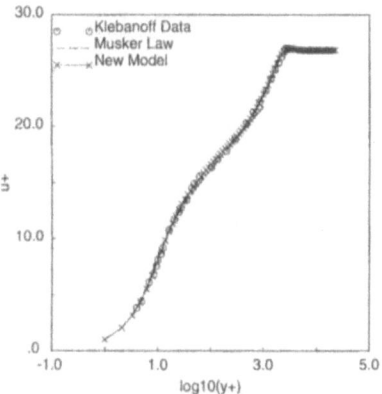

Figure 8: Velocity profile of Klebanoff zero pressure gradient flow (TC3-1)

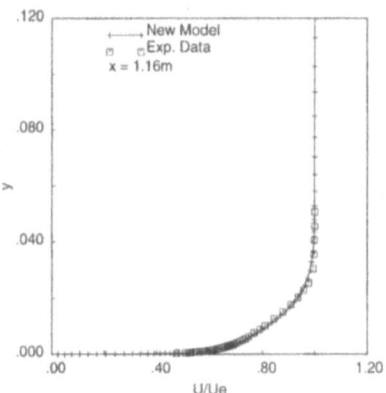

Figure 11: Velocity profile of Samuel-Joubert flow at x = 1.16m (TC3-2)

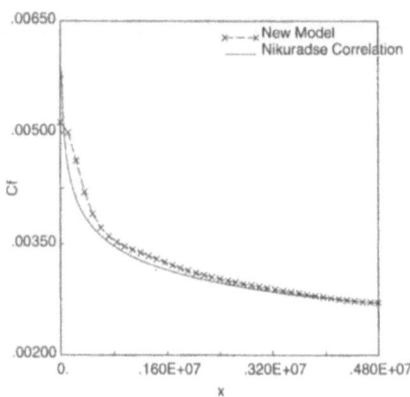

Figure 9: Wall friction coefficient distribution for zero pressure gradient flow (TC3-1)

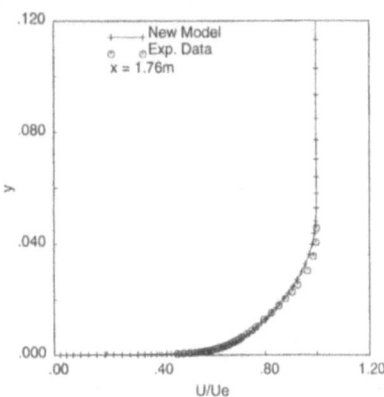

Figure 12: Velocity profile of Samuel-Joubert flow at x = 1.76m (TC3-2)

187

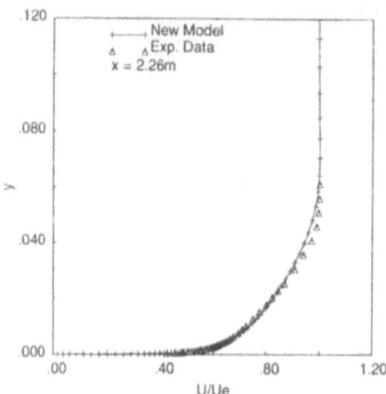

Figure 13: Velocity profile of Samuel-Joubert
flow at x = 2.26m (TC3-2)

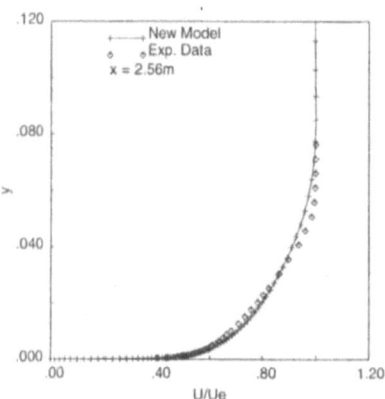

Figure 14: Velocity profile of Samuel-Joubert
flow at x = 2.56m (TC3-2)

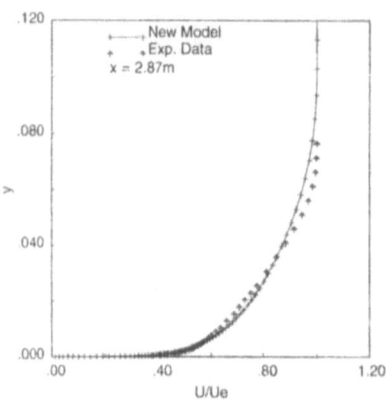

Figure 15: Velocity profile of Samuel-Joubert
flow at x = 2.87m (TC3-2)

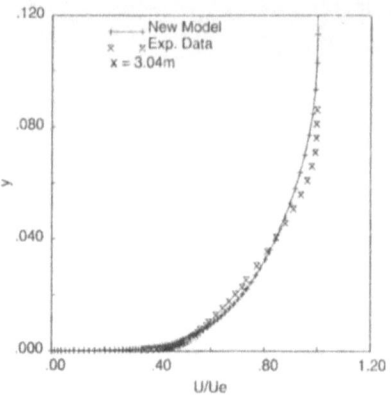

Figure 16: Velocity profile of Samuel-Joubert
flow at x = 3.04m (TC3-2)

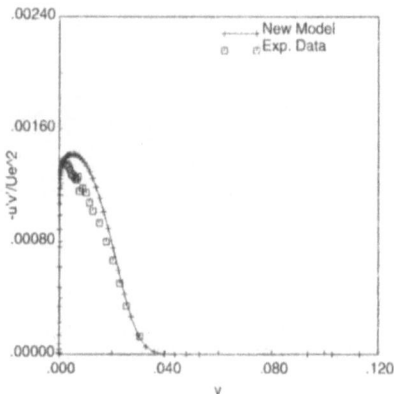

Figure 17: Reynolds stress distribution of
Samuel-Joubert flow at x = 1.14m (TC3-2)

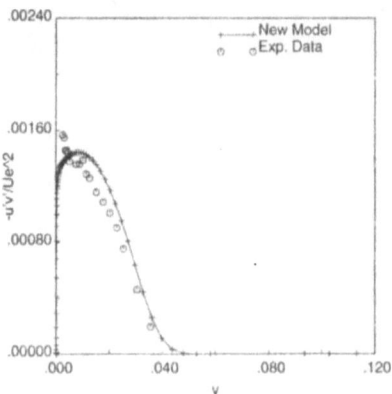

Figure 18: Reynolds stress distribution of
Samuel-Joubert flow at x = 1.79m (TC3-2)

188

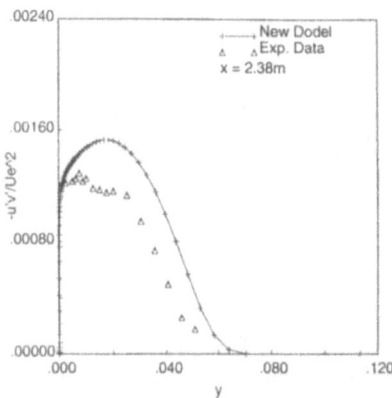

Figure 19: Reynolds stress distribution of Samuel-Joubert flow at x = 2.38m (TC3-2)

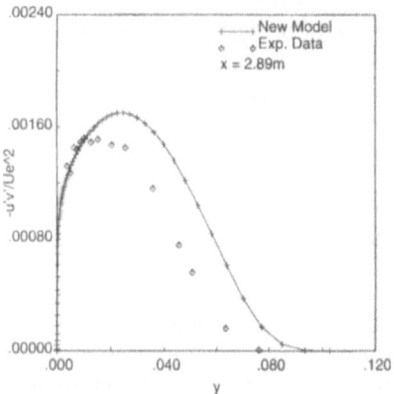

Figure 20: Reynolds stress distribution of Samuel-Joubert flow at x = 2.89m (TC3-2)

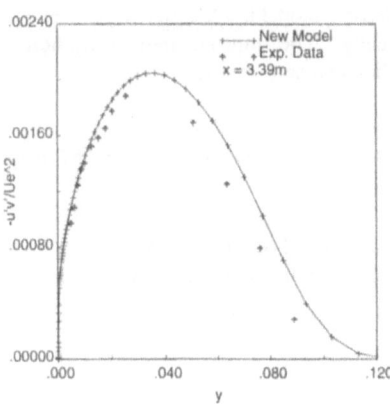

Figure 21: Reynolds stress distribution of Samuel-Joubert flow at x = 3.39m (TC3-2)

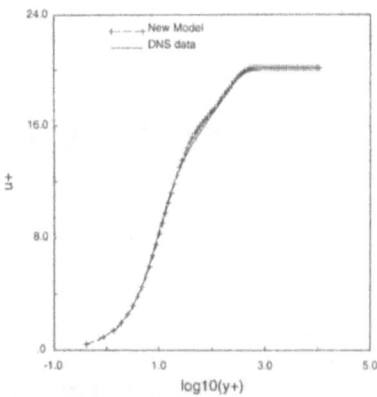

Figure 22: Velocity profile for Spalart sinkflow (TC3-3)

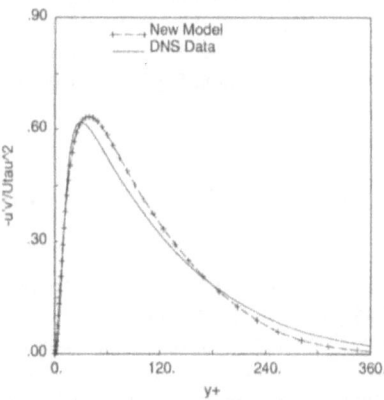

Figure 23: Reynolds stress distribution for Spalart sinkflow (TC3-3)

189

FLAT PLATE BOUNDARY LAYERS

Adil A. Dafa'Alla, Robert D. Harper & Michael M. Gibson
Mechanical Engineering Department, Imperial College, London SW7 2BX

SUMMARY

A new low-Reynolds number, two-equation turbulence model based on the variables $q \equiv \sqrt{k}$ and $\zeta \equiv \tilde{\varepsilon}/2q$ is described, this model and the low-Re model of Launder & Sharma are used to compute three different flat plate boundary layer flows within TC3. The computations are compared with experimental data from the zero pressure gradient boundary layer of Klebanoff [5], the adverse pressure gradient boundary layers of Samuel & Joubert, [6] and also for the favourable pressure gradient boundary layer, with the DNS data of Spalart, [8] and the experimental data of Jones & Launder, [7]. The q-ζ model is seen to perform slightly better than the k-ε model and also exhibits considerable savings in computing time for the flows presented.

INTRODUCTION

The k-ε model of turbulence suffers from the grave defect that its constituent equations are ill-adapted to integration to the wall. Consequently its application is effectively restricted to high Reynolds number turbulence outside the viscous layers, which have to be bridged by wall functions. The lack of a natural wall boundary condition for the ε equation presents problems which may only be surmounted by the introduction of terms which require an excessively fine grid for their resolution. These difficulties are avoided by changing to variables which are zero at the wall and vary linearly with distance close to it. The q-ζ model is a k-ε model replacement designed primarily to improve computational efficiency. The improvement in predictive accuracy shown in the present results is an unexpected but welcome bonus.

This report documents the results of several boundary layer calculations using the new q-ζ turbulence model and compares them with k-ε model predictions and experimental or DNS data, where applicable.

Calculations have been performed for comparison with the hot-wire measurements in a constant-pressure boundary layer [4, 5], TC3-1, with the hot-wire measurements in a decelerated boundary layer [6], TC3-2 and also with TC3-3, accelerated sink-flow boundary layer measurements [7] and DNS [8].

THE q-ζ TURBULENCE MODEL

In the first of the equations of the new model, the turbulent kinetic energy, k, is replaced by its square root, $q \equiv \sqrt{k}$, whose equation is readily obtained from the k-equation via:

$$\frac{Dq}{Dt} = \frac{1}{2k^{1/2}} \frac{Dk}{Dt} \; .$$

The result of this transformation is the transport equation:

$$U_i \frac{\partial q}{\partial x_i} = \frac{\partial}{\partial x_j} \left\{ \left(\nu + \frac{\nu_t}{\sigma_q} \right) \frac{\partial q}{\partial x_j} \right\} + \frac{\Pi}{2q} + \frac{P}{2q} - \frac{\varepsilon}{2q} + \frac{\nu}{q} \left(\frac{\partial q}{\partial x_j} \right) \left(\frac{\partial q}{\partial x_j} \right) .$$

P and ε are the production and dissipation rates of k, and Π is the pressure-diffusion term. The sum of the last two terms is the dissipation rate of q, which we designate by ζ. Then, letting $Q \equiv P/2q$ denote the production rate of q, the q-equation is written as:

$$U_i \frac{\partial q}{\partial x_i} = \frac{\partial}{\partial x_j} \left\{ \left(\nu + \frac{\nu_t}{\sigma_q} \right) \frac{\partial q}{\partial x_j} \right\} + \frac{\Pi}{2q} + Q - \zeta \; .$$

The second equation, for ζ, is obtained by applying the following transformation to the equations of the 'standard' k-ε model:

$$\frac{D\zeta}{Dt} = \frac{1}{2k^{1/2}} \frac{D\varepsilon}{Dt} - \frac{\varepsilon}{4k^{3/2}} \frac{Dk}{Dt} \; .$$

The result is:

$$U_i \frac{\partial \zeta}{\partial x_i} = \frac{\partial}{\partial x_j} \left\{ \left(\nu + \frac{\nu_t}{\sigma \zeta} \right) \frac{\partial \zeta}{\partial x_j} \right\} + \frac{\zeta}{q} \left(C_{\zeta 1} f_{\zeta 1} Q - C_{\zeta 2} f_{\zeta 2} \zeta \right) + \psi$$

where the C_ζ and f_ζ are constants and low-Reynolds-number damping functions related to the equivalent quantities in the parent ε equation, and ψ stands for possible additional source terms which some investigators have found to be necessary to account for viscous effects in near-wall turbulence. Additionally:

$$Q \equiv \gamma_t \frac{\partial U_i}{\partial x_j} \left(\frac{\partial U_i}{\partial x_j} + \frac{\partial U_j}{\partial x_i} \right), \qquad \gamma_t \equiv \frac{C_\mu f_\mu}{4} \frac{q^2}{\zeta}$$

and the eddy viscosity is,

$$\nu_t \equiv C_\mu f_\mu \frac{q^3}{2\zeta} \; .$$

Both q and ζ vary linearly in y as $y \rightarrow 0$ and the wall boundary conditions are $q = 0$, $\zeta = 0$.

Functions and Constants

The q-ζ model uses established constants and functions from the k-ε model to which they are related by:

$$C_\zeta f_\zeta \equiv 2C_\varepsilon f_\varepsilon - 1 \; .$$

For the present calculations the constants correspond to those of the 'standard' high-Re k-ε model $(C_{\varepsilon 1}, C_{\varepsilon 2}) = (1.44, 1.92)$ from which $(C_{\zeta 1}, C_{\zeta 2}) = (1.88, 2.84)$ and $\sigma_\zeta = \sigma_\varepsilon = 1.3$. The viscous-layer damping functions are taken from Launder and Sharma [1]:

$$f_\mu = \exp\left\{\frac{-A_\mu}{\left(1 + Re_t/50\right)^2}\right\}, \qquad f_{\varepsilon 1} = 1.0, \qquad f_{\varepsilon 2} = 1 - 0.3\exp\left(-Re_t^2\right)$$

where Re_t is the turbulence Reynolds number $q^3/2\nu\zeta$. We set $A_\mu = 6.0$ instead of 3.4, as in Launder and Sharma, so as to improve agreement with direct simulations of channel flow. The secondary mean-flow production term is also borrowed from the Launder and Sharma model as:

$$\psi = 2\nu\gamma_t\left(\frac{\partial^2 U_i}{\partial x_j \partial x_k}\right)\left(\frac{\partial^2 U_i}{\partial x_j \partial x_k}\right)$$

and, following time honoured practice, we ignore the pressure diffusion $\Pi/2q$ for the time being.

Note that although model constants and damping functions are required, these need not necessarily be those of Launder and Sharma which are used only because (a) the local Reynolds number Re_t is a better parameter for general flow calculations than the distance from the wall used in the generality of such models and (b) the Launder-Sharma model is sufficiently well-established to form a suitable benchmark for comparison. A limited number of calculations with alternative constants and functions have been performed in order to calibrate the model [2, 3] these results will not be presented here.

COMPUTATIONAL DETAILS

All calculations were performed using a code developed by Dr B.A. Younis which is based on the Patankar-Spalding GENMIX code. The code utilises a 2D parabolic solver with space marching in the streamwise direction, the pressure gradient is defined explicitly and the cross-stream velocity is obtained via continuity. At each step the computational domain expands or contracts in the wall-normal direction in order to accommodate the flow width.

The grids that were used were not those specified in the test case description, although grid independent results were obtained for each case. Approximately 100 non-uniformly distributed nodes were used across each boundary layer with the first point at no more than $y^+ = 1$. The inlet profiles prescribed in the test case specification were used throughout. Zero gradient boundary conditions were applied at the boundary layer edge for velocity, however, it was necessary to apply an arbitrary free stream turbulence level of approximately 1% at the edge of the boundary layer for the turbulence quantities.

RESULTS

TC3-1: Zero Pressure Gradient Boundary Layer

Figures 1a and 1b show the calculated distributions of mean velocity (in wall co-ordinates) and the variation of the local skin-friction coefficient compared with the data of Klebanoff [5] from a uniform-pressure boundary layer respectively. The results from the two low-Re models are indistinguishable for all practical purposes. But the q-ζ model results were obtained in nearly half the time [2].

TC3-2: Adverse Pressure Gradient Boundary Layer

Figures 2a-2d show results for the decelerated boundary layer of Samuel and Joubert [6]. the initial conditions supplied for this calculation were not entirely satisfactory, as is indicated by the slight inconsistency in the initial values of the local skin-friction coefficient c_f (figure 2a). The decline in c_f in the approach to separation, which is usually difficult to predict accurately, is here somewhat better predicted with the q-ζ model, possibly because this model is associated with smaller length scales which two-equation models in general tend to overpredict in such circumstances. Results for the velocity profiles at x between 1.16m and 3.39m are shown for the q-ζ model only. Initially the agreement with the data is good. However, as dp/dx increases downstream the agreement deteriorates, especially in the outer region. The measured Reynolds-stress profiles (figures 2c, 2d) are poorly predicted by both models, the q-ζ results again being marginally superior.

TC3-3: Favourable Pressure Gradient Boundary Layer

Calculations of a sink-flow boundary layer are compared in figures 3a and 3b with the measurements of Jones and Launder [7] and the DNS of Spalart [8] for the same conditions. The mean-velocity profiles show reasonable agreement, as do the cross-stream distributions of the Reynolds stress in three instances, only in the experimental data is the high near-wall peak in $-\overline{uv}$ not recovered. In both cases the q-ζ results are slightly superior.

COMPARISON OF THE q-ζ AND k-ε MODELS

Some of the test cases discriminate between the two low-Reynolds-number models to a surprising degree, see also TC2 (incompressible recirculating flows) within these proceedings [9, 10]. At first sight it might be supposed that the models should give identical results, derived as they are from the same source and employing in this case the same damping functions. In fact the models differ by the omission of certain terms which arise in the $\varepsilon \rightarrow \zeta$ transformation.

When the result is written out in full it will be found to be (in high Reynolds-number form for simplicity):

$$\frac{D\zeta}{Dt} = \frac{\partial}{\partial x_j}\left(\frac{v_t}{\sigma_\varepsilon}\frac{\partial \zeta}{\partial x_j}\right) + \left(C_{\varepsilon 1} - \frac{1}{2}\right)\frac{\zeta}{k}P - \left(C_{\varepsilon 2} - \frac{1}{2}\right)\frac{\zeta}{k}\varepsilon + \frac{1}{2}\left(\frac{1}{\sigma_k} + \frac{1}{\sigma_\varepsilon}\right)\frac{v_t}{k}\frac{\partial k}{\partial x_j}\frac{\partial \zeta}{\partial x_j}$$

$$+ \left(\frac{1}{\sigma_\varepsilon} - \frac{3}{2\sigma_k}\right)\frac{\zeta}{k^2}v_t\frac{\partial k}{\partial x_j}\frac{\partial k}{\partial x_j} - \frac{1}{2}\left(\frac{1}{\sigma_k} - \frac{1}{\sigma_\varepsilon}\right)\frac{\zeta}{k}v_t\frac{\partial^2 k}{\partial x_j \partial x_j} \ .$$

If all of this were to be programmed the results would be identical to those of the k-ε model - for this is only the ε-equation in another guise. But for the present calculations only the first three terms on the right-hand side have been retained: diffusion, production and dissipation, and therein lies the difference. All three of the discarded terms are negligibly small in the logarithmic layer and it may be argued that the only one which is non-negligible in the sublayers is the cross-diffusion term:

$$\frac{1}{2}\left(\frac{1}{\sigma_k} + \frac{1}{\sigma_\varepsilon}\right)\frac{v_t}{k}\frac{\partial k}{\partial x_j}\frac{\partial \zeta}{\partial x_j} \ .$$

In an interesting discussion of this point, Wilcox [11] ascribes to a similar term the differences which arise in k-ε and k-ω modelling. Viewed in another light, the difference is that simple gradient diffusion of ε and ζ is assumed in their respective equations. The assumption is equally justifiable in each case but introduces slightly different physics.

CONCLUSIONS

These calculations demonstrate the feasibility of general turbulent-flow calculations with a new two-equation model related to, but not the same as, the k-ε model of turbulence. The chief merit of the new model is that, unlike the equations for k and ε, the q and ζ equations are well conditioned near the wall, where the boundary conditions are specifically defined, and the dependent variables vary linearly with distance from the wall when the wall is close. The equations prove to be less stiff numerically, while linear variations in the sublayer permit the use of a coarser grid with consequent savings in computer time. The economy achieved is substantial in the flows considered here - up to 50% of the low-Re k-ε CPU time.

The k-ε and q-ζ equations do not give exactly the same results when integrated through the sublayers to the wall because the ε and ζ equations are not exact transforms of each other. Gradient diffusion of ζ is not exactly the same process as gradient diffusion of ε. Thus the results from the new form are not exactly the same as those of the k-ε model from which it was derived.

The q-ζ model allows one of the principal problems of current CFD research to be approached from a new angle, the potential of the established models being almost exhausted. It is regarded as a useful vehicle for further development, not least in the possibility of using the length-scale-determining ζ-equation in higher-order closures. But it is likely that the model is some way from the optimum form. The constants and damping functions chosen for these demonstration calculations have known limitations in k-ε modelling, and the search for improvements has become a major research activity, thus far largely fruitless. This is a research area demanding further intensive study, extended well beyond the scope of fitting curves to the results of channel flow simulations. In continuing work on the model we plan to explore the possibility of inertial damping in the outer sublayers $y^+ > 5$ say, and that of recasting the source terms of the ζ-equation in more physically realistic and robust form.

REFERENCES

1. Launder, B.E. & Sharma, B.I. 1974 Application of the energy-dissipation model of turbulence to the calculation of the flow near a spinning disk, Letters in Heat and Mass Transfer 1, 131-138.
2. Gibson, M.M. & Dafa'Alla, A.A. A two-equation model for turbulent wall flow, AIAA J (to appear).
3. Gibson, M.M. & Dafa'Alla, A.A. 1994 The q-ζ model for turbulent wall flow, Fluid Dynamics Division, American Physical Society, 47th Annual Meeting, Atlanta, Georgia.
4. Gibson, M.M., Verriopoulos, C.A. and Vlachos, N.S. 1984 Turbulent boundary layer on a mildly curved convex surface: 1. Mean flow and turbulence measurements, Expts. Fluids 2, 17-24.
5. Klebanoff, P.S. 1954 Characteristics of turbulence in a boundary layer with zero pressure gradient, NACA TN 3178.
6. Samuel, A.E. & Joubert, P.N. 1974 A boundary layer developing in an increasing adverse pressure gradient, J. Fluid Mech. 66, 481-493.
7. Jones, W.P. & Launder, B.E. 1972 Some properties of sink-flow turbulent boundary layers, J. Fluid Mech. 56, 337-351.
8. Spalart, P.R. 1986 Numerical study of sink-flow boundary layers, J. Fluid Mech. 172, 307-328.
9. Harper, R.D. & Gibson, M.M. 1995 ETMA Workshop, TC2: Incompressible recirculating flows. TC2-A Low-Re Backward-Facing Step, TC2-B High-Re Backward-Facing Step.
10. Harper, R.D. & Gibson, M.M. 1995 ETMA Workshop, TC2: Incompressible recirculating flows. TC2-C Fence-on-a-Wall, TC2-D Obstacle-in-Channel.
11. Wilcox, D.C. 1993 Turbulence Modeling for CFD, DCW Industries Inc., La Canada, California.

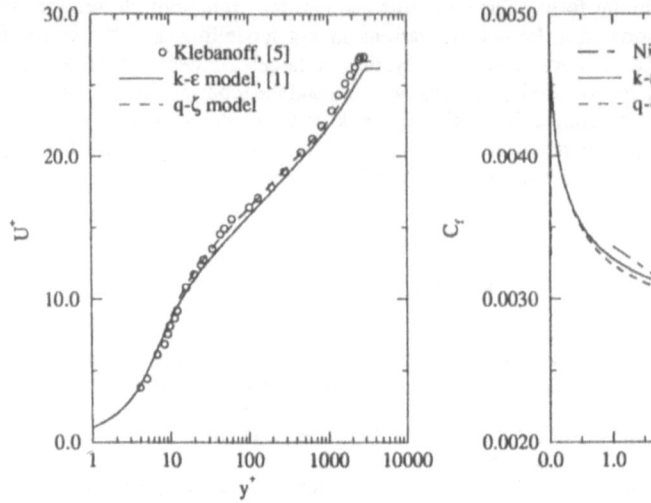

Figure 1a. Mean velocity in a flat plate boundary layer at $Re_x=4.2\times10^6$.

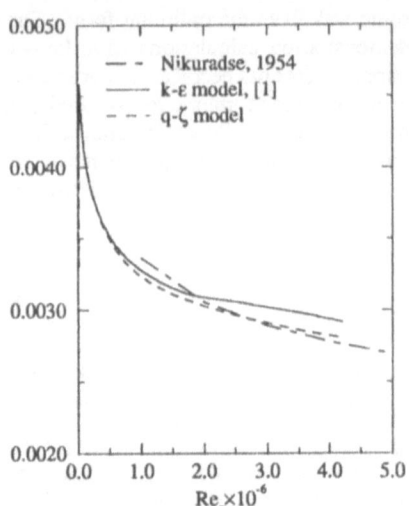

Figure 1b. Skin friction coefficient in a flat plate boundary layer.

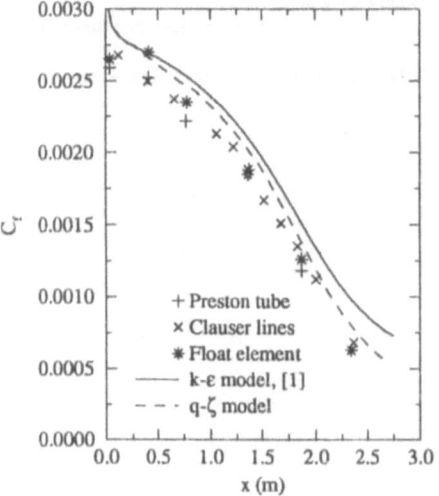

Figure 2a. Skin friction coefficient in adverse pressure gradient boundary layer.

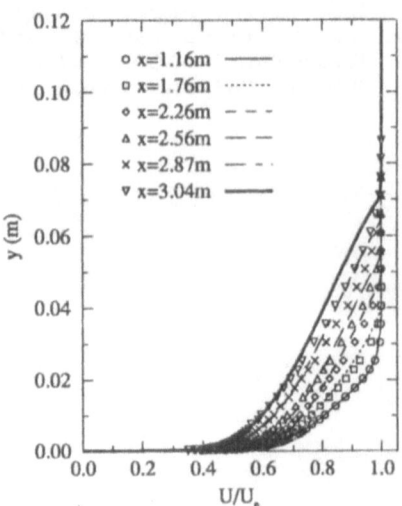

Figure 2b. q-ζ prediction of mean velocity in an adverse pressure gradient boundary layer.

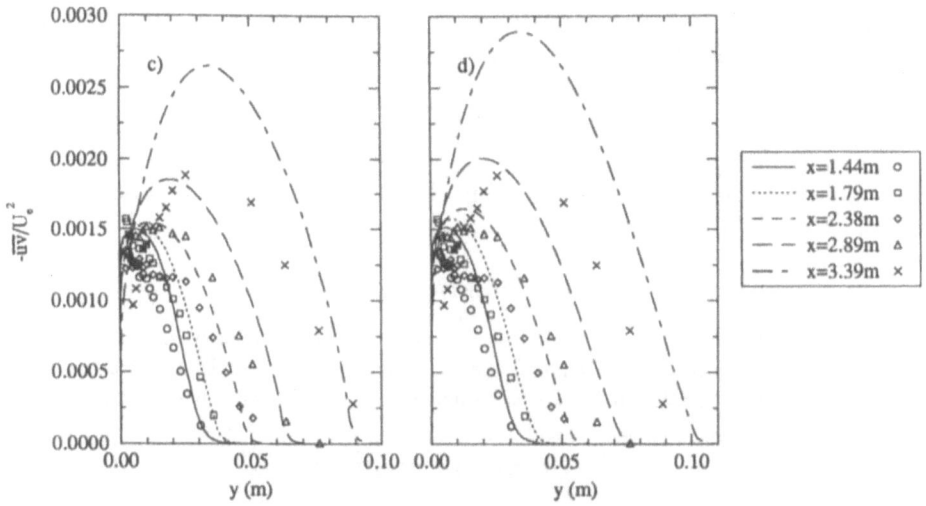

Figures 2c & 2d. q-ζ and k-ε predictions respectively. Shear stress in an
adverse pressure gradient boundary layer.

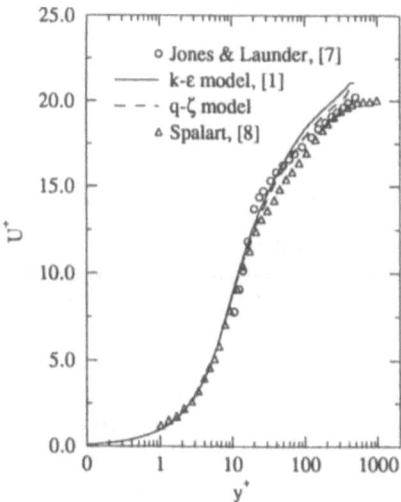

Figure 3a. Mean velocity in a sink flow
boundary layer at R_θ=680,
K=1.5×10^{-6}.

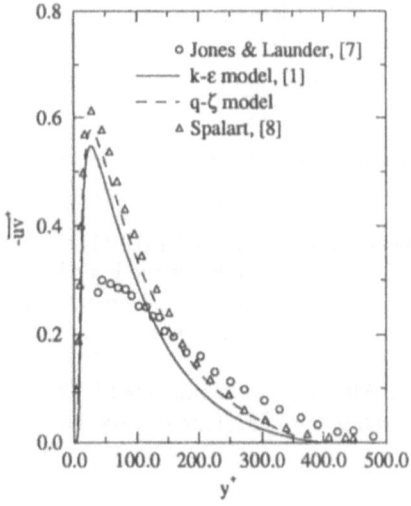

Figure 3b. Turbulence shear stress in a sink
flow boundary layer at R_θ=680
K=1.5×10^{-6}.

Application of Turbulence Models to Incompressible Boundary Layers in Aeronautics

R.A.W.M. Henkes

Faculty of Aerospace Engineering, Delft University of Technology,
P.O. Box 5058, 2600 GB Delft, The Netherlands

SUMMARY

Different turbulence models are applied to the incompressible boundary layer along a flat plate, with a zero, favourable or adverse pressure gradient (test case TC3). The algebraic model of Cebeci & Smith and the $k - \omega$ models give the best overall performance.

INTRODUCTION

This paper summarizes our contribution to test case TC3 of the ETMA-workshop. Nine different models (including an algebraic model, $k - \epsilon$ models, $k - \omega$ models, and a Differential Stress Model) were taken from the literature and applied to the test case. No trial was made to modify and optimize these models.

As the Reynolds number in test case TC3 is high, and as no flow reversal occurs, the Reynolds-averaged Navier-Stokes equations can be simplified to the turbulent boundary-layer equations:

$$\frac{\partial u}{\partial x} + \frac{\partial v}{\partial y} = 0 \tag{1}$$

$$u\frac{\partial u}{\partial x} + v\frac{\partial u}{\partial y} = U_e\frac{dU_e}{dx} + \nu\frac{\partial^2 u}{\partial y^2} - \frac{\partial}{\partial y}\overline{u'v'} \,. \tag{2}$$

The Reynolds stress $\overline{u'v'}$ is modelled by different turbulence models: the algebraic, $k - \epsilon$ and $k - \omega$ models apply the turbulent-viscosity approach, $i.e.$ $-\overline{u'v'} = \nu_t\frac{\partial u}{\partial y}$, whereas the Differential Stress Model (DSM) solves a partial differential equation for $\overline{u'v'}$.

At the outer-edge of the boundary layer the Dirichlet condition $u = U_e$ is prescribed, whereas a homogeneous Neumann condition is applied for all turbulent quantities. The parabolic equations are numerically solved by a straightforward implicit, marching procedure.

All results to be presented in the figures and tables (except for table 1) were made on a 160×160 grid. This grid was checked to give almost grid-independent results.

TURBULENCE MODELS

Algebraic model of Cebeci & Smith (1974)

The turbulent viscosity is modelled as

$$\nu_t = \begin{cases} l^2 \left|\frac{\partial u}{\partial y}\right| \gamma & \text{if} \quad y < y_c \\ \alpha \, U_e\delta^*\gamma & \text{if} \quad y \geq y_c. \end{cases} \tag{3}$$

Here y_c is chosen such that ν_t is continuous in $y = y_c$. The following terms appear in (3):

$$l = \kappa y[1 - \exp(-y^+/A^+)]; \quad \kappa = 0.415; \quad y^+ = \frac{y u_\tau}{\nu}; \quad u_\tau = \sqrt{\nu \left(\frac{\partial u}{\partial y}\right)_w};$$

$$A^+ = \frac{26}{\sqrt{1 - 11.8 p^+}}; \quad p^+ = \frac{\nu U_e}{u_\tau^3} \frac{dU_e}{dx}; \quad \delta^* = \int_0^\infty \left(1 - \frac{u}{U_e}\right) dy;$$

$$\alpha = 0.0168 \frac{1.55}{1 + \Pi}; \quad \Pi = 0.55(1 - \exp(-0.243\sqrt{z_1} - 0.298 z_1));$$

$$z_1 = \frac{Re_\theta}{425} - 1, \quad \text{with} \ Re_\theta > 425; \quad \gamma = \left[1 + 5.5 \left(\frac{y}{y_0}\right)^6\right]^{-1};$$

y_0 is the y − position where $u/U_e = 0.995$.

Low-Reynolds-number $k - \epsilon$ models

Two differential equations are solved for the turbulent kinetic energy (k) and the dissipation rate of turbulent kinetic energy (ϵ), respectively. Applying boundary-layer simplifications, these equations read

$$u\frac{\partial k}{\partial x} + v\frac{\partial k}{\partial y} = \frac{\partial}{\partial y}\left(\nu + \frac{\nu_t}{\sigma_k}\right)\frac{\partial k}{\partial y} + P_k - \epsilon + D, \tag{4}$$

$$u\frac{\partial \epsilon}{\partial x} + v\frac{\partial \epsilon}{\partial y} = \frac{\partial}{\partial y}\left(\nu + \frac{\nu_t}{\sigma_\epsilon}\right)\frac{\partial \epsilon}{\partial y} + (c_{\epsilon 1} f_1 P_k - c_{\epsilon 2} f_2 \epsilon)\frac{\epsilon}{k} + E, \tag{5}$$

with

$$P_k = \nu_t \left(\frac{\partial u}{\partial y}\right)^2, \quad \nu_t = c_\mu f_\mu \frac{k^2}{\epsilon}.$$

Besides the high-Reynolds-number constants c_μ, σ_k, σ_ϵ, $c_{\epsilon 1}$ and $c_{\epsilon 2}$, the model contains the low-Reynolds-number functions f_μ, f_1, f_2, D and E. These functions may depend on y^+, $Re_t = k^2/\nu\epsilon$, or $Re_k = y\sqrt{k}/\nu$.

The following low-Reynolds-number $k - \epsilon$ models are considered:

(i) *Chien* (1980)
$c_\mu = 0.09$, $c_{\epsilon 1} = 1.35$, $c_{\epsilon 2} = 1.8$, $\sigma_k = 1.0$, $\sigma_\epsilon = 1.3$,
$f_\mu = 1 - \exp(-0.0115 y^+)$, $f_1 = 1.0$, $f_2 = 1 - \frac{2}{9}\exp(-(Re_t/6)^2)$,
$D = -2\nu \frac{k}{y^2}$, $E = -2\frac{\nu\epsilon}{y^2}\exp(-0.5 y^+)$ and $k = \epsilon = 0$ at the wall.

(ii) *Lam & Bremhorst* (1981)
$c_\mu = 0.09$, $c_{\epsilon 1} = 1.44$, $c_{\epsilon 2} = 1.92$, $\sigma_k = 1.0$, $\sigma_\epsilon = 1.3$,
$f_\mu = (1 - \exp(-0.0165 Re_k))^2(1 + 20.5/Re_t)$, $f_1 = 1 + \left(\frac{0.05}{f_\mu}\right)^3$, $f_2 = 1 - \exp(-Re_t^2)$,
$D = E = 0$, and $k = 0$, $\frac{\partial \epsilon}{\partial y} = 0$ at the wall.

(iii) *Launder & Sharma* (1974)
$c_\mu = 0.09$, $c_{\epsilon 1} = 1.44$, $c_{\epsilon 2} = 1.92$, $\sigma_k = 1.0$, $\sigma_\epsilon = 1.3$,

$$f_\mu = \exp\left(\frac{-3.4}{(1 + Re_t/50)^2}\right), \ f_1 = 1.0, \ f_2 = 1 - 0.3 \exp(-Re_t^2),$$

$$D = -2\nu\left(\frac{\partial\sqrt{k}}{\partial y}\right)^2, \ E = 2\nu\nu_t\left(\frac{\partial^2 u}{\partial y^2}\right)^2, \ \text{and } k = \epsilon = 0 \text{ at the wall.}$$

(iv) *Launder & Sharma & Yap*

Same as Launder & Sharma model, except for $E = 2\nu\nu_t\left(\frac{\partial^2 u}{\partial y^2}\right)^2 + E_Y$, where E_Y is Yap's

correction: $0.83\left(\frac{k^{3/2}}{2.5y\epsilon} - 1\right)\left(\frac{k^{3/2}}{2.5y\epsilon}\right)^2\frac{\epsilon^2}{k}$

(v) *Lien & Leschziner (1993)* [1]

$c_\mu = 0.09, \ c_{\epsilon 1} = 1.44, \ c_{\epsilon 2} = 1.92, \ \sigma_k = 1.0, \ \sigma_\epsilon = 1.3,$

$$f_\mu = \frac{1 - \exp(-0.016 Re_k)}{1 - \exp(-0.263 Re_k)}, \ f_1 = 1 + \frac{P_k'}{P_k}, \ P_k = \nu_t\left(\frac{\partial u}{\partial y}\right)^2,$$

$$P_k' = \frac{1.92[1 - 0.3\exp(-Re_t^2)]k^{3/2}}{3.53y[1 - \exp(-0.263y)]}\exp(-0.00222 Re_k^2), \ f_2 = 1 - 0.3 \exp(-Re_t^2)$$

$$D = 0, \ E = 0.83\left(\frac{k^{3/2}}{2.44y\epsilon} - 1\right)\left(\frac{k^{3/2}}{2.44y\epsilon}\right)^2\frac{\epsilon^2}{k} \ \text{(Yap's correction)},$$

and $k = 0, \ \epsilon = \lim\limits_{y\to 0} 1.56\dfrac{\nu k}{y^2}$ at the wall.

$k - \omega$ models

In this model the equation for the dissipation rate in the $k - \epsilon$ model is replaced by an equation for the reciprocal turbulent time scale ω. Considered is the low-Reynolds-number $k - \omega$ model of Wilcox (1993) [2]:

$$u\frac{\partial k}{\partial x} + v\frac{\partial k}{\partial y} = \frac{\partial}{\partial y}\left(\nu + \frac{\nu_t}{\sigma_k}\right)\frac{\partial k}{\partial y} + P_k - \epsilon, \tag{6}$$

$$u\frac{\partial\omega}{\partial x} + v\frac{\partial\omega}{\partial y} = \frac{\partial}{\partial y}\left(\nu + \frac{\nu_t}{\sigma_\omega}\right)\frac{\partial\omega}{\partial y} + \alpha\frac{\omega}{k}P_k - \beta\omega^2, \tag{7}$$

with

$$P_k = \nu_t\left(\frac{\partial u}{\partial y}\right)^2, \quad \nu_t = \alpha^*\frac{k}{\omega}, \quad \epsilon = \beta^* k\omega.$$

The low-Reynolds-number functions are defined by

$$\alpha^* = \frac{\alpha_0^* + Re_t/R_k}{1 + Re_t/R_k}, \quad \alpha_0^* = \beta/3, \quad Re_k = 6,$$

$$\alpha = \frac{5}{9}\frac{\alpha_0 + Re_t/Re_\omega}{1 + Re_t/R_\omega}\frac{1}{\alpha^*}, \quad \alpha_0 = 0.1, \quad Re_\omega = 2.7,$$

$$\beta^* = 0.09\frac{5/18 + (Re_t/Re_\beta)^4}{1 + (Re_t/Re_\beta)^4}, \quad Re_\beta = 8.$$

The other coefficients in the model are $\beta = 3/40, \ \sigma_k = \sigma_\omega = 2$. The turbulence-based Reynolds number Re_t is defined here as $k/\omega\nu$. The wall boundary conditions are set to:

$k = 0, \ \omega = \lim\limits_{y\to 0}\dfrac{6\nu}{\beta y^2}$.

200

This low-Reynolds-number model can be transformed into a high-Reynolds-number model by taking the limit $Re_t \to \infty$, which gives $\alpha^* = 1$, $\alpha = 5/9$, $\beta^* = 0.09$.

Differential Stress Model (DSM)

In this model differential equations for all Reynolds-stresses are solved. In addition a differential equation for ϵ is solved. The formulation of Hanjalić et al. (1992) [3] is followed here:

$$u\frac{\partial \overline{u_i'u_j'}}{\partial x} + v\frac{\partial \overline{u_i'u_j'}}{\partial y} = \frac{\partial}{\partial y}\nu\frac{\partial \overline{u_i'u_j'}}{\partial y} + D_{ij}^t + P_{ij} + \Phi_{ij} - \epsilon_{ij}, \tag{8}$$

$$u\frac{\partial \epsilon}{\partial x} + v\frac{\partial \epsilon}{\partial y} = \nu\frac{\partial^2 \epsilon}{\partial y^2} + D_\epsilon^t + c_{\epsilon 1}P_k\frac{\epsilon}{k} - c_{\epsilon 2}f_\epsilon\frac{\epsilon\tilde{\epsilon}}{k} + c_{\epsilon 3}\nu\frac{k}{\epsilon}\overline{v'^2}\left(\frac{\partial^2 u}{\partial y^2}\right)^2, \tag{9}$$

with

$$P_{ij} = -\left(\overline{u_i'u_k'}\frac{\partial u_j}{\partial x_k} + \overline{u_j'u_k'}\frac{\partial u_i}{\partial x_k}\right), \quad P_k = -\overline{u_i'u_k'}\frac{\partial u_i}{\partial x_k},$$

$$D_{ij}^t = \frac{\partial}{\partial x_k}c_s\frac{k}{\epsilon}\overline{u_k'u_l'}\frac{\partial \overline{u_i'u_j'}}{\partial x_l}, \quad D_\epsilon^t = \frac{\partial}{\partial x_k}c_\epsilon\frac{k}{\epsilon}\overline{u_k'u_l'}\frac{\partial \epsilon}{\partial x_l},$$

$$\epsilon_{ij} = f_s\epsilon_{ij}^* + (1 - f_s)\frac{2}{3}\delta_{ij}\epsilon, \quad \epsilon_{ij}^* = \frac{\epsilon}{k}\left[\frac{\overline{u_i'u_j'} + f_d(\overline{u_i'u_k'}n_jn_k + \overline{u_j'u_k'}n_in_k + \overline{u_k'u_l'}n_kn_ln_in_j)}{1 + \frac{3}{2}\frac{\overline{u_p'u_q'}}{k}n_pn_qf_d}\right],$$

$$\Phi_{ij} = \Phi_{ij,1} + \Phi_{ij,1}^w + \Phi_{ij,2} + \Phi_{ij,2}^w,$$

$$\Phi_{ij,1} = -c_1\epsilon a_{ij}, \quad \Phi_{ij,1}^w = c_1^w\frac{\epsilon}{k}f_w(\overline{u_k'u_m'}n_kn_m\delta_{ij} - \frac{3}{2}\overline{u_i'u_k'}n_kn_j - \frac{3}{2}\overline{u_k'u_j'}n_kn_i),$$

$$\Phi_{ij,2} = -c_2(P_{ij} - \frac{2}{3}P_k\delta_{ij}), \quad \Phi_{ij,2}^w = c_2^wf_w(\Phi_{km,2}n_kn_m\delta_{ij} - \frac{3}{2}\Phi_{ik,2}n_kn_j - \frac{3}{2}\Phi_{kj,2}n_kn_i),$$

$$A_2 = a_{ij}a_{ij}, \quad A_3 = a_{ij}a_{jk}a_{ki}, \quad A = 1 - \frac{9}{8}(A_2 - A_3), \quad a_{ij} = \frac{\overline{u_i'u_j'}}{k} - \frac{2}{3}\delta_{ij}.$$

The above expressions are further simplified by assuming that the boundary-layer approximations apply, i.e. $\left|\frac{\partial \phi}{\partial x}\right| << \left|\frac{\partial \phi}{\partial y}\right|$ and $\left|\frac{\partial v}{\partial y}\right| << \left|\frac{\partial u}{\partial y}\right|$. Further, $\underline{n}$ is the unit-vector normal to the wall: $n_2 = 1$, $n_1 = n_3 = 0$.

The high-Reynolds-number constants and the low-Reynolds-number functions (depending on $Re_t = k^2/\nu\epsilon$) in the model are given by: $c_s = 0.22$; $c = 2.58AA_2^{1/4}[1 - \exp(-(0.0067Re_t)^2)]$; $c_1 = 1 - f_s + c$; $c_2 = 0.75A^{1/2}$; $c_1^w = -\frac{2}{3}(1 + c) + 1.67$; $c_2^w = \max\left\{(\frac{2}{3}c_2 - \frac{1}{6})/c_2, 0\right\}$; $c_\epsilon = 0.18$; $c_{\epsilon 1} = 1.44$; $c_{\epsilon 2} = 1.92$; $c_{\epsilon 3} = 0.5$; $f_d = f_s = (1 + 0.1Re_t)^{-1}$; $f_\epsilon = 1 - \frac{0.52}{1.92}\left\{\exp\left[-\left(\frac{Re_t}{6}\right)^2\right]\right\}$, $f_w = \frac{k^{3/2}}{2.5\epsilon y}$, and $\tilde{\epsilon} = \epsilon - 2\nu\left(\frac{\partial\sqrt{k}}{\partial y}\right)^2$.

TEST CASE RESULTS

Test case TC3-1: zero-pressure gradient

Experiments for this configuration with an almost constant free-stream velocity $U_e = 15.25\ m/s$ were performed by Klebanoff (1954). Computations were made in the domain $x_{min} \leq x \leq x_{max}$, and $0 \leq y \leq y_{max}$, with $x_{min} = 0.1\ m$, $x_{max} = 4.9465\ m$, $y_{max} = 0.6183\ m$. The last x-station, x_{max}, corresponds to the Reynolds number $Re_x = U_e x/\nu = 4.8 \times 10^6$ (taking $\nu = 1.5705 \times 10^{-5}$).

The calculation for the u-velocity is started at x_{min} with the common wall-function and defect law for the fully turbulent boundary layer. Turbulence is initiated with the mixing length model, assuming local equilibrium of turbulent energy production and dissipation. A strongly nonuniform grid is used, concentrating grid points close to the wall and close to x_{min}. Increasing the number of grid points from 40×40 to 320×320 reveals a strong grid dependence close to x_{min}: this initial station is too low to sustain the initiated turbulent state, and the flow relaminarizes before undergoing a transition to the fully turbulent branch. Fortunately the turbulent solution fastly becomes almost grid-independent (see table 1).

A calculation was made on the 40×80 grid (uniform in x, nonuniform in y), as recommended in the test-case description. The solution turns out to relaminarize, and remains laminar up to, at least, $Re_x = 4.8 \times 10^6$. Doubling the number of x grid points, however, does give a transition, and the turbulent solution (giving $c_f = 0.00278$ at $Re_x = 4.8 \times 10^6$) is reasonably close to the above-mentioned almost grid-independent solution.

Figure 1a and table 2 compare the experimental wall-shear stress coefficient, $c_f = \dfrac{2\nu}{U_e^2}\left(\dfrac{\partial u}{\partial y}\right)_w$, with the predictions by the different turbulence models. All models are accurate within 8%. Also the velocity profiles (see u^+ vs y^+ in figure 1b) are predicted well; deviations only occur at the outer edge, which are due to small differences in the wall-shear stress.

Test case TC3-2: adverse-pressure gradient

Experiments for this case were performed by Samuel & Joubert (1974) and by Nagano et. al (1991) [4]. For the Samuel & Joubert case we took $x_{min} = 1.04\ m$, $x_{max} = 3.5\ m$ and $y_{max} = 0.2\ m$. The grid is uniform in x direction and nonuniform in y direction. The measured velocity is prescribed at the outer edge of the boundary layer. At x_{min} the measured u profile and turbulence profiles are prescribed. Similar computations were made for the Nagano case, but the initial velocity profile at $x_{min} = 0.1$ was guessed from the wall function and defect law, and the initial turbulence was computed from the mixing-length model assuming local equilibrium.

The predictions for the wall-shear stress are compared with the experiments in table 3 and in figures 2a and 3a. The wall-shear stress coefficient is defined as $c_f = \dfrac{2\nu}{U_{ref^2}}\left(\dfrac{\partial u}{\partial y}\right)_w$ in the Samuel & Joubert case (with $U_{ref} = 26.15\ m/s$), and as $c_f = \dfrac{2\nu}{U_e^2}\left(\dfrac{\partial u}{\partial y}\right)_w$ in the Nagano case. From table 3 (giving c_f at $x = 3.38\ m$ in the Samuel & Joubert case, and at $x = 1.12\ m$ in the Nagano case) it follows that the low-Reynolds-number $k - \omega$ model is closest to the experiments, followed by the algebraic model of Cebeci & Smith. All $k - \epsilon$ models considerably overpredict the wall-shear stress; the Lien & Leschziner model is slightly better than the others. Also the DSM overpredicts the wall-shear stress, in particular for the Nagano case.

Figure 3a, however, shows that the good accuracy of the algebraic model and of $k - \omega$ model does not hold for all measuring points in the Nagano case: the wall-shear stress is underpredicted by up to 19%.

Figure 2b and 3b show that all models give far too large turbulent shear stresses for both cases. The $k - \omega$ model is slightly closer to the experimental levels than the other models.

We also checked the influence of the initial profile in the Nagano case. Starting the calculations with the experimental velocity and turbulence profiles at $x = 0.523$ leaves the wall-shear stress results from figure 3a almost unchanged, whereas the influence on the turbulent shear stress is somewhat larger.

Test case TC3-3: favourable pressure gradient (sink flow)

Calculations were made at a constant reciprocal local-Reynolds number of $Re^{-1} = K = \nu/U_e X = 1.5 \times 10^{-6}$ (here $X = x_0 - x$, with x_0 the position of the sink). The calculations show that the influence of the initial profile on the scaled quantities decays with increasing x, and that a similar, $i.e.$ x–independent, solution is reached.

Comparison with the DNS of Spalart (1986) in table 3 shows that almost all models accurately predict the the wall-shear stress (c_f based on U_e) and Re_θ. The less accurate are the algebraic model and the $k - \epsilon$ model of Launder & Sharma. The accuracy of the maximum shear stress is consistent with the accuracy of the wall-shear stress: that is a too high/low c_f corresponds to a too high/low $\overline{-u'v'}_{max}$. The table also shows that in contrast to the adverse-pressure gradient case, the $k - \omega$ model is no longer the best choice for the favourable-pressure gradient case.

CONCLUSION

Among the models tested, the algebraic model of Cebeci & Smith and the high- or low-Reynolds-number $k - \omega$ models of Wilcox give the best overall performance for incompressible boundary layers (with zero, adverse or favourable pressure gradients). Although not discussed here, however, a disadvantage of the $k - \omega$ model is the large influence of the outer-edge boundary condition for ω. All $k - \epsilon$ models strongly overpredict the wall-shear stress in adverse pressure gradients. The $k - \epsilon$ model of Lien & Leschziner is slightly more accurate than the other $k - \epsilon$ models. Also, the tested Differential Stress Model does not perform well in adverse pressure gradients. It thus seems that the ϵ equation in both the $k - \epsilon$ models and in the DSM needs modifications.

REFERENCES

[1] Lien, F.S. & Leschziner, M.A. 1993 Modelling 2D and 3D separation from curved surfaces with variants of second-moment closure combined with low-Re near-wall formulations. Proc. 9th Symp. on Turbulent Shear Flows, paper 13-1.

[2] Wilcox, D.C. 1993 Turbulence modeling of CFD. DCW Industries Inc.

[3] Hanjalić, K., Jakirlić, Stošić, N., Vasić, S. & Hadžić, I. 1992 Collaborative testing of turbulence models 1990/1992. Report LSTM 352/T/92, Universität Erlangen-Nürnberg.

[4] Nagano, Y., Tagawa, M. & Tsuji, T. 1991 Effects of adverse pressure gradients on mean flows and turbulence statistics in a boundary layer. Proc. 8th Symp. on Turbulent Shear Flows, paper 2-3.

Table 1. Grid dependence for ZPG boundary layer; c_f at $Re_x = 4.8 \times 10^6$ with Launder & Sharma $k - \epsilon$ model.

grid	c_f
40×40	0.00287
80×80	0.00274
160×160	0.00270
320×320	0.00269

Table 2. Deviation from experiments in ZPG boundary layer; experimental: $c_f = 0.00274$ at $Re_x = 4.8 \times 10^6$.

model	% c_f
algebraic	-2
$k - \epsilon$, Chien	$+2$
$k - \epsilon$, Lam & Bremhorst	$+8$
$k - \epsilon$, Launder & Sharma	-2
$k - \epsilon$, Launder & Sharma & Yap	$+8$
$k - \epsilon$, Lien & Leschziner	$+2$
low-Re $k - \omega$	$+2$
high-Re $k - \omega$	$+2$
DSM	$+5$

Table 3. Deviation from experiments in APG boundary layer; $c_f = 0.00063$ at $x = 3.38 \ m$ in experiment of Samuel & Joubert, $c_f = 0.00174$ at $x = 1.12 \ m$ in experiment of Nagano et al..

model	% c_f S.& J.	Nagano
algebraic	-18	-10
$k - \epsilon$, Chien	$+58$	$+74$
$k - \epsilon$, Lam & Bremhorst	$+62$	$+73$
$k - \epsilon$, Launder & Sharma	$+48$	$+61$
$k - \epsilon$, Launder & Sharma & Yap	$+44$	$+56$
$k - \epsilon$, Lien & Leschziner	$+31$	$+40$
low-Re $k - \omega$	-1	-2
high-Re $k - \omega$	$+3$	$+13$
DSM	$+21$	$+51$

Table 4. Deviation from DNS for sink flow; DNS: $c_f = 0.00499$, $Re_\theta = 690$, $\dfrac{(-\overline{u'v'})_{\mathbf{max}}}{U_e^2} = 0.00155$

model	% c_f	% Re_θ	% $\dfrac{(-\overline{u'v'})_{\mathbf{max}}}{U_e^2}$
algebraic	$+13$	18	$+25$
$k - \epsilon$, Chien	-5	-4	-12
$k - \epsilon$, Lam & Bremhorst	-0	$+3$	$+4$
$k - \epsilon$, Launder & Sharma	-13	-13	-23
$k - \epsilon$, Launder & Sharma & Yap	$+2$	$+4$	$+5$
$k - \epsilon$, Lien & Leschziner	-1	$+1$	$+1$
low-Re $k - \omega$	$+8$	$+12$	$+17$
high-Re $k - \omega$	$+5$	$+9$	$+9$
DSM	-9	-9	-19

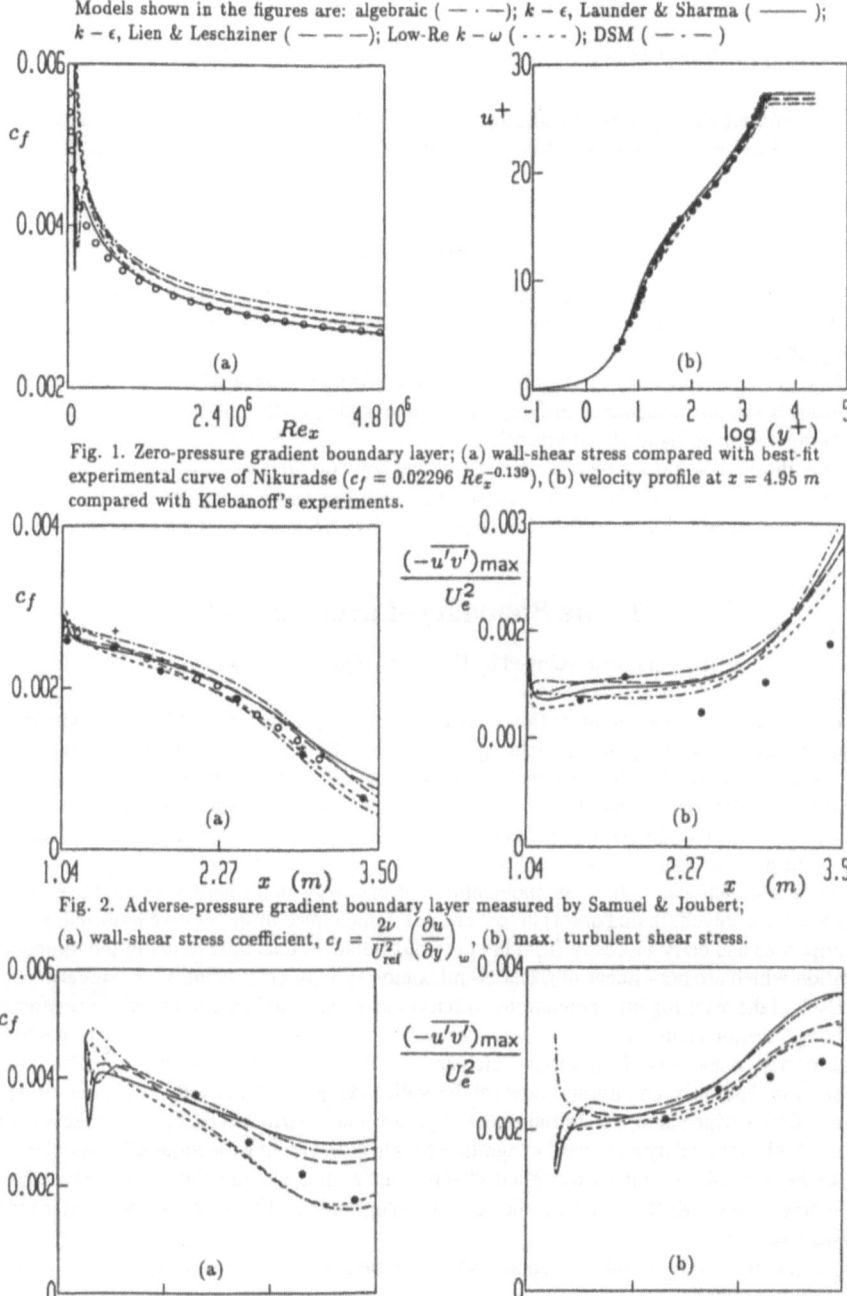

Models shown in the figures are: algebraic (— · —); $k - \epsilon$, Launder & Sharma (——);
$k - \epsilon$, Lien & Leschziner (— — —); Low-Re $k - \omega$ (· · · ·); DSM (— · —)

Fig. 1. Zero-pressure gradient boundary layer; (a) wall-shear stress compared with best-fit experimental curve of Nikuradse ($c_f = 0.02296\ Re_x^{-0.139}$), (b) velocity profile at $x = 4.95\ m$ compared with Klebanoff's experiments.

Fig. 2. Adverse-pressure gradient boundary layer measured by Samuel & Joubert;
(a) wall-shear stress coefficient, $c_f = \dfrac{2\nu}{U_{\mathrm{ref}}^2}\left(\dfrac{\partial u}{\partial y}\right)_w$, (b) max. turbulent shear stress.

Fig. 3. Adverse-pressure gradient boundary layer measured by Nagano et al.;
(a) wall-shear stress coefficient, $c_f = \dfrac{2\nu}{U_e^2}\left(\dfrac{\partial u}{\partial y}\right)_w$, (b) maximum turbulent shear stress.

The Samuel–Joubert Test Case
Computed by a Boundary–Layer Method

Werner Haase, Dornier Luftfahrt GmbH, D–88039 Friedrichshafen, Germany
Harry P. Horton, Queen Mary and Westfield College, London E1 4NS, U.K.

Summary

The boundary layer method for computing two–dimensional, compressible, laminar and turbulent boundary layers has been applied to a non–equilibrium boundary layer with an adverse pressure gradient.

Due to the fact that the boundary layer method provides results on a "free–of–numerical–dissipation" basis, the influence of different turbulence models on the solution is investigated without being swamped by artificial viscosity.

In addition to solutions by other ETMA–workshop contributors – including Navier–Stokes solutions – the results provided in the present contribution are seen as an additional tool to properly describe the boundary–layer flow for the Stanford 80/81, Samuel–Joubert, CTF0141, test case.

1 The Boundary–Layer Method

The code is used in a version developed by H. Horton, Queen Mary and Westfield College, London.

For both, laminar and turbulent flows, the Levy transformation is applied to the Reynolds–averaged boundary layer equations yielding a set of partial differential equations for momentum and energy, respectively. The k–ε equations, used in the 1– and 2–equation models, are similarly transformed. Approximation of the right–hand sides of these equations, containing the streamwise derivatives with three–point backward differences, leads to a coupled, non–linear set of ordinary differential equations to be solved at each streamwise ξ–station.

At each streamwise station, the momentum, mean energy and, where relevant, the equations for ε and k are solved (in that order) in sequence as if uncoupled in an iterative loop, using values of properties and eddy viscosity from the previous iterate to uncouple terms in the momentum equation which are non–linear in f. Under–relaxation is required to achieve convergence.

Each of the resulting linear equations with its associated boundary conditions is transformed, by the application of an invariant embedding algorithm, from a two–point boundary problem into a pair of non–linear sets of equations which are solved by an initial–value method. The first set of equations has initial conditions known at the wall, and is solved by an outward integration pass from $\eta=0$ to an edge value, η_∞. Initial values at η_∞ are then determined for the second set of equations, which is solved by an inward integration pass to $\eta=0$, yielding the required particular solution of the original equations. An implicit 4th–order integration scheme (due to Gear) is used, with a 4th–order Runge–Kutta starter for the outward integration and Gear low–order starter for the inward pass.

The method is fully implicit, second–order accurate in streamwise and fourth–order accurate in the wall–normal direction.

Compared to shooting methods, for which the streamwise intervals must be large, the present procedure offers the advantage that, whilst marching downstream, mesh intervals of any required fineness can be accommodated by the use of sufficiently small integration steps in wall–normal direction.

If the derivatives of the profiles at the boundary layer edge exceed a given value, additional points are added at the end of the boundary layer. In the (wall–normal) η–direction, a geometrical stretching is used on a basis of a prescribed stretching factor. Typically, 200–300 intervals with a stretching factor of about 1.017 are used, in order to achieve the very high resolution of the sublayer which is needed for the 2–equation models. In many circumstances, the present method will be more accurate than a finite difference method for a given number of mesh points.

For the Samuel–Joubert test case presented below, typical mesh parameters for the first and the last profile measured are:

Total number of meshpoints:	265 – 297.
Number of meshpoints within y+=40:	58 – 55.
First wall–nearest step size, y+_wall,:	0.42 – 0.476.

The mean velocity profile at the initial streamwise station is generated by computing the Coles velocity profile corresponding to the measured values of shape factor, H, and momentum thickness Reynolds number R_Θ.

2 Turbulence Model Implementation

Because the numerical method is fully implicit, initial shear stress or eddy viscosity profiles are not required when algebraic models are used, by contrast with Keller box methods such as that of Cebeci. The Johnson–King, 1/2–equation, model as well as the Johnson–Coakley modification is implemented in a fully–implicit manner, using an iterative procedure contained within the loop already used for algebraic models. This does not have any appreciable adverse effect on the rate of convergence of the iteration by comparison with purely algebraic models.

The form of the Johnson–Coakley model incorporated is that using eddy viscosity, rather than mixing length, in the inner region; also hyperbolic–tangent blending is used in place of exponential blending employed in the original Johnson–King model. A Newton iteration is used to compute σ in both Johnson–King and Johnson–Coakley models, but in favourable pressure gradients it has been found necessary to under–relax (by a factor of 0.5) for the Johnson–Coakley model. Lower and upper bounds for σ are set to 0.1 and 1.5, respectively.

For those calculations employing the Hassid–Poreh model, the equation for the turbulent energy, k, is discretised and solved by procedures similar to those used for the mean momentum equation. The momentum and k–equations are solved in succession inside an iterative loop, the eddy viscosity being updated after each iteration. In this case an initial profile of the turbulent energy is required, which is generated by computing an eddy viscosity profile from the van Driest relation in the inner region and a constant mixing length in the outer region. This, in conjunction with the relations given by Hassid and Poreh, provides an implicit expression for k which is solved by iteration at each mesh point in the initial profile.

In the case of employing the 2–equation models by Jones–Launder–Sharma, Lam–Bremhorst and Speziale, all of them in a k–ε form, an initial profile of ε is required. For the first of the models, an iterative procedure is used to construct this profile from the distribution of k, given as above, and the turbulent Reynolds number implied by the assumed eddy viscosity distribution. For the other 2–equation models, ε was derived from the algebraic relation given by Hassid–Poreh. (In the Jones–Launder–Sharma model, ε is the "isotropic part" of the dissipation, whereas in the other models it is the "total" dissipation, hence the need for the two different approaches).

The 2–equation models themselves (with the exception of that of Speziale) are implemented exactly as described by the original authors. Speziale et al. originally used the equation together with a τ equation which was a straightforward combination of the usual k and ε equations. However, difficulties were encountered in applying the inner boundary condition on τ, so that the model was implemented in the standard k–ε form; this is exactly equivalent to the k–τ form and (numerics aside) should yield identical results.

The equations for k and ε are each of second order, so two boundary conditions are required for each of these quantities. For the Lam–Bremhorst and the Speziale models, the inner boundary conditions $k=dk/dz=0$ are applied, but no inner boundary condition on either ε or $d\varepsilon/dz$ is imposed. Instead, the value d^2k/dz^2, obtained at the wall from the k equation, supplies the wall value of ε. In the Jones–Launder–Sharma model, however, ε represents the isotropic part of the dissipation, which vanishes at the wall, thus providing the inner boundary condition $\varepsilon=0$ at $z=0$. Outer boundary conditions on k and ε are applied, these being obtained by streamwise integration of the k–ε equations in the external flow. Initial conditions for these equations are derived from a "standard" pair of values of the quantities $k/U_e=0.0002$ and turbulent Reynolds number $R_t=k^2/(\nu^*\varepsilon)=200$.

3 Results for the Samuel–Joubert Test Case

In the following, results for the Stanford 80/81 test case CTF0141, Samuel–Joubert, are presented using a set of nine different turbulence models ranging from 0–, over 1/2– and 1– to 2–equation models:

(BL)	Baldwin–Lomax	(0)
(CS)	Cebeci–Smith	(0)
(HH)	Horton	(1/2)
(JK)	Johnson–King	(1/2)
(JC)	Johnson–Coakley	(1/2)
(HP)	Hassid–Poreh	(1)
(JLS)	Jones–Launder–Sharma	(2)
(LB)	Lam–Bremhorst	(2)
(SP)	Speziale	(2)

The set of figures presented in the following combine results for the 0–equation models, for the 1/2–equation and the 1– and 2–equation models, respectively.

Fig. 1 presents the U_e/U_∞ distribution versus x used as an input quantity for all calculations performed.

Fig. 2 and Fig. 3 depict skin friction and form parameter for the algebraic turbulence models, CS and BL. In general, the skin friction distributions compared to the measurements are in reasonable agreement although the BL model produces somewhat higher c_f values. Moreover, the deviation increases when the flow is approaching separation where the BL model cannot really predict the flow situation. Compared to the BL model, the CS model behaves much better which might be directly correlated to lower values of eddy viscosity computed in the near wall region, Fig. 10. The predictive accuracy for the form parameter is excellent for the CS model and shows a drastic deviation for the BL model, Fig. 3.

For the 1/2–equation models, Figs. 4 and 5 present (again) skin friction and form parameter. All models involved are in reasonable agreement with the measurements, however, apart from the area x'>2.0m near the last measured station which is closest to separation.

It should be noted at this point that the Samuel–Joubert test case does not reflect a strong non–equilibrium flow. The σ–values for all three 1/2–equation turbulence models, a measure for the equilibrium status of the flow to be investigated, exhibit only moderate minimum values (in the area where flow approaches separation) between 0.7 for the HH and JK model and 0.8 for the JC model.

In Figs. 6 and 7, results for the 1– and 2–equation models are combined. When comparing the performance of the different models, a larger scatter can be detected from the skin friction and form parameter distributions with the (1–equation) HP model as an exception. Comparing the results for all turbulence models tested, the k-ε models depict the largest deviation between calculation and measurements.

To conclude at that point, one can rather easily recognize that the CS model – "although" a "simple algebraic" model – wins that contest, at least on the basis of c_f and H. In the present case, adverse–pressure–gradient flow which (only) approaches separation, the predictive capability of the 2–equation models tested (JLS, LB ans SP) remains poor.

To shed some more light to the performance of the turbulence models, Figs. 8 to 16 present at the last measured station, x=3.4m or x'=2.36m, boundary layer profiles for velocity, Reynolds stress and eddy viscosity, respectively. For the algebraic models (CS and BL), results are presented in Figs. 8, 9 and 10. As suggested in the discussion concerning the wall properties given above, the CS model exhibits a good comparison with the measured velocity and Reynolds stress profile. The BL model produces an eddy viscosity "overshoot" in the outer boundary layer region which is well known, i.e. in that area the BL model does not really produce intermittency at all.

The latter can also be seen for the HH model, Fig. 13. However, for that model, which in general predicts velocity profile shape and Reynolds stresses with good accuracy, the maximum eddy viscosity value is by more than 30% smaller compared to the BL model. Moreover, the "lack" in intermittency provides in the area where the strain rates are already small, an accurate Reynolds stress profile, Fig. 12, and an improved velocity profile shape, Fig. 11.

For the 1–equation HP model and the 2–equation k-ε models, results are presented in Figs. 14 to 16. Apart from the HP model, the use of all other models (as already discussed) do not show very encouraging results, especially in the lower boundary layer region. In that area, the k-ε models produce up to 50% more eddy viscosity, hence preventing "separation" and leading to an over–prediction in Reynolds stresses, Fig. 15.

4 Figures

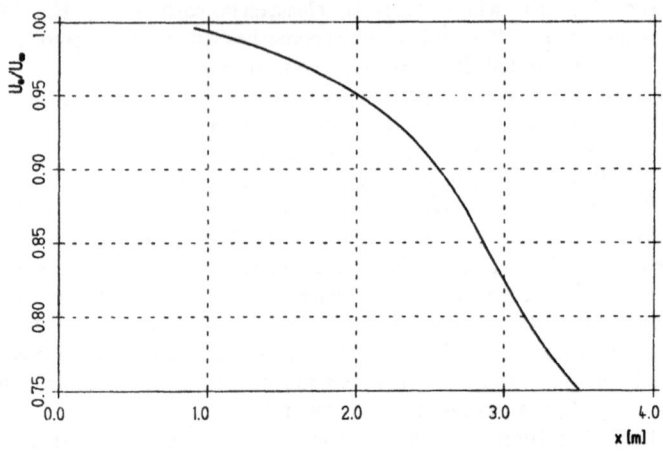

Figure 1 U_e/U_∞ distribution used for all calculations for the Samuel–Joubert, Stanford 80/81, CTF0141, test case

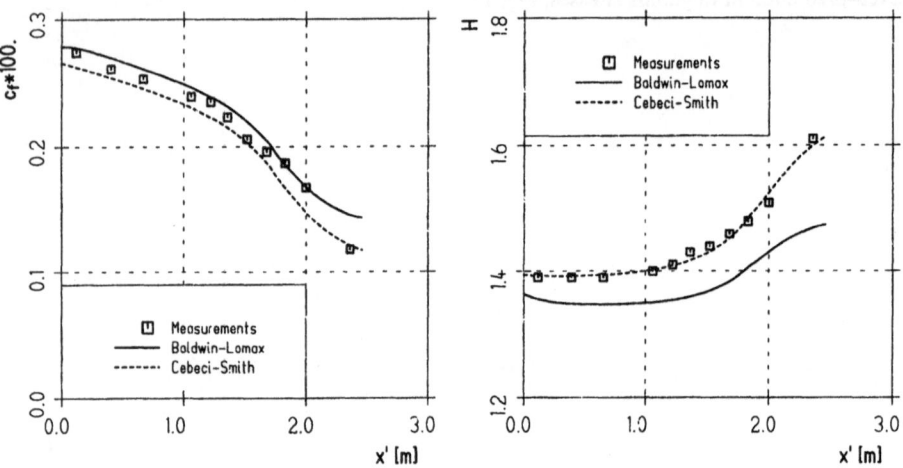

Figure 2 Skin friction for 0–equation models

Figure 3 Form parameter for 0–equation models

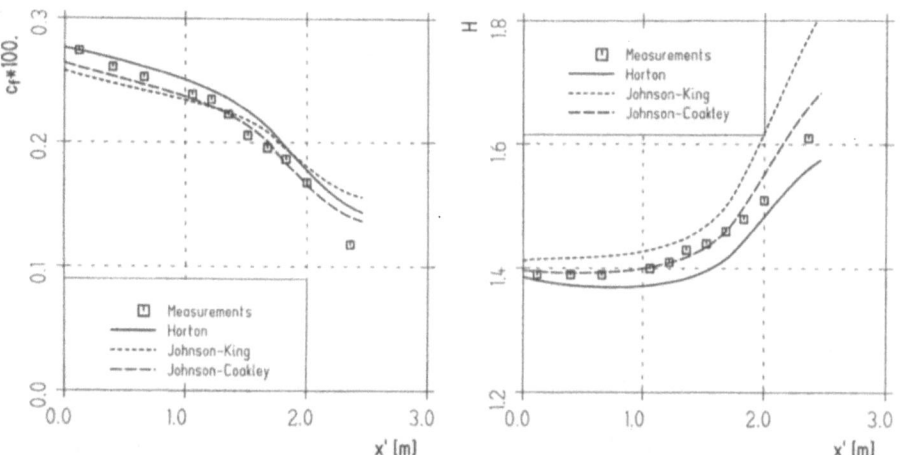

Figure 4 Skin friction for 1/2–equation models

Figure 5 Form parameter for 1/2–equation models

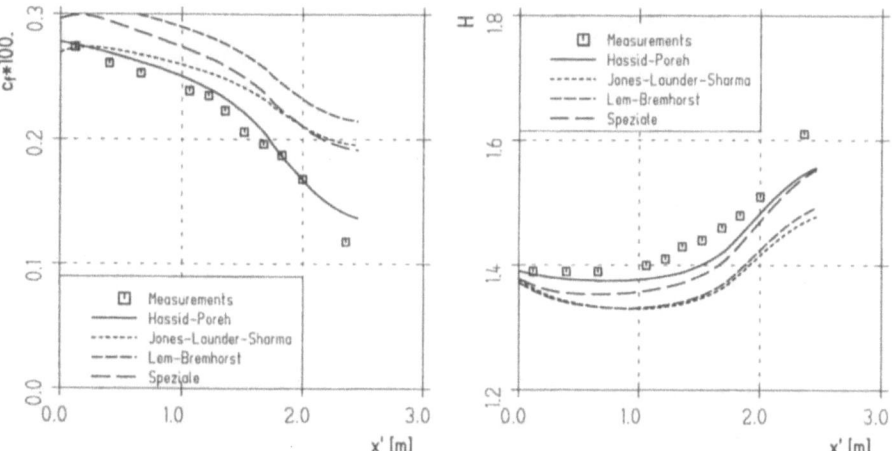

Figure 6 Skin friction for 1– and 2–equation models

Figure 7 Form parameter for 1– and 2–equation models

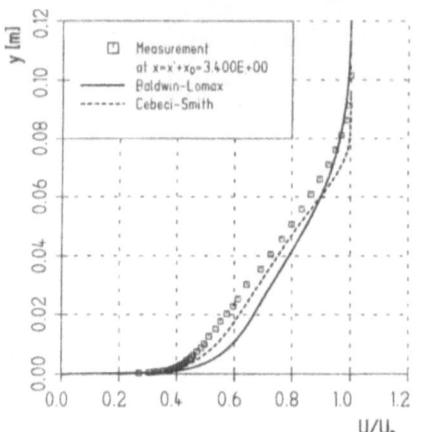

Figure 8 Velocity profiles for 0–equation models

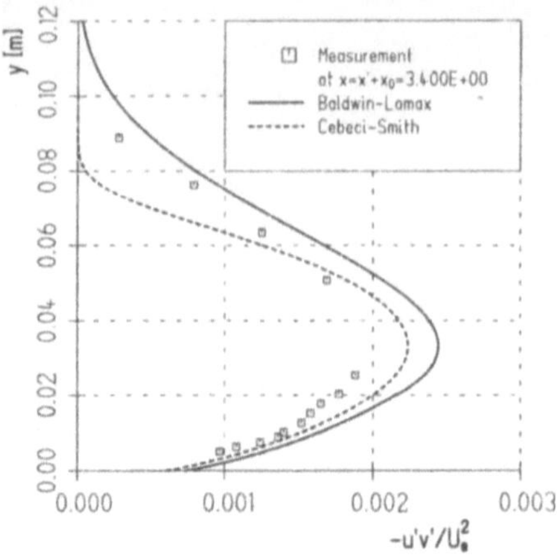

Figure 9 Reynolds–stress profiles for 0–equation models

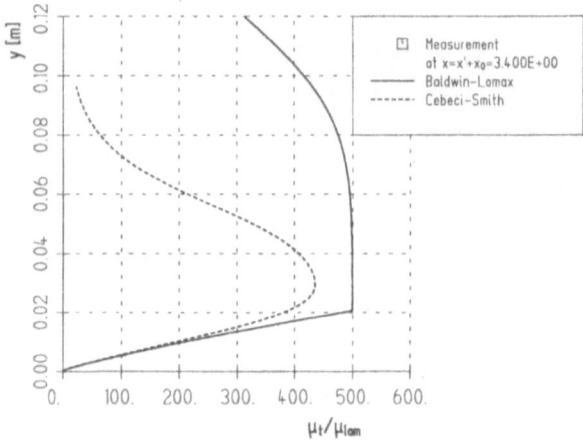

Figure 10 Eddy–viscosity profiles for 0–equation models

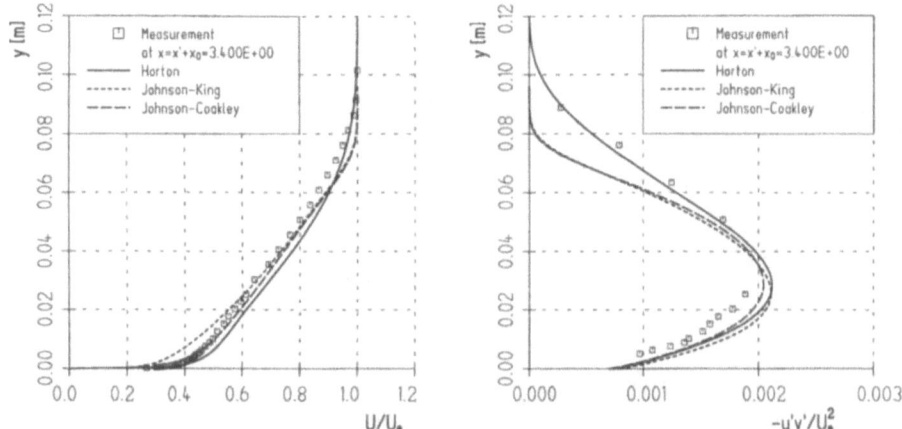

Figure 11 Velocity profiles for 1/2–equation models

Figure 12 Reynolds–stress profiles for 1/2–equation models

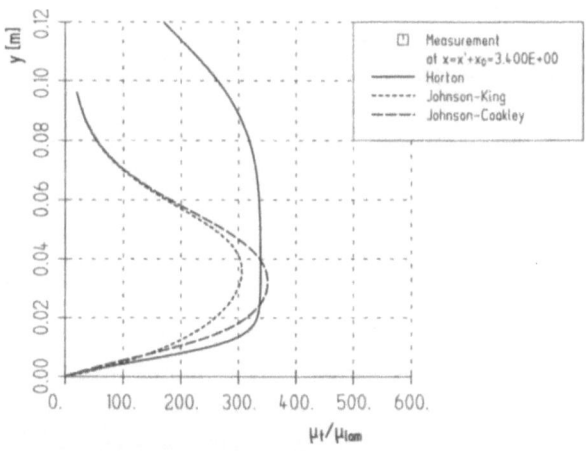

Figure 13 Eddy–viscosity profiles for 1/2–equation models

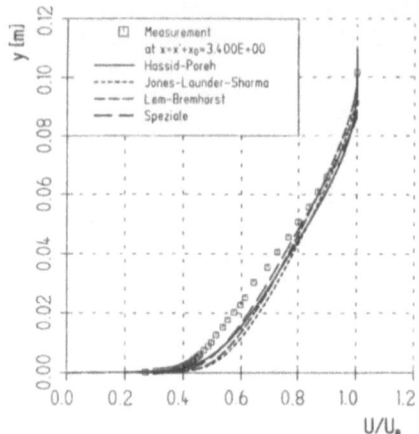

Figure 14 Velocity profiles for 1– and
2–equation models

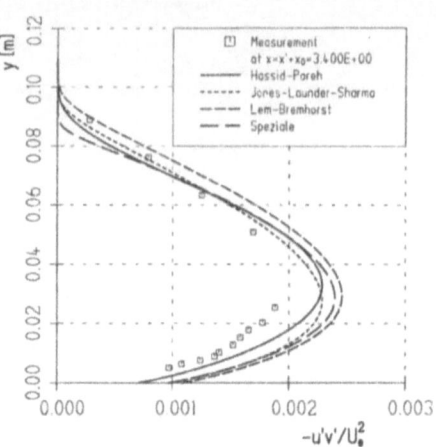

Figure 15 Reynolds–stress profiles for 1–
and 2–equation models

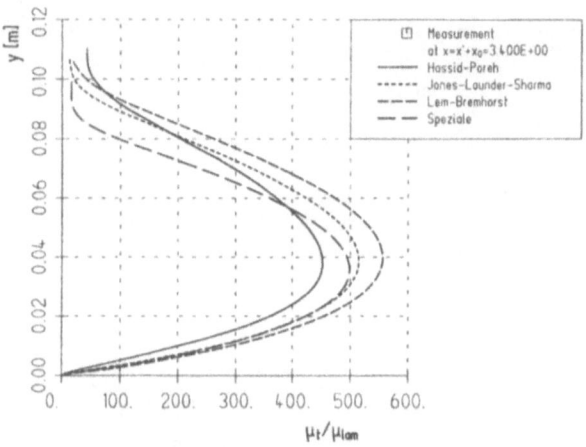

Figure 16 Eddy–viscosity profiles for 1– and
2–equation models

214

Applications of the Finite Element Method to the Reynolds-Averaged Navier-Stokes Equations

Thomas J. R. Hughes, Kenneth Jansen and Guillermo Hauke
Stanford University, Stanford, California USA

SUMMARY

A Galerkin/least-squares finite element method is applied to the Reynolds-averaged Navier-Stokes (RANS) equations. The turbulence model utilizes an eddy viscosity calculated from a velocity scale and a length scale. The velocity scale, q, is evolved via an additional transport equation which is appended to the RANS equations and solved together as a complete system. The length scale, l, is given by algebraic relations involving flow quantities and distance to the wall. The model is shown to give good results on a variety of problems over a broad range of resolution.

INTRODUCTION

A Galerkin/least-squares finite element method is applied to the (RANS) equations. The turbulence model utilizes an eddy viscosity calculated from a velocity scale and a length scale. The velocity scale, q, is evolved via an additional transport equation which is appended to the RANS equations and solved together as a complete system. The length scale, l, is given by algebraic relations involving flow quantities and distance to the wall. The particular model chosen is the Norris-Reynolds (1975) incompressible channel flow model. The model has been extended for application to compressible flows containing wall-bounded shear layers and wakes.

Because the model is posed in terms of a turbulent velocity scale rather than the turbulent kinetic energy k, it enjoys several beneficial properties. First, the q equation combined with the RANS equations can be transformed into a symmetric advective-diffusive system. (Systems of this form inherit many important mathematical properties including discrete entropy production, see Jansen et al. (1993) and references therein). The second benefit is that q is linear near the wall. Therefore the model may be integrated to the wall with one element in the viscous sublayer ($y^+ \leq 5.0$). The use of larger near wall elements results in a reduction of the number elements required to resolve the flow. The increase in the element size also affords larger time steps, which accelerates convergence to steady state.

The model is incorporated into a Galerkin/least-squares finite element program which avoids the use of monotone schemes which are questionable for the RANS equations given the strong influence of source terms. This method combines good stability properties without loss of accuracy. The accuracy of the method has been verified on a wide variety of problems.

215

RANS EQUATIONS

The modeled Reynolds-averaged Navier-Stokes equations are

$$\bar{\rho}_{,t} + [\bar{\rho}\tilde{u}_i]_{,i} = 0$$

$$[\bar{\rho}\tilde{u}_i]_{,t} + [\bar{\rho}\tilde{u}_i\tilde{u}_j]_{,j} = \left[-(\bar{p} + \boxed{\frac{2}{3}\bar{\rho}k})\delta_{ij} + [(\mu^{\text{visc}} + \boxed{\mu_T^{\text{visc}}})S_{ij}(\tilde{u})]\right]_{,j}$$

$$[\bar{\rho}\tilde{e}_{\text{tot}}]_{,t} + [\bar{\rho}\tilde{u}_i\tilde{e}_{\text{tot}}]_{,i} = \left[-(\bar{p} + \boxed{\frac{2}{3}\bar{\rho}k})\tilde{u}_i + (\kappa + \boxed{\kappa_T})\tilde{T}_{,i}\right.$$

$$\left. + (\mu^{\text{visc}} + \boxed{\mu_T^{\text{visc}}})\tilde{u}_jS_{ij}(\tilde{u}) + \boxed{(\mu^{\text{visc}} + \frac{\mu_T^{\text{visc}}}{Pr_k})k_{,i}}\right]_{,i} \tag{1}$$

where

$$S_{ij}(\tilde{u}) = \tilde{u}_{i,j} + \tilde{u}_{j,i} - \frac{2}{3}\delta_{ij}\tilde{u}_{k,k} \qquad \tilde{e}_{\text{tot}} = \tilde{e} + \frac{\tilde{u}_i\tilde{u}_i}{2} + k \qquad \kappa_T = c_p\frac{\mu_T^{\text{visc}}}{Pr_T}. \tag{2}$$

μ_T^{visc} is the eddy viscosity and $k = q^2/2$ is the turbulent kinetic. Here an overbar indicates an ensemble average and a tilde indicates a density-weighted ensemble average. The q equation is given by

$$[\bar{\rho}q]_{,t} + [\bar{\rho}\tilde{u}_iq]_{,i} = [(\mu^{\text{visc}} + \frac{\mu_T^{\text{visc}}}{Pr_k})q_{,i}]_{,i} + S_q \tag{3}$$

$$S_q = \frac{\bar{\rho}}{q}(\mathcal{P} - \epsilon) + (\mu^{\text{visc}} + \frac{\mu_T^{\text{visc}}}{Pr_k})\frac{q_{,i}q_{,i}}{q}. \tag{4}$$

NORRIS-REYNOLDS MODEL

The preceding section is applicable to any q equation model but is not yet closed. To close the model we must define the eddy viscosity, μ_T^{visc}, the turbulent thermal conductivity, κ_T, and the turbulent kinetic energy dissipation, ϵ. For the Norris-Reynolds model (1975) we have:

$$\epsilon = C_4f_\epsilon\frac{q^3}{l} \qquad\qquad \mu_T^{\text{visc}} = C_1f_\mu\bar{\rho}ql$$

$$f_\epsilon = 1 + \frac{C_5}{R_l} \qquad\qquad f_\mu = 1 - e^{-C_3R_y}$$

$$R_l = \frac{\bar{\rho}ql}{\mu^{\text{visc}}} \qquad\qquad R_y = \frac{\bar{\rho}qy}{\mu^{\text{visc}}} \tag{5}$$

$$\kappa_T = \frac{c_p\mu_T^{\text{visc}}}{Pr_T} \qquad\qquad l = C_0\delta[1 - (1 - \frac{y}{\delta})^{C_2}]$$

where y is the normal distance to the wall, δ is the boundary layer thickness and the constants are defined in Table 1. These constants are determined by limiting behavior at walls, log layer equilibrium and performance on a variety of problems. The model was developed for boundary layers around solid boundaries but the length scale can be altered to accommodate wakes as well, see Jansen *et al.* (1993).

Table 1. Various model constants for the Norris-Reynolds model.

Model	C_0	C_1	C_2	C_3	C_4	C_5
Original	0.133	0.39	3.0	0.014	0.0593	2.697
Modified	0.11	0.39	3.73	0.014	0.0593	2.835

RESULTS

The results presented in this section where obtained two years ago during thesis work of the second author. The results were *not* recomputed on the meshes suggested by the workshop organizers. The meshes we used are somewhat coarser. For details concerning solution see Jansen *et al.* (1993).

4.1 Flat Plate Boundary Layer: Incompressible Case

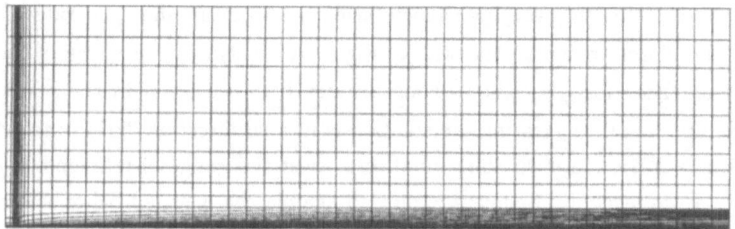

Figure 1. Nearly incompressible flow over a flat plate. The discretization shown here contains 80x60 elements.

The first problem considered is a nearly incompressible high Reynolds number ($Re_L = 12$ million) flow over a flat plate. By running our compressible code at $M = 0.01$ we can approximate incompressible flow. The discretization used for this calculation is shown in Figure 1. There are 60 elements normal to the wall with the first point off the wall at $y^+ = 0.5$. There are 80 elements in the streamwise direction with a concentration of points at the sharp leading edge of the plate. Here a laminar boundary layer begins its growth along the length of the plate until at some point the flow undergoes a transition to turbulence. The exact point of transition depends on many factors. To circumvent this difficulty turbulence modelers often compare results based on variables other than x; see White (1974). For example the skin friction C_f is usually plotted versus Re_θ, the Reynolds number based on the momentum thickness, θ, which is defined as

$$\theta = \int_0^{y_e} \frac{\overline{\rho}\tilde{u}_1(\tilde{u}_1^e - \tilde{u}_1)}{\overline{\rho}^e(\tilde{u}_1^e)^2} dy$$

$$Re_\theta = \frac{\overline{\rho}\tilde{u}_1^e\theta}{\mu^{\mathrm{visc}}}$$

(6)

217

where the superscript e denotes the edge of the boundary layer.

The skin friction is plotted in this fashion in Figure 2. The computed turbulent solution follows the Blasius solution, which is exact for laminar flow, until the point of transition after which it quickly joins the empirical branch. This line is a plot of the Karman-Schönherr correlation

$$\frac{1}{C_f} = 17.08(\log_{10} Re_\theta)^2 + 25.11 \log_{10} Re_\theta + 6.012 \ . \tag{7}$$

To ascertain the quality of the velocity profile, the law of the wall and the log law are plotted with the computed solution at various values of Re_θ in Figure 3.

4.2 Refinement Studies

The mesh shown in Figure 1 is viewed as moderately fine. To study the effect of mesh spacing, the closest point to the wall is allowed to vary while adjusting the remaining nodes to insure a smoothly varying element size (i.e., a new mesh is generated for each change of the closest point to the wall). Figure 4 illustrates the effect of near-wall refinement on the velocity profiles. There is very little variation in the results from $y^+ = 0.25$ to $y^+ = 2.0$, suggesting the solution is normal direction "grid independent". The coefficient of friction is similarly unaffected over this range of refinements as seen in Figure 5. Each level of refinement is plotted against the Karman-Schönherr relation given in (7). The laminar branch of the solution is cut from the figure for the sake of clarity.

Coarse near-wall solution accuracy can be attributed to the choice of q as a turbulent variable which is linear until $y^+ \approx 5$. Similarly, the length scale l is also linear in this region. Thus the scales responsible for turbulence modeling are accurately resolved by linear elements with at least one point at $y^+ \leq 5$. Finally, with the correct turbulence effects, linear elements can also accurately resolve the velocity profile since it too is linear in this region.

4.3 Increasingly Adverse Pressure Gradient

A much more challenging problem is presented by Samuel and Joubert (1975). The changing area duct, sketched in Figure 6, creates an increasingly adverse pressure gradient which in turn dramatically affects the boundary layer and the turbulence within. The coefficient of friction plot is compared with experimental data in Figure 7. The turbulence model properly accounts for the pressure gradient which drives the flow toward separation.

218

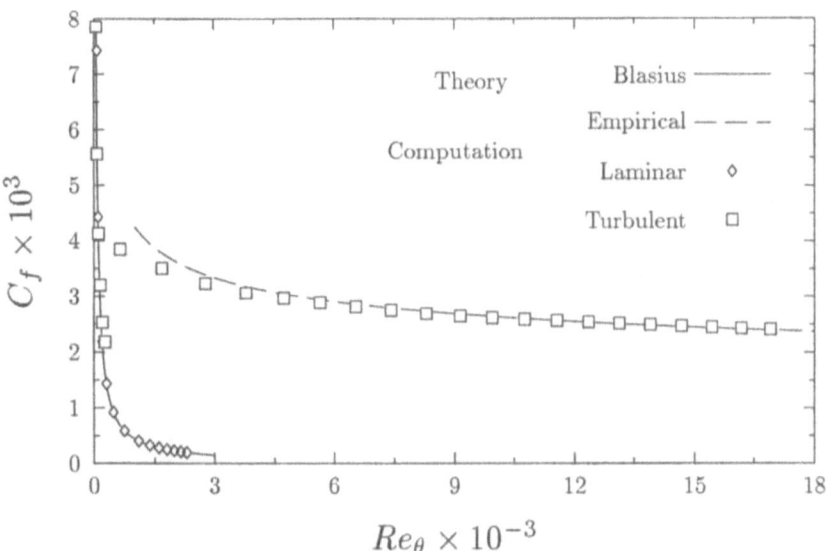

Figure 2. Nearly incompressible flow over a flat plate. Coefficient of friction vs. momentum thickness Reynolds number.

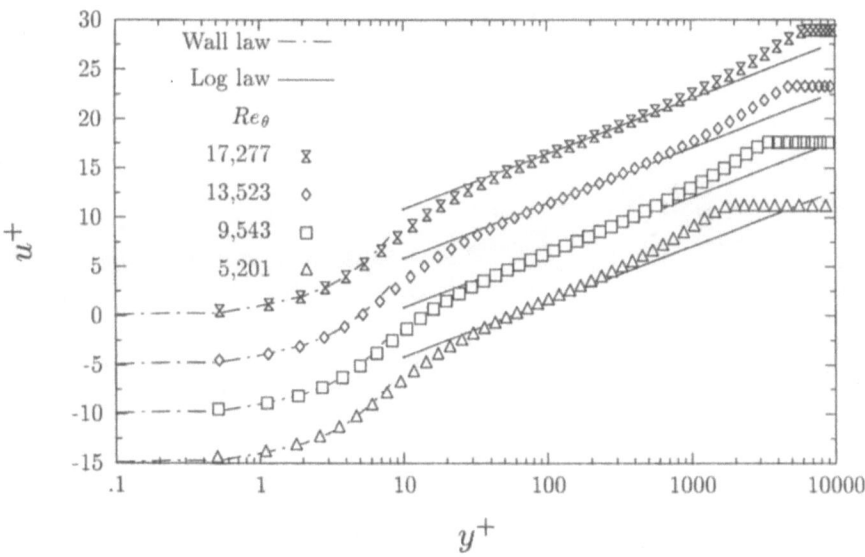

Figure 3. Nearly incompressible flow over a flat plate. Velocity profiles in logarithmic wall coordinates at various values of momentum thickness Reynolds numbers.

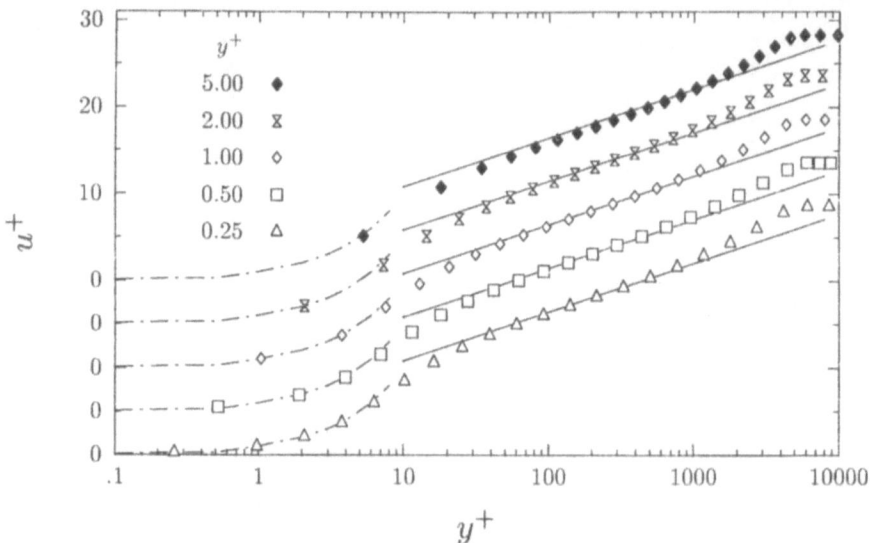

Figure 4. Nearly incompressible flow over a flat plate. The effect of varying the normal direction element size on the boundary layer velocity profile.

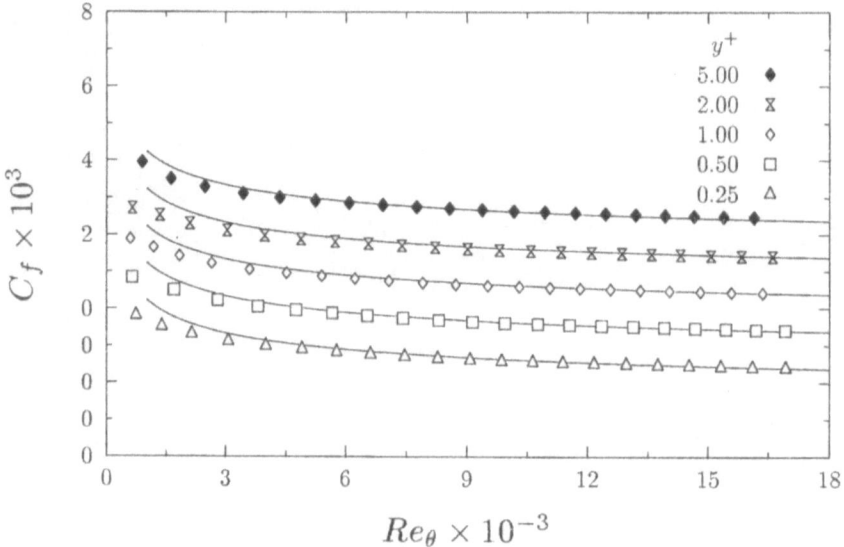

Figure 5. Nearly incompressible flow over a flat plate. The effect of varying the normal direction element size on the coefficient of friction.

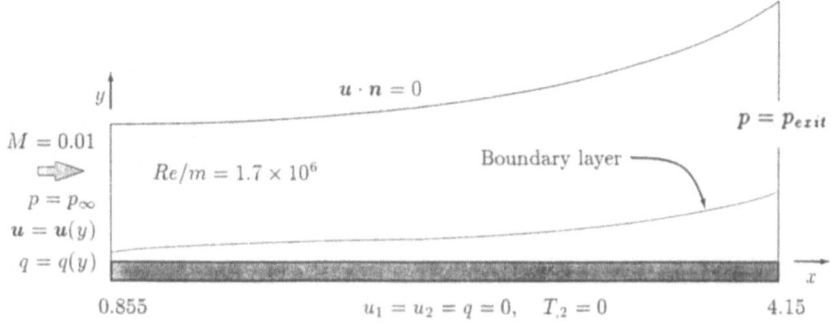

Figure 6. Increasingly adverse pressure gradient flow.

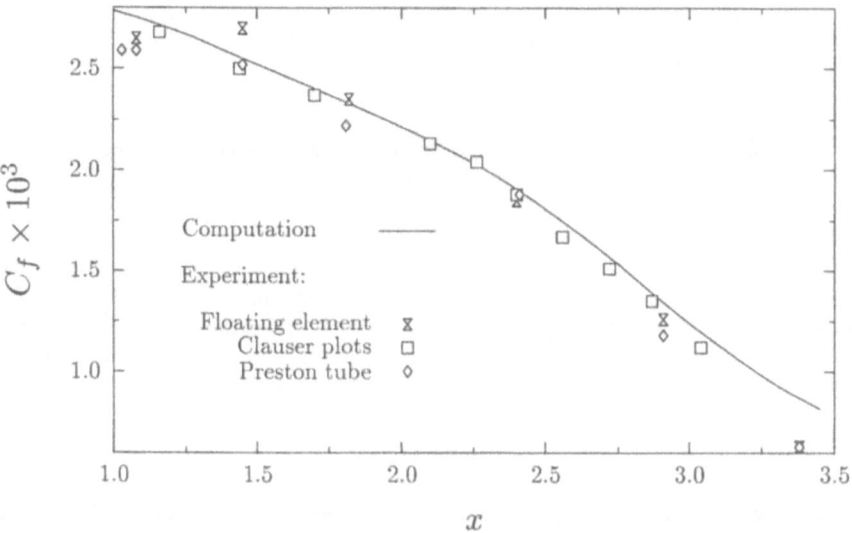

Figure 7. Increasingly adverse pressure gradient flow. The coefficient of friction vs. position in channel. The computation is compared with various methods of experimental measurement.

4.4 Flat Plate Boundary Layer: Compressible Case

To study the effect of compressibility on turbulence the flat plate problem was repeated at various Mach numbers. The results of the study can be observed in Figure 8 where the coefficient of friction at various Mach numbers is compared to Van Driest's approximation and to results submitted to the Collaborative Testing of Turbulence Models (CTTM) study for $Re_\theta = 10,000$; see Bradshaw *et al.* (1991). It is important to note that the constants in Table 1 where not altered for this problem or any other problem in this paper.

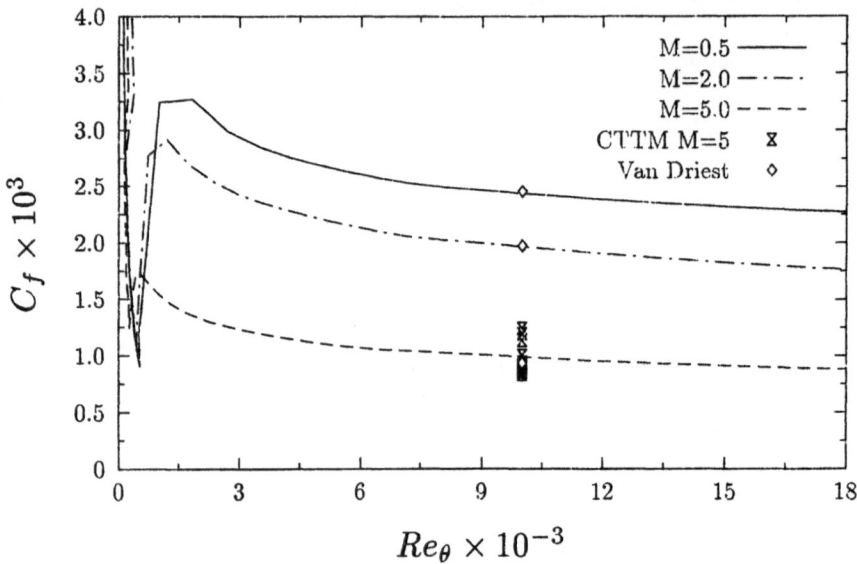

Figure 8. Compressible flow over a flat plate. The effect of varying Mach number on the coefficient of friction.

CONCLUSIONS

The model is seen to perform well over a broad range of Mach numbers. It is also seen to perform very well in adverse pressure gradient flows. Probably the most attractive feature of this model is its ability to be integrated to the wall without a large increase in near wall resolution requirements resulting in a modest computational cost.

REFERENCES

Bradshaw, P., Launder B.E., and Lumley, J.L., 1991, "Collaborative testing of turbulence models", AIAA paper 91-0215.

Jansen, K., Johan, Z., and Hughes, T.J.R., 1993, "Implementation of a one-equation turbulence model within a stabilized finite element formulation of a symmetric advective-diffusive system", *Computer Methods in Applied Mechanics and Engineering*, **105**, pp. 405-433.

Norris, L.H., and Reynolds, W.C., 1975, "Channel flow with a moving wavy boundary," Report FM-10, Stanford University, Department of Mechanical Engineering, Stanford.

Samuel, A.E., and Joubert, P.N., 1975, "A boundary layer developing in an increasingly adverse pressure gradient", *Journal of Fluid Mechanics*, **6**, p. 481.

White, F., 1974, *Viscous fluid flow*. McGraw Hill, New York.

Simulations of Compressible Turbulent Boundary Layer using a Low Reynolds $k - \epsilon$ Model

I. Yudiana[a], and M. Buffat[b]

LMFA, CNRS UMR 5509, [a]ECL, [b]UCB Lyon I

36, avenue Guy de Collongue

69131, Ecully cedex France

Summary

This paper describes the simulation of a compressible turbulent boundary layer with zero pressure gradient, performed using an incompressible low Reynolds $k - \epsilon$ model combined with compressibility modeling. It appears that this model behaves well for such flows.

Introduction

Supersonic boundary layers play an important role in aeronautics. The present study gives some results of the numerical simulations of such flows. For the test case, we focus the attention on the experiments of Mabey [4]. In these experiments, the Mach number is high enough to induce important compressibility effects. The difficulty of this test case is to predict accurately important quantities such as the wall friction coefficient and heat transfer. As different factors influence the precision of the numerical results (the mesh, the inlet conditions, the compressibility terms, the near wall modeling,...), we have used the prescribed mesh and inlet conditions.

Turbulence models

We consider the standard $k - \epsilon$ model, improved with some modifications to take into account the compressibility and the low Reynolds effects.

Dilatation-dissipation models

To take into account an increase of the dissipation with the turbulent Mach number $Ma_t = \frac{\sqrt{2k}}{c}$, an additional compressible dissipation ϵ_c is added into the k equation. Different models, based on direct numerical simulations results, have been proposed for ϵ_c (Sarkar [7], Zeman [10], Wilcox [9]) in term of the incompressible dissipation ϵ_s and the turbulent Mach number Ma_t:

$$\epsilon_c = F(Ma_t)\epsilon_s . \tag{1}$$

The difference between the models are the expression of the dependence on the turbulent Mach number $F(Ma_t)$:

$$F(Ma_t) = 0.5 * Ma_t^2 \quad \text{(Sarkar)} \tag{2}$$

$$F(Ma_t) = \begin{cases} 1.5 * (Ma_t^2 - 0.25^2) & \text{if } Ma_t \geq 0.25 \\ 0 & \text{if } Ma_t \leq 0.25 \end{cases} \quad \text{(Wilcox)} \tag{3}$$

$$F(Ma_t) = \begin{cases} 1 - e^{-\frac{1.2(Ma_t - 0.25)^2}{0.66^2}} & \text{if } Ma_t \geq 0.25 \\ 0 & \text{if } Ma_t \leq 0.25 \end{cases} \quad \text{(Zeman)}. \tag{4}$$

Pressure-dilatation models

For compressible flow, the pressure fluctuation induces a decrease of the production and the dissipation through the pressure-dilatation correlation. Sarkar proposed the following model for the pressure-dilatation term in the turbulent kinetic energy equation:

$$\overline{p'\frac{\partial u_i'}{\partial x_i}} = \alpha_1 Ma_t^2 \bar{\rho}\widetilde{u_i'u_j'}\frac{\partial \tilde{u}_i}{\partial x_j} + \alpha_2 Ma_t^2 C_\mu \bar{\rho}\epsilon \tag{5}$$

$$\text{with } \alpha_1 = 0.4 \text{ and } \alpha_2 = 0.2 . \tag{6}$$

Low Reynolds models

To improve the modeling near the wall, the low Reynolds models of Shih Lumley [8] and Lam Bremhost [5] have also been tested. A first series of computations has been done with the Shih Lumley model. Whereas good results have been obtained for incompressible flows (channel flow), the application to compressible flow was not satisfactory. Computations of compressible boundary layers have shown that the Lam Bremhorst model performs better. Thus the Lam Bremhorst model has been used to perform the test case computations.

To take into account the decrease of the turbulence near the wall, a damping function is introduced in the turbulent viscosity μ_t:

$$\mu_t = C_\mu \rho f_\mu \frac{k^2}{\epsilon} \tag{7}$$

$$\text{with } f_\mu = (1 - e^{-0.0165 Ry})^2 (1 + 20.5/Re_t) \tag{8}$$

$$Re_t = \frac{k^2}{\nu\epsilon} \quad Ry = \frac{\sqrt{k}\,y}{\nu}. \tag{9}$$

Damping functions are also introduced in the turbulence dissipation equation to ensure the right behavior of ϵ at the wall:

$$\frac{\partial \rho\tilde{\epsilon}}{\partial t} + div(\overline{\rho}\tilde{u}\tilde{\epsilon}) = \frac{\partial}{\partial x_i}(\mu + \frac{\mu_t}{\sigma_\epsilon})\frac{\partial \epsilon}{\partial x_i} + C_{\epsilon 1}f_1\rho\frac{\epsilon}{k}P - C_{\epsilon 2}f_2\rho\frac{\epsilon^2}{k} \tag{10}$$

$$f_1 = 1 + (0.05/f_\mu)^3 \tag{11}$$

$$f_2 = 1 - e^{-Re_T^2} \tag{12}$$

$$P = -\widetilde{u_i'u_j'}\frac{\partial \tilde{u}_i}{\partial x_j}. \tag{13}$$

Boundary conditions for k and ϵ, obtained by asymptotic expansion at the wall, are used:

$$k = 0 \qquad \epsilon = 2\nu(\frac{\partial \sqrt{k}}{\partial y})^2_{y=0}. \tag{14}$$

Summary

The final modeled transport equations for k and ϵ are:

$$\frac{\partial \rho k}{\partial t} + div(\overline{\rho}\tilde{u}k) = \frac{\partial}{\partial x_i}(\mu + \frac{\mu_t}{\sigma_k})\frac{\partial k}{\partial x_i} + (1 - \alpha_1 Ma_t^2)\rho P$$
$$- (1 + F(Ma_t) - \alpha_2 Ma_t^2)\rho\epsilon \tag{15}$$

$$\frac{\partial \rho\tilde{\epsilon}}{\partial t} + div(\overline{\rho}\tilde{u}\tilde{\epsilon}) = \frac{\partial}{\partial x_i}(\mu + \frac{\mu_t}{\sigma_\epsilon})\frac{\partial \epsilon}{\partial x_i} + C_{\epsilon 1}f_1\rho\frac{\epsilon}{k}P - C_{\epsilon 2}f_2\rho\frac{\epsilon^2}{k} \tag{16}$$

where

$$C_{\epsilon 1} = 1.44 \quad C_{\epsilon 2} = 1.92 \quad C_\mu = 0.09 \quad \sigma_k = 1.0 \quad \sigma_\epsilon = 1.3. \tag{17}$$

Numerical method

The numerical method uses a mixed finite volume/finite element approximation on non structured grid initially developed by INRIA. The convective fluxes are integrated by a Roe solver with a finite volume approximation and a 2nd order

MUSCL interpolation, whereas the viscous and source terms are integrated using a finite element method [6]. The time integration uses an implicit scheme, with performant preconditioned linear solvers [3]. For the inlet conditions, the fluxes of variables $(\rho, \rho U, \rho V, E, \rho k$ et $\rho \epsilon)$ are imposed. All calculations are carried out on the code "NadiaNG" which is a parallel code using a new object oriented paradigm (C^{++}) and a standard message passing library (PVM). Efficient parallel numerical techniques have been developed, such as an automatic domain partitioning technique and a parallel implicit solver[1].

Computational results

We have used the prescribed mesh and inlet conditions. The data are nondimensionlized by the corresponding values on the boundary layer edge. Numerical results are compared to experimental data, at the stations x=0.623, 0.876, 1.130 and 1.384 m.

The flow in a compressible boundary layer is characterized by large changes in density and temperature which influence the physical properties such as the viscosity μ, (see figures 1—4). To take into account the variation of μ with the temperature, the Sutherland's law has been used in the computations.

The density and temperature changes are a result of compressibility, viscous dissipation and/or heat transfer at the wall. In turbulent flows the transport mechanisms associated with fluctuations quantities such as velocity, temperature, and pressure increase the exchange of momentum and heat considerably. Figures 5, 6 show the variation of the total temperature and the Pitot total pressure. These two quantities are both measured experimentally. It's important to predict accurately those quantities to make sure that the calculation is coherent with the experience. But a significant disagreement between experimental data and computational result is observed for the total temperature. A similar result is obtained for a compressible boundary layer, computed using a turbulent Prandtl number different from unity [2]. Figure 7 shows the variation of the wall friction coefficient Cf compared with the experimental data.

The computation runs on a cluster of DEC Alpha workstations with a FFDI network. Using 4 computer nodes, the parallel computations take about 18 *sec.* elapsed time per iteration, with a local time step and a local CFL of 5. Higher CFL number has not been used because oscillations appear in residuals due to the explicit boundary condition for ϵ. All the residuals have decreased up to a plateau of 10^{-5} in normalized L_2 norm, except for ϵ, which has a higher value (10^{-2}). A converge solution is obtained after 2000 to 3000 iterations, starting from an initial solution.

Conclusion

Initially developed for incompressible flows, the $k - \epsilon$ model of Lam Bremhorst behaves well when used to predict compressible flows. But to obtain a better agreement with experimental data, compressibility terms have to be included in the model. The present results correspond to the Wilcox model.

With the developed numerical method and its parallel implementation, we have been able to obtain good results for this test case. In the future, the boundary condition for ϵ will be imposed implicitly to avoid observed oscillations of residual at high CFL number. We have also noticed that the inlet condition has a strong influence on the predicted values, and future simulation should compute all the development of boundary layer (starting from the stagnation point).

References

[1] M. Buffat. Parallel simulation of compressible turbulent flows using non-structured domains partitioning and object oriented programming. In *Notes on numerical fluid mechanics: CFD on parallel systems*. Vieweg-Verlag, 1994.

[2] Van Driest. Turbulent boundary layer in compressible fluids. *Journal of the Aeronautical Sciences*, 18, 1951.

[3] L. Hallo, JL. Munier, M. Buffat, and G. Brun. Iterative methods for solving implicit non-structured finite volume discretization of euler equations. *International Journal for Numerical Methods in Fluids*, 1994.

[4] P.J. Finley H.H. Fernholz. A critical compilation of compressible turbulent boundary layer data. *AGARDOgraph*, (223), 1977.

[5] Bremhorst K. Lam C.K.G. A modified form of the $k - \epsilon$ model for predicting wall turbulence. *J. Fluids Eng.*, 103:457—460, 1981.

[6] C. Le Ribault. *Simulation des écoulements turbulents compressibles par une méthode mixte éléments finis/volumesfinis*. PhD thesis, Thèse Ecole Centrale de Lyon, 1991.

[7] S. Sarkar, G. Erlabacher, and M.Y. Hussaini. Compressible homogeneous shear: simulation and modeling. In *Eight Symposium on turbulent shear flows, Munich sept. 9-11*, 1991.

[8] John L. Lumley Tsan-Hsing Shih. Kolmogorov behavior of near-wall turbulence and its application in turbulence modeling. *Comp. Fluid Dynamics*, 1:43—46, 1993.

[9] D.C. Wilcox. *Turbulence modeling for CFD*. Griffin Printing, Glendale California, 1993.

[10] O. Zeman. dilatational dissipation: the concept and application in modeling compressible mixing layers. *Physics of Fluids*, pages 178–188, 1990.

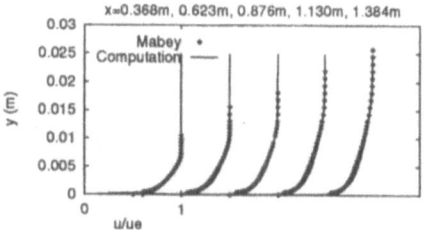

Figure 1: Velocity distribution

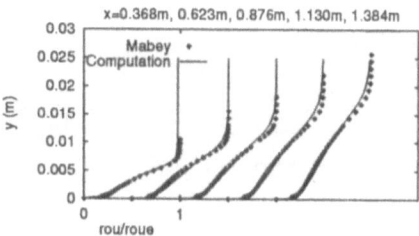

Figure 2: x-momentum distribution

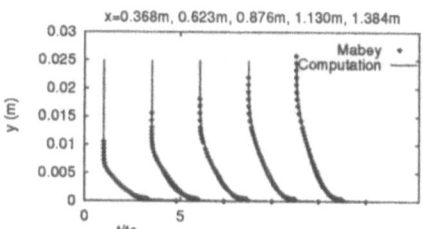

Figure 3: Temperature distribution

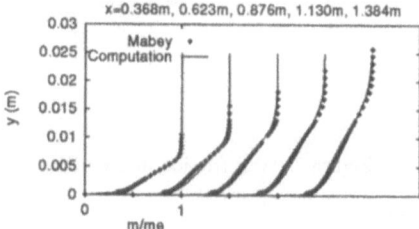

Figure 4: Mach number distribution

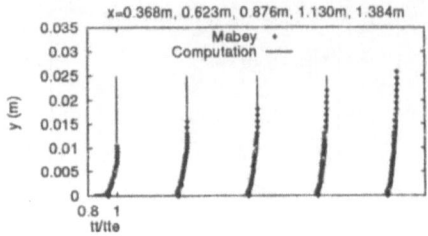

Figure 5: Total temperature distribution

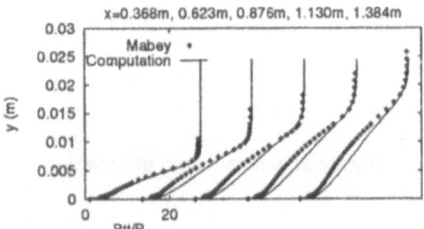

Figure 6: Total pressure distribution

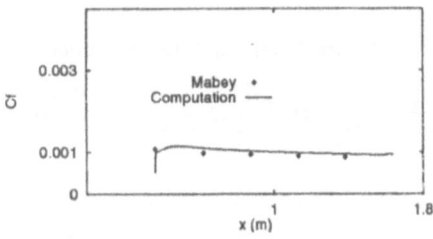

Figure 7: Wall friction distribution

230

SOLUTIONS OF NON-EQUILIBRIUM WALL BOUNDARY LAYERS WITH A LOW-Re-NUMBER SECOND MOMENT CLOSURE

S. Jakirlić[1], K. Hanjalić[2] and I. Hadžić[1]

[1]*LSTM, University of Erlangen, Cauerstr. 4, 91058 Erlangen, Germany*
[2]*Delft University of Technology, Lorentzweg 1, 2628 CJ Delft, The Netherlands*

SUMMARY

The paper outlines the turbulence model, the numerical method and presents major results of computation of wall boundary layer flows with severe adverse and favourable streamwise pressure gradients, as well as the constant pressure boundary layer, which serves as the reference flow. The flows considered served as test cases in the ETMA[1] project. The applied turbulence model is a second-moment (Reynolds stress) closure with new extensions, defined in invariant form, to account for the viscous and wall proximity effects, as well as to suppress the excessive growth of the length scale. The same model reproduced well all major mean flow and turbulence properties in a variety of wall-shear layers within a broad span of pressure gradients, including the boundary layer subjected to increasingly adverse pressure gradient leading to separation, as well as a strong favourable pressure gradient leading to laminarization.

INTRODUCTION

Computational methods and turbulence models are nowadays extensively used to solve complex flows and transport phenomena of practical relevance in many branches of industry, including aeronautics. In view of that, to report on new solutions of the wall boundary layers may sound outdated. Yet, except for simple cases of equilibrium flat plate boundary layers with no or small extra strains, accurate reproduction of these flows poses still a serious challenge to computational methods. This is a major reason why the wind tunnel experiments still serve as indispensable source of information, particularly if high data accuracy and reliability are required.

Wall boundary layers have for long served as basic flow configuration for studying the dynamics of wall turbulence, as well as for validating the analytical and numerical tools for solving the fluid flow problems. Accessibility to most measuring techniques, similarity features pertinent to no other common turbulent flows and parabolic nature which makes them easily tractable by numerical methods, are only some of decisive factors which have for long kept these flows in focus of fluid mechanics research. Equally important is their direct relevance to many branches of industry, particularly to aeronautics: the air flow around an aircraft can be viewed as a collection of boundary layers, subjected to various forms of extra strain rates and exhibiting

---

[1]*Efficient Turbulence Models for Aeronautics, the EC research project within the BRITE-EURAM programme, 1992-1995.*

a variety of features, with only few regions where the flow departs from the classical Prandtl boundary layer assumptions.

Turbulence modelling is in essence an approximation and its development has relied on the availability of experimental or other evidence about flow properties and tubulence structure in different types of flows. A comprehensive experimental data base *for simpler flows*, accumulated over the years and enriched also recently by direct numerical simulation (DNS), played a crucial role in the early stage of designing and validating turbulence models and numerical methods. However, more complex flows are still difficult to measure and almost impossible to solve by DNS, so that the availability of information about the flow and turbulence decreases markedly with the increase in the flow complexity. The lack of reliable data for more complex situations has been one of the major deterrents to further improvement of turbulence models, not only because of inability to verify the model performances and to diagnose its shortcomings, but even more because of a tendency among inexperienced users to employ the models to the prediction of complex cases with features absent in simpler flows for which the models were calibrated.

The situation has improved more recently with the appearance for new experimental techniques and DNS, though any flow complexity usually poses difficulties to all techniques so that the desirable information about flow and turbulence structure in more complex flows, which would substantially help in improving the turbulence models, are still missing.

This brief discussion of some major issues regarding the needs for better turbulence model for aeronautical flows, outlines also major motives for undertaking the European project ETMA, aimed at developing Efficient Turbulence Models for Aeronautics, which encompasses several parallel and mutually complementary experimental and modelling research programmes. The present paper reports on a new second-moment closure model with incorporated low-Re-number and wall vicinity modifications in invariant form, and its application to the computation of a range of attached wall shear flows with different external conditions. More specifically, the paper presents results of computations of wall boundary layers, subjected to strong and increasing adverse and favourable pressure gradients, leading to separation, or laminarization, respectively.

TURBULENCE MODEL

The new version of the low-Re-number second-moment closure, used for present computation (denoted in Figures as HJ model), has been described in more details elsewhere (Hanjalić and Jakirlić, 1994, [5], or Hanjalić et al., 1995, [2]). It will suffice here to state briefly that the model is based on the standard second moment closure with first-order linear models for the rapid and slow part of the pressure strain process (Rotta, Naot), with Gibson and Launder (1978), [3], wall reflections, and with the dissipation rate equation for defining the turbulence scale. This model, which served as the high-Re-number asymptote, was extended to account for the viscosity and wall-proximity effects. These modifications involve several empirical functions, but all of them are defined in terms of scalar parameters, invariant to the wall topography:

turbulence Reynolds number $Re_t = k^2/\varepsilon\nu$ and invariants of the anisotropy of the stress tensor, A_2, A_3, and of the dissipation-rate tensor, E_2 and E_3. The dissipation equation was extended by the addition of the terms suggested by Hanjalic and Launder (1980), [4], - S_{ε_4}, and by Jakirlić and Hanjalić (1994), [5], - S_l. The term S_{ε_4}, which enhances the effect of irrotational strain, can be written in the form $C_{\varepsilon_4}(\overline{v^2} - \overline{u^2})(\partial U/\partial x)\varepsilon/k$ (with $C_{\varepsilon_4} = 1.16$), which now separates the contribution of the shear strain from the irrotational one. This term, found to improve predictions of flows with strong adverse pressure gradients, proved also beneficial in computing the sink flows at higher values of acceleration parameter and in predicting laminarization at appropriate conditions. The term S_l, introduced specifically to deal with the separating and reattaching flows, is explained in more detail in the paper contributed for the computations of the incompressible recirculating flows (these proceedings).

The applied model can be summarized as follows:

- Stress transport equation:

$$\frac{D\overline{u_i u_j}}{Dt} = \frac{\partial}{\partial x_k}\left[\left(\nu + C_s \frac{k}{\varepsilon}\overline{u_k u_l}\right)\frac{\partial \overline{u_i u_j}}{\partial x_l}\right] - \left(\overline{u_i u_k}\frac{\partial U_j}{\partial x_k} + \overline{u_j u_k}\frac{\partial U_i}{\partial x_k}\right) + \dot{\Phi}_{ij} - \varepsilon_{ij} \quad (1)$$

$$\Phi_{ij,1} = -C_1 \varepsilon a_{ij} \qquad \Phi_{ij,2} = -C_2\left(P_{ij} - \frac{2}{3}P_k \delta_{ij}\right) \quad (2)$$

$$\Phi_{ij,1}^w = C_1^w f_w \frac{\varepsilon}{k}\left(\overline{u_k u_m}n_k n_m \delta_{ij} - \frac{3}{2}\overline{u_i u_k}n_k n_j - \frac{3}{2}\overline{u_k u_j}n_k n_i\right) \quad (3)$$

$$\Phi_{ij,2}^w = C_2^w f_w\left(\Phi_{km,2}n_k n_m \delta_{ij} - \frac{3}{2}\Phi_{ik,2}n_k n_j - \frac{3}{2}\Phi_{kj,2}n_k n_i\right) \quad (4)$$

$$C_1 = C + \sqrt{A}E^2 \quad C = 2.5AF^{1/4}f \quad F = \min\{0.6; A_2\} \quad (5)$$

$$f = \min\left\{\left(\frac{Re_t}{150}\right)^{3/2}; 1\right\} \qquad f_w = \min\left[\frac{k^{3/2}}{2.5\varepsilon x_n}; 1.4\right] \quad (6)$$

$$C_2 = 0.8A^{1/2} \quad C_1^w = \max(1 - 0.7C; 0.3) \quad C_2^w = \min(A; 0.3) \quad (7)$$

$$A = 1 - \frac{9}{8}(A_2 - A_3) \quad A_2 = a_{ij}a_{ji} \quad A_3 = a_{ij}a_{jk}a_{ki} \quad (8)$$

$$E = 1 - \frac{9}{8}(E_2 - E_3) \quad E_2 = e_{ij}e_{ji} \quad E_3 = e_{ij}e_{jk}e_{ki} \quad (9)$$

$$a_{ij} = \frac{\overline{u_i u_j}}{k} - \frac{2}{3}\delta_{ij} \quad e_{ij} = \frac{\varepsilon_{ij}}{\varepsilon} - \frac{2}{3}\delta_{ij} \quad Re_t = \frac{k^2}{\nu\varepsilon} \quad (10)$$

$$\varepsilon_{ij} = f_s \varepsilon_{ij}^* + (1 - f_s)\frac{2}{3}\delta_{ij}\varepsilon \quad f_s = 1 - \sqrt{A}E^2 \quad f_d = (1 + 0.1Re_t)^{-1} \quad (11)$$

$$\varepsilon_{ij}^* = \frac{\varepsilon}{k}\frac{[\overline{u_i u_j} + (\overline{u_i u_k}n_j n_k + \overline{u_j u_k}n_i n_k + \overline{u_k u_l}n_k n_l n_i n_j)f_d]}{1 + \frac{3}{2}\frac{\overline{u_p u_q}}{k}n_p n_q f_d} \quad (12)$$

- Dissipation rate transport equation:

$$\frac{D\varepsilon}{Dt} = \frac{\partial}{\partial x_k}\left[\left(\nu + C_\varepsilon \frac{k}{\varepsilon}\overline{u_k u_l}\right)\frac{\partial \varepsilon}{\partial x_l}\right] - C_{\varepsilon_1}\frac{\varepsilon}{k}\overline{u_i u_j}\frac{\partial U_i}{\partial x_j} - C_{\varepsilon_2}f_\varepsilon\frac{\varepsilon\tilde{\varepsilon}}{k}$$

$$+C_{\varepsilon_3}\nu\frac{k}{\varepsilon}\overline{u_j u_k}\frac{\partial^2 U_i}{\partial x_j \partial x_l}\cdot\frac{\partial^2 U_i}{\partial x_k \partial x_l} + S_{\varepsilon_4} + S_l \tag{13}$$

$$f_\varepsilon = 1 - \frac{C_{\varepsilon_2} - 1.4}{C_{\varepsilon_2}}\exp\left[-\left(\frac{Re_t}{6}\right)^2\right] \tag{14}$$

$$S_l = \max\left\{\left[\left(\frac{1}{C_l}\frac{\partial l}{\partial x_n}\right)^2 - 1\right]\left(\frac{1}{C_l}\frac{\partial l}{\partial x_n}\right)^2 ; 0\right\}\frac{\tilde{\varepsilon}\varepsilon}{k}A \quad l = \frac{k^{3/2}}{\varepsilon} \tag{15}$$

$$C_s = 0.22 \quad C_\varepsilon = 0.18 \quad C_{\varepsilon_1} = 1.44 \quad C_{\varepsilon_2} = 1.92 \quad C_{\varepsilon_3} = 0.25 \quad C_l = 2.5 \ .$$

THE NUMERICAL METHOD

Computations were performed by a finite-volume Navier-Stokes numerical solver for two-dimensional flows in an orthogonal coordinate system with colocated variable arrangement, (Obi, Peric and Scheuerer, 1991, [6]). For the flows considered here, the code was parabolized in so far that the pressure gradient was specified explicitly and the V component of the mean velocity calculated from the continuity equation. Apart from these approximations, the method retains all the features of a general code and accounts for the streamwise gradients of all variables. Some tests performed by applying the full (elliptic) form of the equations gave almost identical results for the 2-D boundary layers and sink flows. The typical number of grid points across the flow was about 100 in the case of sink flows and about 150-220 for the zero- and adverse pressure gradient boundary layers. The grid was clustered in the region close to the wall so that at least 30 grid points were within the viscosity-affected zone. The nearest point close to the wall was typically at $y^+ \approx 0.01 - 0.1$. The solutions were obtained by marching downstream and solving the equations at each streamwise positions by the iterative lower-upper (ILU) method after Stone. A typical size of the forward step was 2-5 % of the boundary layer thickness.

RESULTS AND DISCUSSION

We consider first the constant pressure boundary layer, as the reference flow case. Figures 1 to 5 show the computed results for major mean flow and turbulence properties for a wide range of Reynolds number. For comparison, we selected the experimental data of Wiegardt (1944), Klebanoff (1954) and Nagano et al. (1993), as well as DNS of Spalart (1988), each covering a different range of Reynolds numbers. In spite of the fact that these results originated over a time span of more than half a century, they agree remarkably well for the overlapping Reynolds numbers. Computations of the constant pressure boundary layers by different turbulence models have been reported frequently in literature, but it is not rear to find satisfactory agreement only for some, relatively narrow span of Re numbers, either at the high or low Re numbers. What is even more disturbing is that more frequently new models have been reported with applications to some more complex flows, without testing them in simple flows, only to discover later that some of them do not reproduce within a desirable accuracy even the simple constant pressure boundary layers. For that reason we show here an extensive comparison of the modelled solutions with different data and for different Re numbers.

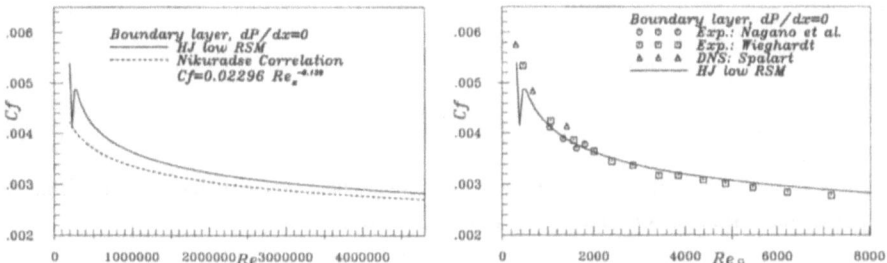

Fig 1 Friction coefficient in the zero pressure gradient boundary layer. Comparison with Nikuradse correlation (left) and with different experimental results (right)

Fig 1 presents the variation of the friction factor with Re number. The computations have been initiated at $Re_\theta \approx 300$ using the initial profiles of mean velocity, turbulent stresses and dissipation rate from DNS of Spalart. Due to incompatibility of the modelled equations and direct numerical solutions (mainly of the dissipation rate ε very close to the wall) there is a steep fall in C_f initially, but the new modelled solutions recover the value of C_f fast and further development of the friction factor along the plate is reproduced well up to very high Re numbers.

Fig 2 Mean velocity profiles in the zero pressure gradient boundary layer

Fig 3 Shear stress component in the zero pressure gradient boundary layer

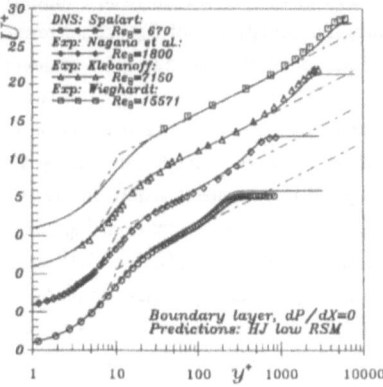

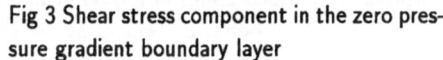

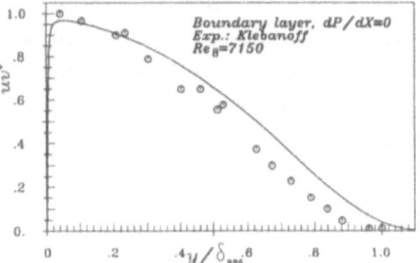

Fig. 2 shows the semi-logarithmic plots of mean velocity for four different Re numbers Re_θ

from 670 to 15570, compared with DNS and three sets of experimental data, showing excellent agreement for all four cases. Comparison with Klebanoff's measurements of turbulent shear stress is shown in Fig. 3 and of the rms of the velocity fluctuations in Fig. 4a for $Re_\theta = 7150$. Apart from some discrepancies in the outer region, where all computed turbulence fluctuations are slightly higher, the agreement is good.

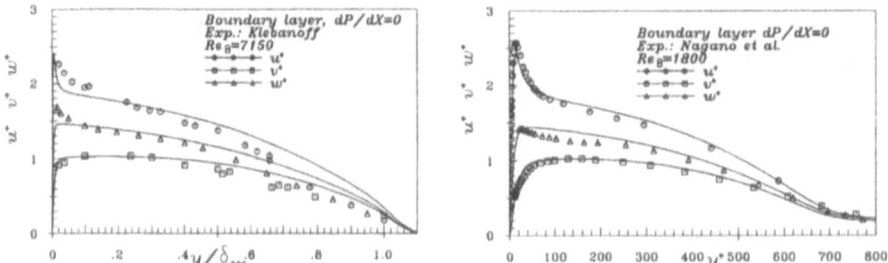

Fig 4 Normal stress components in the zero pressure gradient boundary layer. Comparison with experimental results of Klebanoff (left) and Nagano et al. (right)

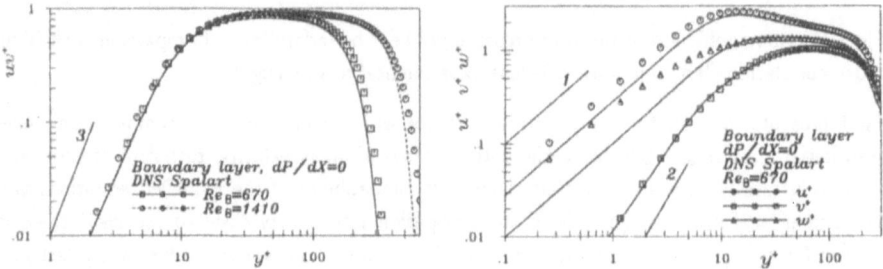

Fig 5 Blow-up of stress components in the zero pressure gradient boundary layer. Comparison with DNS of Spalart

It should be noted that the wall distance on abscissa was normalized with the boundary layer thickness δ_{995} which brings in a dose of uncertainty; an adjustment of δ_{995} could reduce this discrepancy. Even better agreement across the whole boundary layer has been achieved with the experimental results of Nagano et al. for $Re_\theta = 1800$, Fig. 4b. Although the two sets of profiles look similar, they differ notably in the near wall area, where the streamwise velocity fluctuation exhibits a more pronounced peak at low Re numbers (though, admittedly, the near wall resolution for the high Re number is not very good on the linear scale). In order to illustrate further the model performances very close to the wall, we compare in Fig. 5a log-log blow up of turbulent velocity fluctuations and of the shear stress with the DNS of Spalart for $Re_\theta = 670$. These data are selected because no experimental results for all turbulence components are available in such close vicinity of a solid wall. Except for $\tilde{w}^+$, the agreement is very good. Inability to reproduce $\tilde{w}^+$ is a consequence of the simple linear model of the rapid part of the pressure strain term and the cure for this is expected in a more advanced model of this term in general. It should be emphasized that all four stress components show proper asymptotic behaviour as the wall approaches, as illustrated by the slope of the log-log curves.

236

The next case considered is the boundary layer in an increasing adverse pressure gradient. Selected were the experimental results of Samuel and Joubert. This flow served also as one of the test cases in both AFOSR/HTTM Stanford 1980/81 and 1990/91 collaborative testing of turbulence models. Most models are able to reproduce this flow in the initial region while the pressure gradient is moderate, but many failed to reproduce the flow development further downstream. The last station at which the measurements were performed (x=3.4 m) is the most challenging, not only because of a very strong pressure gradient, but also because of an apparently delayed and cumulative action of the upstream variation of the external flow conditions upon turbulence parameters, which most models can not account for.

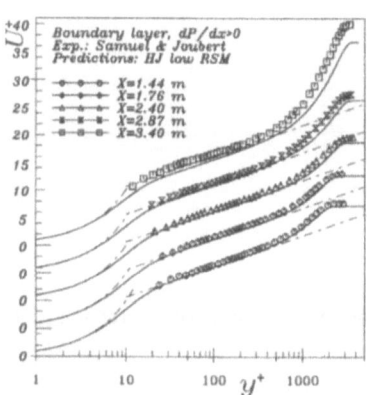

Fig 6 Mean velocity profiles in an increasingly adverse pressure gradient boundary layer

Fig 7 Wall shear stress in an increasingly adverse pressure gradient boundary layer

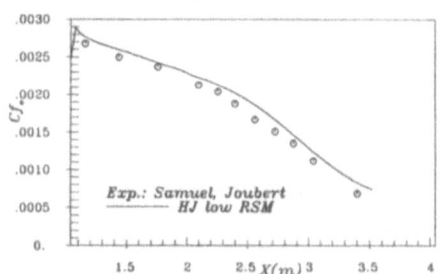

A good reproduction of the wall shear and mean velocity and turbulent stresses close to the wall in the present case indicates that the near-wall modelling in terms of stress- and dissipation-rate anisotropy captures well the whatever effects of the external conditions may penetrate into this region. On the other hand, the computed wake at the last station downstream is too low, what is probably a consequence of inadequacy of the linear pressure strain model and of the scale equation as a whole. Fig. 8 show details of the mean velocity profiles at several stations and Fig. 9 the profiles of the turbulent shear stress at corresponding positions. The agreement with the experimental data can be regarded as fully satisfactory.

The last we discuss the wall boundary layers subjected to severe acceleration. As a representative test case we selected the sink flow in a range of pressure gradients, expressed in terms of acceleration parameter $K = \nu/U_e^2(dU_e/dx)$, where U_e is the free stream velocity. Because of self-similarity features, the sink flow at a particular acceleration parameter retains constant integral properties (shape factor, friction coefficient) independent of the initial conditions and as such is very convenient for testing modelling concepts and numerical methods. The computations were performed for four values of K between 1.5×10^{-6} and 3.2×10^{-6}, for which Spalart (1986) performed direct numerical simulations. As it is well known, at $K \approx 3.0 \times 10^{-6}$ the boundary layer laminarizes irrespective of the initial level of turbulence. Solutions of sink flows at moderate pressure gradients have been reported earlier, but to the authors awareness, no model was reported which could predict equally well strongly decelerated and strongly accelerated boundary layers, and, particularly the laminarization of the latter at appropriate acceleration parameter.

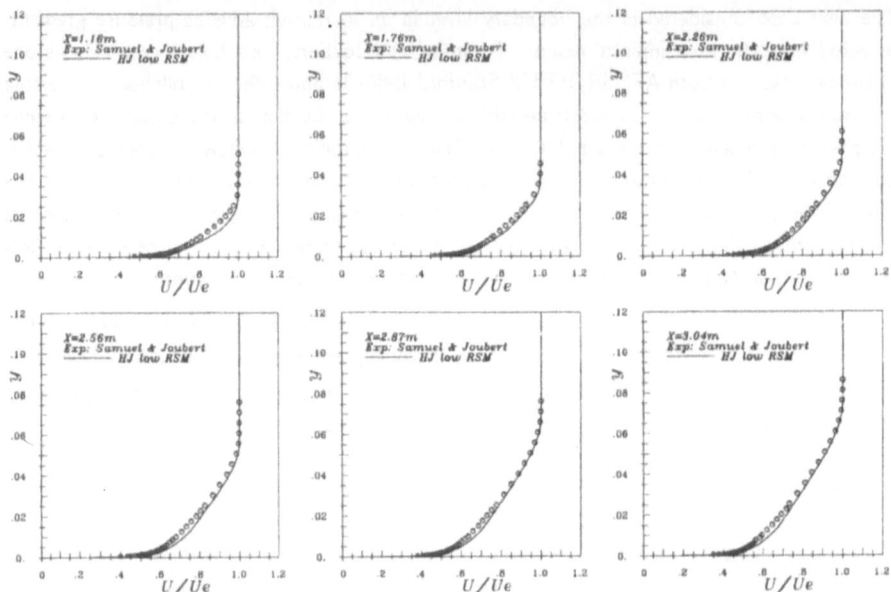

Fig 8 Mean velocity profiles at selected stations in an increasingly adverse pressure gradient boundary layer

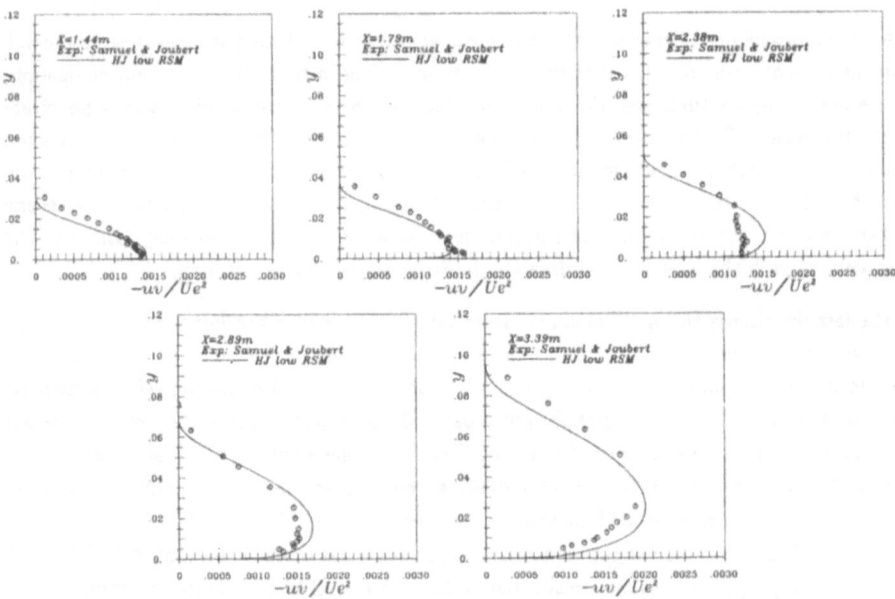

Fig 9 Reynolds shear stress profiles at selected stations in an increasingly adverse pressure gradient boundary layer

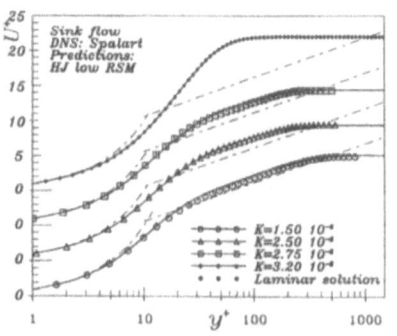

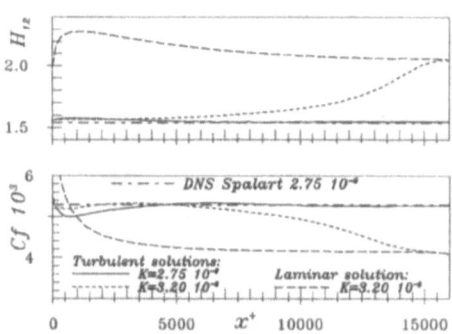

Fig 10 Semi-log plots of mean velocity profiles in a sink flow

Fig 11 Shape factor and friction coefficient in a sink flow

Fig. 10 shows the semi-logarithmic plot of the mean velocity for four values of K. The three profiles corresponding to the lower three values of K show excellent agreement with the DNS data, while the fourth profile for $K=3.2\times10^{-6}$ is the laminar solution, obtained by the model with initially turbulent flow. In order to illustrate further the model performances we concentrate on the values of K around the laminarization threshold by considering the flow evolution at $K=2.7\times10^{-6}$ (just below the critical value) and at $K=3.2\times10^{-6}$ (just above the critical value). Both computations started with initial profiles taken from DNS for $K=2.7\times10^{-6}$. If the K parameter was kept unchanged, after some adjustement both the shape parameter H and the friction factor C_f return to the initial value and remained constant, in full agreement with the values obtained by DNS, Fig. 11. If K was increased to 3.2×10^{-6}, the turbulence decayed steadily, and the flow laminarized, as shown both by the velocity profile in Fig. 10, and by the typically laminar values of both H and C_f, Fig. 11. For comparison, the solutions of purely laminar sink flow for the same K were also shown in Fig. 11.

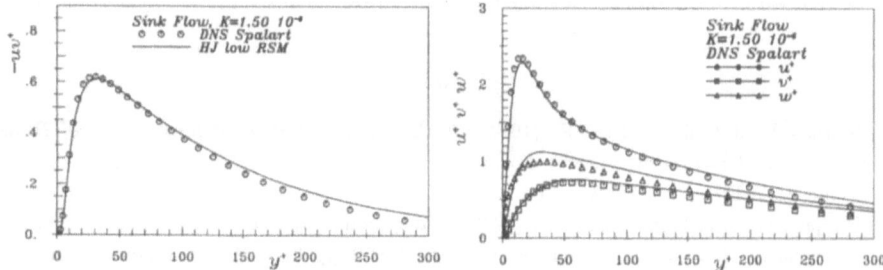

Fig 12 Stress components in a sink flow

Figure 12 compares the computed profiles of the turbulent shear stress and components of the velocity fluctuations with DNS results for $K=1.5\times10^{-6}$, showing excellent agreement except for a marginal discrepancy in $\tilde{v}^=$ component. As can be noted from Fig. 12., strong acceleration of the mean flow is reflected in a strong damping of the velocity fluctuations normal to the wall, causing an increase in the normal stress anisotropy, than in zero pressure gradient equilibrium flow. This behaviour seems well reproduced by the considered model.

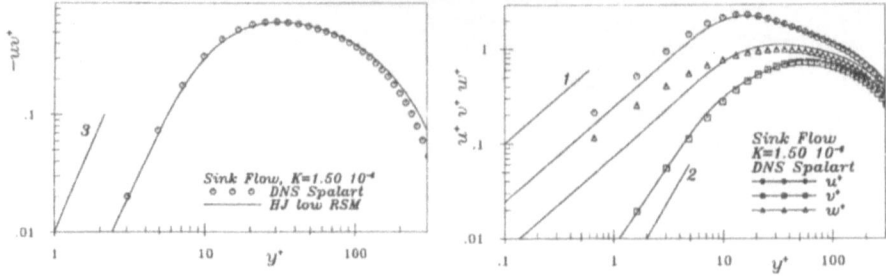

Fig 13 Blow-up of stress components in a sink flow

In spite of very strong pressure gradient, the wall asymptotic behaviour of stress components retain the same pattern as in the constant pressure gradient boundary layer, complying with the same limiting slopes, as the wall is approached, and this behaviour is also well captured by the model, Fig. 13.

CONCLUSIONS

A new version of the second moment closure model is presented which proved to reproduce well non-equilibrium wall-parallel turbulent flows at a range of pressure gradients and Reynolds numbers. The model contains the modifications for viscosity and wall-proximity effects in invariant form in terms of the turbulence Reynolds number and second and third invariants of the stress and dissipation-rate anisotropy tensors. The same model reproduced well decelerated boundary layer at increasing adverse pressure gradient approaching separation and the sink flow at a range of favourable pressure gradients including the case of laminarization at the appropriate value of the acceleration parameter. The model was also applied to constant pressure gradient wall boundary layer at a broad range of Reynolds number showing excellent agreement with several sets of experimental data and results of direct numerical simulation.

REFERENCES

[1] Hanjalić K., Jakirlić S., Durst F., (1994.): *A Computational Study of Joint Effects of Transverse Shear and Streamwise Acceleration on Three- Dimensional Boundary Layers*, Int. J. of Heat and Fluid Flow, Vol. 15, No.4, pp 269-282,

[2] Hanjalić K., Jakirlić S., Hadžić I., (1995.): *Computation of Oscillating Turbulent Flows at Transitional Re-Numbers*, Turbulent Shear Flows , Vol. 9, pp 323-342, Eds. F.Durst et al.,

[3] Gibson M.M., Launder B.E., (1978): *Ground effects on pressure fluctuations in the atmospheric boundary layer*, J. of Fluid Mech., Vol. 86, pp 491-511,

[4] Hanjalić K., Launder B.E., (1980): *Sensitizing the Dissipation Equation to Irrotational Strains*, J. of Fluids Engineering, Vol. 102, pp 34-40,

[5] Jakirlić S., Hanjalić K., (1994.): *On the Performance of the Second-Moment High- and Low-Re-Number Closures in Reattaching Flows*, Proc. Int. Symposium on Turbulence, Heat and Mass Transfer, Lisbon, 9-12 August,

[6] Obi S., Perić M., Scheuerer G., (1991): *Second-moment calculation procedure for turbulent flows with colocated variable arrangement*, AIAA J. Vol. 29, pp 585-590.

SYNTHESIS ON TEST CASE TC3 OF ETMA WORKSHOP

(FLAT PLATE BOUNDARY LAYERS)

Ch. Hirsch and Erbing Shang

Dept. of Fluid Mechanics, Vrije Universiteit Brussel

Pleinlaan 2, 1050 BRUSSEL

INTRODUCTION

The flat plate boundary layer is a classical wall flow against which any model of general use has to be validated. The primary character of the flow arises from the presence of the wall. Because of the no-slip condition, a low Reynolds number viscous layer is distinct from the high Reynolds number outer part of the boundary layer. This situation remains challenging for turbulence computations, since the asymptotic modelling of shear flows at high Reynolds number cannot be used everywhere in the flow. Similarly, in numerical methods, this flow requires large number of grid points to capture the small scales near the wall. The simple geometry relives us of all grid generation problems so that the physics of the models can be fully investigated. Furthermore, the case of boundary layers subjected to pressure gradients is very common in practice and has also to be examined. On the other hand, up to Mach number of 5, many aspects of supersonic boundary layer on adiabatic wall look very much the same as at low speeds, with variable density. A straightforward extension of the subsonic work can then be performed. The four sub-test cases are proposed and tested:

- *TC3-1*. Klebanoff's zero pressure gradient boundary layer flow;

- *TC3-2*. Samuel & Joubert adverse pressure gradient flow;

- *TC3-3*. Spalart sink flow;

- *TC3-4*. Mabey's supersonic flow on adiabatic wall (M = 4.5).

The solutions have been contributed by:

- DORNIER   –   Dornier Luftfart GmbH *(W. Haase and H. Horton)*

- DUT   –   Delft University of Technology *(R. Henkes)*

- IC   –   Imperial College *(A. Dafa'Alla, R. Harper and M. Gibson)*

- ECL   –   Ecole Centrale de Lyon *(I. Yudiana and M. Buffat)*

- LSTM   –   Erlangen *(S. Jalirlic, I. Hadzic and K. Hanjalic)*

- STANFORD –   Stanford University *(T. Hughes, K. Jansen and G. Hauke)*

- VUB   –   Vrije Universiteit Brussel *(Ch. Hirsch and E. Shang)*.

Totally seventeen turbulence models are tested covering algebraic eddy viscosity models, half equation models, one equation models, different two equation models (k-ε, k-ω, q-ζ, and k-τ), and the full Reynolds stress models. Some modified version of the above based models have been also checked. The new models arising from this ETMA project, VUB's algebraic eddy viscosity model and IC's q-ζ two equation model, are also examined.

This report is divided the into following parts: the overview of contributions, turbulence models, grids, computational results and conclusions.

AN OVERVIEW OF CONTRIBUTIONS

Since many turbulence models are applied, the following abbreviations are used

- BL   Baldwin-Lomax   (0)
- C   Chien   (2)
- CS   Cebeci-Smith   (0)
- CS–M   Cebeci-Smith Modified   (0)
- DHG   Dafa'Alla-Harper-Gibson $(q$–$\zeta)$
- H   Horton   (1/2)
- HJ   Hanjalic-Jakirlic   (RSM)
- HJ-M   Hanjalic-Jakirlic Modified   (RSM)
- HP   Hasid-Poreh   (1)
- HS   Hirsch--Shang   (0)
- JC   Johnson-Coakley   (1/2)
- JK   Johnson-King   (1/2)
- LB   Lam-Bremhorst   (2)
- LB-M   Lam-Bremhorst Modified   (2)
- LL   Lien-Leschziner   (2)
- LS   Launder-Sharma   (2)
- LSY   Launder-Sharma-YAP   (2)
- NR   Norris-Reynolds   (1)
- S   Speziale   (2)
      W   Wilcox   (2)

Table 1. Contributors and cases computed

Contributors	Cases	Turbulence Models
DORNIER	TC3-2	BL, CS, JK, JC, H, HP, LB, LS, S
DUT	TC3-1, TC3-2, TC3-3	C, CS–M, HJ, LL, LS, LSY, W
IC	TC3-1, TC3-2, TC3-3	DHG, LS
ECL	TC3-4	LB-M
LSTM	TC3-1, TC3-2, TC3-3	HJ-M
STANFORD	TC3-1, TC3-2, TC3-4	NR
VUB	TC3-1, TC3-2, TC3-3, TC3-4	HS, BL

The number in the brackets stands for type of turbulence models, i.e., "0" representing zero equation model. In Table 1, all the contributions are summarised.

TURBULENCE MODELS

Among these models, two of them arises from the ETMA project. One is 'HS', a new algebraic turbulence model reformulation developed by VUB. The model is independent of y^+, taking into account the correct near wall behaviour of turbulent shear flows. It is a one formula model covering both inner and outer regions. For the details of the model, one can refer to Hirsch and Shang [1].

The other is q-ζ model, a new two equation model. The chief merit of this model is that, unlike the equations for k and ε, the q and ζ equations are well conditioned near the wall, where the boundary conditions are specially defined, and the dependent variables vary linearly with distance from the wall when the wall is close. The details of the model formulation is given in Dafa'Alla [2].

DUT uses the modified version of 'CS' model, 'CS–M', which taking account the pressure gradient effect on the near wall damping function coefficient, A^+. While DORNIER applies the bending function on the original *Cebeci-Smith* model for calculating the eddy viscosity, which is still represented by 'CS' in this report.

One modified version on *Lam-Bremhorst k-ε* model, 'LB-M' model, is proposed by ECL [3], which taking into account an increase of the dissipation with the turbulent Mach number, $Ma_t = \sqrt{2k}/c$, by introducing an additional compressible dissipation ε_c into the k – equation.

In the Reynolds stress model, Hanjalic and Jakirlic [4] proposed a new version of their model, 'HJ–M'. The dissipation equation is extended by the addition of two terms, S_{ε_4} and S_l. The term S_{ε_4}, enhances the effect of irrotational strain, while S_l is introduced to deal with the separating and reattaching flow.

GRIDS

The grids used by the contributors are listed in Table 2. For test case TC3-1, TC3-2 and TC3-3, only VUB use the prescribed grids, 41x81, which has constant interval in the streamwise direction and is clustered in the normal direction by a stretching parameter of 1.1. The majority of the contributors added more or even double the grid points in both directions. With 1–Eq. and 2–Eq. models, the grids have to be clustered towards the leading edge of the flat plate to capture the transition phenomenon. Only STANFORD [5] use the less grid points, 60, in the normal direction to the wall while testing with Norris–Reynolds q model. All contributors put their first grid point at $y_1^+ \leq 1$.

For test case TC3-4, all contributors used the grids prescribed by VUB. The mesh proposed contains 113x81 grid points. The mesh spacing is determined by a constant stretching parameter of 1.11 in the normal direction. In the streamwise direction, a parabolic distribution is applied with clustering at the leading edge.

The solutions to be shown in this report are all grid independent. Therefore the question on the grid quality is avoided.

Table 2. Grids used

Contributors	Cases	Grids
DORNIER	TC3-2	$NJ=297$, $y_1^+ \approx 0.42 \sim 0.476$
DUT	TC3-1,TC3-2,TC3-3	160x160
IC	TC3-1,TC3-2,TC3-3	$NJ=100$, $y_1^+ \leq 1$
ECL	TC3-4	113x81, $y_1^+ \approx 0.1 \sim 1$
LSTM	TC3-1,TC3-2,TC3-3	$NJ=150$, $NJ=200$, $NJ=100$, $y_1^+ \approx 0.01 \sim 0.1$
STANFORD	TC3-1, TC3-2,	80x60, $y_1^+ = 0.5$
	TC3-4	113x81, $y_1^+ \approx 0.1 \sim 1$
VUB	TC3-1,TC3-2,TC3-3	41x81, $y_1^+ \approx 0.1 \sim 1$
	TC3-4	113x81, $y_1^+ \approx 0.1 \sim 1$

COMPUTATIONAL RESULTS

TC3-1: KLEBANOFF ZERO PRESSURE GRADIENT BOUNDARY LAYER FLOW

Klebanoff's test case is a zero pressure gradient incompressible flow over a flat plate [6]. The incoming boundary layer is artificially thickened. The free stream velocity is $u_e = 15.24$ (m/s). At the measuring station, the boundary layer thickness is $\delta = 76.2$ mm. The wall friction parameter [7] is $\sigma = u_\tau/u_e = 0.037$. The Reynolds number is $Re = u_e X_0/v = 4.2 \times 10^6$ based on a turbulent boundary layer development length of $X_0 = 4.32816$ m.

Five contributors calculate this test case, DUT, IC, LSTM, STANFORD, and VUB. Fig.1 to 3 present the results of wall friction coefficient distribution compared with the Nikuradse correlation and the experimental data. All models are accurate within 8%. Fig.4 to 6 show the velocity profiles compared with Klebanoff data. The deviations only occur at the outer edge, which are due to small differences in the wall shear stress.

STANFORD [5] has made a mesh refinement studies for 'NR' model, showing that the first grid point to the wall could be put at $y^+ = 5$. This can be attributed to the choice of q as a turbulent variable which is linear until $y^+ = 5$. Similarly, the length scale l is also linear in this region. Thus the scales responsible for turbulence modelling are accurately resolved by linear elements with at least one point at $y^+ \leq 5$. Finally, with the correct turbulence effects, linear elements can also accurately resolve the velocity profile since it too is linear in this region.

In contrast to the STANFORD [5] computation, for q–ζ model, IC has made the same exercise by using only 54 points in the cross-stream direction, with the first point located at $y^+ \approx 3$, and the remaining points being at $y^+ > 5$. The saving in nodes is also due to the linear variation of both q and ζ in the near wall region. The economy achieved is substantial in the flow considered here – up to 50% of 'LS' model CPU time.

TC3-2: SAMUEL–JOUBERT ADVERSE PRESSURE GRADIENT FLOW

Samuel & Joubert's experiment is performed in a long region of a flexible-roof tunnel with a finely controlled streamwise pressure distribution. Mean flow and fluctuating quantities are obtained in the incompressible turbulent boundary layer developing on a smooth wall in a pressure domain where both dp/dx and d^2p/dx^2 are positive (increasingly adverse pressure gradient). The boundary layer is tripped by pins center spacing at a streamwise position of x = 0.161m. This flow has a strong enough pressure gradient that it appears to be approaching separation at the end of the tested region, but no separation is present. The centerline velocities are obtained within the working region 0.8m ≤ x ≤ 3.5m. For the details one may refer to Samuel & Joubert [8] and the STANFORD 80-81 database (Case f0141a) [9].

Six contributors calculate this test case, DORNIER, DUT, IC, LSTM, STANFORD and VUB. Fig. 7 to 10 depict the wall friction coefficient distribution. For the algebraic models, see Fig.7 and Fig10, the 'HS' model predicts the wall friction very well. But 'BL' model shows slightly higher values when the flow approaches separation. The 'CS' model results obtained by Dornier show a very good agreement with the experimental data, while in contrast, the DUT underpredicts C_f by 18% with the 'CS–M' model, as shown in Fig.10. The difference might be due to the pressure gradient modification used in 'CS–M', which only affects the near wall region behaviour of the model.

For the 1/2–Eq. models, all models involved are in a good agreement with the measurements, however, apart from the area x > 3m near the last measured station which is closest to separation, as shown in Fig.8. DORNIER reported that at that the present test case does not reflect a strong non-equilibrium flow. The σ–values for all three 1/2–Eq. models, a measure for the equilibrium status of the flow, exhibit only moderate minimum values (in the area where flow approaches separation) between 0.7 for the 'H' and 'JK' model and 0.8 for the 'JC' model [10].

Fig.8 also demonstrates the C_f results from 1–Eq. models, 'HP' and 'NR' models. These two models show the same tendency, and behave like the 1/2–Eq. models.

The 2–Eq. model results are shown in Fig.9 and 10. The k–ω model result is the closest to the experimental data. The q–ζ model does a better job compared to the other k–ε model numerical results, possibly because q–ζ model is associated with smaller length scales which 2–Eq. models in general tend to overpredict in such circumstances.

For the Reynolds stress models, LSTM use the modified version of 'HJ' model, 'HJ–M', by adding two new terms, while DUT [11] applies the version without modified, 'HJ'. But the results from two versions are almost identical, showing some overprediction on the wall friction coefficient globally, see Fig.7 and 10.

The Reynolds stress results at streamwise position, x = 1.79m, where the pressure gradient is not so strong, are plotted in Fig, 11 to 13. The reasonable agreement can be achieved among all the models results and the measurements.

In the stronger pressure gradient region, x = 3.39m, 'BL' and 'CS' models overpredict the Reynolds stress, while 'HS' shows a very promising result, as shown in Fig.14. Still the difference is shown between 'CS' model and 'CS–M' model results. This may attribute to the bending function combining the inner and outer eddy viscosity formulae used in 'CS' model by DORNIER.

In general, all the 1/2-Eq. and 1-Eq. models, predict reasonably well the Reynolds stress, as seen in Fig.15 for x = 3.39m. The 'H' model and 'HP' model are superior to the 'JK', 'JC" and 'NR' models, 'JK' and 'JC' models underpredicting the Reynolds stress in the outer region and 'NR' model slightly overpredicting in the majority of region within the boundary layer.

All the 2-Eq. models used do not show very encouraging results on Reynolds stress, especially in the lower boundary layer region, as shown in Fig.16 and 17 for x = 3.39m. DORNIER [10] has investigated more details by plotting the eddy viscosity distribution, and shows that 2-Eq. models produce up to 50% more eddy viscosity, hence preventing "separation" and leading to an over-prediction in Reynolds stress.

The RSM model, 'HJ' model, predicts higher value on the Reynolds stress, as seen from Fig.17 where the DUT result is shown for x = 3.39m. In contrast, the modified version of 'HJ' model, 'HJ-M' model demonstrates improved results even though a slight underprediction close to the edge of the boundary layer can be observed. This might be attributed to the addition of an extra term, $S_{\epsilon 4}$, which enhances the effect of irrotational strain. This term is found to improve predictions of flows with strong adverse pressure gradients.

The velocity profiles are shown in Fig.18 to 20 for the measuring station x= 1.76m, where the pressure gradient is not strong enough. Since all the models predict the Reynolds stress distribution reasonably well, good agreement can be achieved between the numerical results and the measurements.

Fig.21 to 24 depict the velocity distribution at the position x= 3.04m located in the stronger pressure gradient region. For the algebraic models, the 'CS' model does the good job. The new model, 'HS', does demonstrate a reasonable result. Comparing to the other 1/2-Eq., 1-Eq. and 2-Eq. models, the 'H' and 'HP' models behave best. The modified version of 'HJ' does improve the numerical results, as seen in Fig. 18 and Fig.20.

TC3-3: SPALART SINK FLOW

The incompressible flow in a two dimensional convergent channel is called sink flow. In the sink flow the boundary layer edge velocity is inversely proportional to the distance from the sink. The statistical quantities of the turbulent sink flow are self-similar. The experiments show that the self-similarity is to be reached asymptotically as one approaches the sink [12]. In such a solution all the non-dimensional quantities like the acceleration parameter $K = \nu/U_e^2 (dU_e/dx)$, wall friction coefficient C_f, shape factor H, and thickness Reynolds number $Re_{\delta*}$ and Re_θ are independent of the streamwise coordinate x. The entrainment is zero, in the sense that the edge of the boundary layer is also a streamline. The sink flow boundary layer is the purest example of an 'equilibrium' turbulent boundary layer, a boundary layer with a shape that is invariant in the streamwise direction [13]. The Spalart DNS test case with $K = 1.5 \times 10^{-6}$ is used for the numerical simulation. For the details, one may refer to the file 'Sink_K.150' in the STANFORD 80-81 database [9] and Spalart [13].

Four contributors calculate this test case, IC, DUT, LSTM and VUB. Since the flow is self similar, the results at the middle of the flat plate are compared.

The numerical results of Reynolds stress and velocity distribution are compared with the Spalart DNS data in Fig. 25 ~ 26 and Fig.27 ~ 28. The VUB new algebraic model, "HS' model, still demonstrates very promising results for both Reynolds stress distribution and velocity profile. The 'CS–M' model, shows the overprediction on the Reynolds stress value, causing an underprediction on the velocity. Since in the near wall region, 'CS-M' gives

$-\overline{u'v'}^+ = o(y^{+4})$, it does not follow the correct near-wall region behaviour $-\overline{u'v'}^+ = o(y^{+3})$. Also the results show that the pressure gradient modification on the wall damping function coefficient, A^+, does not demonstrate the expected effect.

Within the 2–Eq. models used, the 'LL' model gives the better result. The q–ζ model is superior to 'LS' model even though it slightly underpredicts the Reynolds stress. In contrast to the adverse pressure gradient case, the k–ω model is no longer showing the best performance over the k–ε models.

The best performance is achieved by the modified version of *Hanjalic–Jakirlik* model, 'HJ–M' model, while the unmodified version of *Hanjalic–Jakirlik* model, gives slightly underpredicted Reynolds stresses, which in turn result in overprediction on the velocity. The difference is also due to the addition of the additional term S_{ε_4}, which enhances the effect of irrotational strain.

TC3-4: MABEY SUPERSONIC FLOW ON ADIABATIC WALL (M = 4.5)

Mabey's test case considers the development of the turbulent boundary layer on an adiabatic flat plate with zero pressure gradient at a freestream Mach number of 4.5. The experiment is performed on a flat plate with a length of 1.65m. Surface roughness is within 0.64 μm except for the transition strip from x = 2.584 to 5.08 mm where small glass spheres of 0.28 mm diameter are distributed sparsely. The profiles are measured with a combined Pitot and total temperature probe at streamwise positions x = 0.368, 0.623, 0.876, 1.130 and 1.384m. The good agreement of experimental data with both the logarithmic law and the outer law proves that the boundary layer is fully developed in this experiment [14]. The working fluid is treated as a perfect gas with constant specific heats. The flow Reynolds number per unit length is $Re/L = 2.82 \times 10^7$. For the details, one may refer to the test case ID No.7402 in AGARDograph No. 223 [15].

Three contributors calculate this test case, ECL, STANFORD and VUB. Fig. 29 presents the wall friction distribution for the models, 'LM–R', 'NR', and 'BL'.

The numerical results have been compared with the measurements at x = 0.623, 0.876, 1.130 and 1.384m. Since all the quantities to be compared at four measuring stations have similar behaviour, here only the numerical results at the last measuring station x = 1.384m are presented. Fig.30 to 34 present the numerical results of velocity, temperature, Mach number, Pitot total pressure and total temperature, respectively. Compared to the other two models, the 'NR' model does best. Even the other two also show reasonable results, these two models' results lie aside the experimental data. If one model overpredict one quantity, the other will underpredict it, vice versa.

CONCLUSIONS

The number of turbulence models tested is the largest compared to the other test cases within the ETMA workshop. Totally seventeen turbulence models are applied covering algebraic eddy viscosity models, half equation models, one equation models, different kinds two equation models (k-ε, k-ω, k-τ *and* q-ζ), and the full Reynolds stress models. Some modified version of the above based models have been also checked. The new models arising from this ETMA project, VUB's algebraic eddy viscosity model and IC's q-ζ two equation model, are also examined.

The grids used by the contributors are much finer than the prescribed grids. Only STANFORD (Hughes 1995) uses less grid points, 60, in the normal direction to the wall while testing with *Norris–Reynolds q*–model. When 1–Eq. and 2–Eq. models are applied, the grids have to be clustered towards the leading edge of the flat plate to capture the transition phenomenon. All contributors put their first grid point at $y^+ \leq 1$.

For the algebraic turbulence models, *Cebeci–Smith* model ('CS') gives best results with boundary layer codes. The pressure gradient modification on the wall damping function parameter, A^+, does not improve the near wall region behaviour of the *Cebeci–Smith* model. *Baldwin–Lomax* model performs reasonably well when the pressure gradient is not too strong to separate. The VUB new model shows very promising results.

For the 1/2–Eq. models, all models give reasonable results, but the *Horton* model always does a better job than the *Johnson–King* and *Johnson–Coakley* models because the "lack" in intermittency of the model provides, in the area where the strain rates are already small, an accurate stress profile.

The 1–Eq. models, *Hassid–Poreh* ('HP') model and *Norris–Reynolds* ('NR') model are all superior to the 2–Eq. models tested. This means that the ε –equation needs to be modified in future investigations.

The q-ζ model, and the well known k-ω model demonstrate better performance than the k-ε models applied with only an exception for the sink flow, where the *Lien–Leschziner* model shows results superior to the q-ζ and k-ω models.

While *Norris–Reynolds* ('NR') and q-ζ models are applied, since q is taken as a turbulent variable which is linear until $y^+ = 3$, and the length scale l (for 'NR' model) and the new variable ζ (for q-ζ model) are also linear in this region, the scales responsible for turbulence modelling are accurately resolved by linear elements with at least one point at $y^+ \leq 3$.

For the *Hanjalic and Jakirlic* Reynolds stress model, good results can be obtained if the additional term, $S_{\varepsilon4}$, is added, which enhances the effect of irrotational strain.

REFERENCES

[1] Hirsch, Ch. and Shang E., 1995, "A Numerical Evaluation of A New Algebraic Turbulence Model", see this Proceeding.

[2] Dafa'Alla, A., Harper, R and Gibson, M., 1995, "ETMA Workshop TC3: Flat Plate Boundary Layers", see this Proceeding.

[3] Yudiana, I. and Buffat, M., 1995, 'Simulations of Compressible Turbulent Boundary Layer using a Low Reynolds Model, Proceeding of ETMA Workshop (See this proceeding).

[4] Jakirlic, S., Hadzic, I and Hanjalic, K., "TC3: Flat Plate Boundary Layers", Proceeding of ETMA Workshop (See this proceeding).

[5] Hughes, T.J.R., Jansen, K. and Hauke, G., 1995, "Applications of the Finite Element Method to the Reynolds–Averaged Navier-Stokes Equations", see this Proceeding.

[6] Klebanoff., P. S., 1954, "Characteristics of Turbulence in a boundary layer with zero pressure gradient", NACA Technical Note 3178.

[7] Hinze, J. O., 1959, "Turbulence", McGRAW-HILL BOOK COMPANY, pp.493.

[8] Samuel, A.E. and Joubert, P.N., 1974, "A Boundary Layer Development in an Increasingly Adverse Pressure Gradient", Journal of Fluid Mech., Vol.66, pp.481-505.

[9] STANFORD 80-81 database (Case f0141a).

[10] Haase, W. and Horton, H., "The Samuel-Joubert Test Case Computed by a Boundary -Layer Method", Proceeding of ETMA Workshop (See this proceeding).

[11] Henkes, R.A.W.M., "Application of Turbulence Models to Incompressible Boundary Layers in Aeronautics", Proceeding of ETMA Workshop (See this proceeding).

[12] Jones, W.P. and Launder, B.E., 1972, 'Some Properties of Sink flow Turbulent Boundary Layers', J. of Fluid Mech., Vol.56, pp337-351.

[13] Spalart, P. R., 1986, 'Numerical Study of Sink flow Boundary Layers', J. of Fluid mech., Vol.172, pp.307-328.

[14] Fernholz, H.H. and Finley, P.J. 1980, "A Critical Commentary on Mean Flow Data for Two-dimensional Compressible Turbulent Boundary Layers", AGARDograph 253.

[15] Fernholz, H.H. and Finley, P.J. 1977, "A Critical Compilation of Compressible Turbulent Boundary Layer Data", AGARDograph 223.

249

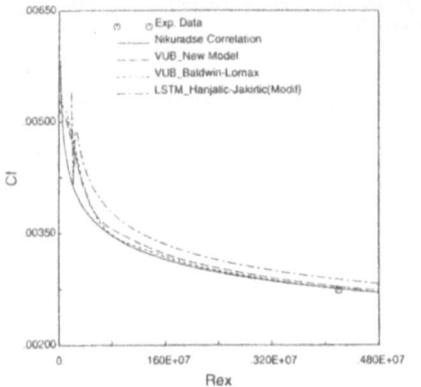

Figure 1: Wall friction coefficient distribution for zero pressure gradient flow (TC3-1)

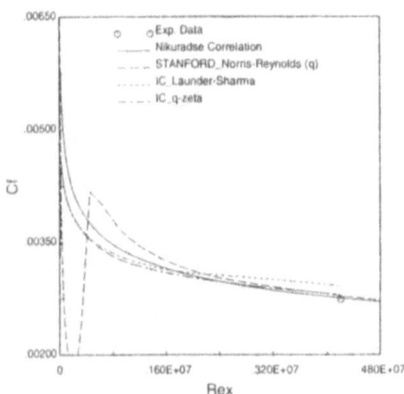

Figure 2. Wall friction coefficient distribution for zero pressure gradient flow (TC3-1)

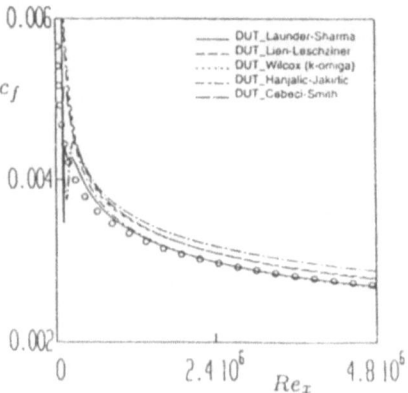

Figure 3: Wall friction coefficient distribution for zero pressure gradient flow (TC3-1)

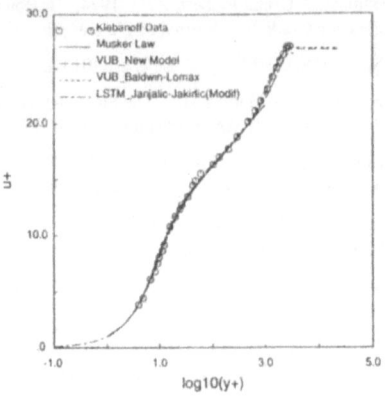

Figure 4: Velocity profile of Klebanoff zero pressure gradient flow (TC3-1)

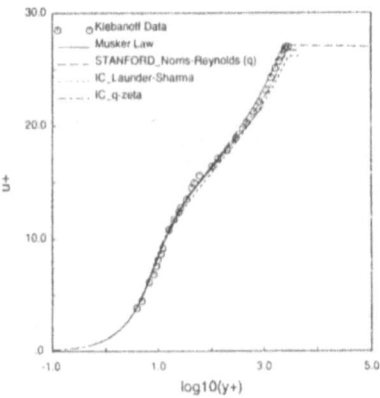

Figure 5: Velocity profile of Klebanoff zero pressure gradient flow (TC3-1)

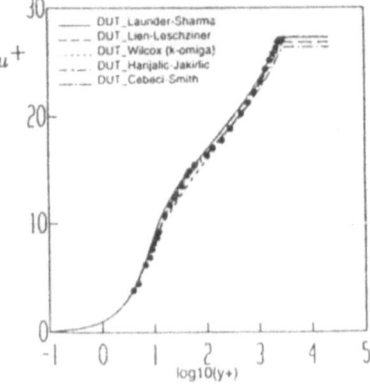

Figure 6: Velocity profile of Klebanoff zero pressure gradient flow (TC3-1)

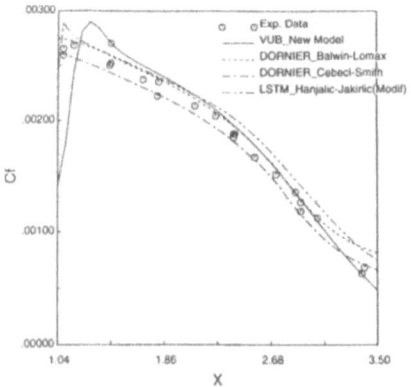

Figure 7: Wall friction coefficient distribution for Samuel-Joubert flow (TC3-2)

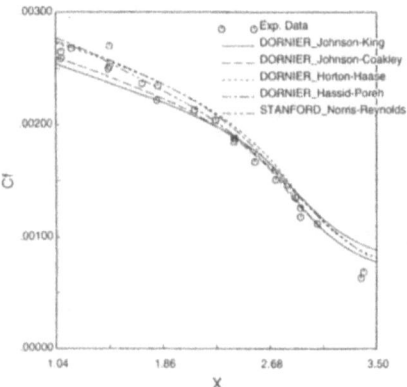

Figure 8: Wall friction coefficient distribution for Samuel-Joubert flow (TC3-2)

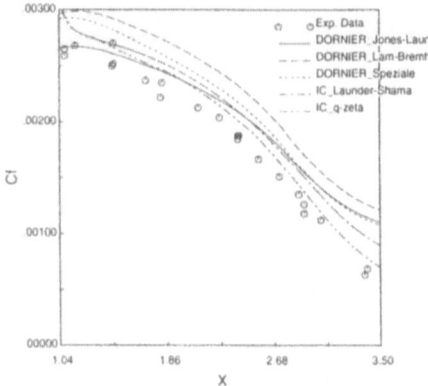

Figure 9: Wall friction coefficient distribution for Samuel-Joubert flow (TC3-2)

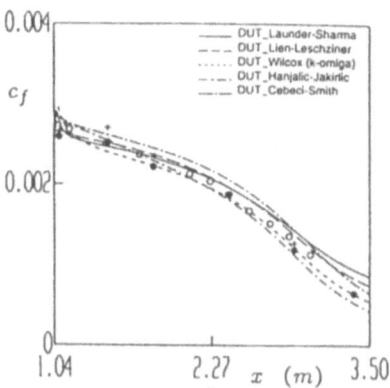

Figure 10: Wall friction coefficient distribution for Samuel-Joubert flow (TC3-2)

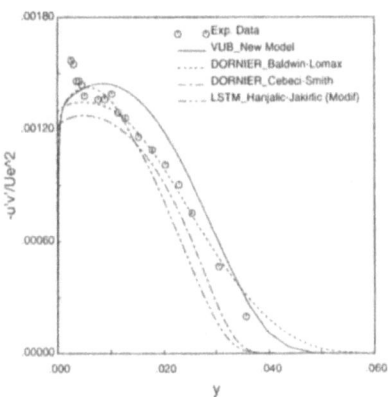

Figure 11: Reynolds stress distribution of Samuel-Joubert flow at x = 1.79m (TC3-2)

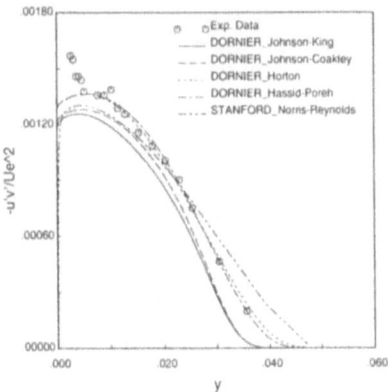

Figure 12: Reynolds stress distribution of Samuel-Joubert flow at x = 1.79m (TC3-2)

251

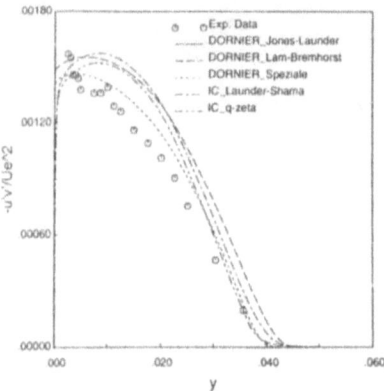

Figure 13: Reynolds stress distribution of Samuel-Joubert flow at x = 1.79m (TC3-2)

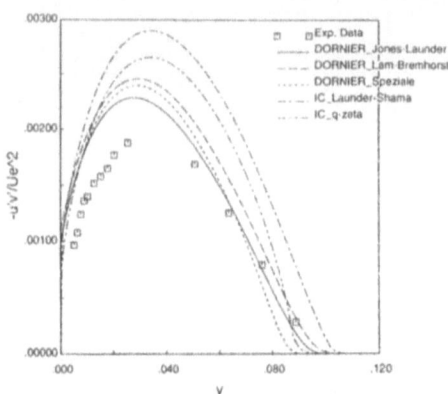

Figure 16: Reynolds stress distribution of Samuel-Joubert flow at x = 3.39m (TC3-2)

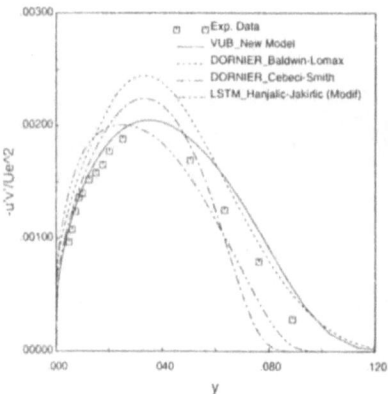

Figure 14: Reynolds stress distribution of Samuel-Joubert flow at x = 3.39m (TC3-2)

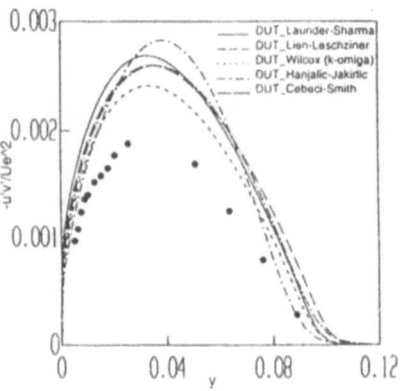

Figure 17: Reynolds stress distribution of Samuel-Joubert flow at x = 3.39m (TC3-2)

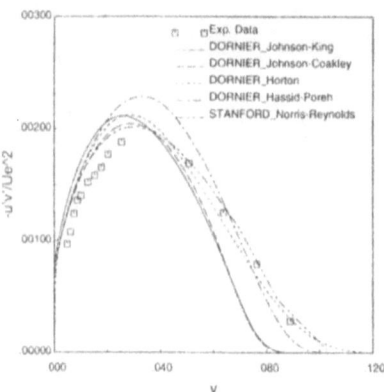

Figure 15: Reynolds stress distribution of Samuel-Joubert flow at x = 3.39m (TC3-2)

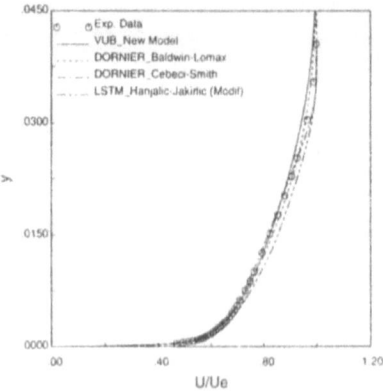

Figure 18: Velocity distribution of Samuel-Joubert flow at x = 1.76m (TC3-2)

252

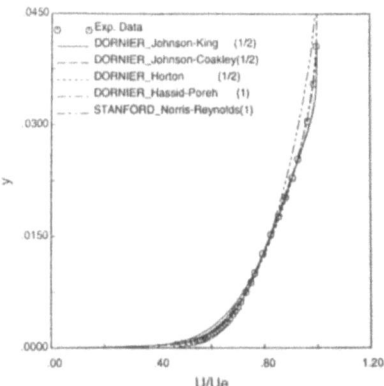

Figure 19: Velocity distribution of Samuel-Joubert flow at x = 1.76m (TC3-2)

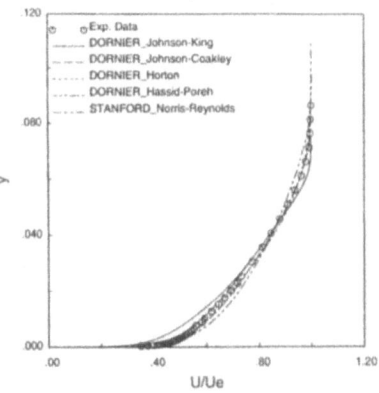

Figure 22: Velocity distribution of Samuel-Joubert flow at x = 3.04m (TC3-2)

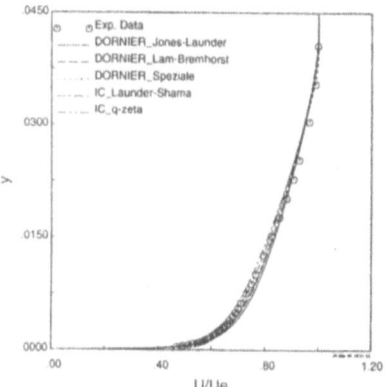

Figure 20: Velocity distribution of Samuel-Joubert flow at x = 1.76m (TC3-2)

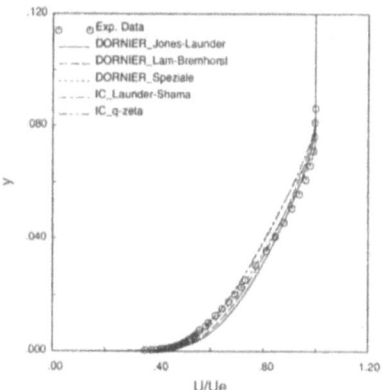

Figure 23: Velocity distribution of Samuel-Joubert flow at x = 3.04m (TC3-2)

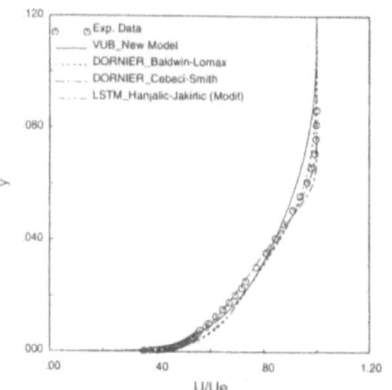

Figure 21: Velocity distribution of Samuel-Joubert flow at x = 3.04m (TC3-2)

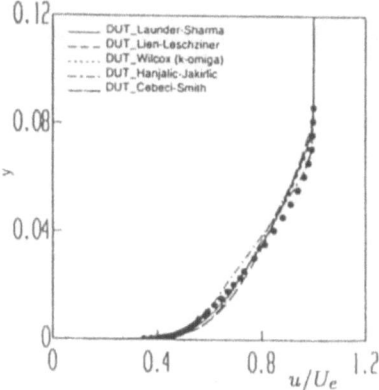

Figure 24: Velocity distribution of Samuel-Joubert flow at x = 3.04m (TC3-2)

253

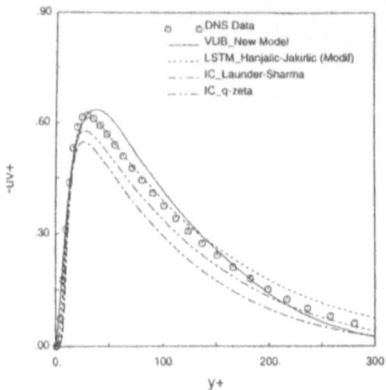

Figure 25: Reynolds stress distribution of Spalart sink flow (TC3-3)

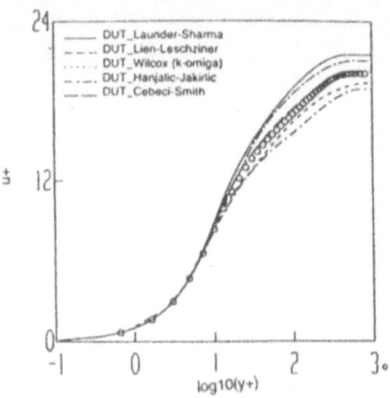

Figure 28: Velocity distribution of Spalart sink flow (TC3-3)

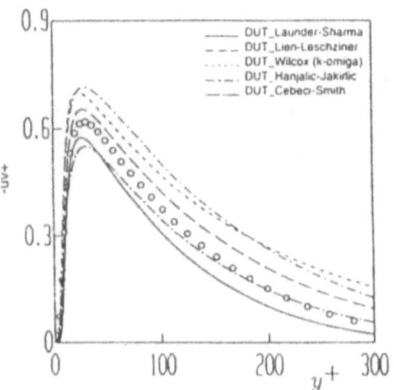

Figure 26: Reynolds stress distribution of Spalart sink flow (TC3-3)

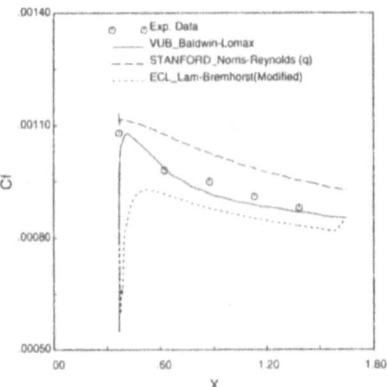

Figure 29: Wall friction coefficient distribution for Mabey supersonic flow (TC3-4)

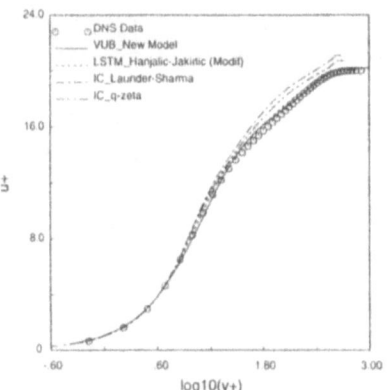

Figure 27: Velocity distribution of Spalart sink flow (TC3-3)

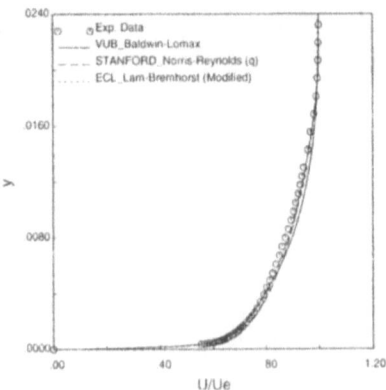

Figure 30: Velocity distribution at x = 1.384m for Mabey supersonic flow (TC3-4)

254

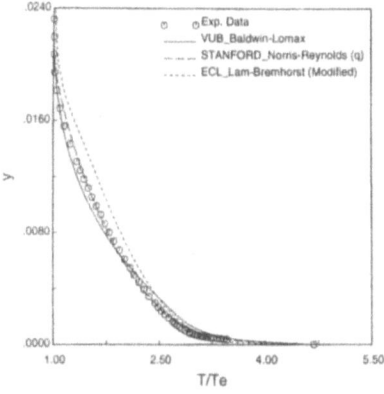

Figure 31: Temperature distribution at x = 1.384 m for Mabey supersonic flow (TC3-4)

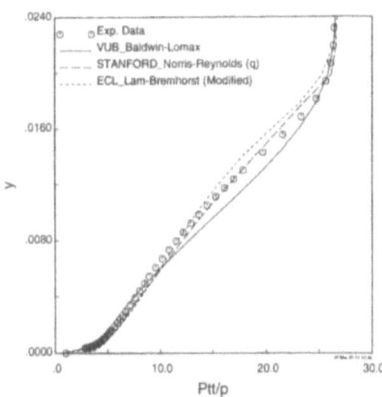

Figure 33: Pitot total pressure distribution at x = 1.38 4m for Mabey supersonic flow (TC3-4)

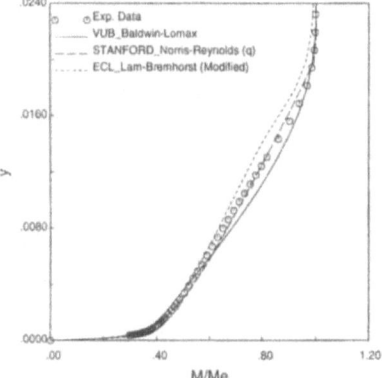

Figure 32: Mach number distribution at x = 1.384 m for Mabey supersonic flow (TC3-4)

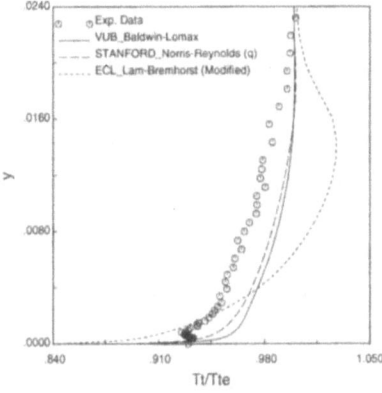

Figure 34: Total temperature distribution at x = 1.384 m for Mabey supersonic flow (TC3-4)

CHAPTER 5 : SHOCK REFLECTION

ETMA Test Case 6
SHOCK REFLECTION ON A FLAT PLATE
DESCRIPTION OF THE TEST CASE

R. Arina

Dipartimento di Ingegneria Aeronautica e Spaziale

Politecnico di Torino

Corso Duca degli Abruzzi, 24 - Torino (Italy)

SUMMARY

In this report we analyze the two-dimensional flow field generated by the reflection of an incident oblique shock wave on a flat wall, and its interaction with the turbulent boundary layer developing on the plate. The general features of the flow field are described, and the experimental measurements of Delery [1] are presented. Appropriate boundary conditions for the numerical calculation of this case are given.

1. INTRODUCTION

The interaction of a shock wave with a boundary layer is an important problem in aerodynamics, both theoretically and for practical applications. Its prediction is a challenge for the turbulence model community. The interaction with a shock wave imparts to a boundary layer a so rapid deceleration that the present models are unable to represent correctly. None of the most commonly used turbulence models are entirely satisfactory, lacking in the description of the behavior of turbulent structures interacting with a shock.

The interaction due to the reflection of an oblique shock over a flat wall is one of the basic configurations involving interactions between a shock wave and a turbulent boundary layer in supersonic flows. In the present case the incident shock is not excited by the incoming turbulent boundary layer, as in the case of the shock/boundary layer interaction over a compression ramp. Since a cause of unsteadiness is removed, the mechanism of generation of turbulence by compressibility effects may differ from the mechanism present in the interaction on a corner, at least for the incident wave. Moreover, being the wall straight, there are no influences due to geometric singularities.

For the present configuration, there is a modest amount of experiments available in the literature. The main concern in all these experiments, is to minimize the three-dimensional perturbations due to side walls. Side effects can be avoided employing an axisymmetric geometry. Some results for axisymmetric tests are reported in the Proceedings of the 1980 HTTM Stanford Conference [2], and in the report of Fernholz and Finley [3]. For planar geometries, the only available complete set of experimental data referring to an essentially two-dimensional flow field, is provided by Delery [1]. The present Test Case is focused on this set of experiments.

2. FLOW FIELD DESCRIPTION

When passing throughout the oblique incident shock, the incoming flow field is deflected toward the wall, and the necessity for the downstream flow field to remain parallel to the wall, entails the formation of a reflected shock. This reflected shock issues from the point where the incident shock interacts with the wall (Figure 1). The presence of the wall boundary layer makes the reflection quite complex. The incident shock reduces it strength as it penetrates into the boundary layer, and it progressively curves becoming almost normal in proximity of the boundary-layer sonic line. In this region the shock vanishes and the induced pressure rise tends to propagate upstream in the subsonic region of the boundary layer, causing a rapid thickening of the boundary layer just upstream the point of impingement. This tickening, causes in many circumstances the separation of the boundary-layer, and the generation of outgoing compression waves which rapidly collapse into the leading reflected shock.

The reflected shock is generated upstream of the point of impingement of the incident shock, therefore there will be an interaction between them. The interaction may happen inside the outer part of the boundary layer, or outside it, in the essentially inviscid part of the flow field. In the first case this crossing generates reflected waves which, interacting with the boundary layer along its sonic line, give rise to an outgoing expansion fan, which will interact with the reflected shock, weakening it. In the second case, at the crossing point a slip line will be generated. Also in this last case there is an outgoing expansion fan, similar to the one previously described, but being generated by the weaking of the impinging shock.

The expansion fan causes a strong deviation of the flow towards the wall, such that the outer stream is accelerated towards the wall. The boundary layer, downstream the point of impingement, decreases its thickness, and eventually reattaches. The external flow field deviates to become parallel to the wall, and compression waves, collapsing into another reflected shock, are genrated. This second shock, usually quite weak, having a slower incoming flow field, is less oblique of the leading reflected shock, therefore it will merge in it, giving rise to the reflected shock predicted by the inviscid theory far away from the wall.

From the above description, it is evident how the presence of strong pressure gradients, adverse in the upstream part, and favorable downstream, can give rise to very complex turbulent structures. This aspect is analyzed in the next sections.

3. EXPERIMENTAL SETUP

For planar geometries, the only available complete set of experimental data referring to a essentially two-dimensional flow field, is provided by Delery [1]. In the present Test Case we focus the attention on these experiments. The experiments have been performed at the ONERA S5CH continuos wind tunnel, with a test section of 300 mm span, and 150 mm height, equipped with a contoured nozzle producing a uniform supersonic free stream with Mach number M_∞ equal to 2.4, Reynolds number per meter Re_m equal to 7.01×10^6 m^{-1}, total pressure p^o equal to 74.9 kPa and total temperature T^o 317 K.

The straight wall was equipped with 66 pressure taps and eight orifices allowing the passage of pressure and temperature probes. To ensure a fully turbulent boundary layer in the interaction zone, the transition is triggered by a wire located at 430 mm upstream the working region. The incident oblique shock is generated by a flat plate shock generator, with sharp leading edge placed above the wall, with incidence with respect to the free stream direction (figure 1).

The experimental data include several flow situations, namely the cases of adiabatic and heated wall (T_{wall}/T_{rec}=2) with two different strengths of the mpinging oblique shock, corresponding to two angles of incidence of the shock generator (5 deg and 8.75 deg). In both cases, adiabatic and heated wall, with the weaker shock the boundary layer reaches the state of incipient separation, while with the strongest shock, it is subjected to extended separation. Being interested in the effects of strong pressure gradients on the turbulence structure, in the present Test Case we will focus our attention to the adiabatic case, with extended separation. In this case the incident oblique shock generated by the shock generator with incidence of 8.75 deg with respect to the free stream, has a slope of 31.8 deg.

The two-dimensionality of the measured flow field was checked by means of surface flow visualizations.

The region of interaction has been investigated by means of several experimental techniques, providing a detailed survey of the mean flow field and of the turbulence quantities. The mean pressure has been determined by means of static pressure probes and wall pressure measurements. The mean total temperature distribution has been measured with thermocouples. Mean and fluctuating velocity measurements were performed with a two-component laser Doppler velocimeter (LDV). The probe volume was of 0.2 mm diameter, enabling measurements up to 0.3 mm from the wall. The main limitation of accuracy comes from the flow tracking by particles, and from the bias produced by inappropriate seeding. This can be important in the shock region itself. Outside this region, the accuracy is of the order of 1-3% for the mean velocity and 5% for the rms fluctuating velocity, with a maximum of 15/20 % for the shear stress.

An important remark is that all the flow quantities (static pressure, total pressure, velocity, Reynolds stress components) have been directly determined.

4.   FLOW FIELD PROPERTIES

The measured wall pressure distribution is shown in figure 2. This curve exhibits three inflection points, typical of an interaction with separation. The existence of an extended separated region is indicated by the presence of a plateau, following the rapid pressure rise responsible of the separation process. The pressure distribution goes toward a maximun, nearly equal the to level predicted by the perfect fluid, well downstream the region of interaction.

In figure 3 the experimental isomach contours are presented. The local Mach number is obtained by the LDV measurements of the mean velocity, and the total temperature measurements. The countours in the external inviscid part, clearly show the crossing between the incident shock, and the leading reflected shock, as well as the weak secondary reflected shock, as explained above. The isomach lines in the boundary layer indicate how the subsonic part of the boundary layer is rapidly thickened by the interaction process.

In figures 4-10 we report the measured distributions along the wall normal, of mean flow quantities (streamwise velocity component, temperature, Mach number and static pressure), and of the turbulent kinetic energy, for some of the stations in the interval $80 \leq x \leq 220$ where the measurement were performed (figure 1). Analyzing the streamwise velocity profiles it can be argued that the boundary layer separates at $x \approx 130$ mm, and reattaches at $x \approx 180$ mm. The upper kink in the profile at x=120 mm is the trace of the incident shock, while at x-140 mm, we have the trace of the leading reflected shock, and at x=160 mm the distorsion of the upper part of the profile is induced by the downstream expansion fan.

The rise of the static pressure near the wall, for x=120 mm, is an effect of the upstream influence. In the pressure profile x=130 mm we can see the two traces of the incident (upper) and of the reflected (lover) shocks.

261

The experimental turbulent kinetic energy profiles, which have been obtained from the measured velocity fluctuations as

$$k = \frac{1}{2} \left(\overline{u'^2} + \overline{v'^2} + \frac{\overline{u'^2} + \overline{v'^2}}{2} \right) \ ,$$

show that the interaction with the shock entails a large increase in the fluctuation level. The maximum of velocity fluctuations, becomes very intense and is well detached from the wall. The Reynolds stress component measurements, not reported here, reveals also that a strong anisotropy develops. Moreover the interaction effects, very important in the interaction region, persist far downstream.

5. BOUNDARY CONDITIONS FOR THE NUMERICAL CALCULATIONS

An appropriate computational domain for the present test-case is a rectangle with lower boundary the wall.

The incoming shock wave can be directly enforced throughout the boundary conditions at the inlet boundary. The apex of the shock generator is placed outside the computational domain (figure 1). The intersection of the shock wave with the inlet boundary is at $x = 0$ and $y = 93.2$ mm. Below this point the free stream flow conditions must be imposed ($M_\infty = 2.4$, $Re_m = 7.01 \times 10^6 \, m^{-1}$, $p_\infty^\circ = 74.9 \, kPa$, $T_\infty^\circ = 317 \, K$), while above it, the flow behind the shock has the following properties: $M = 2.0494$, $p = 1.7036 \times p_\infty$, $\rho = 1.4567 \times \rho_\infty$, and the flow is deflected of 8.75 deg with respect to the x-axis.

The length of the flat plate is equal to 410 mm., and for the inviscid theory the incident shock wave should impinge the wall at $x \approx 150$. mm.

At the inlet boundary, near the wall, it is necessary to specify the incoming turbulent boundary layer profile at the beginning of the computational domain ($x = 0$ m). In [1] it is reported that at $x = -14$. mm the boundary layer is fully turbulent, and it has a thickness $\delta \approx 8$ mm, and the following integral characteristics: displacement thickness $\delta^* = 1.78$ mm, momentum thickness $\theta = 0.454$ mm, incompressible shape factor $H = 1.40$ and Reynolds number made with the displacement thickness $Re_{\delta^*} = 1.25 \times 10^4$.

REFERENCES

[1] Delery J., *Etude experimentale de la reflection d'une onde de choc sur une paroi chauffe en presence d'une couche limite turbulente*, La Recherche Aerospatiale, 1, 1992.

[2] Kline et. al.(ed.s), *Proceedings of the 1980 HTTM Stanford Conference on Complex Turbulent Flows*, Stanford University, 1981.

[3] Fernholz H.H., Finley P.J., *A further compilation of compressible boundary-layer data with a survey of turbulence data*, AGARDograph 263, 1981.

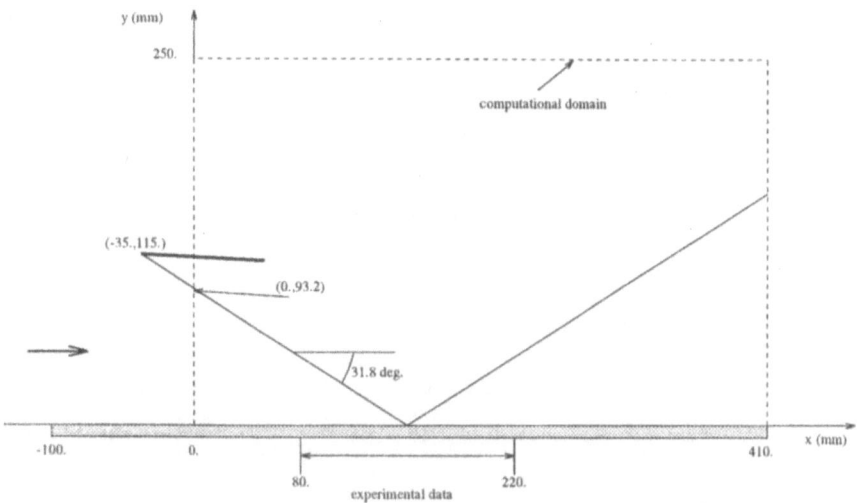

Figure 1: Test Case geometry

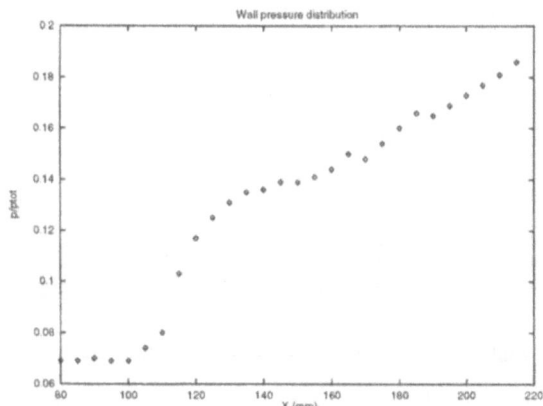

Figure 2: TC6 - Experimental wall pressure

263

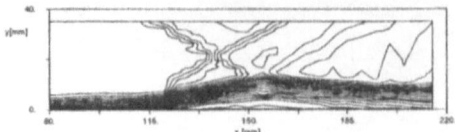

Figure 3: TC6 - Experimental isoMach contours

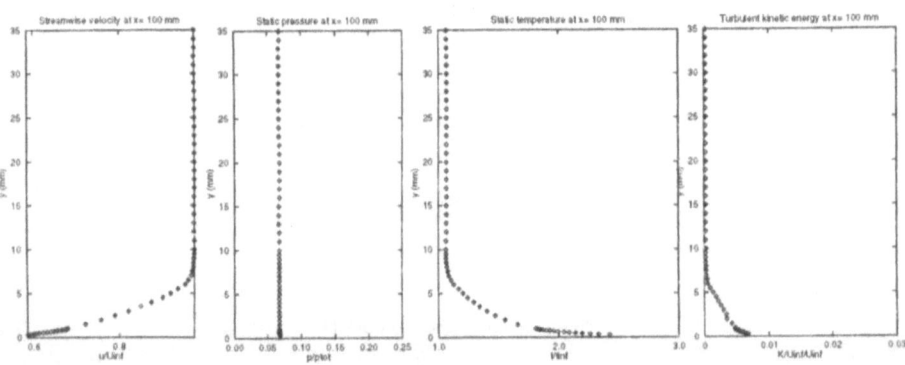

Figure 4: — experimental profiles x = 100 mm

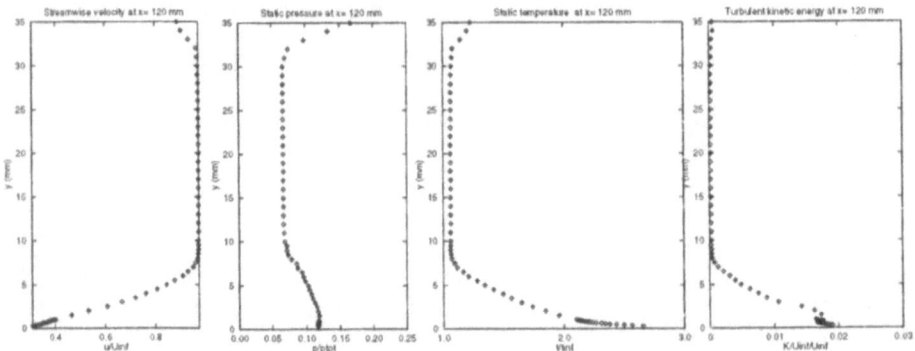

Figure 5: — experimental profiles x = 120 mm

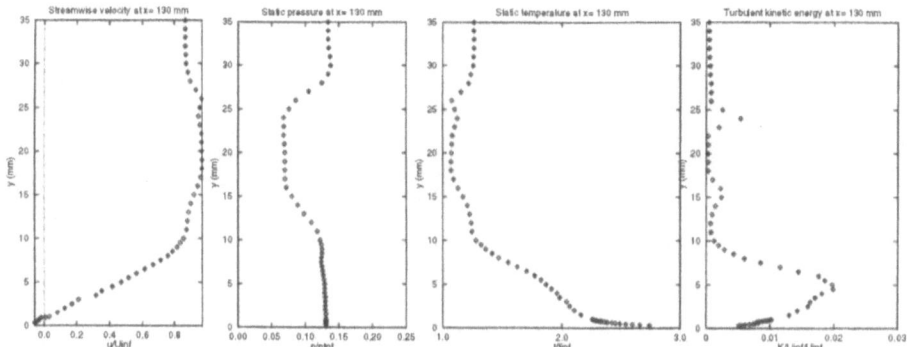

Figure 6: — experimental profiles x = 130 mm

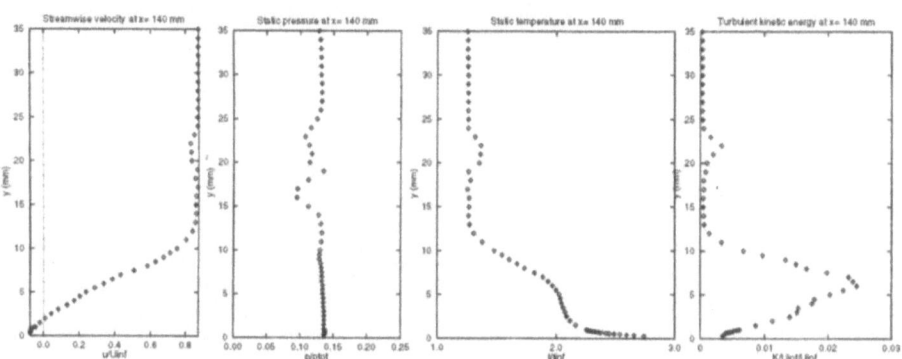

Figure 7: — experimental profiles x = 140 mm

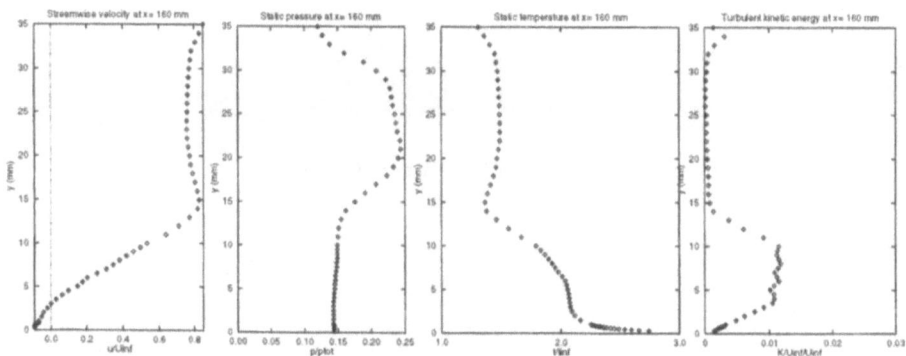

Figure 8: — experimental profiles x = 160 mm

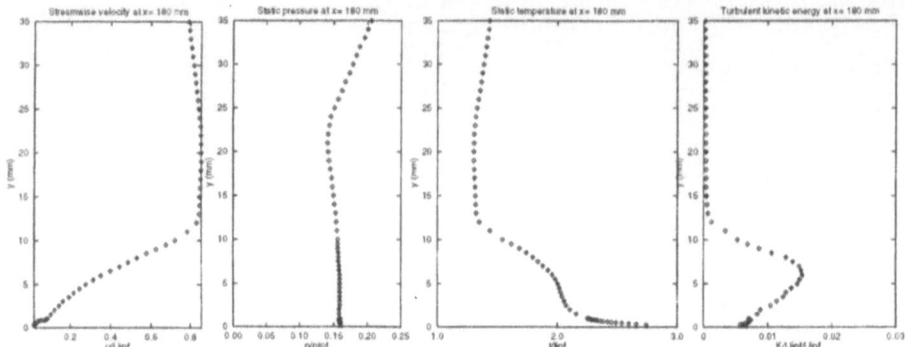

Figure 9: — experimental profiles x = 180 mm

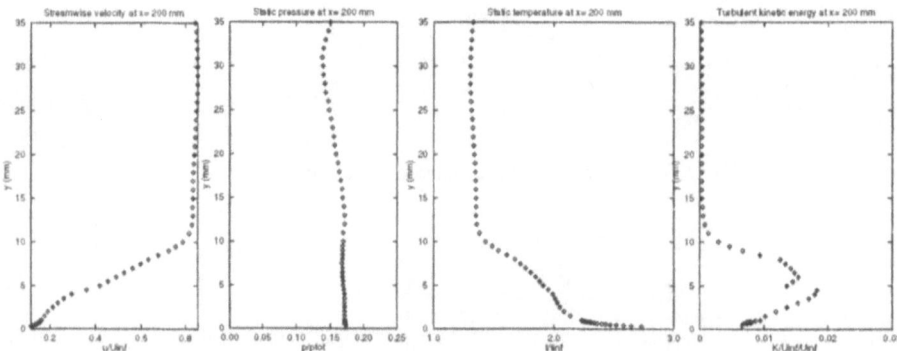

Figure 10: — experimental profiles x = 200 mm

NUMERICAL SIMULATION OF SHOCK REFLECTION
WITH A COMPRESSIBLE $K - \varepsilon$ MODEL

R. Arina and F. Ramella
Dipartimento di Ingegneria Aeronautica e Spaziale
Politecnico di Torino
Corso Duca degli Abruzzi, 24 - Torino (Italy)

SUMMARY

In this paper we calculate the flow field generated by the reflection of an incident oblique shock wave on a flat wall. The numerical solution is obtained using the Favre-averaged Navier-Stokes equations and a $k - \varepsilon$ turbulence model with compressibility corrections and near-wall modeling. The compressibility terms in the transport equation for the turbulent kinetic energy are based on the Sarkar et al. model [6,7], while the compressibility effects in the equation for the dissipation rate are based on the model of Coleman and Mansour [2]. The results are compared with the experimental data of Delery [3].

1. INTRODUCTION

The interaction due to the reflection of an oblique shock over a flat wall is one of the basic configurations involving interactions between a shock wave and a turbulent boundary layer in supersonic flows. In this case the incident shock is not excited by the incoming turbulent boundary layer, as in the case of the shock/boundary layer interaction over a compression ramp. Therefore the compressibility effects on turbulence may be different. Moreover there are no influences due to geometric singularities.

Due to the strong pressure gradients, the turbulent structure may be affected by compressibility effects, therefore low-speed turbulence models may fail in modeling the correct behavior of the fluctuating field, leading to wrong predictions of the mean flow field. In the recent years several models for the compressibility effects have been proposed. Most of them have been applied to free shear layers. Very few is known about their behavior in the case of boundary layers. Therefore it is interesting to check their validity in cases such as the shock reflection flow field, in order to gain new informations and new hints for future improvements. The purpose of this paper is to test some recent ideas in the case of a compressible turbulent shear layer developing along a wall and subjected to strong pressure gradients inducing separation.

For a correct simulation, it is a crucial issue to specify the shape of the incoming boundary layer as close as possible to the experimental one. The shock going inside the boundary layer become weaker and vanish. The pressure signal carried by the shock is necessarily transmitted in the upstream direction through the subsonic inner part of the boundary layer. The deep of the upstream influence appears as being essentially a function of the height of this subsonic part, therefore the fullness of the incoming boundary layer profile plays a determinant role in the interaction mechanism. The available experimental data provide only integral parameters for the upstream incoming boundary layer, therefore an iterative procedure is necessary to generate profiles with appropriate fullness. In the present case, the incoming mean velocity and temperature profiles, have been obtained by the Coles' method, extended to compressible flows [4], based on the compressible logarithmic wall law. In this way the calculated profiles exactly reproduce the measured momentum thickness.

2. GOVERNING EQUATIONS

The gas is assumed perfect (air, $\gamma = 1.4$, $Pr = 0.72$), and the Shuterland law for the molecular viscosity is adopted. The equations for the mean flow quantities, are obtained from the Navier-Stokes equations by Favre averaging. The closure for the Reynolds stresses is obtained by a $k - \varepsilon$ two-equation model, with near-wall treatment. Their form is described in the next section.

The governing equations for the mean flow field $U = (\bar{\rho}, \bar{\rho}\tilde{u}_i, \bar{\rho}\tilde{E})^T$, take the form

$$\frac{\partial U}{\partial t} + \frac{\partial F_j}{\partial x_j} = \frac{\partial V_j}{\partial x_j} \ , \tag{1}$$

where the inviscid fluxes F_j are

$$F_j = \begin{pmatrix} \bar{\rho}\tilde{u}_i \\ \bar{\rho}\tilde{u}_i\tilde{u}_j + \bar{p}\delta_{ij} \\ (\bar{\rho}\tilde{E} + \bar{p})\tilde{u}_i \end{pmatrix} \ ,$$

and V_j are the viscous fluxes

$$V_j = \begin{pmatrix} 0 \\ 2(\bar{\mu} + \mu_t)\left(\tilde{S}_{ij} - \frac{1}{3}\delta_{ij}\frac{\partial \tilde{u}_k}{\partial x_k}\right) - \frac{2}{3}\bar{\rho}k\delta_{ij} \\ \tilde{u}_i\left[2(\bar{\mu} + \mu_t)\left(\tilde{S}_{ij} - \frac{1}{3}\delta_{ij}\frac{\partial \tilde{u}_k}{\partial x_k}\right) - \frac{2}{3}\bar{\rho}k\delta_{ij}\right] + \gamma(\frac{\mu}{Pr} + \frac{\mu_t}{Pr_t})\frac{\partial \tilde{e}}{\partial x_i} \end{pmatrix} \ .$$

$\bar{\rho}\tilde{E} = \bar{\rho}\tilde{e} + \bar{\rho}\frac{\tilde{u}_i\tilde{u}_i}{2} + \bar{\rho}k$, $k = \frac{\widetilde{u_i''u_i''}}{2}$ is the turbulent kinetic energy, $\tilde{S}_{ij} = \frac{1}{2}\left(\frac{\partial \tilde{u}_i}{\partial x_j} + \frac{\partial \tilde{u}_j}{\partial x_i}\right)$ is the Favre-averaged rate of strain tensor. The equations are discretized in space by a finite volume method. For the convective terms an improved version of the AUSM flux-vector splitting formulation [5] with MUSCL extrapolation for second order accuracy, is employed, while the viscous terms are discretized by a central scheme. The time discretization is first order and implicit, and the linear system of equations formed at each time-integration step, is solved by a nonlinear Krylov subspace projection method (preconditioned GMRES in Jacobian-free formulation).

The computational domain, is parameterized by a structured grid stretched near the wall.

On the upper boundary, as well as on the inflow boundary, the supersonic inviscid flow is fully specified. The incident shock wave is specified at the inlet boundary, and the incoming boundary layer is fully described using the compressible extension of the Coles' method for the mean flow quantities [4]. Local equilibrium conditions and a mixing length model are assumed for the specification of the turbulence quantities. At the outlet boundary, we impose characteristic boundary conditions in the inviscid part, and second order extrapolation in the boundary-layer region. The wall is considered adiabatic.

3. COMPRESSIBLE $k - \varepsilon$ MODEL

In the present work we use a near-wall two-equation turbulence model of the $k - \varepsilon$ type. In the model we introduce a variable density extension of the near-wall treatment proposed by Chien for incompressible flows [1], combined with the implementation of the model for the compressibility terms proposed by Sarkar, Speziale et al. [6,7], for the k equation, and by Coleman and Mansour [2] for the ε equation.

Consistent with [6], the dissipation is decomposed into solenoidal and compressible parts: $\varepsilon = \varepsilon_s + \varepsilon_c$. Here, ε_s represents the dissipation associated with the energy cascade. The length and time scales are built up from ε_s, which is obtained from a modeled transport equation. Sarkar et al. model the compressible dissipation in the form $\varepsilon_c = \alpha M_t^2 \varepsilon_s$, where $M_t = (\widetilde{u_i'' u_j''}/\gamma R \tilde{T})^{0.5}$ is the turbulence Mach number, and $\alpha = 0.5$.

The Reynolds stress tensor is modeled in the standard eddy viscosity form,

$$R_{ij} = -\frac{2}{3}\bar{\rho}k\delta_{ij} + 2\mu_t\left(\tilde{S}_{ij} - \frac{1}{3}\tilde{S}_{kk}\delta_{ij}\right) \quad ,$$

with the eddy viscosity

$$\mu_t = \bar{\rho}C_\mu f_\mu \frac{k^2}{\varepsilon_s} \quad ,$$

where $C_\mu = 0.09$, and f_μ is a wall damping function.

The modeled transport equation for k is, neglecting the turbulent fluctuations in the viscosity

$$\frac{\partial}{\partial t}(\bar{\rho}k) + \frac{\partial}{\partial x_i}(\bar{\rho}\tilde{u}_i k) = R_{ij}\frac{\partial \tilde{u}_i}{\partial x_j} - \bar{\rho}\varepsilon + \overline{p'\frac{\partial u_i'}{\partial x_i}}$$

$$-\overline{u_i''}\frac{\partial \bar{p}}{\partial x_i} + \overline{u_i''}\frac{\partial \overline{\tau_{ij}}}{\partial x_j} + \frac{\partial}{\partial x_i}\left[\left(\bar{\mu} + \frac{\mu_T}{\sigma_k}\right)\frac{\partial k}{\partial x_i}\right] \quad , \qquad (2)$$

where $\sigma_k = 0.75$, and

$$\overline{\tau_{ij}} = -\frac{2}{3}\mu\frac{\partial \tilde{u}_k}{\partial x_k}\delta_{ij} + \mu\left(\frac{\partial \tilde{u}_i}{\partial x_j} + \frac{\partial \tilde{u}_j}{\partial x_i}\right) \quad ,$$

is the mean viscous stress.

Following Sarkar et al. [6], the pressure dilatation correlation is modeled as follows

$$\overline{p'\frac{\partial u_i'}{\partial x_i}} = -\alpha_1 \mathcal{P} M_t^2 + \alpha_2 \bar{\rho}\varepsilon_s M_t^2 \quad ,$$

where $\mathcal{P} = R_{ij}\frac{\partial \tilde{u}_i}{\partial x_j}$ is the turbulence production, $\alpha_1 \approx 0.4$ and $\alpha_2 \approx 0.2$. The mass flux term is modeled by the gradient transport hypothesis [7]

$$\overline{u''}_i = \frac{C_\mu f_\mu}{\bar{\rho}\sigma_\rho}\frac{k^2}{\varepsilon_s}\frac{\partial \bar{\rho}}{\partial x_i} \quad .$$

with $\sigma_\rho = 0.5$.

The dissipation rate equation is assumed of the form

$$\frac{\partial}{\partial t}(\bar{\rho}\tilde{\varepsilon}_s) + \frac{\partial}{\partial x_i}(\bar{\rho}\tilde{u}_i\tilde{\varepsilon}_s) = -\left[\frac{1}{3} + n(\gamma - 1)\right]\bar{\rho}\tilde{\varepsilon}_s\frac{\partial \tilde{u}_i}{\partial x_j}$$

$$C_{\varepsilon 1}f_1\frac{\tilde{\varepsilon}_s}{k}R_{ij}\left(\frac{\partial \tilde{u}_i}{\partial x_j} + \frac{1}{3}\frac{\partial \tilde{u}_k}{\partial x_k}\delta_{ij}\right)$$

$$-C_{\varepsilon 2}f_2\bar{\rho}\frac{\tilde{\varepsilon}_s^2}{k} + \frac{\partial}{\partial x_i}\left[\left(\bar{\mu} + \frac{\mu_T}{\sigma_\varepsilon}\right)\frac{\partial \tilde{\varepsilon}_s}{\partial x_i}\right] + \bar{\rho}\mathcal{E} \quad , \qquad (3)$$

Where $C_{\varepsilon 1} = 1.5$, $C_{\varepsilon 2} = 1.83$, $\sigma_{\varepsilon} = 1.3$. The damping functions f_{μ}, f_1, f_2, as well as the wall dissipation term $\mathcal{E}$, and the wall boundary condition D for $\varepsilon \, (= \tilde{\varepsilon}_s + D)$, are modeled as proposed by Chien [1]. The first member of the rhs, takes into account the variation of the viscosity along a streamline [2]. For an isentropic compression $n \approx 0.7$.

The solution of the turbulence transport equations is decoupled from the solution of the equations governing the mean flow field. At each iteration the mean flow quantities are kept frozen and the turbulence transport equations are solved with an implicit, first order accurate, discretization in time. In space, second order upwind discretization (MUSCL) of the convective terms, and centred discretization for the diffusive ones, are employed. The block matrix system is solved by Gaussian elimination.

4.   NUMERICAL RESULTS AND DISCUSSION

The flow field, with free-stream conditions $M_{\infty} = 2.4$, Reynolds number per meter $Re_m = 7.01 \times 10^6$ m^{-1}, total pressure $p^0 = 74.9$ kPa and total temperature $T^0 = 317$ K has been discretized with 121×121 grid points, and a clustering near the wall such that , the first point is always placed at $y^+ \leq 1$. In this way the solution can be considered grid independent.

The rectangular computational domain is $[0., .4] \times [0., .25]$ meters, In figure 1 we show the isopressure contours. They clearly show the typical structure of the shock reflection over a wall, with a leading reflected shock arising upstream the point of impingement of the incident shock, an expansion fan just downstream the *inviscid* point of reflection, followed by a weak compression fan merging outside into the leading reflected shock.

However this scenario is not typical of a strongly separated flow. This impression is confirmed by the pressure wall distribution (figure 2) where the calculated distribution does not have the typical plateau, present instead in the experiments. From the skin friction coefficient (figure 3) it is evident that the region interested by a separated flow is quite small.

The next figures 4-10, comparing the computed profiles along the wall normal of streamwise velocity component, pressure and turbulent kinetic energy with the experimental ones, confirm that the boundary layer thickening is not properly simulated.

In spite of the good agreement with the experimental inlet profiles, obtained by matching the experimental momentum thickness at $x = 0$, it is evident that the present turbulence model is not able to correctly mimic the flow physics, as soon as the upstream influence starts it effects. In conclusion we can argue that the $k - \varepsilon$ model, in the present form does underestimate the strong increase of the turbulent fluctuations induced by the shock, and therefore underpredicted the separation region.

REFERENCES

[1] Chien, K.-Y.,*Predictions of Channel and Boundary-Layer Flows with a Low-Reynolds-Number Turbulence Model*, AIAA J., 20, 1, pp. 33-38, 1982.

[2] Coleman, G.N. and Mansour, N.N., *Simulation and Modeling of Homogeneous Compressible Turbulence under Isotropic Mean Compression*, Eighth Symposium on Turbulent Shear Flows, 1991, Munich, Germany.

[3] Delery J., *Etude Experimentale de la Reflection d'une Onde de Choc Sur une Paroi Chauffee en Presence d'une Couche Limite Turbulente*, La Recherche Aerospatiale, 1, pp. 1-23, 1992.

[4] Huang P.G., Bradshaw P. and Coakley T.J., *Skin Friction and Velocity Profile Family for Compressible Turbulent Boundary Layers*, AIAA J., 31, 9, pp. 1600-1604, 1993.

[5] Liou M.-S. and Steffen C. J., *A New Flux Splitting Scheme*, J. Comput. Phys., 107, pp. 23-39, 1993.

[6] Sarkar S., Erlebacher G., Hussaini M.Y. and Kreiss H.O., *The Analysis and Modeling of Dilatational Terms in Compressible Turbulence*, J. Fluid Mech., 227, pp. 473-493, 1991.

[7] Speziale, C.G. and Sarkar S., *Second-Order Closure Models for Supersonic Turbulent Flows*, AIAA Paper 91-0217, 1991.

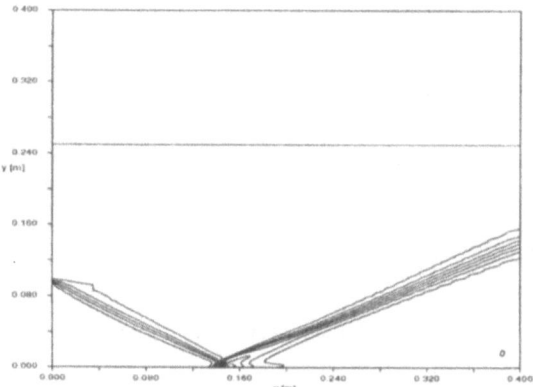

Figure 1: Isopressure contours

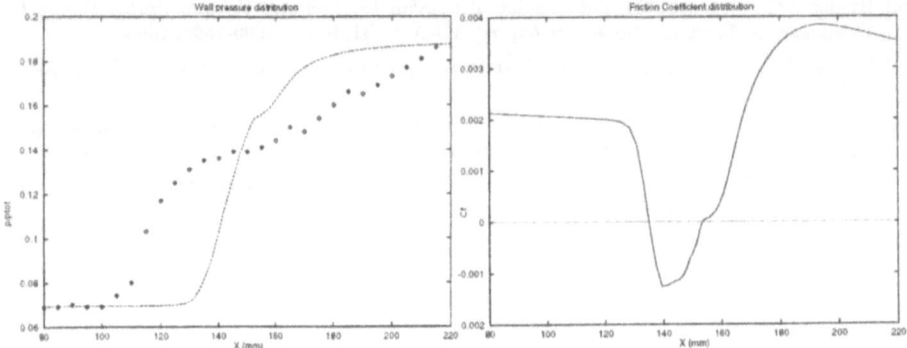

Figure 2: — present calculation, ◇ experimental data [3]

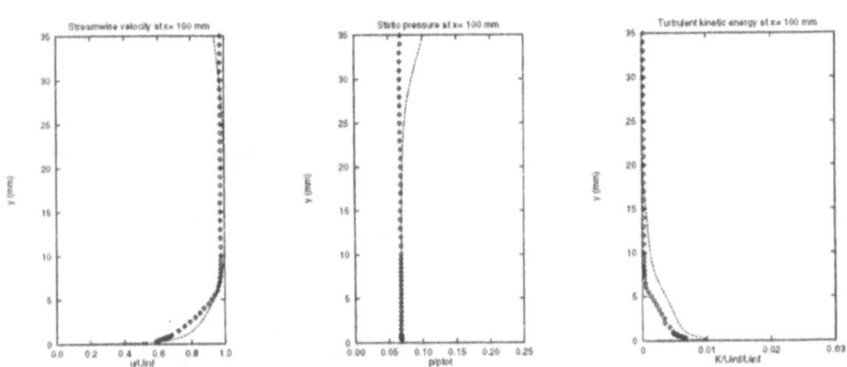

Figure 3: — present calculation, ◇ experimental data [3]

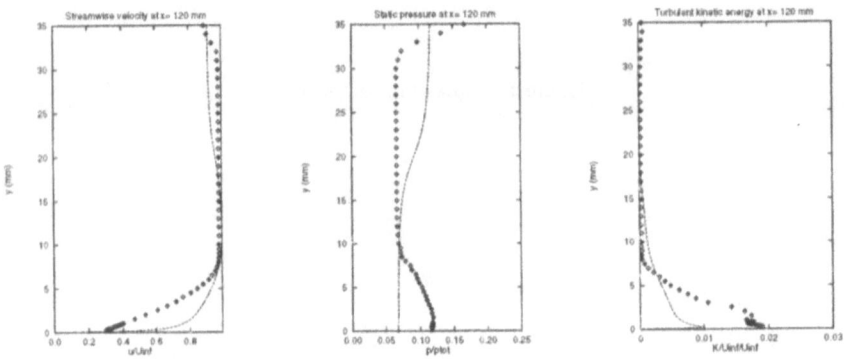

Figure 4: — present calculation, ◇ experimental data [3]

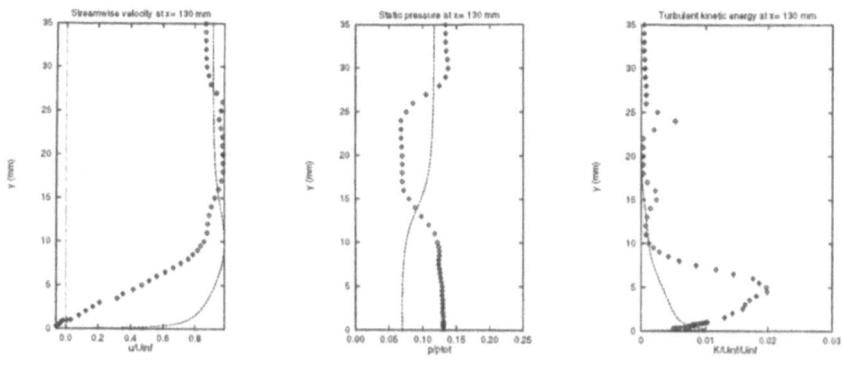

Figure 5: — present calculation, ◇ experimental data [3]

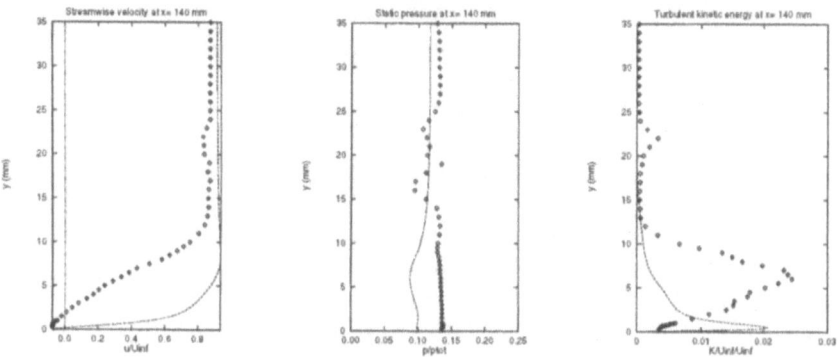

Figure 6: — present calculation, ◇ experimental data [3]

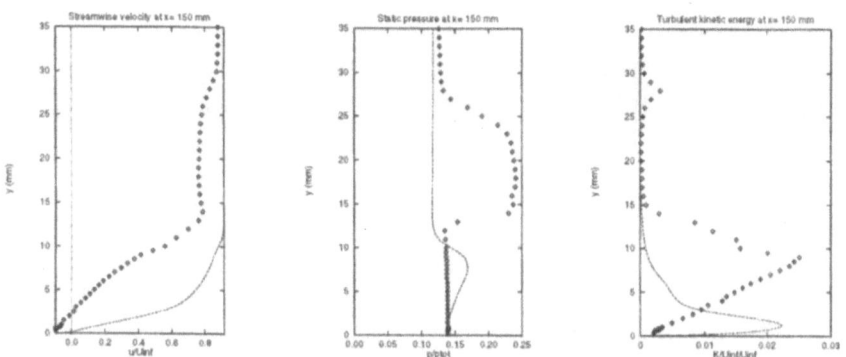

Figure 7: — present calculation, ◇ experimental data [3]

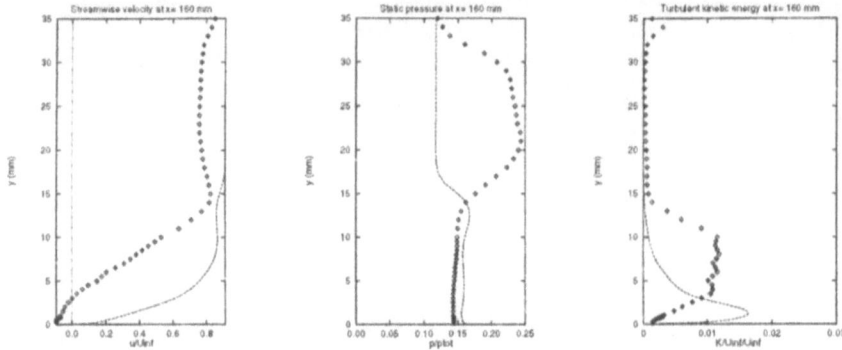

Figure 8: — present calculation, ◇ experimental data [3]

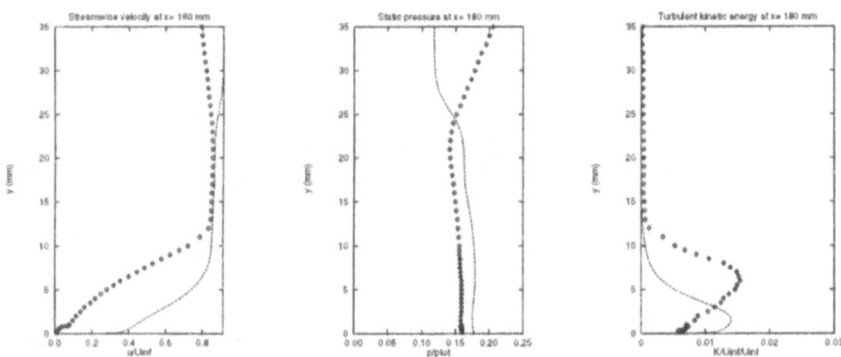

Figure 9: — present calculation, ◇ experimental data [3]

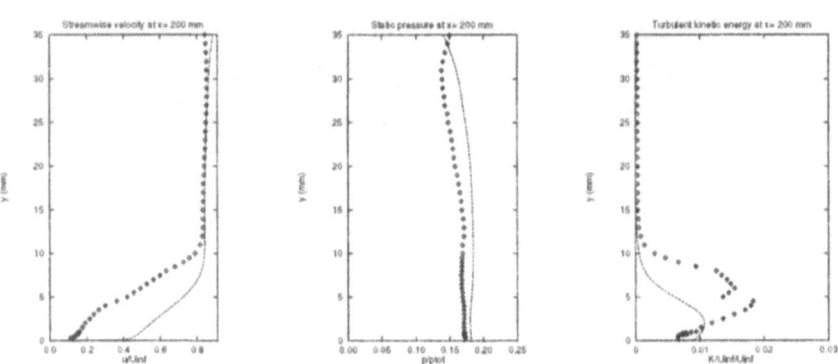

Figure 10: — present calculation, ◇ experimental data [3]

Simulations of Shock Reflection on Flat Plate using a Low Reynolds $k - \epsilon$ Model

I. Yudiana, M. Buffat

LMFA,CNRS URA 263, ECL, UCB Lyon I

36, avenue Guy de Collongue

69131 Ecully cedex France

Abstract

This paper describes the simulation results of a shock reflection on a flat plate performed using an incompressible low Reynolds $k - \epsilon$ model combined with compressibility modeling. It appears that this model behaves well for such flows.

Introduction

The shock reflection on a flat plate is a complex test case which embodies all the difficulties of compressible turbulent flows, with a shock/boundary layer interaction and compressibility effects. For the present test case, we focus the attention on the experiments of Delery [3]. The interaction between the shock and the boundary layer induces a large recirculation bubble and a thickening of the boundary layer. Two intensities of the incident shock wave were experimentally considered. Here, only the configuration corresponding to the formation of an extended separated region will be simulated. The difficulty of this test case is to predict accurately the separation. Different factors influence the precision (the mesh,the inlet conditions, the compressibility terms, the near wall modeling,...). To minimize numerical dependencies between the contributors, the prescribed inlet profiles have been used and the incoming shock wave has been enforced by the boundary conditions.

Turbulence models

We consider the standard $k - \epsilon$ model, improved with some modifications to take into account the compressibility and the low Reynolds effects.

Dilatation-dissipation models

To take into account an increase of the dissipation with the turbulent Mach number $Ma_t = \frac{\sqrt{2k}}{c}$, an additional compressible dissipation ϵ_c is added into the k equation. Different models, based on direct numerical simulations results, have been proposed for ϵ_c (Sarkar [7], Zeman [10], Wilcox [9]) in term of the incompressible dissipation ϵ_s and the turbulent Mach number Ma_t:

$$\epsilon_c = F(Ma_t)\epsilon_s . \tag{1}$$

The difference between the models are the expression of the dependence on the turbulent Mach number $F(Ma_t)$:

$$F(Ma_t) = Ma_t^2 \text{ (Sarkar)} \tag{2}$$

$$F(Ma_t) = \begin{cases} 1.5 * (Ma_t^2 - 0.25^2) & \text{if } Ma_t \geq 0.25 \\ 0 & \text{if } Ma_t \leq 0.25 \end{cases} \text{ (Wilcox)} \tag{3}$$

$$F(Ma_t) = \begin{cases} 1 - e^{-\frac{1.2(Ma_t - 0.25)^2}{0.66^2}} & \text{if } Ma_t \geq 0.25 \\ 0 & \text{if } Ma_t \leq 0.25 . \end{cases} \text{ (Zeman)} \tag{4}$$

$$\tag{5}$$

Pressure-dilatation models

For compressible flow, the pressure fluctuation induces a decrease of the production and the dissipation through the pressure-dilatation correlation. Sarkar proposed the following model for the pressure-dilatation term in the turbulent kinetic energy equation:

$$\overline{p' \frac{\partial u_i'}{\partial x_i}} = \alpha_1 Ma_t^2 \overline{\rho} \widetilde{u_i' u_j'} \frac{\partial \tilde{u}_i}{\partial x_j} + \alpha_2 Ma_t^2 C_\mu \overline{\rho} \epsilon \tag{6}$$

$$\text{with } \alpha_1 = 0.4 \text{ and } \alpha_2 = 0.2 . \tag{7}$$

Low Reynolds models

To improve the modeling near the wall, the low Reynolds models of Shih Lumley [8] and Lam Bremhost [5] have also been tested. A first series of computations has been done with the Shih Lumley model. Whereas good results have been obtained for incompressible flows (channel flow), the application to compressible flow was not satisfactory. Computations of compressible boundary layers have shown that the Lam Bremhorst model performs better. Thus the Lam Bremhorst model has been used to perform the test case computations.

To take into account the decrease of the turbulence near the wall, a damping function is introduced in the turbulent viscosity μ_t:

$$\mu_t = C_\mu \rho f_\mu \frac{k^2}{\epsilon} \tag{8}$$

with $f_\mu = (1 - e^{-0.0165 Ry})^2 (1 + 20.5/Re_t)$ \hfill (9)

$$Re_t = \frac{k^2}{\nu \epsilon} \quad Ry = \frac{\sqrt{k}\, y}{\nu}. \tag{10}$$

Damping functions are also introduced in the turbulence dissipation equation to ensure the right behavior of ϵ at the wall:

$$\frac{\partial \rho \tilde{\epsilon}}{\partial t} + div(\bar{\rho} \tilde{u} \tilde{\epsilon}) = \frac{\partial}{\partial x_i}(\mu + \frac{\mu_t}{\sigma_\epsilon})\frac{\partial \epsilon}{\partial x_i} + C_{\epsilon 1} f_1 \rho \frac{\epsilon}{k} P - C_{\epsilon 2} f_2 \rho \frac{\epsilon^2}{k} \tag{11}$$

$$f_1 = 1 + (0.05/f_\mu)^3 \tag{12}$$

$$f_2 = 1 - e^{-Re_T^2} \tag{13}$$

$$P = -\widetilde{u_i' u_j'}\frac{\partial \tilde{u}_i}{\partial x_j}. \tag{14}$$

Boundary conditions for k and ϵ, obtained by asymptotic expansion at the wall, are used:

$$k = 0 \qquad \epsilon = 2\nu(\frac{\partial \sqrt{k}}{\partial y})^2_{y=0}. \tag{15}$$

Summary

The final modeled transport equations for k and ϵ are:

$$\frac{\partial \rho k}{\partial t} + div(\bar{\rho} \tilde{u} k) = \frac{\partial}{\partial x_i}(\mu + \frac{\mu_t}{\sigma_k})\frac{\partial k}{\partial x_i} + (1 - \alpha_1 Ma_t^2)\rho P$$
$$- (1 + F(Ma_t) - \alpha_2 Ma_t^2)\rho \epsilon \tag{16}$$

$$\frac{\partial \rho \tilde{\epsilon}}{\partial t} + div(\bar{\rho} \tilde{u} \tilde{\epsilon}) = \frac{\partial}{\partial x_i}(\mu + \frac{\mu_t}{\sigma_\epsilon})\frac{\partial \epsilon}{\partial x_i} + C_{\epsilon 1} f_1 \rho \frac{\epsilon}{k} P - C_{\epsilon 2} f_2 \rho \frac{\epsilon^2}{k} \tag{17}$$

where

$$C_{\epsilon 1} = 1.44 \quad C_{\epsilon 2} = 1.92 \quad C_\mu = 0.09 \quad \sigma_k = 1.0 \quad \sigma_\epsilon = 1.3. \tag{18}$$

Numerical method

The numerical method uses a mixed finite volume/finite element approximation on non structured grid initially developed by INRIA. The convective fluxes are integrated by a Roe solver with a finite volume approximation and a 2nd order

MUSCL interpolation, whereas the viscous and source terms are integrated using a finite element method [6]. The time integration uses an implicit scheme, with performant preconditioned linear solvers [4]. For the inlet conditions, the fluxes of variables $(\rho, \rho U, \rho V, E, \rho k$ et $\rho\epsilon)$ are imposed. All calculations are carried out on the code "NadiaNG" which is a parallel code using a new object oriented paradigm (C^{++}) and a standard message passing library (PVM). Efficient parallel numerical techniques have been developed, such as an automatic domain partitioning technique and a parallel implicit solver [2].

Computational results

From the specification of the test case, the incident shock wave induced by a shock generator, with an incident angle of 8.75 degrees and an inlet Mach number of 2.4, has an angle of 31.8 degrees. But the experimental angle measured afterwards by Benay using strioscopy [1] is 34.4 degrees, which corresponds to a shock generator angle of 11 degrees. Considering these remarks, we have computed two configurations, the first corresponding to an angle of 8.75 degrees (1) and the second to an angle of 11 degrees (2).

The figure 1 shows the computed and measured wall pressure distributions for the two runs. We notice that, with the first configuration, there is no separation observed. Whereas, the second computation predicts a separation of the boundary layer. It seems that an incident angle of 8.75 degrees is not strong enough to induce a recirculation zone. In the following, we will consider the case corresponding to an angle of 11 degrees.

We have presented the isomach contours and the wall friction coefficient distribution respectively in the figure 2 and 3. We see that the separated region is located between x=120 and 170 mm. This is in a good agreement with the experimental result. However the pressure distribution after the shock is higher than the experimental data.

The velocity distributions are plotted on figures 4 and 5. Whereas the model predicts with a good accuracy the velocity field upstream of the shock foot, it underestimates the thickness of the boundary layer downstream.

We have compared the predicted turbulent kinetic energy with the measurements on figures 6 and 7. We observe a significant disagreement, especially in the region located between x=160 and 200 mm. It seems that in the reattachment region, the $k - \epsilon$ model is not able to predict correctly the turbulence [1].

The computation runs on a cluster of DEC Alpha workstations with a FFDI network. Using 4 computer nodes, the parallel computations take about 20 $sec.$ elapsed time per iteration, with a local time step and a local CFL of 5. Higher CFL number has not been used because oscillations appear in residuals due to the explicit boundary condition for ϵ. All the residuals have decreased up to a plateau of 10^{-5} in normalized L_2 norm. ¸A converge solution is obtained after 5000 iterations, starting from an initial solution.

Conclusion

Initially developed for incompressible flows, the $k - \epsilon$ model of Lam Bremhorst behaves well when used to predict compressible flows involving separation. But to obtain a good agreement with experimental data, compressibility terms have to be included in the model. The present results correspond to the Wilcox model.

With the developed numerical method and its parallel implementation, we have been able to obtain good results for this test case, without special tunning of the turbulence model. In the future, the boundary condition for ϵ will be imposed implicitly to avoid observed oscillations of residual at high CFL number.

References

[1] R. Benay. Modélisation de la turbulence dans une interaction choc-couche limite sur une paroi chauffée. *La Recherche Aérospatiale*, (no. 5):pages 45—68, Septembre—Octobre 1991.

[2] M. Buffat. Parallel simulation of compressible turbulent flows using non-structured domains partitioning and object oriented programming. In *Notes on numerical fluid mechanics: CFD on parallel systems*. Vieweg-Verlag, 1994.

[3] J. Delery. Etude experimentale de la réflexion d'une onde choc sur une paroi chauffée en présence d'une couche limite turbulente. *La Recherche Aérospatiale*, 1992.

[4] L. Hallo, JL. Munier, M. Buffat, and G. Brun. Iterative methods for solving implicit non-structured finite volume discretization of euler equations. *International Journal for Numerical Methods in Fluids*, 1994.

[5] Bremhorst K. Lam C.K.G. A modified form of the $k - \epsilon$ model for predicting wall turbulence. *J. Fluids Eng.*, 103:457—460, 1981.

[6] C. Le Ribault. *Simulation des écoulements turbulents compressibles par une méthode mixte éléments finis/volumesfinis*. PhD thesis, Thèse Ecole Centrale de Lyon, 1991.

[7] S. Sarkar, G. Erlabacher, and M.Y. Hussaini. Compressible homogeneous shear: simulation and modeling. In *Eight Symposium on turbulent shear flows, Munich sept. 9-11*, 1991.

[8] John L. Lumley Tsan-Hsing Shih. Kolmogorov behavior of near-wall turbulence and its application in turbulence modeling. *Comp. Fluid Dynamics*, 1:43—46, 1993.

[9] D.C. Wilcox. *Turbulence modeling for CFD*. Griffin Printing, Glendale California, 1993.

[10] O. Zeman. dilatational dissipation: the concept and application in modeling compressible mixing layers. *Physics of Fluids*, pages 178–188, 1990.

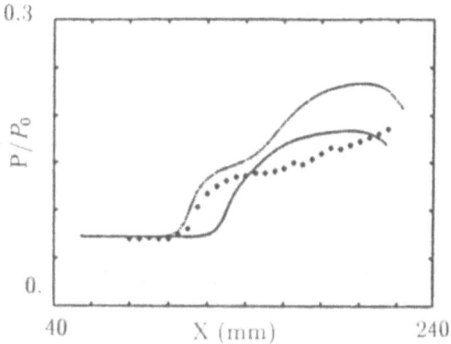

Figure 1: Wall Pressure distribution

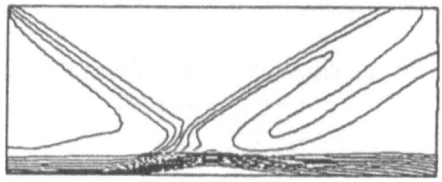

Figure 2: Isomach contours

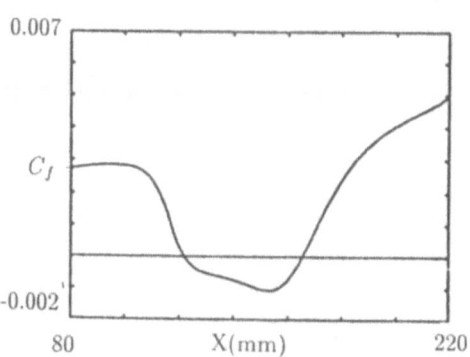

Figure 3: Wall friction coefficient distribution

281

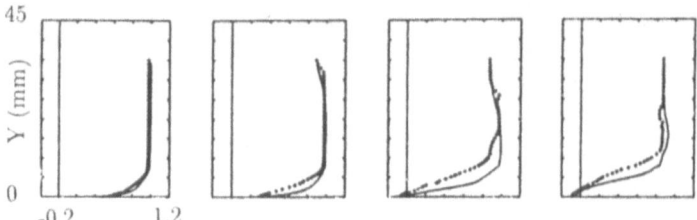

Figure 4: Velocity distribution at x=100,120,130 and 140 mm

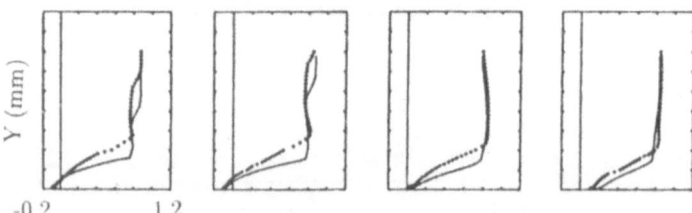

Figure 5: Velocity distribution at x=150,160,180 and 200 mm

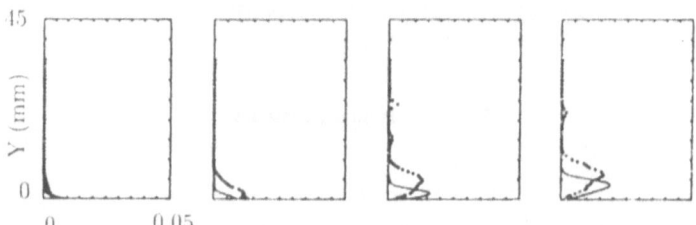

Figure 6: Turbulent kinetic energy distribution at x=100,120,130 and 140 mm

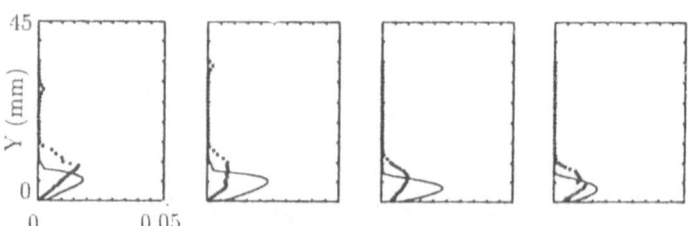

Figure 7: Turbulent kinetic energy distribution at x=150,160,180 and 200 mm

ASSESSMENT OF A ONE-EQUATION POINTWISE TURBULENCE MODEL FOR COMPRESSIBLE FLOW. - TESTCASE TC6.

Nicola Ceresola

ALENIA - Ingegneria Velivoli Difesa
C.so Marche,41 - 10146 Torino Italy

Summary.

A pointwise version of the one-equation Baldwin-Barth model for the turbulence Reynolds number R_T is proposed. The y^* dependence of the wall damping terms is replaced with a dependence on R_T, that was obtained by comparison with the results of a direct simulation on a flat plate. The model is applied to the supersonic shock-boundary layer interaction (testcase TC6). Both the velocity and the temperature fields are satisfactorily predicted, except in the recovery region. In particular, a quite good near-wall behaviour is evidenced, obtaining a first assessment of the validity of the present approach.

1 INTRODUCTION.

The present needs of modern industrial R&D require to make CFD simulations past more and more complex 3D geometries and in a variety of physical conditions. To fulfill the requests of the development engineers, it is therefore mandatory to look for numerical methods ensuring the maximum possible flexibility and generality in practical applications. The quick turnaround times that are required for a practical usage are, in fact, incompatible with setting up adhoc modifications when one has to do with different geometries or with changed physical conditions.

In that sense, the main difficulty that is associated with classical first order, two equations methods is the necessity to explicitly determine lengthscales, as the distance from solid walls, and velocity profiles. As an example, if multiple shear layers are present, such an evaluation may be very cumbersome, often compromising a correct prediction of the turbulence quantities. The presence of wall functions, moreover, requires the knowledge of the distance from the solid surfaces, precluding any general-purpose treatment of flows past arbitrary geometries, expecially when we have to do with structured multiblock or unstructured meshes.

In a recent work, Baldwin and Barth [1] proposed a self-consistent one-equation model for the turbulence Reynolds number Re_t avoiding the need to supply the algebraic length scales. A fully pointwise implementation of the model was later proposed by Goldberg [2], who substituted the y^* dependence of the damping factors with a dependence on Re_t itself. In the present paper, a form of the damping function is tried, which is mutuated both from [1] and from direct simulation data.

The scope of the present work is to carry on an evaluation of the model, as a possible candidate to be applied to geometrically complex industrial problems.

2 THE TURBULENCE MODEL.

2.1 The field equation.

The Baldwin-Barth model consists in a field equation for the turbulence Reynolds number

$$R_t = \frac{\kappa^2}{\nu\epsilon} .$$

The governing equation was derived from the $\kappa - \epsilon$ equations under the main hypothesis that the production equals the dissipation in the outer region of the boundary layer. It is given by

$$\frac{\partial \Phi}{\partial t} + V \cdot \nabla \Phi = (c_{\epsilon_2} f_2 - c_{\epsilon_1})\sqrt{(\Phi P)} + \left(v + \frac{v_t}{\sigma_\epsilon}\right)\nabla^2 \Phi - (\nabla v_t) \cdot \nabla(\Phi)/\sigma_\epsilon \qquad (1)$$

where $\Phi = v R$, f_2 is a damping function and

$$P = v_t \left[\left(\frac{\partial u_i}{\partial x_j} + \frac{\partial u_j}{\partial x_i}\right)\frac{\partial u_i}{\partial x_j} - \frac{2}{3}\left(\frac{\partial u_k}{\partial x_k}\right)^2\right]$$

is the turbulence production.

The eddy viscosity $v_t = c_\mu v k^2/\epsilon$ is given by, from (1), by

$$v_t = c_\mu f_\mu (v R_T)$$

where f_μ is a wall damping function.

During the preliminary testing on flat plate flow with zero pressure gradient, it was found that correct velocity profiles could only be obtained by halving the value of the dissipation coefficient σ_ϵ with respect to the one suggested in [1]. As it will be pointed out in the following, this is probably due to the fact that the model predicts a lack of dissipation with respect to the production in the outer part of the boundary layer.

2.2 Near-wall treatment.

In [1], the damping function f_μ is calibrated by comparison with the Cebeci-Smith model for an incompressible flat-plate flow, supposing that a linear relation exists between R_t and y^* in the near-wall region:

$$y_* = \frac{c_\mu}{k}R_t \qquad (2)$$

The damping factor is given by

$$f_\mu = \left(1 - e^{-\frac{y^*}{A^*}}\right)\left(1 - e^{-\frac{y^*}{A_2^*}}\right).$$

In the present work, a fully pointwise model was adopted, simply substituting () in (), to eliminate the explicit dependence of f_μ on the wall distance. In addition, correcting factors were introduced, to empirically fit the behavior of f_μ according with the prediction of direct simulations on a flat plate [3]. The final formula is

$$f_\mu = \left(1 - 0.5e^{-\frac{R_t c_\mu}{kA^*}}\right)\left(1 - 0.5e^{-\frac{R_t c_\mu}{kA_2^*}}\right) + 0.07\left(1 - e^{-(R_t/25)^2}\right)$$

The standard Launder's formula was kept for f_2:

$$f_2 = \left(1 - 0.22e^{-0.1R_t^2}\right).$$

3 BASIC NUMERICAL ALGORITHM.

The thin-layer Reynolds-averaged Navier-Stokes equations are solved with a finite differences, space-centered, implicit algorithm, with second and fourth-order numerical dissipation. A different treatment of the nonlinear fourth-order dissipation term was needed in solving Eq.() with respect to the Navier-Stokes equations. In the latter case, the scaling factor is taken to be proportional to the Euler spectral radius $\Phi = (|u|+a)/\Delta x$, while, in solving the R_t equation, it was redefined as $\Phi = a/\Delta x$. Besides this difference, the resolution schemes were identical, and the same numerical damping coefficients and CFL number were used.

Fully converged solution were obtained in all cases, with a five orders of magnitude drop in the residual in about 5000 iterations.

An implicit treatment of the turbulence production term was also attempted, following the identity

$$\sqrt{(\nu R_t P)} = \sqrt{\left(c_\mu f_\mu \frac{P}{\nu_t}\right)}(\nu R_t)$$

but no particular improvements in the rate of convergence were evidenced.

4 RESULTS.

The shock-boundary layer interaction in a supersonic flow past a flat plate (testcase TC6) may be a first workbench to assess the performance of the model on a physically complex flow. The critical phenomenon to be predicted is the shock-induced separation, which location and extension strongly depends upon the characteristics of the turbulence model.

The computations were made on a 151x81 grid (fig.1). At inflow, velocity and temperature profiles were specified, based on the Coles' family of profiles with the Van Driest correction for compressibility. The initial eddy viscosity profile was then computed, solving eq.(1) with the convection terms set to zero.

No information was given to the turbulence model about the presence and location of the solid wall. The main purpose of the present investigation was, in effect, to assess if the model have correctly "felt" the presence of the shear layers, only using mean-flow informations.

In Fig. 2 the experimental Mach number contours are depicted. The incident and reflected shock are clearly seen, as well as the expansion fan over the top of the separation bubble. A strong thickening of the boundary layer also occurs during the interaction. Fig.3 shows the computed Mach number contours, representing quite well the qualitative structures of the flow.

From the wall pressure values in Fig.4 a substantial agreement between the computed and measured shock locations are evidenced.

The velocity and Mach number profiles at a station immediately upstream of the interaction ($X = 80$ mm) is shown in fig.5.

The profiles at a station at the beginning of the recirculation region ($X = 120$ mm) are depicted in fig.6 and 7. The height of the separation bubble, as well as of the velocity and Mach number profiles, are well predicted by the computation. The same holds true for the profiles at $X = 130$ mm, in figs.8, 9, and at $X = 140$ mm (figs.10, 11).

In the reattaching flow ($X = 160$ mm, in fig.12, 13, and $X = 180$ mm, in fig.14, 15) the computation predicts a rapid recovery to a stable attached boundary layer flow, while the experimental profiles are still thick and less energetic. It results in a clear underestimation of the boundary layer thickness. As it was already noted, the flow is here characterized by a gradual return to the equilibrium between the mean flow and the turbulence, that cannot be simulated by equilibrium models.

The wall skin friction distribution is reported in fig.16.
A global view of the boundary layer features is given in figs. 17,18 and 19, where the displacement thickness, momentum thickness and incompressible shape factor distributions are reported.

5 CONCLUSIONS.

The pointwise version of the one-equation Baldwin-Barth model was applied to the TC5 testcase. A damping function was tried, which seems to predict well enough the near-wall behaviour of the velocity profiles. The results seemed to be at least comparable whith those obtained with the standard versions of $\kappa - \epsilon$. In our view, this makes the model to be potentially attractive for engineering applications on complex geometries and in multiblock environments.

In addition, its operation count is about half than the one of two-equations models, and it can be implemented in a fully implicit way, enhancing the well conditioning of the matrices and the robustness of the code.

The main problem that came out from the present validation seemed to be a substantial overprediction of the $\frac{\rho}{\epsilon}$ ratio in the wake region of the profile. Further theoretical and numerical work is thought to be required to overcome this difficulty.

6 BIBLIOGRAPHY.

[1]   Baldwin, B.S., and Barth, T.: *A one-equation turbulence transport model for high Reynolds number wall-bounded flows.*
       NASA TM-102847, 1990.

[2]   Goldberg, U.C., and Ramakrishnan, S.V.: *A pointwise version of the Baldwin-Barth turbulence model.*
       Rockwell International Science Center, 1993.

[3]   Gilbert, N., and Kleiser, L.: *Turbulence model data derived from direct numerical simulations.*
       Advances in Turbulence 2, H.H. Fernholz and H.E.Fiedler, Editors, Springer-Verlag, 1989.

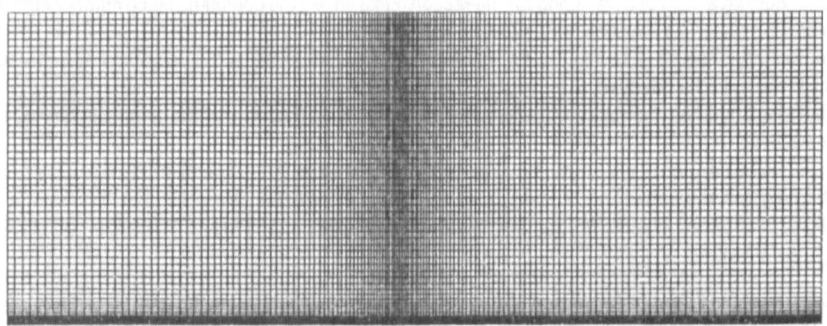

Fig.1 - Computational grid.

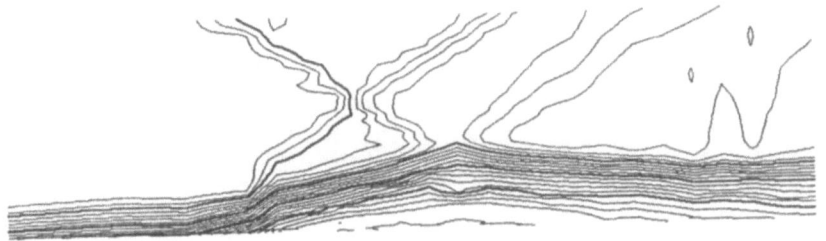

Fig.2 - Mach contours - experiment.

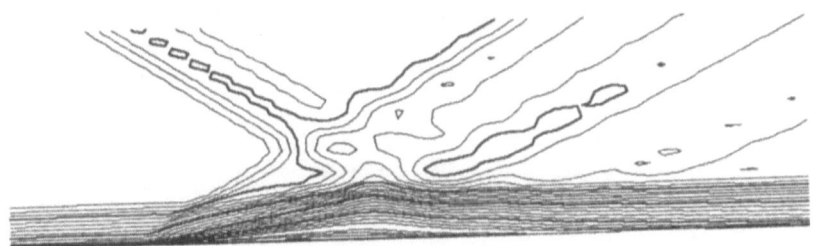

Fig.3 - Mach contours - present calculation

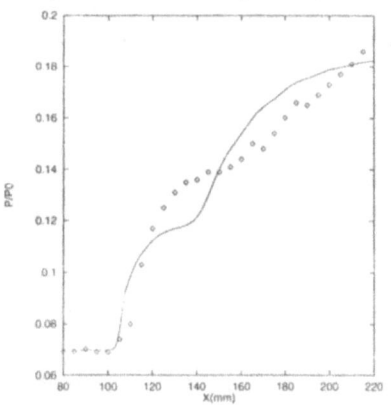

Fig.4 - Wall pressure distribution.

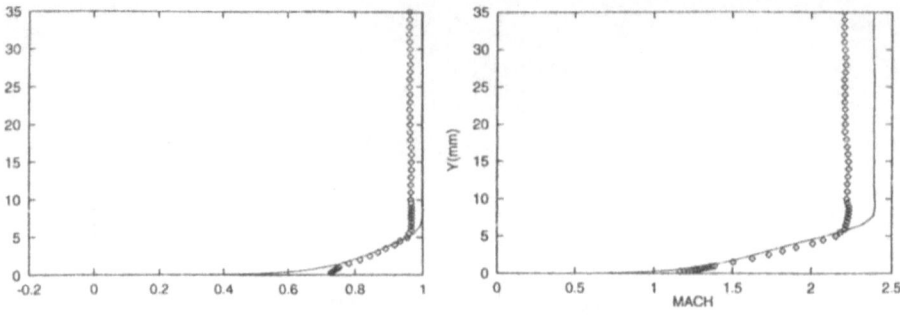

Fig.5 - Velocity profiles at x = 80 mm

Fig.6 - Mach profiles at x = 80 mm

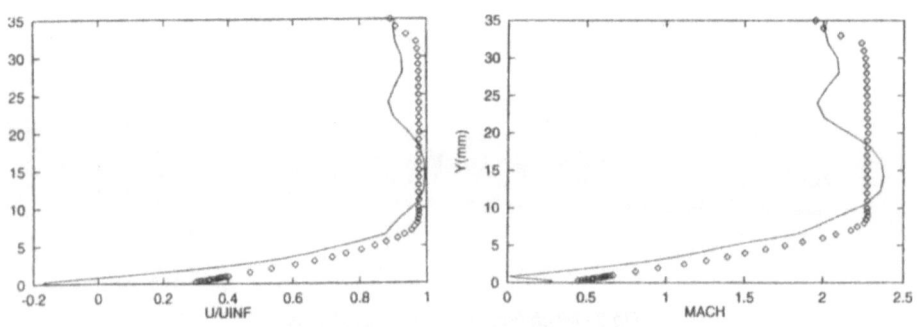

Fig.7 - Velocity profiles at x = 120 mm

Fig.8 - Mach profiles at x = 120 mm

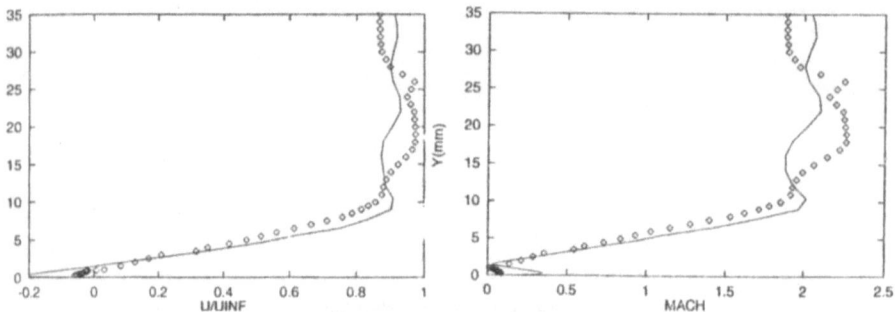

Fig.9 - Velocity profiles at x = 130 mm

Fig.10 - Mach profiles at x = 130 mm

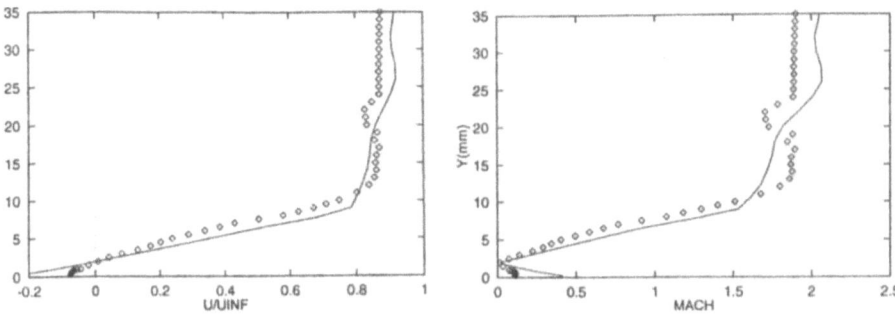

Fig. 11 - Velocity profiles at x = 140 mm   *Fig. 12 - Mach profiles at x = 140 mm*

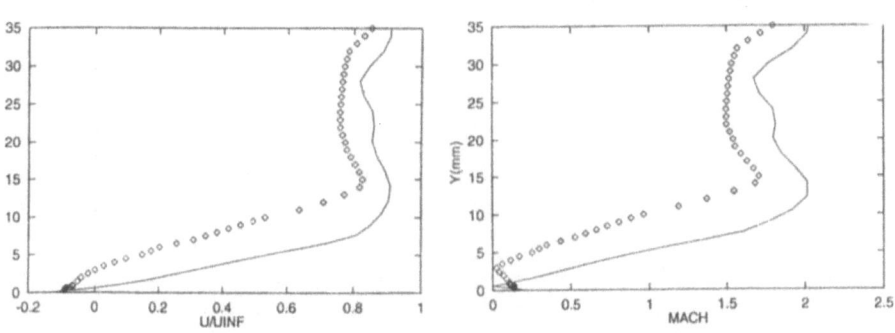

Fig. 13 - Velocity profiles at x = 160 mm   *Fig. 14 - Mach profiles at x = 160 mm*

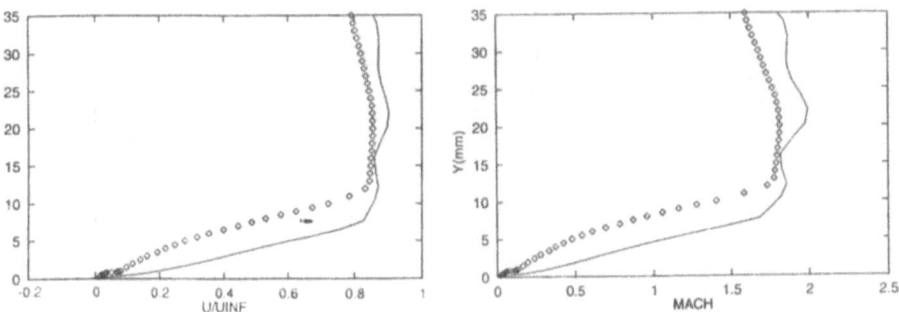

Fig. 15 - Velocity profiles at x = 180 mm   *Fig. 16 - Mach profiles at x = 180 mm*

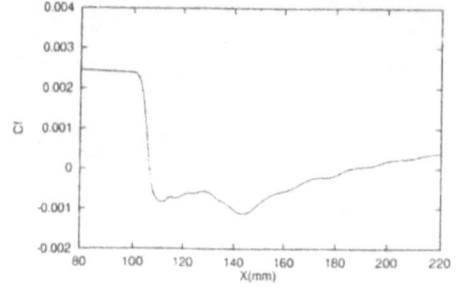

Fig. 17 - Wall skin friction

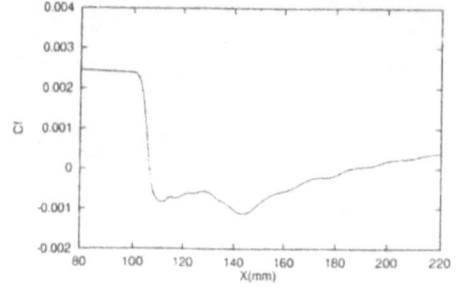

Fig. 18 - Displacement thickness

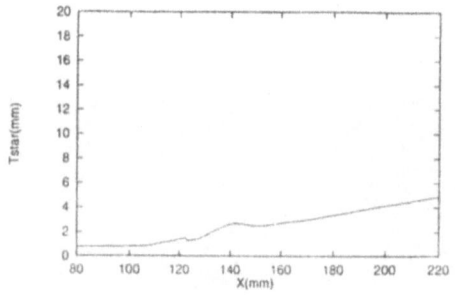

Fig. 19 - Momentum thickness

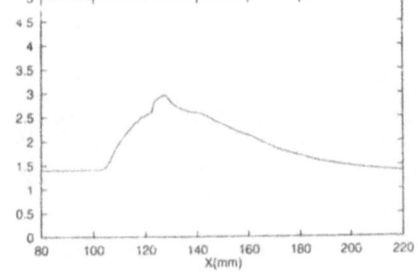

Fig. 20 - Shape parameter

ETMA Test Case 6
SHOCK REFLECTION ON A FLAT PLATE
SYNTHESIS OF THE CALCULATIONS

R. Arina

Dipartimento di Ingegneria Aeronautica e Spaziale
Politecnico di Torino
Corso Duca degli Abruzzi, 24 - TORINO (Italy)

SUMMARY

The calculations of Test Case 6 are critically compared. The turbulence models, as well as the the numerical methods employed are discussed, considering the main differences in the numerical results, with respect to the experimental data.

1. INTRODUCTION

Three contributors have performed calculation of Test Case 6, the shock reflection on a flat plate. One of them, Ceresola (ALENIA), has used a variant of the one equation turbulence model of Baldwin and Barth, while the other two, Arina and Ramella (TORINO) and Yudiana and Buffat (ECL) have employed the $k - \varepsilon$ model with compressibility corrections.

The main concern of ALENIA, is the application of turbulence models in complex geometries, therefore the focus has been on the capability of predicting accurately and without excessive amount of computational cost shock/boundary layer interaction phenomena. For this reason a one equation model has been selected, and implemented into a thin layer numerical solver. The numerical algorithm, for structured grids, is an implicit time integration of the averaged Navier-Stokes equations, with the thin-layer approximation, combined with a second order centred discretization with forth order artificial viscosity. A second fundamental issue for ALENIA is the possibility of avoiding an explicit knowledge of the distance normal to the wall, therefore a pointwise formulation of the turbulence models is mandatory. The novelty of the turbulence model consists into a new pointwise formulation of the Baldwin-Barth model.

The other two contributors have studied the effects of existing compressibility corrections for the $k - \varepsilon$ turbulence model. In both cases a low-Re form of the turbulence model has been adopted. ECL solves the Favre averaged Navier-Stokes equations on an unstructured mesh, with a mixed finite volume/finite element formulation. The convective fluxes are discretized by the Roe flux-difference splitting and second order MUSCL extrapolation, diffusive terms are discretized by a Galerkin finite element method. The time integration is implicit with preconditioned linear solvers. The numerical solution of the Favre averaged Navier Stokes equations done by TORINO is based on a finite volume discretization on structured grids. The convective terms are discretized by an upwind AUSM flux-vector splitting, with second order MUSCL extrapolation, and the diffusive terms are discretized with a centred scheme. The time integration is implicit with nonlinear GMRES solver.

2. PRESCRIPTION OF THE INCOMING BOUNDARY LAYER

When a shock wave penetrates inside a boundary layer, it encounters a flow with lower and lower Mach number, therefore it must adapt itself and it becomes vanishingly weak, at the boundary-layer sonic line. The pressure signal carried by the shock is necessarily trasmitted in the upstream direction through the subsonic part of the boundary layer. The pressure rise induces a thickening, and eventually separation, of the boundary layer. From this simple description of the physical mechanism of upstream influence, it is evident that the *form* of the boundary layer profile plays a determinant role in the interaction.

For this reason it is very important to prescribe at the inlet boundary a boundary layer which is as close as possible to the experimental one. In order to reduce the degree of uncertainess all the contributors where provided with a prescribed profile of the boundary layer.

The available experimental data of Delery [2], provide only integral parameters for describing the inlet boundary layer. Between them, the most significative is the momentum thickness, which gives a measure of the fulness of the boundary layer. And as previously remarked, the fullness is a determinant factor for the interaction mechanism. The construction of a boundary layer profile, for flat plate, with prescribed momentum thickness, can be obtained by an iterative procedure, as explained by Huang et al. [4]. The method is based on the extension of the incompressible similarity laws of the wall by means of the Van Driest's transformation.

In Appendix we report the boundary layer profile, and its integral parameters, adopted by the contributors. The turbulent kinetic energy and the dissipation rate, have been obtained from the velocity and temperature profiles, employing the Baldwin-Lomax algebraic model and assuming local equilibrium between production and dissipation of turbulent kinetic energy.

3. COMPARISON OF THE RESULTS

All the calculations seem to be grid independent, each contributor having checked them on different grids. Moreover comparing the velocity profile on the flat plate just before the beginning of the interaction for x=80 and x=100 (figure 3), the solutions are identical.

The main differences come out comparing the pressure wall distribution (figure 1) and the skin friction coefficient distribution (figure 2). Considerable differences are present. Except for the results of ALENIA, the other two do not fit with the experimental data.

As a first remark, it is possible to see that the Baldwin Barth model of ALENIA is able to capture the essential features of the interaction, with the presence of a strong separation region. Actually the length of the separation bubble is overestimated, as can be seen from the comparisons of the calculated profiles with the experimental ones. The separation point is predicted a little more upstream.

The results obtained with the compressible $k - \varepsilon$ of TORINO and ECL, seems very different but the disagreement is due to a different strenght of the incident shock employed in the two simulation.

Indeed also ECL has found almost identical results to those of TORINO, employing the incident shock strenght given in [2]. This is not surprising because the two turbulence models are very similar, except for some minor changes in the dissipation rate equation in the case of TORINO. This agreement confirms that the solution are unaffected from numerical influences, and that the two different low-Re treatments behave in the same way. In both cases the compressible $k - \varepsilon$ model is not able to take into account the increase of turbulence production

292

induced by the strong pressure gradient, as can be seen from the profiles of turbulent kinetic energy k of TORINO. The maximun of k is always underestimated, as well as it position. The resulting wall pressure distribution presents the three typical inflexion points, but there is no plateau, which indicates the presence of extended separation [3]. Therefore it could be argued that the compressible $k - \varepsilon$ is not able to model the increase of turbulence production entrained by the interaction with the shock.

Referring to the work of Benay [1], who measured afterwards the angle of the incident shock in the same experimental set up of [2], and found that it should be of 34.4 deg, instead of 31.8 deg, ECL applied the new boundary conditions, corresponding to a stronger incident shock, and found a massive separation. Looking at the profiles at different x-stations, it is possible to note that now the flow field is much better represented with respect to the results of TORINO and ECL (with the *standard* shock intensity). The skin friction distribution indicates that the separation region is between $x \approx 130$ and $x \approx 170$, as in the experiment of Delery [2]. Some discrepancies can be found in the reattachment region. If we could assert that the stronger incident shock is the right one, then we could conclude that the $k - \varepsilon$ model, with the compressibility corrections, models fairly well the physics. However the resulting pressure wall distribution is quite far from the experimental one. This difference is hard to be explained. The higher level of predicted pressure seems to be due to an overshoot. As explained by Delery and Marvin [3], in some cases the pressure wall distribution presents such kind of overshoot. It would have been interesting to see the behavior of the pressure further downstream.

4. CONCLUSIONS

From the above considerations we can conclude that the Baldwin Barth model (ALENIA) gives a quite good description of the phenomenon. Moreover we can argue that the thin-shear layer approximation, employed by ALENIA, is appropriate.

The results of TORINO and ECL, concerning the $k - \varepsilon$ model, do not enable us a definite conclusion. On the contrary they open the problem of the real value of the intensity of the experimental incident shock. The quite good agreement with the experimental data, of the results of ECL, would carry to the conclusion that the stronger shock is the good one. However the results of TORINO show that downstream the wall pressure recover the value predicted by the invisid theory. Moreover the good results obtained by ALENIA, with the weak shock, also contradict the previous conclusion. Therefore further investigations are necessary in order to clarify this point.

REFERENCES

[1] Benay R., *Modelisation de la turbulence dans une interaction onde de choc-couche limite sur une paroi chauffee*, La Recherche Aerospatiale, 5, 1991.

[2] Delery J., *Etude experimentale de la reflection d'une onde de choc sur une paroi chauffe en presence d'une couche limite turbulente*, La Recherche Aerospatiale, 1, 1992.

[3] Delery J. and Marvin J.G., *Shock-Wave Boundary Layer Interactions*, AGARDograph 280, 1986.

[4] Huang P.G., Bradshaw P., Coakley T.J., *Skin Friction and Velocity Profile Family for Compressible Turbulent Boundary Layers*, AIAA J., **31**, 9, 1600-1604, 1993.

APPENDIX: Incoming Boundary Layer Profiles

Table I - External Flow Field

Re/L	Mach	Total Temp
7.01e6 m^1	2.4	317 K

Table II - Integral parameters

δ	δ^*	θ_k	H_{12}	H_k	c_f
6.9683874E-03 m	2.2525215E-03 m	6.5851392E-04 m	4.961501	1.301726	1.5003645E-03

Table III - Boundary Layer Profiles ($\theta = .454e - 3$ m)

y/δ	u/u_{ext}	T/T_{ext}	K/u_{ext}^2	$\varepsilon/u_{ext}^3/\delta$
0.00000E+00	0.00000E+00	0.21520E+01	0.00000E+00	0.00000E+00
0.82341E-05	0.82121E-03	0.21520E+01	0.22207E-12	0.66443E-11
0.17962E-04	0.17914E-02	0.21520E+01	0.50236E-11	0.15030E-09
0.29453E-04	0.29374E-02	0.21520E+01	0.36283E-10	0.10856E-08
0.43029E-04	0.42914E-02	0.21520E+01	0.16506E-09	0.49386E-08
0.59067E-04	0.58909E-02	0.21520E+01	0.58522E-09	0.17509E-07
0.78013E-04	0.77804E-02	0.21519E+01	0.17776E-08	0.53183E-07
0.10040E-03	0.10013E-01	0.21519E+01	0.48649E-08	0.14555E-06
0.12684E-03	0.12650E-01	0.21518E+01	0.12362E-07	0.36987E-06
0.15808E-03	0.15765E-01	0.21517E+01	0.29734E-07	0.88959E-06
0.19498E-03	0.19445E-01	0.21516E+01	0.68579E-07	0.20517E-05
0.23857E-03	0.23792E-01	0.21514E+01	0.15307E-06	0.45792E-05
0.29008E-03	0.28928E-01	0.21511E+01	0.33285E-06	0.99567E-05
0.35092E-03	0.34994E-01	0.21507E+01	0.70860E-06	0.21194E-04
0.42280E-03	0.42158E-01	0.21502E+01	0.14823E-05	0.44325E-04
0.50771E-03	0.50618E-01	0.21493E+01	0.30553E-05	0.91325E-04
0.60802E-03	0.60606E-01	0.21482E+01	0.62165E-05	0.18569E-03
0.72652E-03	0.72391E-01	0.21466E+01	0.12501E-04	0.37296E-03
0.86652E-03	0.86282E-01	0.21443E+01	0.24853E-04	0.73979E-03
0.10319E-02	0.10262E+00	0.21411E+01	0.48806E-04	0.14468E-02
0.12273E-02	0.12178E+00	0.21366E+01	0.94432E-04	0.27781E-02
0.14581E-02	0.14411E+00	0.21305E+01	0.17909E-03	0.51960E-02
0.17308E-02	0.16984E+00	0.21221E+01	0.32996E-03	0.93403E-02
0.20529E-02	0.19898E+00	0.21109E+01	0.58288E-03	0.15822E-01
0.24335E-02	0.23118E+00	0.20966E+01	0.97151E-03	0.24658E-01
0.28830E-02	0.26561E+00	0.20788E+01	0.15055E-02	0.34594E-01
0.34141E-02	0.30114E+00	0.20579E+01	0.21520E-02	0.43183E-01
0.40416E-02	0.33653E+00	0.20345E+01	0.28414E-02	0.48091E-01
0.47828E-02	0.37077E+00	0.20094E+01	0.34975E-02	0.48466E-01

Table III - continued

y/δ	u/u_{ext}	T/T_{ext}	K/u_{ext}^2	$\varepsilon/u_{ext}^3/\delta$
0.56584E-02	0.40314E+00	0.19834E+01	0.40651E-02	0.45084E-01
0.66929E-02	0.43327E+00	0.19573E+01	0.45189E-02	0.39488E-01
0.79149E-02	0.46106E+00	0.19315E+01	0.48575E-02	0.33135E-01
0.93586E-02	0.48658E+00	0.19064E+01	0.50933E-02	0.27009E-01
0.11064E-01	0.51001E+00	0.18822E+01	0.52441E-02	0.21617E-01
0.13079E-01	0.53160E+00	0.18589E+01	0.53278E-02	0.17129E-01
0.15459E-01	0.55162E+00	0.18364E+01	0.53610E-02	0.13521E-01
0.18271E-01	0.57031E+00	0.18146E+01	0.53573E-02	0.10683E-01
0.21593E-01	0.58793E+00	0.17934E+01	0.53284E-02	0.84787E-02
0.25517E-01	0.60470E+00	0.17727E+01	0.52841E-02	0.67766E-02
0.30153E-01	0.62084E+00	0.17522E+01	0.52327E-02	0.54638E-02
0.35630E-01	0.63651E+00	0.17317E+01	0.51812E-02	0.44480E-02
0.42100E-01	0.65187E+00	0.17112E+01	0.51358E-02	0.36569E-02
0.49744E-01	0.66703E+00	0.16905E+01	0.51017E-02	0.30355E-02
0.58774E-01	0.68208E+00	0.16694E+01	0.50831E-02	0.25423E-02
0.69441E-01	0.69710E+00	0.16479E+01	0.50841E-02	0.21473E-02
0.82043E-01	0.71214E+00	0.16259E+01	0.51094E-02	0.18294E-02
0.96931E-01	0.72726E+00	0.16034E+01	0.51659E-02	0.15737E-02
0.11452E+00	0.74252E+00	0.15801E+01	0.52641E-02	0.13701E-02
0.13530E+00	0.75800E+00	0.15560E+01	0.51327E-02	0.11473E-02
0.15984E+00	0.77381E+00	0.15309E+01	0.44376E-02	0.85761E-03
0.18884E+00	0.79010E+00	0.15045E+01	0.38691E-02	0.65195E-03
0.22309E+00	0.80704E+00	0.14764E+01	0.34075E-02	0.50573E-03
0.26356E+00	0.82488E+00	0.14462E+01	0.30352E-02	0.40133E-03
0.31137E+00	0.84387E+00	0.14133E+01	0.27343E-02	0.32588E-03
0.36785E+00	0.86430E+00	0.13771E+01	0.24853E-02	0.26967E-03
0.43457E+00	0.88638E+00	0.13370E+01	0.22646E-02	0.22484E-03
0.51339E+00	0.91015E+00	0.12927E+01	0.20396E-02	0.18448E-03
0.60651E+00	0.93520E+00	0.12448E+01	0.17653E-02	0.14247E-03
0.71652E+00	0.96035E+00	0.11953E+01	0.13876E-02	0.95201E-04
0.84648E+00	0.98324E+00	0.11492E+01	0.88680E-03	0.46846E-04
0.10000E+01	0.10000E+01	0.11147E+01	0.60651E-03	0.32039E-04

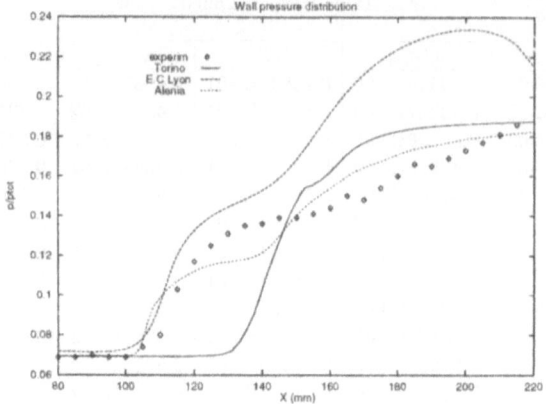

Figure 1: TC6 - Wall pressure distribution

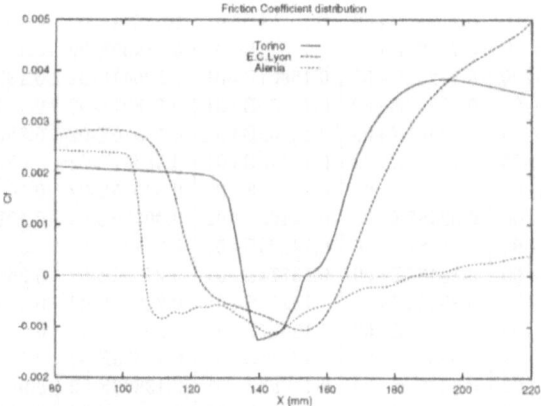

Figure 2: TC6 - Skin Friction Coefficient distribution

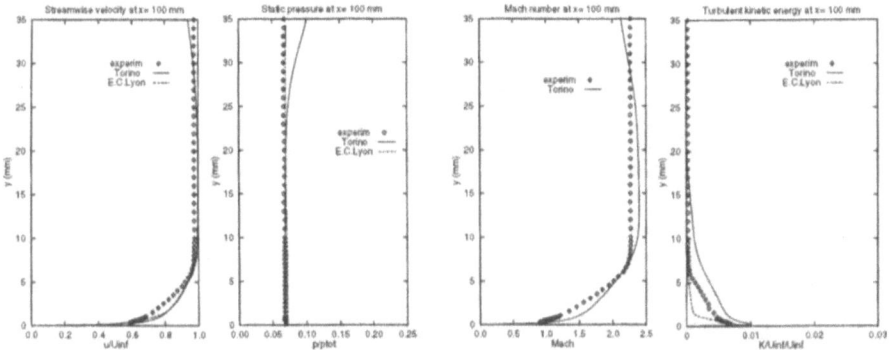

Figure 3: x = 100 [mm]

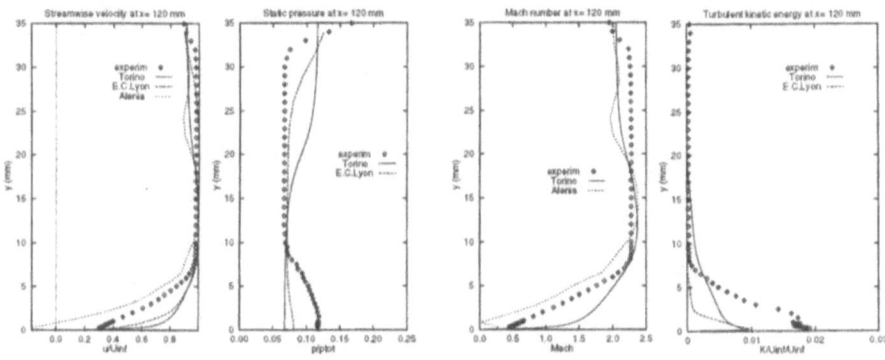

Figure 4: x = 120 [mm]

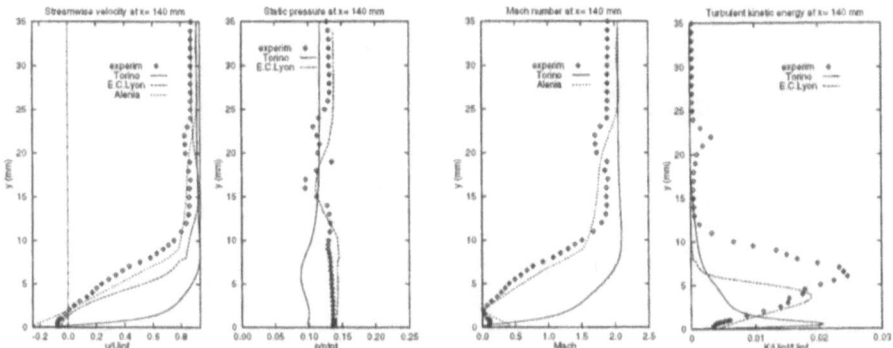

Figure 5: x = 140 [mm]

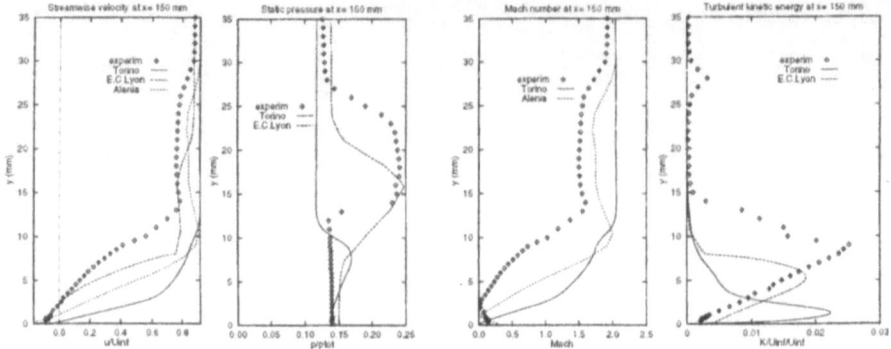

Figure 6: x = 150 [mm]

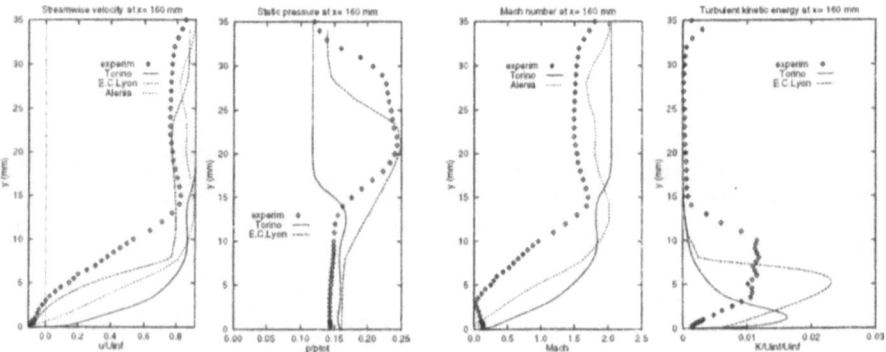

Figure 7: x = 160 [mm]

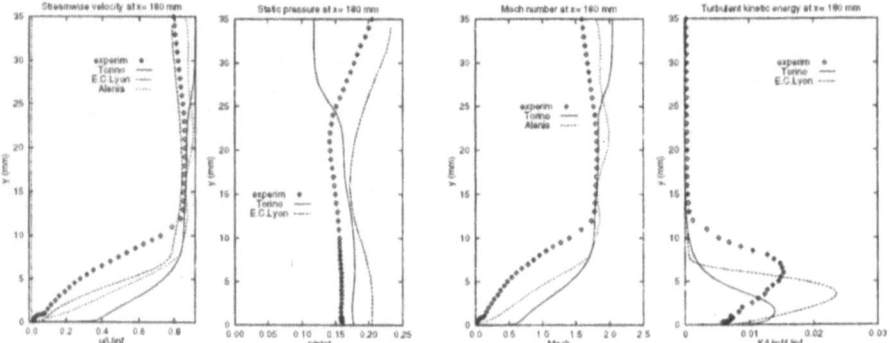

Figure 8: x = 180 [mm]

CHAPTER 6 : RAMP FLOW

Supersonic Compression Ramp Flow. Synthesis of Results.

Kyriakos C. Giannakoglou
National Technical University of Athens,
Lab. of Thermal Turbomachines
P.O. Box 64069, 157 10 Athens, Greece
kgianna@central.ntua.gr

SUMMARY

This text concerns the numerical study of the supersonic flow over a two-dimensional ramp at two corner angles. The text is split into two parts. In the first part, specifications concerning data and the format of the requested results are provided. In the second part, numerical results predicted by various contributors to the ETMA workshop, using a variety of turbulence models, are presented and compared with the available experimental data.

INTRODUCTION

The compression ramp flow constitutes one of the most interesting problems in fluid mechanics, involving the shock wave- turbulent boundary layer interaction which gives rise to flow separation. The compression ramp flow was examined at the 1980-81 AFOSR-HTTM-Stanford Conference on Complex Turbulent Flows, under code number 8631, with an infinite Mach number equal to 2.85 and at different corner angles of 8°, 16°, 20° and 24°. Detailed analyses of the experimental data and flow features involved can be found in the cited references, [9], [10], [11] and [12].

Numerical results obtained using seven Navier-Stokes solvers, where closure is effected through different turbulence models, are presented and compared with experimental data. Contributors , describe their own methods in different articles in the same volume. These papers have not been included in the list of references in the present paper, for the sake of economy in space.

TEST CASE SPECIFICATION

The cases analyzed during the Workshop are those corresponding to corner angles of 20° and the 24°. The other two milder cases (8° and the 16°) have been examined only by a small number of contributors and will not be included in the present synthesis. A 99x99 structured grid for the 24° case, generated and firstly tested at the Institut de Mecanique de l' INSA, has been distributed to all participants. This grid was appropriate for low-Reynolds k- ϵ calculations, ensuring a non-dimensional distance y^+ of the first grid-node off the wall less than 1, when a cell centered discretization was used. For participants working with vertex-centered discretizations on structured grids, the aforementioned grid was enriched by adding two more grid lines, locally, in the vicinity of the solid wall. With the aforementioned modifications the y^+ of the first node off the wall remained within the aforesaid limits. For the 20° case, a 99x101 structured grid was prepared by the National Technical University of Athens (NTUA), by merely retaining the stretching law of the 24° grid. Contributors working with unstructured grids had to generate their own grids.

Preliminary runs, as well as experience gained during previous calculations, revealed the dependency of the numerical predictions upon the proper definition of the supersonic inlet profile. Consequently, an inlet velocity profile and the corresponding turbulence data have been distributed among the partners with the intention to minimize possible calculation discrepancies due to different inlet conditions. The inlet profiles were neither mandatory nor of proven reliability and accuracy. They were in the form of recommended velocity, density, pressure, turbulent energy dissipation and turbulent kinetic energy distribution which, on the basis of existing experience, may fit to the measured data.

In the cases to be examined in the Workshop, the inlet boundary layers, were in equilibrium condition, with zero pressure gradient and near-adiabatic wall conditions. For the proposed cases the infinite conditions were defined in terms of the boundary layer edge at X=-0.5m (where the origin is set to the ramp corner) . These initial conditions are given in Table 1.

The recommended inlet velocity profile was obtained through the Kunh and Nielsen [5] expression for turbulent incompressible velocity profiles and transformed to compressible flows using the van-Driest transformation [2] . The corresponding temperature profile was built as described in [14] and the static pressure at the inlet was assumed constant and equal to the infinite one. Turbulence quantities at the inlet were described as in [14].

REPORTING REQUIREMENTS

The list of aerodynamic quantities, which contributors have been asked to submit, were in accordance with the measurements by Settles et al, as published in the 1980-81 Stanford Conference Proceedings. Other flow quantities distributions and two-dimensional plots, in the vicinity of the shock wave boundary layer interaction, were

302

Table 1: Initial Conditions for the 20° and the 24° cases

	20° case	24° case
Infinite Mach Number	2.79	2.84
Reynolds Number per Unit Length	6.3×10^7	6.3×10^7
Total Temperature	258 K	262 K
Wall Temperature	274 K	276 K
Infinite Temperature	101 K	100 K
Infinite Velocity	562 m/s	569 m/s
Infinite Static Pressure	26000 Pa	24000 Pa
Boundary Layer Thickness	0.025 m	0.023 m
Displacement Thickness	0.0066 m	0.0061 m
Momentum Thickness	0.0013 m	0.0012 m

Table 2: Contributors and Turbulence Models Used

	NAME	INST.	TURB. MODEL
Code1	LeRibault and Buffat	ECL, Lyon	Low-Re $k - \epsilon$, compr. corr.
Code2	Vassilopoulos, Giannakoglou and Papailiou	NTUA, Athens	Low-Re $k - \epsilon$, A.R.S.M.
Code3	Arina and Ramella	Pol. Torino	Low-Re $k - \epsilon$, compr. corr.
Code4	Wallin	F.F.A., Sweden	Low-Re $k - \epsilon$, E.A.R.S.M.
Code5	Mohammadi	INRIA, France	Two-Layer $k - \epsilon$
Code6	Freskos and Koschel	RWTH, Aachen	Low-Re $k - \epsilon$
Code7	Loyau	INSA, Rouen	Low-Re $k - \epsilon$

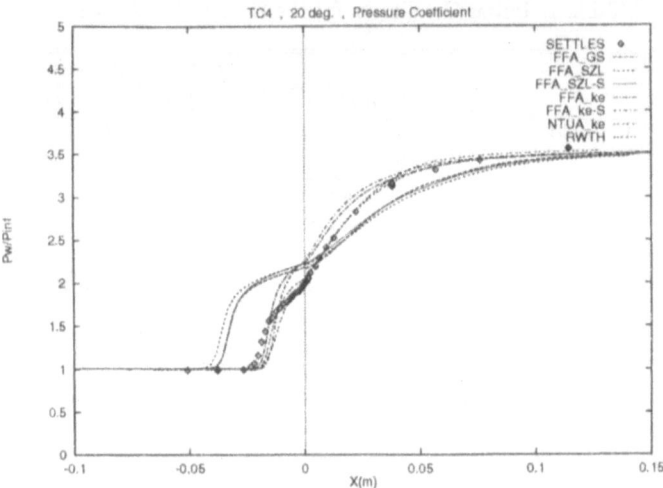

Figure 1: Pressure Coefficient Distribution along the solid wall.Ramp flow, 20°.

additionally requested.

In order to keep a common style in the presentation of results and to facilitate their comparison, contributors were requested to follow the guidelines provided below:

1. For all graphs, the X-coordinate is defined in the streamwise direction along the surface of the solid wall (negative X-values correspond to locations upstream of the corner). The Y-coordinate is zero on the solid walls and its orientation should be as follows: (a) for the 20° case and for $X \leq 0.0127m$ then Y is vertical, otherwise Y is normal to the ramp surface, (b) for the 24° case and for $X \leq 0.0102m$, then Y is vertical, otherwise Y is normal to the ramp surface.

2. The wall pressure distribution will be normalized on the infinite static pressure and plotted versus the X-coordinate.

3. The wall friction coefficient based on infinite conditions will be plotted versus the X-coordinate.

4. During the workshop, different profiles (Mach number, static pressure, non-dimensional streamwise velocity, turbulent kinetic energy) versus the Y-coordinate have been presented and compared. This corresponds to a huge amount of plots which will be omitted here, for the sake of economy in space.

COMPARISON OF THE NUMERICAL PREDICTIONS

All seven contributions to the Ramp Flow Test Case in the ETMA Workshop are listed in Table 2. The turbulence models used are also tabulated. All turbulence models

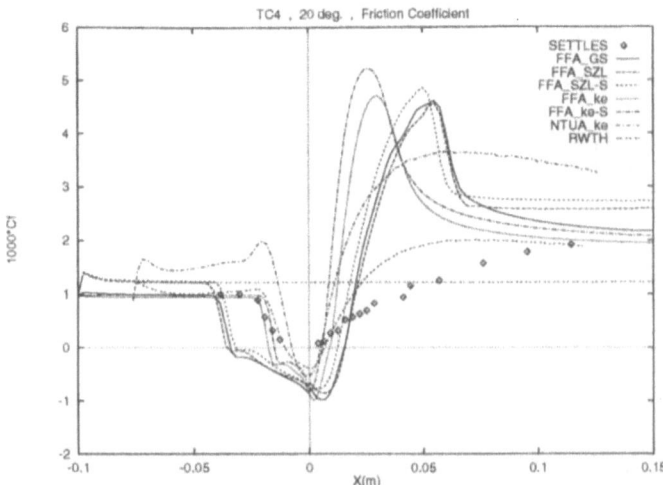

Figure 2: Friction Coefficient Distribution along the solid wall.Ramp flow, 20°.

used are based on the two differential equations governing the turbulent kinetic energy k and the turbulent energy dissipation ϵ and can be distinguished by differences in the near wall treatment, differences in the treatment of source terms and the addition or not of an Algebraic Reynolds Stress Model.

All participants made their calculations using structured grids. Only two of them, *Code1* and *Code5* made use of unstructured grids.

The variety of the turbulence models as well as the discretization and solution schemes tested by the different contributors in the framework of the ETMA Workshop, gave the possibility of:

1. Investigating the behaviour of the widely used two-equation k-ϵ model, in various forms, with or without corrections to account for the compressibility effects, in separated flows where separation is induced from the shock wave.

2. Investigating the behaviour of Algebraic Reynolds Stress models used in order to overcome the turbulence anisotropy in modelling the shock induced separation. In the present text, the numerical difficulties faced during the practical implementation of the ARSM and the remedies adopted in order to overcome the problems will be reported in brief.

3. Assessing the various compressibility corrections tested, for the dilatational dissipation or/and the pressure- dilatation correlation, in the examined flow case.

Figures 1 and 2 for the 20° case and figures 3 and 4 for the 24° one summarize the predictions of some of the contributors to the ETMA Workshop. The curves can be

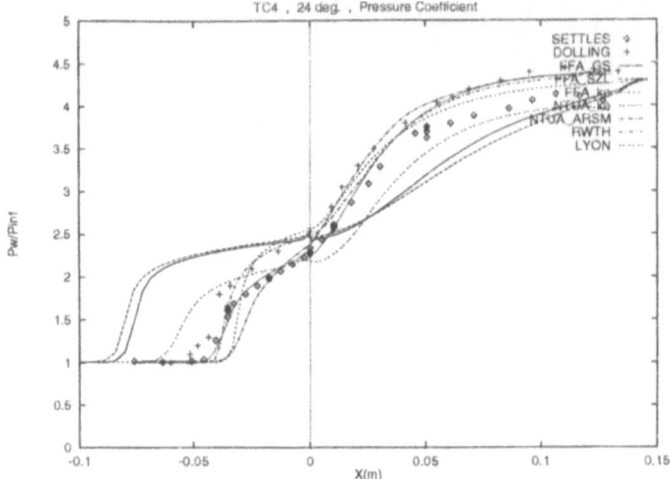

Figure 3: Pressure Coefficient Distribution along the solid wall.Ramp flow, 24°.

distinguished by the initials of the Institution providing the data, see Table 2 and an extension showing the type of models used. For the 20° case, the experimental data of Settles are included, while for the 24° case, experimental data from both Settles and Dolling are illustrated.

As mentioned in a previous section, the state of the incoming turbulent boundary layer is crucial, if the predictions are to be compared with available measurements. The sensitivity to the selection of the initial profiles and the freestream turbulence values has been proved (among others) by numerical tests performed using *Code6*.

The compressibility corrections in the low-Reynolds number $k - \epsilon$ model have been introduced in two ways. In the first way, the compressible dissipation term ϵ_c is introduced in the k-equation, in order to correct the deficiency for the compressible mixing layer. This is in general given by

$$\epsilon_c = F(M_t)\epsilon \tag{1}$$

where ϵ is the incompressible dissipation term and M_t is the turbulent Mach number. The latter is defined as

$$M_t = \frac{2k}{a^2} \tag{2}$$

where a is the speed of sound and is used in order to make active the extra dilatational dissipation term ϵ_c, in the compressible flow regime. The second turbulence equation retains its form and it is still solved in terms of the incompressible or solenoidal dissipation

306

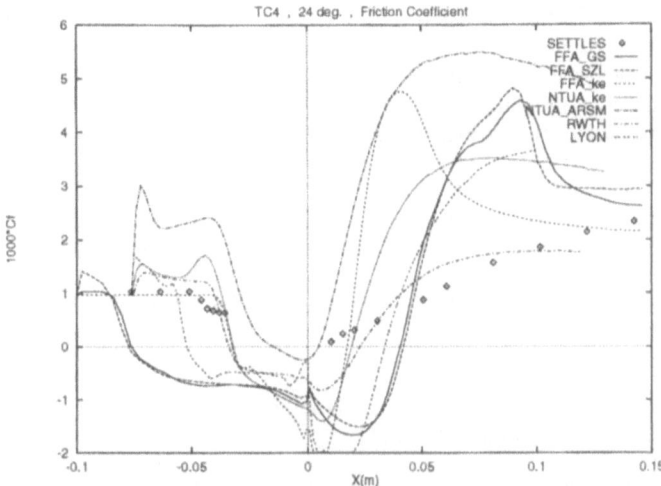

Figure 4: Friction Coefficient Distribution along the solid wall.Ramp flow, 24°.

ϵ. Models proposed by Sarkar [8], Wilcox [15] or Zeman [16] have been alternatively tested using *Code1*, *Code3* and *Code4*. The distinguishing feature between the three models is the way the functional $F(M_t)$ is defined. These models will be not repeated here and can be found in the cited references. The functional term $F(M_t)$ in the Sarkar model is proportional to M_t^2 and becomes negligible at low Mach number flows, whereas in the Zeman and Wilcox models ϵ_c is explicitly set to zero below a certain M_t threshold ($M_t = 0.25$), that is , below this threshold the compressibility effect is negligible. The Wilcox model combines the simplicity of the functional form used in the Sarkar model with the threshold existing in the Zeman model.

The second modification introduced in order to account for the compressibility effects was that related to the modelling of the pressure fluctuation term. In compressible flows, the pressure dilatation correlations lead to a decrease in turbulence production and dissipation. As a consequence, following the modification proposed by Sarkar [8], the pressure dilatation correlation gives rise to extra production and dissipation terms, which are scaled by an appropriate factor being proportional to M_t^2. Their signs are such that they decrease the overall production and dissipation terms, as introduced in the standard models, when added to them.

The aforementioned models for the compresible dissipation and the Sarkar model for the pressure dilatation have been tested in *Code1*. The compressibility terms lead to a large recirculation length. From the pressure distribution along the wall, it can be seen that the compressibility models by Sarkar, combined with the wall function technique, improves the quality of the solution at separation.

The Sarkar model for the pressure dilatation term was also used in *Code3*. For the 24° case, the plateau in pressure is visible. in the results presented during the Workshop. These results are not included in the figures. The low Reynolds k- ϵ model of Chien [1]

307

was also enhanced with Sarkar compressibility corrections, in *Code4*. The Sarkar model moves the shock induced separation point quite upstream, with respect to the standard k- ϵ model and this is also clear from the plots showing the friction coefficient distribution along the wall. The previously discussed results demonstrate that the compressibility corrections, as used in the various codes, lead to controversial results.

In *Code6* the dilatational dissipation model gave no improvement to the results and this is why the model was finally abandoned. On the contrary, the pressure dilatation model of Zeman seemed to have an important effect on the shock topology.

As a conclusion of the previous analysis of results, the following remarks can be given, in brief:

1. Different behaviour of the models can be found if combined with the wall function technique, or integrated to the wall.

2. Different behaviour of the models can be found when the models are combined with different "basic" two-equation models.

3. Their role needs to be re-examined when used in separated compressible flows.

4. A careless usage of modifications existing in the literature may deteriorate, unexpectedly, the numerical results.

As far as Algebraic Reynolds stress modifications to the k- ϵ are concerned, two contributors have tested two different ARSM techniques. In *Code4*, following the original work by Pope [6], the Reynolds stress anisotropy tensor is expressed as a tensor polynomial of the mean flow stress and the rotation tensors. The coefficient in the polynomial expression are functions of the mean flow stress, rotation tensor invariants and the turbulent Reynolds number. The Algebraic Reynolds Stress models are explicitly used along with a two-equation turbulence model. In this way, numerical problems related to the stiffness of the system of the Reynolds stress equations are overcome. The viscous terms in the mean flow equations are modified through the Explicit ARSM (EARSM) but the k- ϵ equations are still solved in their standard form.

The EARSM is tested in the form of two existing variants, the first having constant coefficients, as proposed by Speziale [13] (denoted by SZL in the figures) and the second incorporating variable coefficients which are functions of the mean flow stress and vorticity tensor invariants, as proposed by Gatski and Speziale [4] (denoted by GS in the figures).

In *Code 2*, the ARSM is combined with an explicit Navier-Stokes algorithm where an implicit treatment is incorporated in order to stabilize the solution the part of the numerical solution dealing with the k- ϵ source terms. The ARSM model was set up as a variant of the original scheme proposed by Rodi [7]. The algebraic system set for the Reynolds stresses was explicitly solved by (a) decoupling the normal from the non-orthogonal Reynolds stresses, in order to overcome stability problems and (b) underrelaxing the algebraic equations in order to ensure the diagonal dominality of the scheme.

The conclusion from the application of the EARSM model through *Code 4* is that it predicts a separation point much earlier than the k- ϵ model (for the 20° case) while for the case of 24° corner angle, the EARSM model separated early, at the inflow.

The conclusions from the application of the ARSM model, through *Code 2* is that it fails to reproduce the correct separation zone and leads to higher friction coefficient distributions; on the contrary, the standard k- ϵ model (the version previously combined with the ARSM) reproduces the experimental data close to the wall and predicts the correct separation zone. Nevertheless, it must be pointed out that, by examining the velocity profiles at the various axial positions, the ARSM model provides velocity distributions which are very close to the experimental ones.

REFERENCES

[1] *Chien, K. Y.* , "Predictions of Channel and Boundary Layer Flows with a Low-Reynolds Number Turbulence Model" , AIAA Journal **20**, No 1, (1982).

[2] *Cousteix, J.* , "Turbulence et Couche Limite" ,Cepandues Edition, Toulouse, (1973).

[3] *Dolling, D. S.* and *Murphy, M. T.* , "Unsteadiness of the Separation Shoch Wve Structure in a Supersonic Compression Ramp Flowfield" , AIAA Journal **21**, No 12, (1983).

[4] *Gatski T. B. and Speziale, C. G.*, "On Explicit algebraic stress models for complex turbulent flows" , J. Fluid Mech., **254**, (1993).

[5] *Kuhn, G. D.* and *Nielsen, J. N.*, "Prediction of Turbulent Separated Boundary Layers" , AIAA Paper **73-663**, (1973).

[6] *Pope, S. B.*, "A More General Effective-Viscosity Hypothesis" , J. Fluid Mech, **72**, Part 2, (1975).

[7] *Rodi, W.*, "A New Algebraic Relation for Calculating the Reynolds Stresses' , Mechanics of Fluid, **56**, 219-221, (1976).

[8] *Sarkar, S., Erlebacher, G.* , *Hussaini, M. Y.* and *Kreiss, H. O.* , "The Analysis and Modelling of Dilatational Terms in Compressible Turbulence" , J.Fluid Mech, **227**, (1991).

[9] *Settles, G. S.*, "An Experimental Study of Compressible Turbulent Boundary Layer Separation at High Reynolds Numbers" ,Ph.D. Thesis, Princeton University, (1975).

[10] *Settles, G. S., Vas, I. E.* and *Bogdonoff, S. M.* , "Details of a Shock Separated Turbulent Boundary Layer at a Compression Corner" , AIAA Journal **14**, No 12, (1976).

[11] *Settles, G. S., Fitzpatrick, T. J.* and *Bogdonoff, S. M.* , "Detailed Study of Attached and Separated Compression Corner Flow Fields in High Reynolds Number Supersonic Flow" , AIAA Journal **17**, No 6, (1979).

[12] *Settles, G. S., Gilbert, R. B.* and *Bogdonoff, S. M.* , "Data Compilation for Shock Wave Turbulent Boundary Layer Interaction Experiments on 2-D Compression Corners" , Princeton Univ. MAE Rep. 1489, (1980).

[13] *Speziale, C. G.*, "On non-linear k-l and k-ϵ models of turbulence", , J. Fluid Mech., **178**, (1987).

[14] *Vassilopoulos, C., Giannakoglou, K. C.* , and *Papailiou K. D.*, "Study of the Supersonic Compression Ramp Flow Using an Explicit Fractional-Step Technique and the $k - \epsilon$ Turbulence Model" , ETMA Workshop, Manchester, U.K. (1994).

[15] *Wilcox, D. C.*, "Turbulence Modeling for CFD" , Griffin Printing, Glendale, California, (1993).

[16] *Zeman, O.*, "Dilatational Dissipation: The Concept and Application in Modelling Compressible Mixing Layers" , Physics of Fluids A, **2**, No 2, (1990).

STUDY OF THE SUPERSONIC COMPRESSION RAMP FLOW
USING THE k-ε TURBULENCE MODEL WITH AND WITHOUT
ALGEBRAIC REYNOLDS STRESS MODIFICATIONS

C. Vassilopoulos, K.C. Giannakoglou and K.D. Papailiou
Lab. of Thermal Turbomachines
National Technical University of Athens
P.O. Box 64069, 157 10 Athens, Greece.

INTRODUCTION

In this work, the low-Reynolds k-ε model, with or without Algebraic Reynolds Stress modifications will be used for the numerical modelling of the two-dimensional shock wave / turbulent boundary layer interaction problem, over compression ramps. Two different compression ramps will be used, with corner angles equal to 20 and 24 degrees, according to the experiments performed in the Princeton University [1], [2].

In the present study two turbulence models will be used in combination with the same numerical kernel. The first turbulence model is the standard Jones and Launder k-ε one [3] for low Reynolds numbers while the second is an Algebraic Reynolds Stress modification of the aforementioned k-ε model. Both models are implemented in the same basic numerical kernel used, that of code ATHENA (A Turbulent Hyperbolic Explicit Navier-Stokes Algorithm) developed in the Lab. of Thermal Turbomachines of NTUA [4], [5]. It is an explicit, time-marching fractional-step method which solves the unsteady Navier-Stokes equations through the successive solution of simple one-dimensional operators. The advantage of such a fractional-step algorithm is that it allows greater time-steps, compared to other explicit solvers. There are two major numerical difficulties that need to be overcome when the above models are integrated in such an explicit solver. The first is related to the stiff low-Re source terms which appear in the right-hand side of the k-ε equations and render the method prone to instabilities. The use of a semi-implicit treatment of source-terms is the proposed remedy to this problem. When the ARS model is used, there is an additional problem that of succesfully solving the algebraic set of equations in terms of the Reynolds stress components. Special techniques like an implicit residual smoothing of the so calculated Reynold stresses and an underelaxation mechanism have been proved necessary in order to obtain a converged Reynolds stress field.

In the interest of space, the mean flow equations and the fractional-step technique will not be elaborated here. For a detailed presentation of the explicit numerical method, the reader should turn to a recent publication by the authors [6]. In the present paper, our aim is to convey to the reader information about the turbulence models used as well as their performance in the examined cases. Furthermore, techniques which are necessary in order to overcome numerical problems, due to the coupling of these models with an explicit scheme, will be discussed.

311

TURBULENCE MODELING

THE LOW-REYNOLDS k-ε MODEL

The differential equations for the turbulent kinetic energy k and the turbulent energy dissipation ε are written in the form [3]:

$$\frac{\partial}{\partial t}(\rho k) + \frac{\partial}{\partial x}(\rho k u) + \frac{\partial}{\partial y}(\rho k v) = \frac{\partial}{\partial x}\left[\left(\mu + \frac{\mu_t}{Pr_k}\right)\frac{\partial k}{\partial x}\right] + \frac{\partial}{\partial y}\left[\left(\mu + \frac{\mu_t}{Pr_k}\right)\frac{\partial k}{\partial y}\right] + S_k \tag{1}$$

$$\frac{\partial}{\partial t}(\rho \varepsilon) + \frac{\partial}{\partial x}(\rho \varepsilon u) + \frac{\partial}{\partial y}(\rho \varepsilon v) = \frac{\partial}{\partial x}\left[\left(\mu + \frac{\mu_t}{Pr_\varepsilon}\right)\frac{\partial \varepsilon}{\partial x}\right] + \frac{\partial}{\partial y}\left[\left(\mu + \frac{\mu_t}{Pr_\varepsilon}\right)\frac{\partial \varepsilon}{\partial y}\right] + S_\varepsilon \ . \tag{2}$$

The involved source terms read

$$S_k = P_k - \rho \varepsilon - L_k \qquad , \qquad L_k = 2\mu\left(\frac{\partial \sqrt{k}}{\partial n}\right)^2$$

$$S_\varepsilon = c_{\varepsilon_1} P_k \frac{\varepsilon}{k} - c_{\varepsilon_2} f_2 \rho \frac{\varepsilon^2}{k} + L\varepsilon \quad , \qquad L_\varepsilon = 2\frac{\mu \mu_t}{\rho}\left(\frac{\partial^2 V_t}{\partial n^2}\right)^2$$

where V_t denotes the velocity component tangential to the solid walls and n is the normal to the wall direction. The Prandtl numbers are set equal to $Pr_k = 1.0$ and $Pr_\varepsilon = 1.3$. The molecular viscosity μ is obtained using the Sutherland's law. The production term P_k, results directly from the Boussinesq hypothesis and reads

$$P_k = \mu_t\left(\frac{\partial u}{\partial y} + \frac{\partial v}{\partial x}\right)^2 + 2\mu_t\left[\left(\frac{\partial u}{\partial x}\right)^2 + \left(\frac{\partial v}{\partial y}\right)^2\right] - \frac{2}{3}\mu_t\left(\frac{\partial u}{\partial x} + \frac{\partial v}{\partial y}\right)^2 - \frac{2}{3}\rho k\left(\frac{\partial u}{\partial x} + \frac{\partial v}{\partial y}\right) \tag{3}$$

while the turbulent viscosity μ_t is obtained from the Prandtl-Kolmogorov relation

$$\mu_t = c_\mu f_\mu \frac{\rho k^2}{\varepsilon} \ . \tag{4}$$

The constants c_μ, $c_{\varepsilon 1}$ and $c_{\varepsilon 2}$ are set equal to 0.09, 1.45 and 1.92 respectively and the functions f_μ and f_2 are given by

$$f_2 = 1 - 0.3\exp(-Re_t^2) \quad , \quad f_\mu = \exp\left[\frac{-3.4}{\left(1 + \frac{Re_t}{50}\right)^2}\right] \quad , \quad Re_t = \frac{\rho k^2}{\mu \varepsilon} \ .$$

THE k-ε MODEL WITH ALGEBRAIC REYNOLDS STRESS MODIFICATIONS

The starting point for the development of the proposed Algebraic Reynolds Stress Model is the Favre-averaged Reynold stress equations which can be written in the form

$$\frac{\partial}{\partial t}(\rho R_{ij}) + \frac{\partial}{\partial x_k}(\rho u_k R_{ij}) - D_{ij}^1 = P_{ij} + \Phi_{ij} + D_{ij}^2 \tag{5}$$

where

312

$$R_{ij} = \widetilde{u_i'' u_j''} .$$
(6)

The diffusion term D^1_{ij}, the turbulence production term P_{ij}, the pressure-velocity fluctuations correlation Φ_{ij} and the dissipation term D^2_{ij} are given respectively by

$$D^1_{ij} = -\frac{\partial}{\partial x_k}\left(\rho \, \widetilde{u_k'' u_i'' u_j''} + \widetilde{pu_j''}\delta_{ik} + \widetilde{pu_i''}\delta_{jk} - \mu\widetilde{S_{ik}u_j''} - \mu\widetilde{S_{jk}u_i''}\right)$$
(7)

$$P_{ij} = -\rho\left(R_{ki}\frac{\partial u_j}{\partial x_k} + R_{kj}\frac{\partial u_i}{\partial x_k}\right)$$
(8)

$$\Phi_{ij} = \overline{p\left(\frac{\partial u_i''}{\partial x_j} + \frac{\partial u_j''}{\partial x_i}\right)}$$
(9)

$$D^2_{ij} = -\mu \overline{S_{ik}\frac{\partial u_j''}{\partial x_k}} - \mu \overline{S_{jk}\frac{\partial u_i''}{\partial x_k}}$$
(10)

where δ_{ij} is the delta of Kronecker and S_{ij} denotes the rate-of-strain tensor

$$S_{ij} = \frac{\partial u_i}{\partial x_j} + \frac{\partial u_j}{\partial x_i} - \frac{2}{3}\delta_{ij}\frac{\partial u_k}{\partial x_k} .$$
(11)

By neglecting the effect of density fluctuations on turbulence quantities as well as any term involving viscosity fluctuations, all terms in (5) can be modelled [7]. Thus, the pressure-velocity fluctuations correlation Φ_{ij} is approximated by three additive quantities which represent the quadratic, the linear and the wall influence terms, yielding

$$\Phi_{ij} = -c_1\frac{\rho\varepsilon}{k}\left(R_{ij} - \frac{2}{3}\delta_{ij}k\right) - B_1\left(P_{ij} - \frac{2}{3}\delta_{ij}P_k\right) - B_2\left(D_{ij} - \frac{2}{3}\delta_{ij}P_k\right) -$$
$$- B_3\rho k S_{ij} + \frac{\rho\varepsilon}{k}R_{ij}c_3\frac{k^{3/2}}{\varepsilon n} - \frac{2}{3}\delta_{ij}\rho\varepsilon c_3\frac{k^{3/2}}{\varepsilon n} + c_4\frac{k^{3/2}}{\varepsilon n}(P_{ij} - D_{ij})$$
(12)

where n stands from the distance from the wall and

$$B_1 = \frac{c_2 + 8}{11} \quad , \quad B_2 = \frac{8c_2 - 2}{11} \quad , \quad B_3 = \frac{30c_2 - 2}{55}$$

while

$$c_1 = 1.5 \quad , \quad c_2 = 0.4 \quad , \quad c_3 = 0.125 \quad , \quad c_4 = 0.015 .$$

The dissipation term D_{ij}^2 is related to the dissipation tensor and is approximated by the sum of a scalar and an anisotropic part, namely

$$D^2_{ij} = -\rho(1 - f_s)\frac{2}{3}\delta_{ij}\varepsilon - \rho f_s\frac{\varepsilon}{k}R_{ij}$$
(13)

where

$$f_s = \frac{1}{1 + \frac{Re_t}{10}} \quad . \tag{14}$$

The modelled Reynolds stress transport equations are finally written in the form

$$\frac{\partial}{\partial t}(\rho R_{ij}) + \frac{\partial}{\partial x_k}(\rho u_k R_{ij}) - \frac{\partial}{\partial x_k}\left(C_s \rho \frac{k}{\varepsilon} R_{kl} \frac{\partial R_{ij}}{\partial x_k} + \mu \frac{\partial R_{ij}}{\partial x_k}\right)$$

$$= P_{ij} - C_1 \rho \frac{\varepsilon}{k}\left(R_{ij} - \frac{2}{3}\delta_{ij}k\right) - C_2\left(P_{ij} - \frac{2}{3}\delta_{ij}P\right) \tag{15}$$

$$+ \frac{k^{\frac{3}{2}}}{\varepsilon n}\left[C_3 \rho \frac{\varepsilon}{k}\left(R_{ij} - \frac{2}{3}\delta_{ij}k\right) + C_4(P_{ij} - D_{ij})\right] - (1-f_s)\frac{2}{3}\delta_{ij}\rho\varepsilon - f_s \rho \frac{\varepsilon}{k} R_{ij}$$

where

$$D_{ij} = -\rho\left(R_{ki}\frac{\partial u_k}{\partial x_j} + R_{kj}\frac{\partial u_k}{\partial x_i}\right).$$

According to the Rodi assumption [8], we may also assume that the convection and diffusion part in equations (12) remains proportional to the corresponding part of the turbulent kinetic energy equation (1), the proportionality factor being the ratio R_{ij}/k. By combining equations (1) and (15) the aforementioned assumption leads to the following algebraic equation which may provide the normal and cross-components of the Reynolds stresses

$$\left(1 + B_1 + C_4 \frac{k^{3/2}}{\varepsilon n}\right)P_{ij} + \left(-B_2 - C_4 \frac{k^{3/2}}{\varepsilon n}\right)D_{ij} - B_3\rho k S_{ij} + \frac{2}{3}\delta_{ij}P_k(B_1 + B_2 - 1) +$$

$$+ \frac{2}{3}\delta_{ij}\rho\varepsilon\left(C_1 - 1 + f_s + \frac{P_k}{\rho\varepsilon} - C_3 \frac{k^{3/2}}{\varepsilon n}\right) = \frac{\rho\varepsilon}{k} R_{ij}\left(C_1 - 1 + f_s + \frac{P_k}{\rho\varepsilon} - C_3 \frac{k^{3/2}}{\varepsilon n}\right). \tag{16}$$

Details about the numerical solution of system (16) are given in a following section.

NUMERICAL SOLUTION

THE FRACTIONAL-STEP ALGORITHM

In the fractional step algorithm, mean flow and turbulence equations are all included in the same array of dependent variables Q and solved through the successive application of one-dimensional operators. The fractional-step procedure could be expressed in the symbolic form

$$Q^{n+2} = L_\xi^H L_\eta^H L_\xi^P L_\eta^P L^{ST} L^{ST} L_\eta^P L_\xi^P L_\eta^H L_\xi^H Q^n. \tag{17}$$

The double and inverse sequence of the one-dimensional operators ensures second order accuracy in time. In the computational domain (ξ,η), the one-dimensional operators L are appropriately subscribed, to indicate whether sweeps are performed along the η or ξ=constant grid lines. Superscripts H, P and ST distinguish the nature of the part of the equations which is resolved by each operator. Thus, the L^H operator corresponds to the inviscid part of the equations (H=Hyperbolic), the L^P operator solves for the viscous part (P=Parabolic), while the L^{ST} operator handles the source terms which appear in the turbulence equations. Second order accuracy in space is obtained using the predictor-corrector

314

MacCormack scheme for the hyperbolic and the parabolic operators. The source terms appearing in the turbulence equations influence the code stability, mainly during the early stages of the computation. In order to overcome convergence problems, a semi-implicit treatment is used in the solution of the L^{ST} operators, [4], [6].

The governing equations are discretized using centered finite-difference schemes on a collocated grid. Mean flow and turbulence equations are all stored at the same nodes which are located at the intersection of grid lines. Thus, in order to prevent odd-even uncoupling effects and to avoid oscillations close to shock waves, extra dissipation (a blend of second- and fourth-order derivatives, [9]) is explicitly added to the solution array Q at the end of any iteration. The well known Jameson scheme is further modified and numerical stability is ensured through a smoother ε-distribution close to solid walls. Details about the modified artificial dissipation scheme could be found in [5].

IMPLEMENTATION OF THE ARS MODEL

The algebraic system (16) can be iteratively solved to provide the normal R_{11}, R_{22}, R_{33} and the cross R_{12} Reynolds stresses. Unfortunately, such an iterative scheme suffers from considerable convergence problems. Numerical tests have shown that any iterative scheme which calculates all Reynolds stresses, by exactly satisfying (16) at each time-step of the time-marching algorithm is both unstable and time-consuming. Consequently, equations (16) are solved in a segregated way, by treating separately the orthogonal and the cross Reynolds stresses; according to the proposed solution method, equations (16) are not exactly satisfied at the early stages of the time-marching algorithm. For stability reasons the production term P_k in (16) is calculated using the Boussinesq hypothesis (3), in terms of the gradients of the velocity components and it is only indirectly linked to the normal P_{ij} components provided by (8).

Concerning the normal Reynolds stresses R_{ij}, the solution procedure for (16) will be demonstrated below for the R_{11} component, [10], [11]. The extention to R_{22} and R_{33} is straightforward. So, for $i=j=1$, equation (16) can be written in the synoptic form

$$R_{11} = A_0 + A_1 R_{11} + A_2 \tag{18}$$

where $A_2 = A_2(R_{22}, R_{33})$ is a function of the remaining orthogonal Reynolds stresses. The coefficients A_i are further analyzed into a positive and a non-positive part

$$A_i = A_i^+ + A_i^- \tag{19}$$

giving rise to the final equation which explicitly updates R_{11}

$$R_{11} = \frac{A_0 + A_1^+ R_{11} + A_2^+}{1 - A_1^- - A_2^- \frac{1}{R_{11}^{old}}} \tag{20}$$

using the R_{11}^{old} value which is known from the previous time step.

The so calculated orthogonal Reynolds stresses are then corrected in order to satisfy the equation

$$R_{11} + R_{22} + R_{33} = 2k \tag{21}$$

while maintaining their relative ratios. The non-orthogonal stress R_{12} is simply obtained from (16), by using the R_{11}, R_{22} and R_{33} values resulting from (20) and (21). All Reynolds stresses

R_{ij} are under-relaxed using a relaxation coefficient of $\omega = 0.3$. The under-relaxed values of R_{ij} are finally smoothed using a Laplacian filter of the form

$$R_{ij} + \varepsilon \ (\delta_x^2 + \delta_y^2) \ R_{ij} = 0 \qquad (22)$$

where ε is an appropriate constant ($\varepsilon = 0.1\text{-}1.0$) and δ_x, δ_y the centered finite-difference operators.

The calculated Reynolds stresses are directly lumped to the right-hand-side of the momentum equations. Practically speaking, this means that the eddy viscosity coefficient is zero in these equations; of course, the real static pressure must replace the effective pressure in these equations. In the energy equation, the eddy viscosity coefficient μ_t is also eliminated from the diffusion term and products of the Reynolds stresses with the appropriate velocity components are introduced in the r.h.s. of the equation. On the contrary, the Boussinesq hypothesis is retained in both the k and ε transport equations.

RESULTS AND DISCUSSION

FLOW OVER THE 24 DEF. COMPRESSION RAMP

The computational grid used in the present study is presented in figure (1). It consists of 99×101 grid nodes [12] and it is appropriately stretched close to the ramp surface, where a maximum y^+ distance of approximately 2.0 has been obtained. The fixed inlet profile was obtained through the Kuhn and Nielsen [13] expression for turbulent velocity profiles, combined with the second van Driest transformation [14]

$$\left(\frac{U}{U_\infty}\right)_c = \frac{B + D\sin\left[A\left(\frac{U}{U_\infty}\right)_{INC} - \sin^{-1}\left(\frac{B}{D}\right)\right]}{2A^2}$$

where

$$A = \sqrt{\frac{\gamma-1}{2}M^2\frac{T_\infty}{T_W}} \quad , \quad B = \left(1+\frac{\gamma-1}{2}M^2\right)\frac{T_\infty}{T_W} - 1 \quad , \quad D = \sqrt{B^2 + 4A^2}$$

$$\left(\frac{U(y)}{U_\infty}\right)_{INC} = \frac{U_T}{U_\infty}\left[\frac{1}{\kappa}\ln\left(1+\frac{y|U_T|}{\nu_W}\right) + C - \left(3.39\frac{y|U_T|}{\nu_W} + C\right)e^{-0.37\frac{y|U_T|}{\nu_W}} + \frac{\pi}{\kappa}\left(1-\cos\left(\pi\frac{y}{\delta}\right)\right)\right]$$

where the subscript W indicates quantities at the wall while C and INC characterize equivalent compressible and incompressible flow conditions. The constants κ and C are set equal to .41 and 5.1 respectively. The input data for the velocity profile, are the external Mach number, the infinite temperature, the wall temperature, the friction velocity and the boundary layer thickness. The corresponding temperature profile yields

$$\frac{T(y)}{T_W} = 1 + B\frac{U(y)}{U_\infty} - \left(A\frac{U(y)}{U_\infty}\right)^2.$$

At the inlet, the pressure is assumed uniform and equal to the infinite one.

The turbulence kinetic energy profile is built using the expressions

316

$$k^+ = \frac{k}{U_T^2} = \min\left(\sqrt{c_\mu}\, y^+, 0.035 y^{+2}\right) \ , \ y^+ \leq 10$$

$$k^+ = \frac{k}{U_T^2} = \max\left(3.333\left(\frac{y_{CUT}^+ - y^+}{y_{CUT}^+ - 10}\right), \frac{k_\infty}{U_T^2}\right) \ , \ y^+ > 10$$

where $y^+_{CUT} = 400$. The infinite value of k results from the known turbulence intensity level. The turbulent viscosity is evaluated through an algebraic turbulence model; in the present study the Cebeci-Smith model has been used. Finally, the turbulence energy dissipation is calculated from the Prandtl-Kolmogorov relation. Figure (2) illustrates the non-dimensional inlet profile and compares it with the logarithmic-law theoretical one for incompressible flows. In this figure, the imposed inlet profile has been mapped into an equivalent incompressible one, through the van-Driest transformation. The inlet profile has been built using the following flow data:

Infinite Mach Number	2.84
Reynolds number per unit length	6.3×10^7
Total Temperature	262 K
Boundary layer thickness	0.023 m .

Figure (3) shows the calculated iso-Mach contours. Figures (4) and (5) show the wall pressure and friction coefficient distributions obtained using the standard low-Reynolds k-ε model with or without algebraic modifications, as discussed in Section 2.2. The pressure distribution, calculated using both models, are more close to the measurements by Dolling and Murphy, rather than to those by Settles et al; these differences are more distinct in the region which follows the shock induced separation. The pressure rise within and after the shock is well predicted; better results are obtained using the low-Re k-ε model which, in the shock region are in perfect agreement with the measurements by Settles et al. The use of the ARS model provides a shock wave which lies slightly downstream compared to the standard k-ε model. The pressure recovery in the part starting at the separation point and going up the corner location is more abrupt in the case of the ARS model. It can be noticed that, in a few grid positions after the inlet, the friction coefficient is increased. In the case of the ARS model, this increase is unexpectedly high and could be probably attributed to inconsistencies between the inlet k and ε distributions and the ARS model. Thus, the ARS model leads to a friction coefficient distribution which is shifted to higher values and consequently to a smaller separation zone.

Figure (6) presents the calculated non-dimensional velocity distributions at three locations along the ramp surface, where experimental data are also available. These locations are at the longitudinal distances x=-.0102m, x=.0102m and x=.0305m. The comparison with experiments is satisfactory. As previously discussed, the ARS model underestimates the separation zone. On the contrary, it provides velocity distributions which are very close to experiments in the outer region.

FLOW OVER THE 20 DEG. COMPRESSION RAMP

For the 20 degrees ramp flow case, a similar grid to the one presented in figure (1) has been used; this similarity exists in the way grid nodes have been arranged in the transversal direction. The inlet flow profile was built in an analogous way, using the following data:

Infinite Mach Number	2.79
Reynolds number per unit length	6.3×10^7
Total Temperature	258 K
Boundary layer thickness	0.025 m .

Figures (7) and (8) show the wall pressure and friction coefficient distributions obtained using the standard low-Reynolds k-ε model. The pressure distribution is in excellent agreement with measurements, but the friction coefficient is overestimated. The rise in the friction coefficient starts from the first grid locations after the inlet section. The calculated friction coefficient at the exit is also higher than its measured value, but this has been observed in other numerical studies as well.

Figure (9) presents the calculated non-dimensional velocity distributions at three specific locations along the ramp surface. These are at x = -.0111m, x = 0.0m and x = .0127m and satisfactorily match the experimental profiles.

CONCLUSIONS

The ramp flow problem, at 20 and 24 degrees has been numerically investigated using two turbulence models and an explicit numerical kernel. Firstly, numerical problems associated to the use of en explicit solver along with differential equations having stiff source-terms have been solved. On the other hand, the algebraic system of equations which provides the Reynolds stresses at each node has been successfully solved by decoupling the involving equations and enforcing the diagonal dominality in each equation. Under-relaxation and smoothing of the obtained Reynolds stress fields are necessary. With the imposed inlet profiles, the ARS model fails to reproduce the correct separation zone and provides higher friction coefficients. Nevertheless, the ARS model gives accurate results far from solid walls. The standard k-ε model reproduces the experimental data with satisfactory accuracy and predicts the correct separation zone.

REFERENCES

[1]   Dolling, D.S., Murphy, M.T.: "Unsteadiness of the separation shock wave structure in a supersonic compression ramp flowfield", AIAA Journal, 21, No. 12, December (1983), pp. 1628-1634.

[2]   Settles, G.S., Vas, I.E., Bogdonoff, S.M.: "Details of a shock-separated turbulent boundary layer at a compression corner', AIAA Journal, 14, No. 12, December (1976), pp. 1709-1715.

[3]   Jones, W.P., Launder, B.E.: "The prediction of laminarization with a two-equation model of turbulence", Int. Journal Heat Mass Transfer, 15 (1972), pp. 301-314.

[4]   Simandirakis, G., "Numerical solutions of Navier-Stokes equations for transonic flows inside turbine bladings", Ph.D. Thesis, Athens, February 1992.

[5]   Giannakoglou, K., Simandirakis, G., Papailiou, K.D.: "Turbine cascade calculated through a fractional step Navier-Stokes algorithm", ASME Paper 91-GT-55 (1991).

[6]   Dejean, F., Vassilopoulos, C., Simandirakis, G., Giannakoglou, K., Papailiou, K.D., "Analysis of transonic turbomachinery flows using a 2-D explicit low-Reynolds k-ε Navier-Stokes solver", ASME Paper 94-GT-63 (1994).

[7]   Vandromme, D.: "Turbulence modelling for compressible flows and implementation in Navier-Stokes solvers", Introduction to the modelling of Turbulence, VKI LS 1991-02 (1991).

[8]   Rodi, W.: "A new algebraic relation for calculating the Reynolds stresses", Mechanics of Fluid, 56 (1976), pp. 219-221.

[9]   Jameson, A., Schmidt, W., Turkel, E.: "Numerical solutions of the Euler equations by finite volume methods using Runge-Kutta time stepping schemes", AIAA Paper 81-1259 (1981).

[10]   Stolcis, L.: "Computation of the turbulent flow over a high-lift system using algebraic Reynolds stress model", Applied Mathematics and Simulation, CRS4-APPMATH-93-4, March (1993).

[11]   Chaput, E., Tourrette, L., Undreiner, S.: "Navier-Stokes 2D et 3D pour la conception aérodynamique des avions de transport", 30ème Colloque d'Aérodynamique Appliquée, Ecole Centrale de Nantes, 25-26-27 Octobre (1993).

[12]   Loyau, H.: personal communication .

[13]   Kuhn, G.D., Nielsen, J.N.: "Prediction of turbulent separated boundary layers", AIAA Paper 73-663 (1973).

[14]   Cousteix, J.: "Turbulence et couche limite", Cepandues Edition, Toulouse 1989.

Acknowledgements

Part of the work concerning the development of the ARS model was funded by the ETMA (Efficient Turbulence Models for Aeronautics) Project (BRITE-EURAM-2076/2032). The authors would like to acknowledge Dr. G. Simandirakis' contribution to the development of the fractional step numerical algorithm.

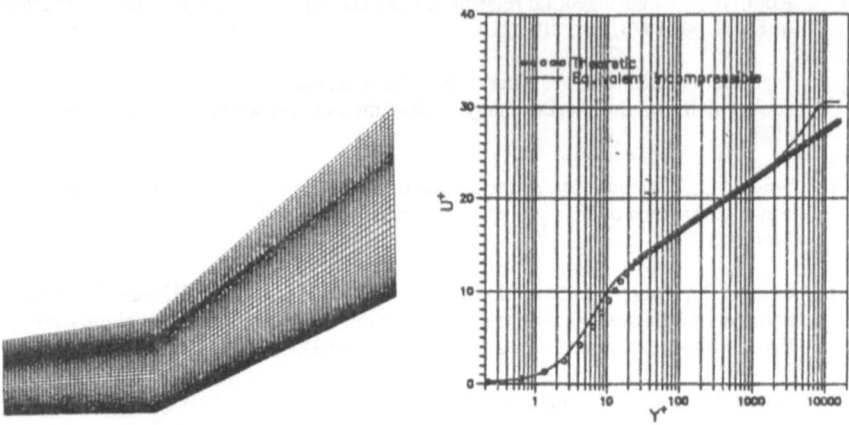

Fig. 1   The computational grid (24 deg.)   Fig. 2   Equivalent incompressible inlet velocity
                                                              profile (24 deg.)

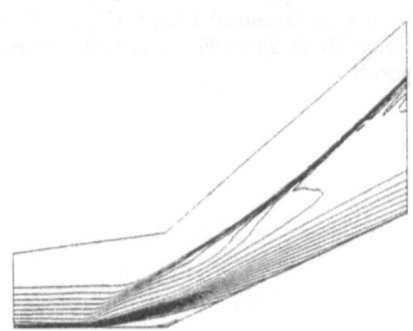

Fig. 3   Calculated iso-Mach contours (24 deg.)

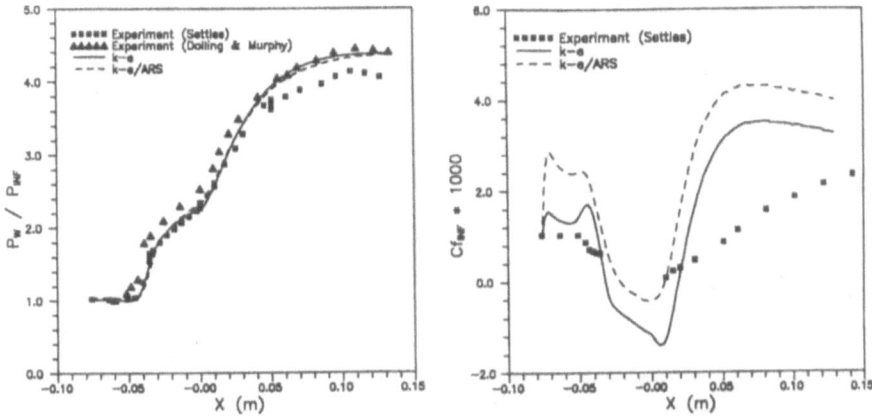

Fig. 4 Pressure Coefficient distribution (24 deg.)

Fig. 5 Friction Coefficient distribution (24 deg.)

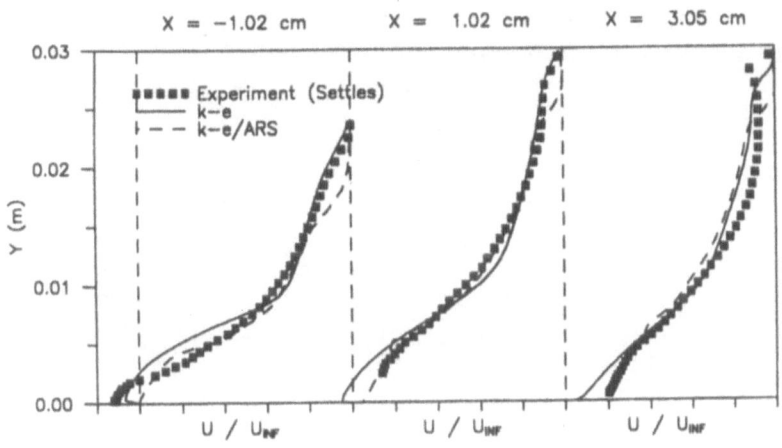

Fig. 6 Velocity Profiles at three axial positions (24 deg.)

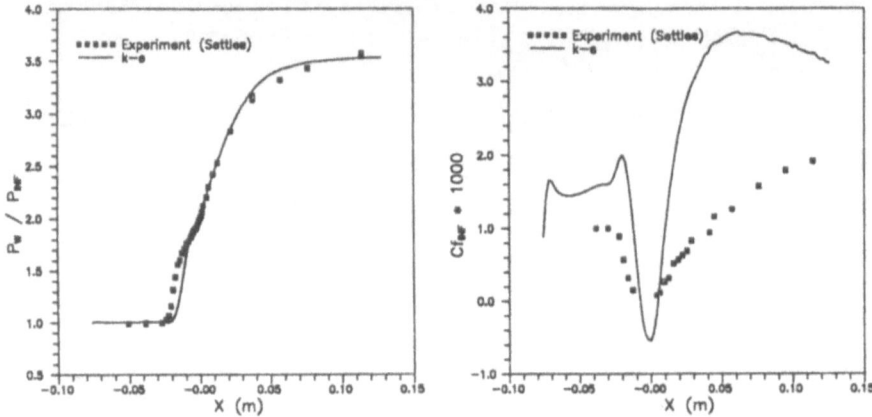

Fig. 7   Pressure Coefficient distribution (20 deg.)

Fig. 8   Friction Coefficient distribution (20 deg.)

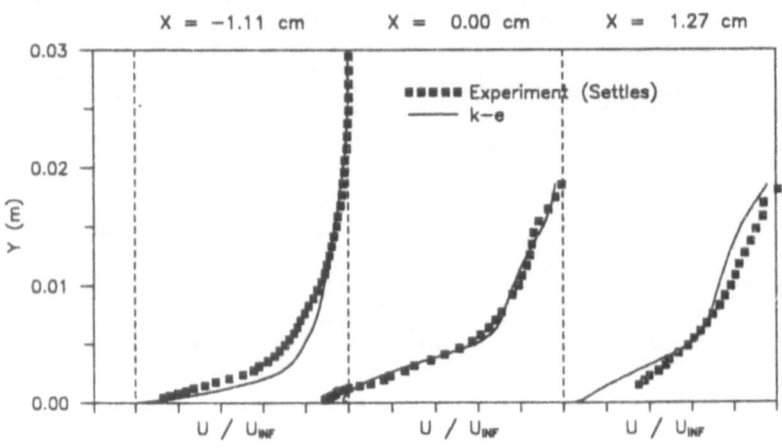

Fig. 9   Velocity Profiles at three axial positions (20 deg.)

NUMERICAL SIMULATION OF COMPRESSION RAMP FLOWS
WITH A COMPRESSIBLE $K - \varepsilon$ MODEL

R. Arina and F. Ramella
Dipartimento di Ingegneria Aeronautica e Spaziale
Politecnico di Torino
Corso Duca degli Abruzzi, 24 - I-10129 Torino (Italy)

SUMMARY

In this paper we present the results of the numerical investigation of the shock wave/turbulent boundary layer interaction on a compression corner. The calculations are performed using the Favre-averaged Navier-Stokes equations and a $k - \varepsilon$ turbulence model with compressibility corrections and near-wall modeling. The compressibility terms in the transport equation for the turbulent kinetic energy are based on the Sarkar et al. model [6,7], while the compressibility effects in the equation for the dissipation rate are based on the model of Coleman and Mansour [2]. Flowfield data are presented for a compression corner angle of 24°, with Mach number 2.84, and compared with the experimental data of Settles, Fitzpatrick and Bogdonoff reported in [3]. A good prediction of the upstream influence has been obtained by a careful specification of the incoming boundary layer matching the available experimental informations.

1. INTRODUCTION

The interaction of a shock wave with a boundary layer on a two-dimensional compression ramp represents a good test case for turbulence models. The strong shock-turbulence interaction entails a large increase in the fluctuation level, with a global large increase of the turbulent mixing rate, and consequent rapid acceleration of the flow field in the downstream part. For these reasons, low-speed turbulence models are not able to correctly predict the separation region. Therefore it is necessary to introduce compressibility effects. Several studies have been proposed in the recent years for modeling the compressible terms. In spite of the efforts, they are not satisfactory yet, and further numerical tests are needed for a better understanding of the behavior of compressible turbulence. The purpose of this paper is to test some recent ideas in the case of a compressible turbulent shear layer developing along a wall.

In order to have a realistic simulation, it is a crucial issue to specify the shape of the incoming boundary layer as close as possible to the experimental ones. The shock going inside the boundary layer become weaker and vanish. The pressure signal carried by the shock is necessarily transmitted in the upstream direction through the subsonic inner part of the boundary layer. The upstream influence length appears as being essentially a function of the height of this subsonic part, therefore the fullness of the incoming boundary layer profile has a determinant role in the interaction mechanism. The available experimental data provide only integral parameters at inlet, therefore an iterative procedure is necessary to generate profiles with appropriate fullness. The inlet mean velocity and temperature profiles, have been obtained by the Coles' method, extended to compressible flows [4], based on the compressible logarithmic wall law. In this way the calculated profiles exactly reproduce the measured momentum thickness.

2. GOVERNING EQUATIONS

The equations for the mean flow quantities, are obtained from the Navier-Stokes equations by Favre averaging. The gas is assumed perfect (air, $\gamma = 1.4$, $Pr = 0.72$), and the Shuterland law for the molecular viscosity is adopted. The closure for the Reynolds stresses is obtained by a $k - \varepsilon$ two-equation model, with near-wall treatment, described in detail in the next section.

The governing equations for the mean flow field $(\bar{\rho}, \tilde{u}_i, \tilde{E})$, take the form

$$\frac{\partial \bar{\rho}}{\partial t} + \frac{\partial}{\partial x_i}(\bar{\rho}\tilde{u}_i) = 0 \ ,$$

$$\frac{\partial}{\partial t}(\bar{\rho}\tilde{u}_i) + \frac{\partial}{\partial x_j}(\bar{\rho}\tilde{u}_i\tilde{u}_j + \bar{p}\delta_{ij}) = \frac{\partial}{\partial x_j}\left[2(\bar{\mu} + \mu_t)\left(\tilde{S}_{ij} - \frac{1}{3}\delta_{ij}\frac{\partial \tilde{u}_k}{\partial x_k}\right) - \frac{2}{3}\bar{\rho}k\delta_{ij}\right] \ ,$$

$$\frac{\partial}{\partial t}(\bar{\rho}\tilde{E}) + \frac{\partial}{\partial x_i}\left[(\bar{\rho}\tilde{E} + \bar{p})\tilde{u}_i\right] =$$
$$\frac{\partial}{\partial x_i}\left[\tilde{u}_i\left(2(\bar{\mu} + \mu_t)\left(\tilde{S}_{ij} - \frac{1}{3}\delta_{ij}\frac{\partial \tilde{u}_k}{\partial x_k}\right) - \frac{2}{3}\bar{\rho}k\delta_{ij}\right) + \gamma(\frac{\mu}{Pr} + \frac{\mu_t}{Pr_t})\frac{\partial \tilde{e}}{\partial x_i}\right] \ .$$

where $\bar{\rho}\tilde{E} = \bar{\rho}\tilde{e} + \bar{\rho}\frac{\tilde{u}_i\tilde{u}_i}{2} + \bar{\rho}k$, $k = \frac{\widetilde{u_i''u_i''}}{2}$ is the turbulent kinetic energy, and $\tilde{S}_{ij} = \frac{1}{2}\left(\frac{\partial \tilde{u}_i}{\partial x_j} + \frac{\partial \tilde{u}_j}{\partial x_i}\right)$ is the Favre-averaged rate of strain tensor. The equations are discretized in space by a finite volume method. For the convective terms an improved version of the AUSM flux-vector splitting formulation [5] is employed, while the viscous terms are discretized by a central scheme. The time discretization is first order and implicit, and the linear system of equations formed at each time-integration step, is solved by a nonlinear Krylov subspace projection method (preconditioned GMRES in Jacobian-free formulation).

The domain over the corner, is parameterized by a structured grid stretched along the wall and around the origin of the ramp.

On the upper boundary, as well as on the inflow boundary, the supersonic inviscid flow is fully specified. The incoming boundary layer is specified using the compressible extension of the Coles' method for the mean flow quantities as previously described. Local equilibrium conditions and a mixing length model are assumed for the specification of the turbulence quantities. At the outlet boundary, we impose characteristic boundary conditions in the inviscid part, and second order extrapolation in the boundary-layer region. The wall is considered adiabatic.

3. COMPRESSIBLE $k - \varepsilon$ MODEL

In the present work we use a near-wall two-equation turbulence model of the $k - \varepsilon$ type. In the model we introduce a variable density extension of the near-wall treatment proposed by Chien for incompressible flows [1], combined with the implementation of the model for the compressibility terms proposed by Sarkar, Speziale et al. [6,7], for the k equation, and by Coleman and Mansour [2] for the ε equation.

Consistent with [6], the dissipation is decomposed into solenoidal and compressible parts: $\varepsilon = \varepsilon_s + \varepsilon_c$. Here, ε_s represents the dissipation associated with the energy cascade. The length and time scales are built up from ε_s, which is obtained from a modeled transport equation. Sarkar et al. model the compressible dissipation in the form $\varepsilon_c = \alpha M_t^2 \varepsilon_s$, where $M_t = (\overline{u_i'' u_j''}/\gamma R\tilde{T})^{0.5}$ is the turbulence Mach number, and $\alpha = 0.5$.

The Reynolds stress tensor is modeled in the standard eddy viscosity form,

$$R_{ij} = -\frac{2}{3}\bar{\rho}k\delta_{ij} + 2\mu_t \left(\tilde{S}_{ij} - \frac{1}{3}\tilde{S}_{kk}\delta_{ij} \right) \quad ,$$

with the eddy viscosity

$$\mu_t = \bar{\rho}C_\mu f_\mu \frac{k^2}{\varepsilon_s} \quad ,$$

where $C_\mu = 0.09$, and f_μ is a wall damping function.

The modeled transport equation for k is, neglecting the turbulent fluctuations in the viscosity

$$
\begin{aligned}
\frac{\partial}{\partial t}(\bar{\rho}k) + \frac{\partial}{\partial x_i}(\bar{\rho}\tilde{u}_i k) = {} & R_{ij}\frac{\partial \tilde{u}_i}{\partial x_j} - \bar{\rho}\varepsilon + \overline{p'\frac{\partial u_i'}{\partial x_i}} \\
& -\overline{u_i''}\frac{\partial \bar{p}}{\partial x_i} + \overline{u_i''}\frac{\partial \overline{\tau_{ij}}}{\partial x_j} + \frac{\partial}{\partial x_i}\left[\left(\bar{\mu} + \frac{\mu_T}{\sigma_k} \right)\frac{\partial k}{\partial x_i} \right] \quad ,
\end{aligned}
\tag{1}
$$

where $\sigma_k = 0.75$, and

$$\overline{\tau_{ij}} = -\frac{2}{3}\mu\frac{\partial \tilde{u}_k}{\partial x_k}\delta_{ij} + \mu\left(\frac{\partial \tilde{u}_i}{\partial x_j} + \frac{\partial \tilde{u}_j}{\partial x_i} \right) \quad ,$$

is the mean viscous stress.

Following Sarkar et al. [6], the pressure dilatation correlation is modeled as follows

$$\overline{p'\frac{\partial u_i'}{\partial x_i}} = -\alpha_1 \mathcal{P} M_t^2 + \alpha_2 \bar{\rho}\varepsilon_s M_t^2 \quad ,$$

where $\mathcal{P} = R_{ij}\frac{\partial \tilde{u}_i}{\partial x_j}$ is the turbulence production, $\alpha_1 \approx 0.4$ and $\alpha_2 \approx 0.2$. The mass flux term is modeled by the gradient transport hypothesis [7]

$$\overline{u''}_i = \frac{C_\mu f_\mu}{\bar{\rho}\sigma_\rho}\frac{k^2}{\varepsilon_s}\frac{\partial \bar{p}}{\partial x_i} \quad .$$

with $\sigma_\rho = 0.5$.

The dissipation rate equation is assumed of the form

$$
\begin{aligned}
\frac{\partial}{\partial t}(\bar{\rho}\tilde{\varepsilon}_s) + \frac{\partial}{\partial x_i}(\bar{\rho}\tilde{u}_i\tilde{\varepsilon}_s) = {} & -\left[\frac{1}{3} + n(\gamma - 1)\right]\bar{\rho}\tilde{\varepsilon}_s\frac{\partial \tilde{u}_i}{\partial x_j} \\
& C_{\varepsilon 1}f_1\frac{\tilde{\varepsilon}_s}{k}R_{ij}\left(\frac{\partial \tilde{u}_i}{\partial x_j} + \frac{1}{3}\frac{\partial \tilde{u}_k}{\partial x_k}\delta_{ij} \right) \\
& -C_{\varepsilon 2}f_2\bar{\rho}\frac{\tilde{\varepsilon}_s^2}{k} + \frac{\partial}{\partial x_i}\left[\left(\bar{\mu} + \frac{\mu_T}{\sigma_\varepsilon} \right)\frac{\partial \tilde{\varepsilon}_s}{\partial x_i} \right] + \bar{\rho}\mathcal{E} \quad ,
\end{aligned}
\tag{2}
$$

Where $C_{\varepsilon 1} = 1.5$, $C_{\varepsilon 2} = 1.83$, $\sigma_\varepsilon = 1.3$. The damping functions f_μ, f_1, f_2, as well as the wall dissipation term $\mathcal{E}$, and the wall boundary condition D for $\varepsilon\,(=\tilde{\varepsilon}_s + D)$, are modeled as proposed by Chien [1]. The first member of the rhs, takes into account the variation of the viscosity along a streamline [2]. For an isentropic compression $n \approx 0.7$.

325

4.   NUMERICAL RESULTS AND DISCUSSION

The numerical simulation of the supersonic compression ramp flow (24 deg, $\frac{Re_\infty}{m} = 6.3 \times 10^7$, $M_\infty =$ 2.84, $T_\infty = 100°K$), has been performed with a stretched grid of 101×101 points. The computational domain is $[-0.5, .18] \times [0., .23]$ meters, with the corner ramp placed at $x = 0$. The isopressure contours are shown in figure 1. Analyzing their shape, it is possible to remark that the pressure rise induced by the shock wave is strong enough to provoke a significant separation, with a considerable upstream influence. Also evident is the interaction between the upstream first shock, causing the separation, and the compression fan induced by the reattachment of the boundary layer.

In figure 2 the wall pressure distribution and the skin friction coefficient are compared with the experimental data [3]. From these results it is possible to note that the separation point is quite well predicted. As pointed out earlier, it is very important to have an incoming boundary layer as close as possible to the experimental one. In the present case the good agreement with the experimental data can be seen in the flat plate part of the skin friction distribution, as well as with the velocity, Mach and pressure distributions at $x = -0.0635m$. It is important to note that the inlet profile has been specified at $x = -0.5m$, matching the experimental momentum thickness. The good agreement in figure 3 confirm the statement in [3] that the upstream incoming boundary layer develops under a zero pressure gradient.

In the next figures 4-11, we compare the computed profiles with the experimental ones. For the same stations we report also the turbulent kinetic energy profiles. It is possible to remark that the turbulence model is able to predict the increase of the fluctuation level. However the velocity and Mach number profiles show that the outer part of the boundary layer is not very well predicted. And this disagreement becomes larger after the reattaching region, where the boundary layer is strongly accelerated. A possible cause could be the incorrect modelization of the compressibility terms. These remarks confirm that current compressibility corrections are still weak, and further improvements are necessary, especially for taking into account the compressibility effects in accelerated boundary layers.

REFERENCES

[1] Chien, K.-Y.,*Predictions of Channel and Boundary-Layer Flows with a Low-Reynolds-Number Turbulence Model*, AIAA J., 20, 1, pp. 33-38, 1982.

[2] Coleman, G.N. and Mansour, N.N., *Simulation and Modeling of Homogeneous Compressible Turbulence under Isotropic Mean Compression*, Eighth Symposium on Turbulent Shear Flows, 1991, Munich, Germany.

[3] Settles G.S. et al., *Proceedings of the 1980 HTTM Stanford Conference on Complex Turbulent Flows*, Kline et. al.(ed.s), Stanford University, 1981.

[4] Huang P.G., Bradshaw P. and Coakley T.J., *Skin Friction and Velocity Profile Family for Compressible Turbulent Boundary Layers*, AIAA J., 31, 9, pp. 1600-1604, 1993.

[5] Liou M.-S. and Steffen C. J.,*A New Flux Splitting Scheme*, J. Comput. Phys., 107, pp. 23-39, 1993.

[6] Sarkar S., Erlebacher G., Hussaini M.Y. and Kreiss H.O., *The Analysis and Modeling of Dilatational Terms in Compressible Turbulence*, J. Fluid Mech., 227, pp. 473-493, 1991.

[7] Speziale, C.G. and Sarkar S., *Second-Order Closure Models for Supersonic Turbulent Flows*, AIAA Paper 91-0217, 1991.

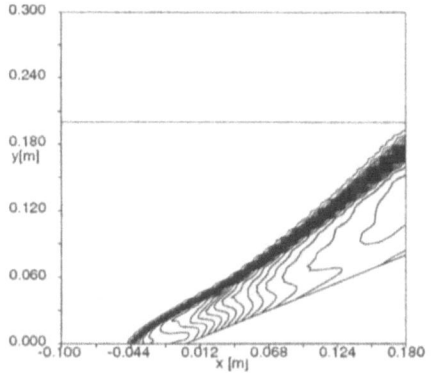

Figure 1: Isopressure contours

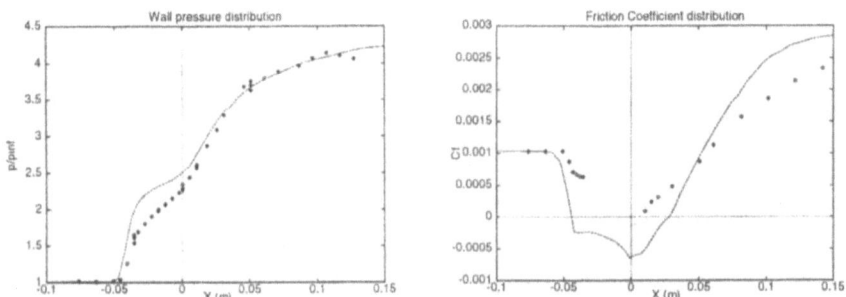

Figure 2: — present calculation, ◇ experimental data [3]

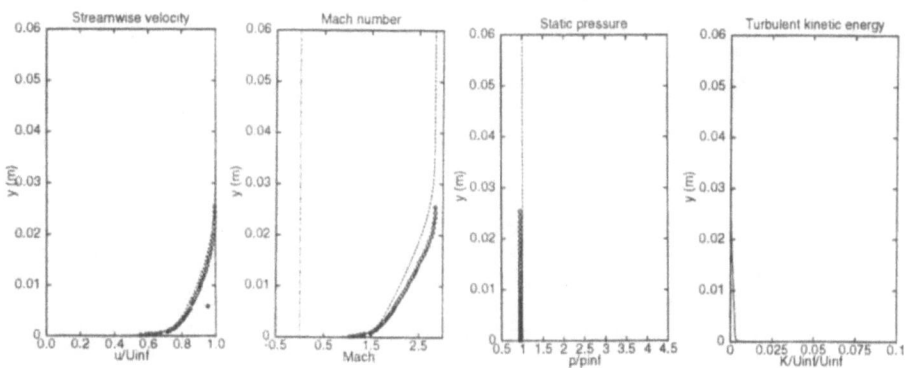

Figure 3: x=-0.0635 m: — present calculation, ◇ experimental data [3]

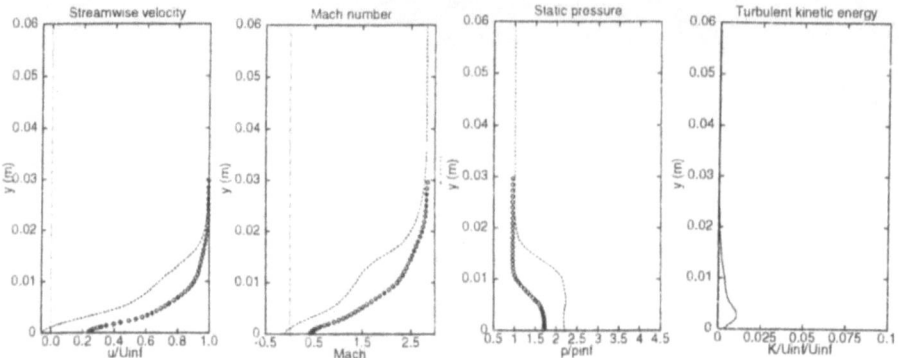

Figure 4: x=-0.0305 m: — present calculation, ◊ experimental data [3]

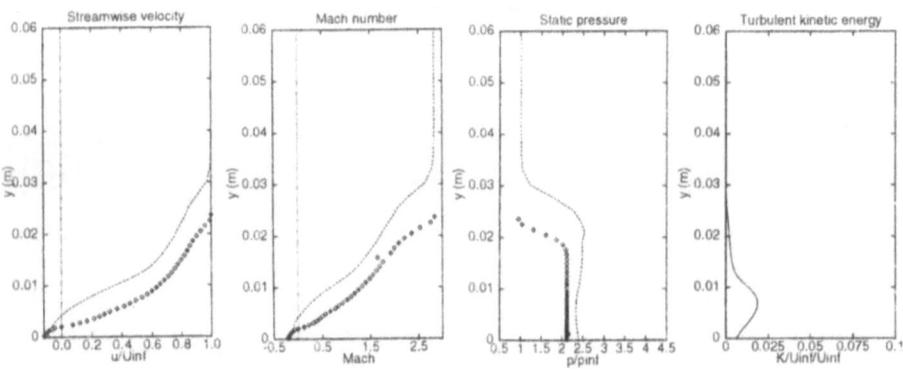

Figure 5: x=-0.0102 m: — present calculation, ◊ experimental data [3]

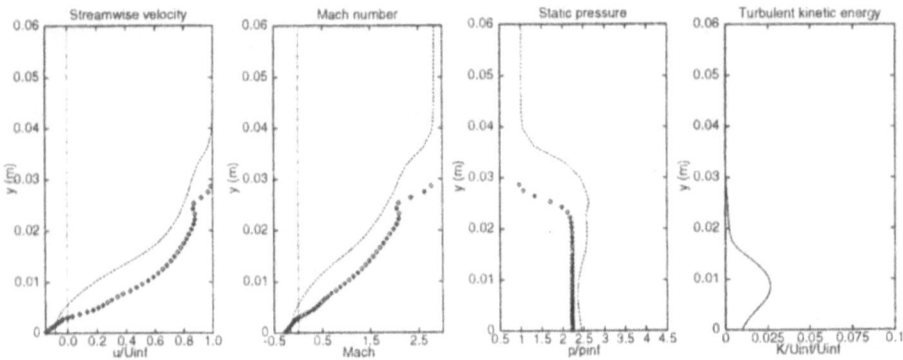

Figure 6: x=0.0 m: — present calculation, ◊ experimental data [3]

328

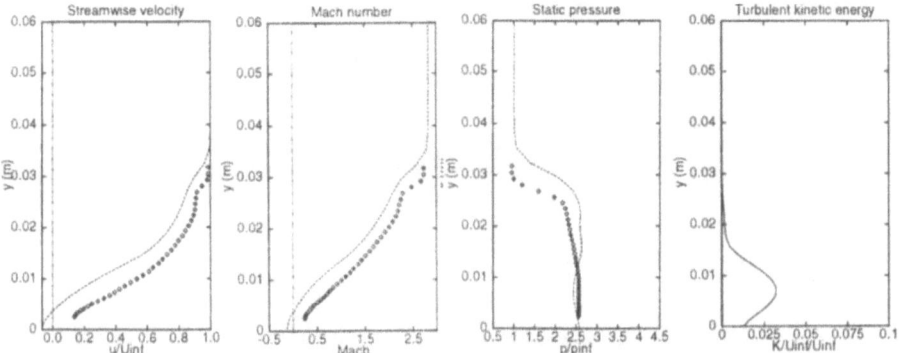

Figure 7: x=0.0102 m: — present calculation, ◇ experimental data [3]

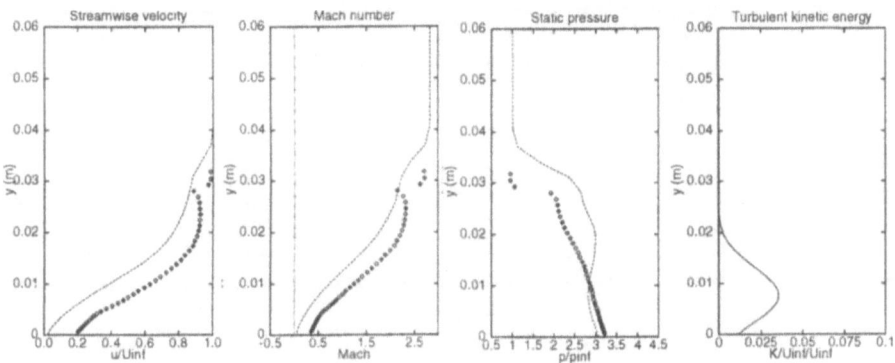

Figure 8: x=0.0305 m: — present calculation, ◇ experimental data [3]

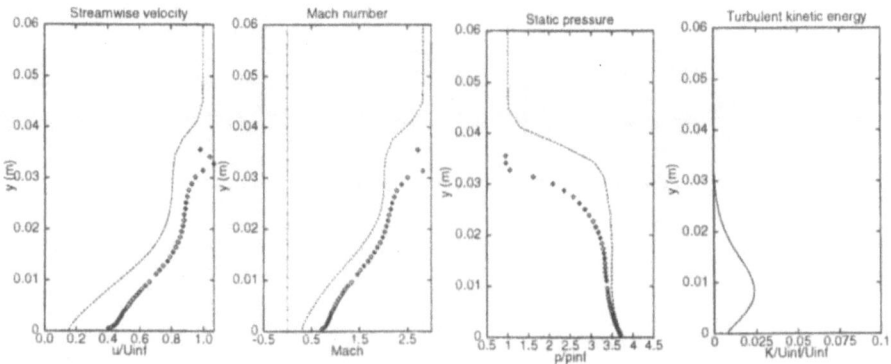

Figure 9: x=0.0610 m: — present calculation, ◇ experimental data [3]

329

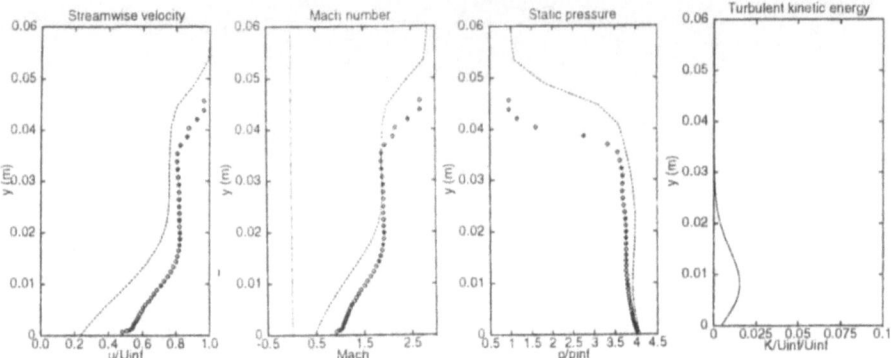

Figure 10: x=0.1016 m: — present calculation, ◊ experimental data [3]

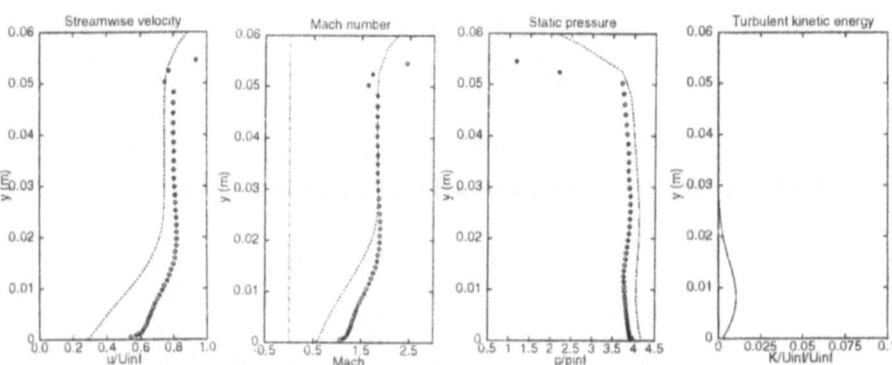

Figure 11: x=0.1422 m: — present calculation, ◊ experimental data [3]

Computation of a shock wave boundary layer interaction on compression ramp flow configuration

H. Loyau,D. Vandromme

LMFN - CORIA URA 230 - INSA de ROUEN
Place E. Blondel, 76130 Mont-Saint-Aignan

SUMMARY

This paper concerns the calculation of test case TC4 of ETMA project, correspondding to the two-dimensional supersonic compression ramp flow. The computation have been performed with a 2D developped for the mass-averaged Navier-Stokes equations in association with a low-Reynolds $k - \epsilon$ turbulence model.

Introduction

Despite the intensive research effort of the past decade to develop more general turbulence models, the two-equation turbulence models still remain the most widely used approach by engineers and scientists for practical flow computations. In this paper, the low-Reynolds number $k - \epsilon$ model of Jones and Launder [1] is performed, on the two-dimensional supersonic compression ramp flow conducted at Princeton University [2], [3] with a deviation angle of 24°. For this case, strong shock wave/boundary layer interaction induces an extended recirculation region.

The results presented have been performed with a 2D MacCormack [4] solver in association with the $k - \epsilon$ model under study.

The governing equations

The formulation of conservation of mass, momentum and energy, together with the physical properties for 2D compressible viscous fluid is expressed in the following form:

$$\frac{\partial U}{\partial t} + \frac{\partial F}{\partial x} + \frac{\partial G}{\partial y} = S_o.$$

where U the conservative variable vector is given by:

$$U = [\rho, \rho u, \rho v, \rho E].$$

and F respectively G, the convective/diffusive fluxes in x, respectively y direction by:

$$F = \begin{bmatrix} \rho u \\ \rho u^2 + p^* + \sigma_x \\ \rho u v + \tau_{xy} \\ (\rho E + p^* + \sigma_x)u + \tau_{xy}v - q_x \end{bmatrix} . \quad G = \begin{bmatrix} \rho v \\ \rho u v + \tau_{xy} \\ \rho v^2 + p^* + \tau_{xy} \\ (\rho E + p^* + \sigma_y)v + \tau_{xy}u - q_y \end{bmatrix} .$$

In this formulation, ρ denotes the averaged density, u and v are the velocity components along x, and y directions, E is the mass-averaged total energy, σ_x and σ_y the deviatoric

331

normal stresses and τ_{xy} the shear stress being given, according to the Stokes hypothesis, by:

$$\sigma_x = 2\mu\frac{\partial u}{\partial x} - \frac{2}{3}\mu\Big(\frac{\partial u}{\partial x} + \frac{\partial v}{\partial y}\Big).$$

$$\sigma_y = 2\mu\frac{\partial v}{\partial y} - \frac{2}{3}\mu\Big(\frac{\partial u}{\partial x} + \frac{\partial v}{\partial y}\Big).$$

$$\tau_{xy} = \mu\Big(\frac{\partial u}{\partial y} + \frac{\partial v}{\partial x}\Big).$$

where μ the molecular viscosity, is expressed as a function of temperature according to the Sutherland's law.

The effective pressure p^* is the sum of the average static pressure p and of the turbulent pressure, which is taken into account in the two-equation turbulence model (see next section)

$$p^* = p + \frac{2}{3}k.$$

The heat flux is expressed following the Fourier's law

$$q = -\lambda\nabla T = -\frac{C_p\mu}{P_r}\nabla T.$$

where P_r the Prandtl number is assumed to be equal to 0.72.

Turbulence model

The turbulence model used here is the Low-Reynolds number $k - \epsilon$ model proposed by Jones and Launder [1. The turbulent transport equations including low-Reynolds number source terms are given by:

$$\frac{\partial \rho k}{\partial t} + \nabla \rho k u + \nabla(\mu_k \nabla k) = S_k = P_k - \rho\epsilon - 2\mu\Big(\frac{\partial\sqrt{k}}{\partial x_n}\Big)^2.$$

$$\frac{\partial \rho\epsilon}{\partial t} + \nabla \rho\epsilon u + \nabla(\mu_\epsilon\nabla\epsilon) = S_\epsilon = C_{\epsilon 1}\frac{\epsilon}{k}P_k - C_{\epsilon 2}f_2\frac{\epsilon}{k}\rho\epsilon + \frac{2\mu\mu_t}{\rho}\Big(\frac{\partial^2 U_t}{\partial x_n^2}\Big)^2.$$

whith

$$P_k = -\rho\widetilde{u''_i u''}_j\frac{\partial u_i}{\partial x_j}.$$

$$\mu_k = \mu + \frac{\mu_t}{\sigma_k} \quad ; \quad \mu_\epsilon = \mu + \frac{\mu_t}{\sigma_\epsilon} \quad ; \quad \mu_t = C_\mu f_\mu \mu R_t.$$

$$R_t = \frac{\rho k^2}{\mu\epsilon} \quad ; \quad f_2 = 1 - 0.3\exp(-R_t^2) \quad ; \quad f_\mu = \exp\Big(\frac{-2.5}{1 + \frac{R_t}{50}}\Big).$$

The following set of constants being adopted for the calculation:

$$C_\mu = 0.09 \quad ; \quad C_{\epsilon 1} = 1.45 \quad ; \quad C_{\epsilon 2} = 1.92 \quad ; \quad \sigma_k = 1.0 \quad ; \quad \sigma_\epsilon = 1.3.$$

By adding these two-supplementary equations to the previous set, the source vector become:

$$S_o = [\, 0\,,\, 0\,,\, 0\,,\, 0\,,\, S_k\,,\, S_\epsilon\,].$$

with

$$S_\epsilon = C_{\epsilon 1}\frac{\epsilon}{k}P_k - C_{\epsilon 2}f_2\frac{\epsilon}{k}\rho\epsilon + \frac{2\mu\mu_t}{\rho}\Big(\frac{\partial^2 U_t}{\partial x_n^2}\Big)^2 \quad ; \quad S_k = P_k - \rho\epsilon - 2\mu\Big(\frac{\partial\sqrt{k}}{\partial x_n}\Big)^2.$$

Numerical method

The preceeding equations are solved on a computational domain of variables ξ and η (transformed coordinates of the physical domain), by the use of finite volume discretisation technique on structured mesh. The new system of equation is solved by using MacCormack [4]'s explicit-implicit finite volume method. This two-step predictor-corrector algorithm is of second-order accuracy in space and time, and the basic discretisation for the convective fluxes is modified in order to take into account the information propagation as done initially by Steger and Warming [5]. The flux splitting is made second order accurate, but is lowered to first order in shock regions. The viscous terms are centered and the source terms are integrated in the center of each control volumes in both ξ and η directional sweeps. The explicit discretisation is complemented with an implicit numerical approximation which is free from stability conditions.

Numerical results

Concerning the compression corner flow computed, the experiment was performed in the Princeton University $203 \times 203mm$, Mach 3 blowdown wind tunnel. The incoming flow boundary-layer makes a natural transition to tubulence in the nozzle and developed along the wall of the tunel. At the location of the compression corner ($1.95m$ downstream of the nozzle throat), the boundary layer is typically a zero pressure gradient, fully turbulent boundary-layer (thickness $\delta_o = 26mm$) in equilibrium and near adiabatic wall conditions.

For the computation of this case, a 100×100 grid size was used and the first streamwise location (about $3\delta_o$) before the corner was initialized with the tubulent flow profile provided by the NTUA, which mean flow characteristic are:

$M_\infty = 2.84$ Infinite Mach numbrer

$T_t = 262\ K$ Total temperature

$T_w = 258\ K$ Wall temperature

$P_\infty = 24000\ Pa$ Infinite static pressure

$\delta = 22\ mm$ Boundary layer thickness.

The results joined to this paper are isomach lines, wall pressure distributions and skin friction ceoficient distribution.

On the isomach lines, one can see that the shock developed in the inviscid part of the flow change progressively when approaching the wall, under the effect of the incoming boundary layer. In case of the configuration of 24° angle, the associated pressure gradient causes the boundary layer to separate upstream, producing an oblic compression fan.

333

A second compression fan occuring after the reattachement point on the inclined wall converges with the first one in order to produce after a deviation point the main shock at the exit of the computational domain, producing a double slope shock structure which is caracteristic of such interaction.

The wall pressure distribution is rather closed to its experimental one but the separation point is not well estimated and we can observed two regions on the pressure distribution where the pressure level is over-estimated.

This tendancy is confirmed by considering the friction coefficient distribution on which these two regions correspond to a too lower friction velocity level, inducing premature boundary layer separation and return to fully turbulent state.

Conclusion

The low-Reynolds $k-\epsilon$ model of Jones and Launder [1] and its application to a complex flow configuration have been presented in this paper. One can say that the prediction given by this incompressible form of $k-\epsilon$ model, on such flow configuration, is still a rather good approximation.

References

[1] JONES W.P. and LAUNDER B.E., (1972): "The calculation of Low-Reynolds-Number phenomena with a two-equation model of turbulence"
J. of Heat and Mass Transfer, vol. 15, No 3, pp. 301-314.

[2] SETTLES G.S., FITZPATRICK T.J. and BOGDONOFF S.M., (1979): "Detaile study of attached and separated compression corner flowfields in high Reynolds number supersonic flows"
AIAA Journal, vol. 17, No 6.

[3] SETTLES G.S., VAS I.E. and BOGDONOFF S.M., (1976): "Turbulent boundary-layer at a compression corner"
AIAA journal, vol. 14, No 3, pp. 1709-1715.

[4] MAC CORMACK R.C., (1985): "Current status of numerical solution of the Navier-Stokes equations"
AIAA paper 85-0032.

[5] STEGER J. and WARMING R.F., (1979): "Flux vector splitting of the inviscid gas dynamics equations with application to finite difference methods"
NASA TM-78605.

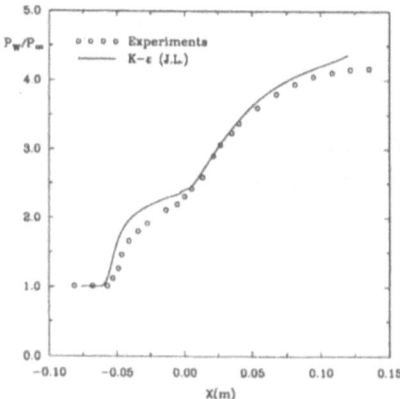

FIG. 1 - Wall pressure distribution

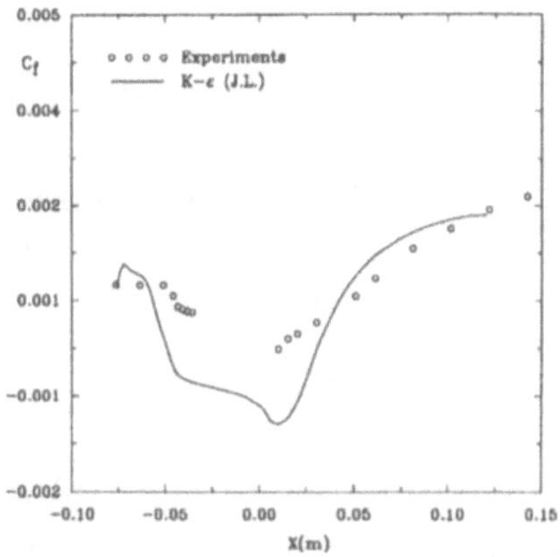

FIG. 2 - Friction coefficient distribution

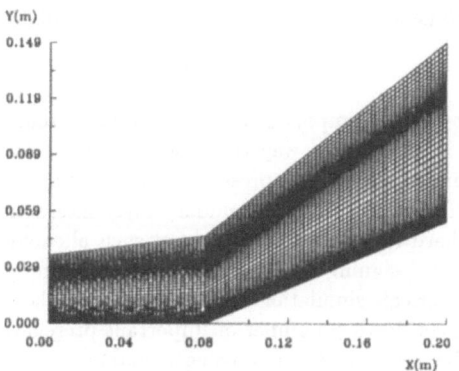

FIG. 3 - Computational grid

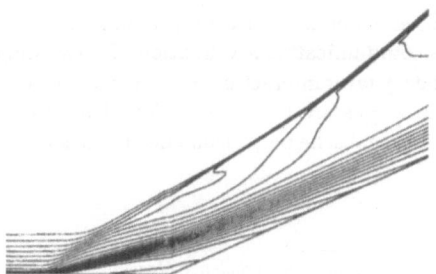

FIG. 4 - Iso-mach lines distribution

335

The shock / turbulent boundary layer interaction over the Princeton 20° and 24° ramps *

G. Freskos [†], W.Koschel [‡]

Institute for Jet Propulsion and Turbomachinery, RWTH Aachen

Templergraben 55, 52056 Aachen, GERMANY

SUMMARY

This paper concerns the calculation of test case TC4 of ETMA project, corresponding to the compression ramp flow geometries. The computation is performed with two numerical schemes developped for the mass-averaged Navier-Stokes equations in association with an algebraic turbulence model and a low-Reynolds number $k - \varepsilon$ model.

Introduction

Among all aircraft engine components, the air-intake system is one of the most complex. It decelerates in an optimal way the flow, before the latter reaches the engine. The flow in mixed compression intakes presents a variety of physical phenomena. Among them, viscous flow effects due to shock-boundary layer interaction. With the advent of powerful computing hardware, numerical study of such phenomena occupies a continuously increasing part in designing and optimizing air-breathing propulsion systems.

The complete -numerical- simulation of the flows in intakes needs 3D codes, which should reply simultaneously to a number of important prerogatives. An efficient treatment of the convective terms of the governing equations should allow good resolution of shock waves, without penalizing the very sensitive boundary layer regions. Moreover, good understanding and modeling of the physical phenomena, present in boundary layer regions, is also necessary, since the existing computing capacities are still several years away from allowing direct simulation of complicated flows.

We propose in this communication a validation of a two-dimensional code on a shock wave / turbulent boundary layer interaction measured at Princeton University [1]. Results obtained for two ramp angles are used here. Complete description of the code can be found in [2], where numerical schemes, turbulence models etc. are presented.

---

*Prepared for the ETMA Workshop on *Turbulence Modeling for Compressible Flows arising in Aeronautics, Manchester, U.K., November 14-17, 1994*

[†]Post-doctorate fellow - Present address: Lab. of Thermal Turbomachines, N.T.U.A., P.O. Box 64069, 157 10 Athens, GREECE

[‡]Professor

The governing equations

In conservative form, the time-dependent Navier-Stokes equation, for 2D planar compressible flow are expressed as follows in a Cartesian coordinate system (x, y):

$$\frac{\partial \vec{U}}{\partial t} + \frac{\partial \vec{F}}{\partial x} + \frac{\partial \vec{G}}{\partial y} = \frac{\partial \vec{F}_v}{\partial x} + \frac{\partial \vec{G}_v}{\partial y}. \tag{1}$$

where $\vec{U}$ is the conservative variables vector:

$$\vec{U} = [\rho, \rho u, \rho v, \rho e_0]^{\mathrm{T}}. \tag{2}$$

Above, ρ is the density, u and v are the velocity components along x, and y directions, and e_0 is the total internal energy.

The convective and diffusive flux vector $\vec{F}$ and $\vec{F}_v$ in x direction are respectively (with p the static pressure) (analogous expressions hold for y direction):

$$\vec{F} = \begin{bmatrix} \rho u \\ \rho u^2 + p \\ \rho u v \\ u(\rho e_0 + p) \end{bmatrix}, \vec{F}_v = - \begin{bmatrix} 0 \\ \sigma'_x \\ \tau_{xy} \\ \sigma'_x u + \tau_{yx} v - q_x \end{bmatrix}.$$

where σ'_i is the deviatoric normal stress $(\sigma'_i = \sigma_i - p)$, $\tau_{ij}, i \neq j$ are the shear stresses and q_i the heat flux along the i direction.

If isotropic, Newtonian fluids, i.e., fluids for which a linear relation between the stress components and those of the rate of strain holds in all directions are considered, the stress components are expressed as follows:

$$\sigma_i = p - \lambda \frac{\partial u_j}{\partial j} - 2\mu \frac{\partial u_i}{\partial i}.$$

$$\tau_{ij} = -\mu \left(\frac{\partial u_i}{\partial j} + \frac{\partial u_j}{\partial i} \right).$$

where λ is the coefficient of bulk viscosity and μ is the ordinary viscosity coefficient. Assuming the mean value of the normal stresses to be equal to the thermodynamic pressure, leads to Stokes' hypothesis: $\lambda + \frac{2}{3}\mu = 0$. The viscosity coefficient μ is expressed as a function of the temperature through Sutherland's law.

The state equation for a perfect gas is

$$p = \rho R T.$$

where T is the temperature and R the air constant $(R = c_p(T) - c_v(T) = 287.04 \mathrm{m}^2\mathrm{s}^{-2}\mathrm{K}^{-1})$. For even low temperature levels $(T \leq 600\mathrm{K})$ the ratio of specific heats can be assumed constant (calorically perfect gas). Then, the kinetic theory of gases gives for air at standard conditions: $\gamma = 1.4$.

The heat flux in each direction is expressed according to Fourier's law, which is written as:

$$\vec{q} = -\chi \vec{\nabla} T.$$

where χ is the coefficient of thermal conductivity. It is computed by assuming the Prandtl number to be equal to 0.72.

The numerical schemes

Two numerical schemes are available. The first is the well-known Mac Cormack scheme combined with the Steger Warming flux-splitting. The other one is a TVD scheme, based on Roe's approximate Rieman solver. An entropy function, proposed by Harten and Hyman and modified by Lafon enforces the entropy condition. An approximate factorization (AF) implicit technique is used in both schemes for convergence acceleration.

The first scheme

MacCormack's predictor-corrector scheme [3] in two-dimensions is given by the following expressions:

$$\text{predictor}: \quad \Delta U_{i,j}^n = -\Delta t \left(\frac{D_+ F_{i,j}^n}{\Delta x} + \frac{D_+ G_{i,j}^n}{\Delta y} \right),$$

$$(3)$$

$$U_{i,j}^{\overline{n+1}} = U_{i,j}^n + \Delta U_{i,j}^n.$$

$$\text{corrector}: \quad \Delta U_{i,j}^{\overline{n+1}} = -\Delta t \left(\frac{D_- F_{i,j}^{\overline{n+1}}}{\Delta x} + \frac{D_- G_{i,j}^{\overline{n+1}}}{\Delta y} \right),$$

$$(4)$$

$$U_{i,j}^{n+1} = \tfrac{1}{2}(U_{i,j}^n + U_{i,j}^{\overline{n+1}} + \Delta U_{i,j}^{\overline{n+1}}).$$

Forward and backward discretizations are alternated during the iterative procedure.

For supersonic flows a shock capturing technique is also needed. Steger-Warming's flux vector splitting [4] helps in resolving the shock waves.

The upwind scheme

The numerical treatment of the convective part of eq.(1) is based on the so-called Total Variation Diminishing approach (TVD).

The above equation can take the following discretized form:

$$\frac{U_i^{n+1} - U_i^n}{\Delta t} + \frac{\tilde{F}_{i+\frac{1}{2},j} - \tilde{F}_{i-\frac{1}{2},j}}{\Delta x} + \frac{\tilde{G}_{i,j+\frac{1}{2}} - \tilde{G}_{i,j-\frac{1}{2}}}{\Delta y} = 0. \tag{5}$$

The numerical convective flux $\tilde{F}$ at $i+\dfrac{1}{2},j$ will be expressed according to Yee's unified formulation (flux difference splitting):

$$\tilde{F}_{i+\frac{1}{2},j} = \frac{1}{2}(F(U_{i+\frac{1}{2},j}^R) + F(U_{i+\frac{1}{2},j}^L) + \hat{R}_{i+\frac{1}{2},j}\Phi_{i+\frac{1}{2},j}). \tag{6}$$

where MUSCL interpolation is assumed. The same analysis also applies to the $i, j + \frac{1}{2}$ interfaces for approximating the $\tilde{G}$ fluxes. Based on scalar considerations and for a fully upwind scheme, the left and right states for the conservative variables are approximated as follows:

$$U^R_{i+\frac{1}{2},j} = U_{i+1,j} - \frac{1}{2}g_{i+1,j}.$$
$$U^L_{i+\frac{1}{2},j} = U_{i,j} + \frac{1}{2}g_{i,j}. \tag{7}$$

where g is the slope limiter function. The behavior of the solution as a function of various limiters can be studied. For the present calculations Roe's minmod limiter was retained. The expression for latter is:

$$g^l_i = \frac{\Delta^l_{i+\frac{1}{2},j}\Delta^l_{i-\frac{1}{2},j} + |\Delta^l_{i+\frac{1}{2},j}\Delta^l_{i-\frac{1}{2},j}|}{\Delta^l_{i+\frac{1}{2},j} - \Delta^l_{i-\frac{1}{2},j}}. \tag{8}$$

where

$$\Delta_{i+\frac{1}{2},j} = U_{i+1,j} - U_{i,j} .$$

Many authors prefer to impose the limiters on the characteristic variables. In our approach limiting was exclusively applied to the conservative variables.

The components of the dissipation term $\Phi_{i+\frac{1}{2},j}$ in (6) are expressed as follows:

$$\hat{\phi}^l_{i+\frac{1}{2},j} = -\psi(\hat{a}^l_{i+\frac{1}{2},j})\hat{\alpha}^l_{i+\frac{1}{2},j},$$
$$\hat{\alpha}_{i+\frac{1}{2},j} = \hat{R}^{-1}_{i+\frac{1}{2},j}(U^R_{i+\frac{1}{2},j} - U^L_{i+\frac{1}{2},j}). \tag{9}$$

where $\hat{a}^l_{i+\frac{1}{2},j}$ are the eigenvalues and $\hat{R}$ is the right eigenvectors matrix of the jacobian A^x of the flux vector F. They are computed at some symmetric average between $U^R_{i+\frac{1}{2},j}$ and $U^L_{i+\frac{1}{2},j}$. Here Roe's average state between $U^R_{i+\frac{1}{2},j}$ and $U^L_{i+\frac{1}{2},j}$ has been chosen [5]. ψ is an entropy function introduced in order to enforce the entropy condition. It is destinated for regions where eigenvalues of the jacobian A^x tend to zero (shock waves or contact discontinuities). It eliminates expansion shocks and ensures a smooth transition from supersonic to subsonic flow. The choice of ψ is very important for viscous flow computations. Indeed, low numerical viscosity in boundary layer regions and stabilization of convective terms are simultaneously required. Harten [6] proposed the following expression for ψ, where a parameter δ adjusts the correction introduced by ψ:

$$\psi(\hat{a}^l_{i+\frac{1}{2},j}) = \begin{cases} |\hat{a}^l_{i+\frac{1}{2},j}|, & |\hat{a}^l_{i+\frac{1}{2},j}| > \delta \\ \dfrac{\hat{a}^{l2}_{i+\frac{1}{2},j} + \delta^2}{2\delta}, & |\hat{a}^l_{i+\frac{1}{2},j}| \leq \delta \end{cases} \tag{10}$$

Furthermore δ can be evaluated locally as [7]:

$$\delta = \delta_1(|u| + |v| + |c|). \tag{11}$$

339

In this way ψ can be switched on by choosing δ_1 large enough in regions where this is necessary (shock capturing), whereas it can be set equal to some small value in zones where it can provoke unwanted effects (boundary layers) [8] .

The turbulence model

The Baldwin-Lomax model

The Baldwin-Lomax model [9] is an algebraic model. It is a two-layer model, in which the turbulent viscosity is given by:

$$\mu_t = \begin{cases} (\mu_t)_{\text{inner}} \, , & y \leq y_c \\ (\mu_t)_{\text{outer}} \, , & y > y_c \end{cases}$$

where y is the normal distance from the wall and y_c is defined as the value for which the outer and inner formulae provide the same value. The Prandtl mixing length concept is used in the inner region:

$$(\mu_t)_{\text{inner}} = \rho l^2 |\omega|$$

in which l is proportional to the wall distance corrected in the near wall region by the Van Driest damping law:

$$l = \kappa y (1 - \exp(\frac{y^+}{A^+}))$$

($|\omega|$ is the modulus of local vorticity).

A different formula is used in the outer region:

$$(\mu_t)_{\text{outer}} = \alpha C_{\text{cp}} \bar{\rho} F_{\text{wake}} F_{\text{kl}}(y)$$

with:

$$F_{\text{wake}} = \min \left(y_{\text{max}} F_{\text{max}} \, , \, C_{\text{wk}} \frac{y_{\text{max}} u_{\text{dif}}^2}{F_{\text{max}}} \right).$$

$F_{\text{max}}, y_{\text{max}}$ are the maximum value and the corresponding distance from the solid wall of the function $F(y)$:

$$F(y) = y|\omega| \left(1 - \exp\left(-\frac{y^+}{A^+} \right) \right).$$

The intermittency factor F_{kl} is expressed as:

$$F_{\text{kl}} = \left(1 + 5.5 \left(\frac{C_{\text{kleb}} y}{y_{\text{max}}} \right)^6 \right)^{-1}.$$

u_{dif} expresses the difference between maximum and minimum velocity in a given profile (it accounts for separation).

The values of the different constants appearing in the model are the following:

$$A^+ = 26 \ , \ C_{cp} = 1.6 \ , \ C_{kl} = 0.3$$

$$C_{wk} = 0.25 \ , \ \kappa = 0.4 \ , \ \alpha = 0.0168 \ .$$

The $k - \varepsilon$ model

The Jones-Launder $k - \varepsilon$ model is used. It has the advantage of not requiring computation of any wall distances, which can be of primary importance for the simplicity in which a turbulence model can be implemented and its extension in three dimensions.

The turbulent viscosity for the Jones-Launder low Reynolds number turbulence model [10], is expressed as follows:

$$\mu_t = c_\mu f_\mu \bar{\rho} \frac{\tilde{k}^2}{\tilde{\varepsilon}}.$$

Two supplementary equations expressing the transport of the turbulent kinetic energy and its dissipation rate are used:

$$\frac{\partial}{\partial t} \left(\bar{\rho} \tilde{k} \right) + \frac{\partial}{\partial x_i} \left(\bar{\rho} \tilde{k} \tilde{U}_i - (\mu + \mu_k) \frac{\partial \tilde{k}}{\partial x_k} \right) = S_k. \tag{12}$$

$$\frac{\partial}{\partial t} \left(\bar{\rho} \tilde{\varepsilon}^* \right) + \frac{\partial}{\partial x_i} \left(\bar{\rho} \tilde{\varepsilon}^* \tilde{U}_i - (\mu + \mu_\varepsilon) \frac{\partial \tilde{\varepsilon}^*}{\partial x_k} \right) = S_\varepsilon. \tag{13}$$

with:

$$\tilde{\varepsilon}^* = \tilde{\varepsilon} - 2\nu \left(\frac{\partial \sqrt{\tilde{k}}}{\partial x_n} \right)^2 .$$

S_k, S_ε are source terms. In the present model they are expressed as follows:

$$S_k = P_k - \bar{\rho} \tilde{\varepsilon}^* - 2\mu \left(\frac{\partial \sqrt{\tilde{k}}}{\partial x_n} \right)^2 .$$

$$S_\varepsilon = C_{\varepsilon_1} \frac{\tilde{\varepsilon}}{\tilde{k}} P_k - C_{\varepsilon_2} f_2 \frac{\tilde{\varepsilon}}{\tilde{k}} \bar{\rho} \tilde{\varepsilon}^* + \frac{2\mu\mu_t}{\bar{\rho}} \left(\frac{\partial^2 \tilde{U}_t}{\partial x_n^2} \right)^2 .$$

where, $\tilde{U}_t$ is the tangential velocity component to the solid boundary(-ies) of the flow, x_n is the normal direction to these boundaries and P_k is the production term:

$$P_k = -\bar{\rho} \widetilde{u_i'' u_j''} \frac{\partial \tilde{U}_i}{\partial x_j}.$$

The wall boundary conditions are $\tilde{k} = \tilde{\varepsilon}^* = 0$, since only the isotropic part of the turbulent dissipation is considered. The constants c_μ, C_{ε_1}, C_{ε_2} are equal to 0.09, 1.45, 1.92 respectively.

The function f_μ allows to damp out the turbulent diffusion mechanisms in wall vicinity for the case of boundary layers:

$$f_\mu = \exp[-\frac{2.5}{1 + \dfrac{R_t}{50}}], \text{ where } : R_t = \frac{\rho \tilde{k}^2}{\mu \tilde{\varepsilon}}.$$

The f_2 is given by the relation:

$$f_2 = 1 - 0.3 \exp(-R_t^2).$$

The diffusivity coefficients are expressed as:

$$\mu_k = \frac{\mu_t}{Pr_k} \; ; \; \mu_\varepsilon = \frac{\mu_t}{Pr_\varepsilon}.$$

with $Pr_k = 1.0, Pr_\varepsilon = 1.3$. The turbulent Prandtl number is $Pr_t = 0.90$.

A shock wave/boundary layer interaction

The flows treated here conern the supersonic flow over a ramp, inclined 20° and 24° with respect to the freestream flow. The flows were investigated experimentally at Princeton University by Settles, Vas and Bogdonoff [1] and Dolling and Murphy [11]. In the first case (20°) the recirculation region is limited, whereas an extended separation zone appears in the second case. (from [1] The viscous shock-b.l. interaction provokes an extended separation zone. The flow reattaches downstream of the corner. According to the results obtained by Settles et al., reverse velocities as high as 16% of the free-stream value are induced. It should be also noted that the induced shock wave originates approximately two boundary layer thicknesses upstream of the corner, turning the outer flow through an initial 10°. Separation and reattachement points were measured to be 1.466 and 0.490 boundary layers thicknesses, upstream and downstream of the corner, respectively.

The free-stream and inflow conditions for the 24° flow are given below:

$$M_\infty = 2.84; P_{t\infty} = 692260 \text{Pa}; T_{t\infty} = 262 \text{K}$$

$$\theta_\infty = 0.0012 \text{m}; Re_\infty = 6.3 \cdot 10^7 \text{m}^{-1}$$

whereas those for the 20° flow are the following:

$$M_\infty = 2.79; P_{t\infty} = 694900 \text{Pa}; T_{t\infty} = 258 \text{K}$$

$$\theta_\infty = 0.0013 \text{m}; Re_\infty = 6.3 \cdot 10^7 \text{m}^{-1}.$$

The numerical experiments

For the calculations presented in this work initialization profiles provided by NTUA were used. For the 24° flow another initialization procedure was also used for comparison reasons. A boundary layer is imposed at the inflow section. It is defined by U_∞, Re_θ and $T_{t\infty}$ and the character of the flow (turbulent or laminar [p.11.23] [12]. The effective velocity approach of Van Driest [p.632] [13] is also integrated .

For the computations based on the Roe-TVD numerical scheme the minmod limiter is used, and the cutoff constants in the entropy function δ_{1x}, δ_{1y} were given the values 10^{-4}. An implicit treatment is combined with both numerical schemes. Approximate factorization technique is used for the resolution of the tri-diagonal linear system. Due to the AF technique an error is introduced in the time relaxation. For elevated CFL numbers, the results may be CFL dependent. (a detailed presentation can be found in [2])

A number of turbulence models can be used: the Baldwin-Lomax algebraic model, the Jones-Launder low Reynolds number $k - \varepsilon$ model, a wall functions model, taking into account compressibility, and finally an Algebraic Stress Model (A.S.M.). We will only use here the first two models. Especially for the $k - \varepsilon$ model, an implicit treatment of the source terms may further enhance the robustness of the numerical approach. An important parameter, which is often neglected in turbulent flow simulations, is the free-stream values of the turbulent quantities. We have noticed, that the flow under consideration, is very sensitive to the profiles and free-stream values of k and ε.

Initialization of the k and ε profiles

The flow proved to be very sensitive to the imposed k and ε profiles at the inlet sections, and more particularly, to the free-stream turbulence. For low levels of the free-stream turbulent kinetic energy values, or, high values of its dissipation, the separation zone propagates upstream, reaching the inlet section.

For the computations presented here, we used the following initialization process. The k-profile is defined as follows:

$$k = \min(c_\mu^{-\frac{1}{2}} \frac{U_\tau^2 y^+}{10}, 0.035 U_\tau^2 y_+^2), y^+ \leq 10$$

$$k = \max(c_\mu^{-\frac{1}{2}} U_\tau^2 (1 - \frac{y^+ - 10}{400 - 10}), k_\infty), y^+ > 10$$

(or, $k = \max(c_\mu^{-\frac{1}{2}} U_\tau^2 (1 - \frac{y^+ - 10}{\frac{\delta_0}{y_\tau} - 10}), k_\infty), y^+ > 10$)

where $c_\mu = 0.09$, $y^+ = \frac{y}{y_\tau}$, $y_\tau = \frac{\mu_w}{\rho_w U_\tau}$, $U_\tau = \sqrt{\frac{\tau_w}{\rho_w}}$ and k_∞ is the imposed free-stream value. It was given the following value: $k_\infty = (\frac{U_\infty}{100})^2$.

The turbulent dissipation is defined by mixing length model considerations: $\varepsilon = c_\mu^{\frac{3}{4}} k^{\frac{3}{2}} \frac{1}{l_t}$

with: $l_t = \min(200 y^+, 0.4 y (1 - e^{-\frac{y^+}{27}}))$.

The flow over the 24° ramp

Discussion of the effect of some numerical factors

Local time step is used, since time accuracy is of no importance, for the flow considered. The -convective- CFL number, we used in most of the computations, was 10. Using Roe's scheme we were able to go up to CFL=30, by employing the implicit treatment of the source terms p.118 [2]. The results remain CFL independent. For a higher CFL value (50) the results become CFL dependent. The Mac Cormack scheme allowed, on the contrary, the use of CFL=50 without inducing any dependency in the results. This should be attributed to the 2nd order accuracy of the scheme, reducing the factorization (i.e. time relaxation) error. The L-norm residual is reduced three orders of magnitude. For this particular computation 720 seconds (600 iterations) were needed on the Siemens-Nixdorf S600/20 vector supercomputer ($120\mu sec$/point/iteration).

The implicit treatment of the source terms was used with CFL=30 and 50. It prevents in both cases divergence of the computations, without altering the induced results. It may effectively be of significant importance in allowing the use of high CFL numbers (even at the first stages of the computation).

Wall pressure, friction coefficient and mean velocity profiles

Fig. 1, 2, present the grid and the iso-Mach lines of a computation where the upwind implicit scheme was used. The $k - \varepsilon$ model takes into account the turbulent character of the flow. The initialization of the velocity and $k - \varepsilon$ profiles are based on the procedure presented in previous paragraphs. The -local- CFL number was equal to 30 and the presented solution was obtained after 1300 iterations.

Results for the wall pressure distribution (fig. 4) and the friction coefficient (fig. 5) were obtained using two different models. For $k - \varepsilon$ model two initialization processes were used, namely an initialization profile proposed by NTUA and the one described above.

The Baldwin-Lomax fails to predict the separation point. Furthermore, a wrong friction coefficient level is deduced in the downstream region. It should be nevertheless noticed, that the upstream free-stream velocity is used in the whole domain, without taking into account the presence of the shock wave.

The wall pressure distribution (fig. 4) obtained with the $k - \varepsilon$ model is in better agreement with the results obtained by Dolling and Murphy. The separation point is well predicted with one noticeable exception. The upwind scheme fails to compute the right separation point when it is combined with the NTUA initialization. It proves to be more sensible to this parameter and shows that the choice of the numerical scheme can be important for the good prediction of phenomena defined primarily by the turbulent character of the flow. As it can be noticed, no data are available for the skin friction coefficient in the separation zone. 'Acceleration' close to the inflow section is provoked by the boundary conditions at this section (a partly subsonic boundary layer is imposed) -for some of the numerical simulations of fig. 5. All predictions give a more important separation zone than the one measured.

The mean velocity profile close to inflow section provided by NTUA matches the experimental data, whereas a small difference can be verified between these data and the RWTH initialization (fig. 6). The mean velocity profiles in subsequent sections show a good agreement with the measurements not very close to the wall. The results obtained using the RWTH profile seem to capture better the experimental distributions. The adjacent to the wall region is not very well resolved (experimentally). In the numerical predictions back flow persists in downstream sections. [1]

The flow over the 20° ramp

Mach contour lines for the flow over the 20° ramp are drawn in fig. 3. In the three last figures the wall pressure and friction coefficient distributions and velocity profiles for the same flow are presented. The $k - \varepsilon$ low-Reynolds turbulence model was used with both numerical schemes, whereas the computations were initialized with profiles provided by NTUA. The first scheme seems to match better the experimental data. Nevertheless influence of the initialization procedure must be further investigated.

Velocity profiles are in good agreement with experience in the region where they are drawn ($x \leq 12.7$mm). More important divergence is to be expected further downstream, as it is suggested by the friction coefficient curve (at least very close to the wall).

Acknowledgements

The development of the two-dimensional code was supported by SNECMA, where the code is already installed. The present research is sponsored by the Commission of the European Communities in the framework of the Human Capital and Mobility Program.

---

[1] Compressibility effects in the turbulence field: Although the free-stream Mach number is not very elevated, significant contribution from the rapid compression part of the pressure-dilatation correlation term is to be expected (see for example [14], [15]). According to other authors even the dilatational dissipation contribution can be of some importance [16]. The models implemented in the code are described in [17]. The dilatational dissipation model did not modify significantly the results. This was expected according to [15]. On the contrary, some first results using the $p - \theta$ model of Zeman, seem to have an important effect on the shock topology.

References

[1] G.S. Settles, I.E. Vas, S.M. Bogdonov. Details of shock-separated turbulent boundary layer at a compression corner. *AIAA Journal*, 14:1709-1715, 1976.

[2] G.O. Freskos. *Physical aspects and numerical simulation of the flows i n supersonic air-intakes.* PhD thesis, INP, Toulouse, France, October 1992.

[3] R.W. Mac Cormack. The effect of viscosity in hypervelocity impact cratering. AIAA paper 69-0354, 1969.

[4] J.L. Steger and R.F. Warming. Flux vector splitting of the inviscid gadynamic equations with applications to finite difference methods. *Journal of Computational Physics*, 40(2):263-293,April 1981.

[5] P.L. Roe. Approximate Riemann solvers, parameter vectors and difference schemes. *Journal of Computational Physics*, 43:357-372, 1981.

[6] A. Harten. On a class of high resolution total-variation-stable finite-difference schemes. *SIAM J. Num. Anal.*, 21:1-23, 1984.

[7] J.L. Montagne, H.C. Yee, G.H. Klopfer, and M. Vinokur. Hypersonic blunt body computations including real gas effects. Technical Report TM100074, NASA, 1988.

[8] A. Lafon. Calcul d'écoulements visqueux hypersoniques. Raport Tecnique OA 32/5005.22, CERT-ONERA, Toulouse, France, Mars 1990.

[9] B.S. Baldwin and H. Lomax. Thin layer approximation and algebraic model for separated turbulent flows. AIAA paper 78-257, 1978.

[10] W.P. Jones and B.E. Launder. The calculation of low Reynolds number phenomena with a two-equation model of turbulence. *Int. J. Heat Transfer*, 16:1119-1130, 1973.

[11] D.S. Dolling, M.T. Murphy. Unsteadiness of the Separation Shock Wave Structure in a Supersonic Compression Ramp Flowfield. *AIAA Journal*, 21:1628-1634, 1983.

[12] J. Cousteix. Aérodynamique en fluide visqueux / Turbulence et Couches Limites. ENSAE, Toulouse, 1988.

[13] F.M. White. Viscous fluid flow McGraw-Hill, New-York, 1974.

[14] O. Zeman. Compressible turbulence subjected to shear and rapid compression, *8th Symposium on Turbulent Shear Flows*, München, Germany, 1991.

[15] D. Vandromme, O. Zeman. Response of the turbulent boundary layer to the compression corner CTR Summer Program Report, Stanford University, 1992.

[16] D.C. Wilcox. Dilatation-Dissipation Corrections for Advanced Turbulence Models, *AIAA Journal*, 30:1311-1320, 1992.

[17] G. Freskos, D. Vandromme and H. Ha Minh, quad On the Numerical simulation of the two-dimensional flow field around a hypersonic Air-Intake - Compressibility effects, XI ISABE, Sept. 20-24,Tokyo, Japan, 1993.

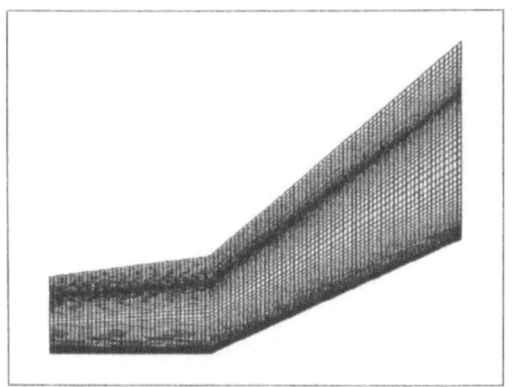

FIG. 1 - Computational grid.

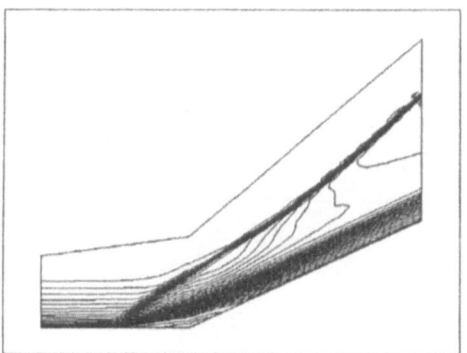

FIG. 2 - Mach contour lines ($k - \epsilon$ model/24°)

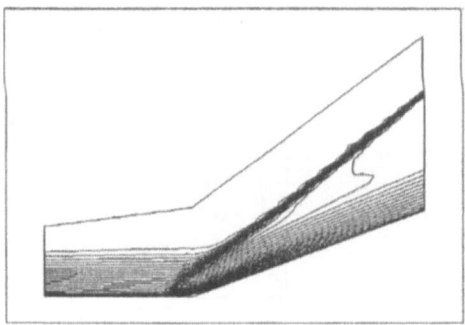

FIG. 3 - Mach contour lines ($k - \epsilon$ model/20°)

347

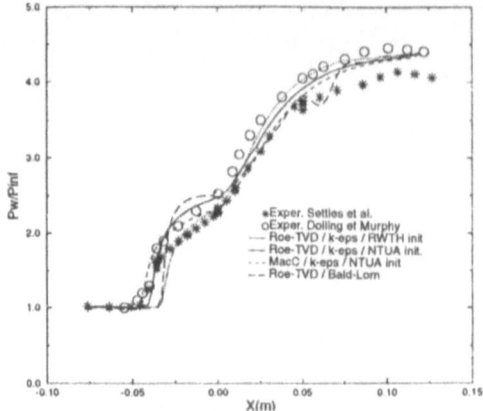

FIG. 4 - Wall pressure distribution (24°)

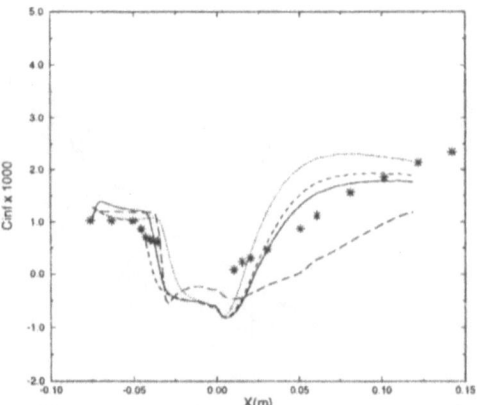

FIG. 5 - Friction coefficient (24°) - Curves as in fig. 4

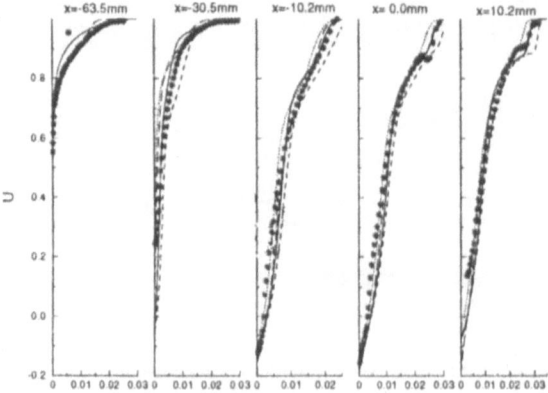

FIG. 6 - Velocity profiles (24°) - Curves as in fig. 4

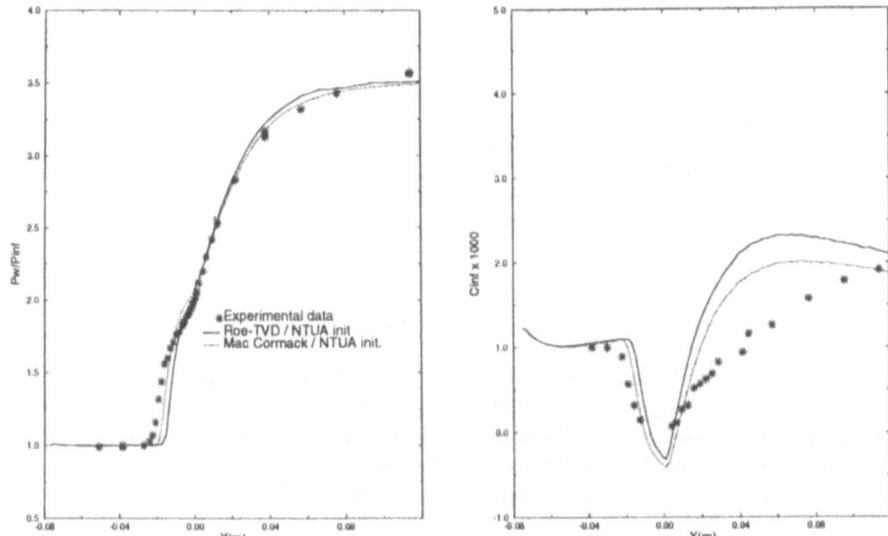

FIG. 7 - Wall pressure distribution 20°.   FIG. 8 -   Friction coefficient (20°) - Curves as
in fig. 7

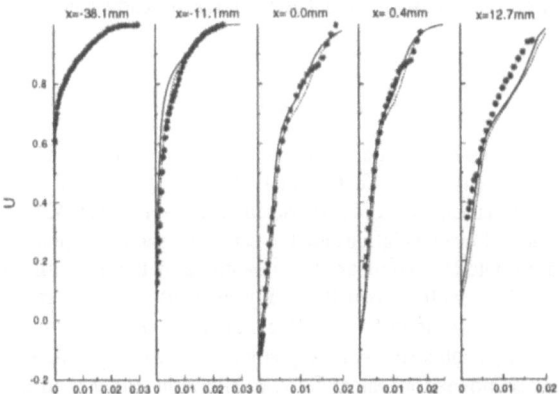

FIG. 9 -   Velocity profiles (20°) - Curves as in fig. 7

COMPUTATIONS ON THE COMPRESSION RAMP USING EXPLICIT ALGEBRAIC REYNOLDS STRESS MODELS

Stefan Wallin
The Aeronautical Research Institute of Sweden, FFA
Box 110 21, S-161 11 Bromma, Sweden

ABSTRACT

The compression ramp flow, test cases TC4 in the ETMA Workshop, have been computed with two different Explicit Algebraic Reynolds Stress Models, EARSM, together with Chien low Reynolds number k-ε model. The effect of the Sarkar compressibility correction of the dissipation has also been studied. The goal of this work is to investigate the possibilities and limitations of the Explicit Algebraic Reynolds Stress Models for this kind of separated flow.

The solutions are, however, not grid independent and one has to be very cautious when drawing any further conclusions from the study of the comparison between computations and experiments. The trends are, however, quite clear and show that the EARSM models gave a much larger separated region and also that the Sarkar compressibility correction of the dissipation gave a slightly larger separation.

The EARSM extension of the k-ε turbulence model did not introduce any further numerical problems and the difference in computational effort to reach steady state was negligible.

1 INTRODUCTION

Most practical calculations of turbulent flow at the present time are based on the averaged Navier-Stokes equations where the effect of the turbulence is given by the averaged Reynolds stresses, hence the designation Reynolds Averaged Navier-Stokes (RANS). The most popular models used in practical flow calculations are the zero, one and two equation models where the Reynolds stresses are related to the mean flow quantities by the generalised Boussinesq eddy viscosity hypothesis. It is well known that eddy viscosity models are unable to properly describe turbulent flows with body forces effects arising from curvature or system rotation. Moreover, the normal Reynolds stresses are predicted to be equal, which is incorrect even in simple cases such as homogeneous shear flow. In situations other than thin boundary layers this may cause significant changes to the mean flow. It is thus a generally accepted fact that the popular k-ε model does not predict the correct size of the separated region behind a backward facing step.

During the 1970s, second-order closure models became popular in which closure was achieved based on the Reynolds stress transport equation (RSTM). History and nonlocal effects are then included in the modelling as well as a more correct description of the turbulent anisotropy. Algebraic Reynolds Stress Models (ARSM) were obtained from the Reynolds stress transport equation where the transport of the Reynolds stress anisotropy was neglected (see Rodi 1976). The Reynolds stresses are here related to the mean flow field by an implicit relation. However,

350

this model has been found to be cumbersome to implement in complex flow, and numerical stiffness problems can result from the need for successive matrix inversion at each iteration.

Pope (1975) developed an algebraic Reynolds stress model (EARSM) by invoking the same equilibrium hypothesis as Rodi (1976) but presented a methodology for obtaining an explicit relation for the Reynolds stresses. The Reynolds stress anisotropy tensor is in his model developed as a tensor polynomial of the mean flow strain and rotation tensors where the coefficients may be functions of the mean flow strain and rotation tensor invariants, and the turbulent Reynolds number.

The functional form of the coefficients can be derived formally from an RST model with the aid of an anisotropy equilibrium hypothesis, Pope presented the solution for a two-dimensional mean flow derived from the RST model by Launder, Reece and Rodi (1975). Gatski & Speziale (1993) extended the results of Pope to three-dimensional turbulent flow and demonstrated a way to remove singular behaviour of the model for large strain rates. It is also possible to find the coefficients from general conditions such as requiring the kinetic energy in each velocity component to remain nonnegative (realizability) and then fit the functions to fundamental turbulence experiments as shown by, e.g. Speziale (1987), Thangham, Abid & Speziale (1991) and Shih, Zhu & Lumley (1992).

These Explicit Algebraic Reynolds Stress Models (EARSM) are used together with a two-equation turbulence model and are simple to implement in an existing Navier-Stokes solver avoiding the numerical stiffness of the implicit stress equation. The number of boundary conditions is also unchanged from the two-equation models. History and nonlocal effects of the anisotropy can not be accounted for in these models, but they have the potential to correctly describe the Reynolds stresses for flows with effects arising from curvature, system rotation and nonequal normal Reynolds stresses.

2 GOVERNING EQUATIONS

2.1 The Favre averaged Navier-Stokes equations

The classical way to decompose a variable into a constant and a fluctuating part is the Reynolds decomposition where the variable $q = \bar{q} + q'$ is split into a time average $\bar{q}$ and the fluctuating part q' where $\overline{q'} = 0$. A favourable decomposition for compressible flow is a mass weighted averaging according to Favre where the variable $q = \tilde{q} + q''$ is split into a mass weighted time average $\tilde{q}$ and the fluctuating part q'' where $\overline{q''} \neq 0$. The time averaged Navier-Stokes equations can be expressed in the following form,

$$\frac{\partial}{\partial t} W + \frac{\partial}{\partial x_j}(Q_j - V_j) = 0 , \tag{2.1}$$

where the state vector, W, is given by

$$W = \left\{ \begin{array}{c} \bar{\rho} \\ \bar{\rho}\,\tilde{u}_i \\ \bar{\rho}\,\tilde{E} \end{array} \right\} \tag{2.2}$$

351

and the convective, Q, and viscous, V, fluxes are

$$
Q_j = \left\{ \begin{array}{c} \overline{\rho \widetilde{u}_j} \\ \overline{\rho \widetilde{u}_i \widetilde{u}_j} + \overline{\rho u''_i u''_j} + \bar{p} \delta_{ij} \\ \overline{\rho \widetilde{E} \widetilde{u}_j} + \overline{\rho u''_i u''_j} \widetilde{u}_i + \bar{p}\, \widetilde{u}_j \end{array} \right\} ,
$$

(2.3)

$$
V_j = \left\{ \begin{array}{c} 0 \\ 2\mu \widetilde{S}_{ij} \\ 2\mu \widetilde{S}_{ij}\widetilde{u}_i + C_p\left(\dfrac{\mu}{Pr}+\dfrac{\mu_t}{Pr_t}\right)\dfrac{\partial \widetilde{T}}{\partial x_j} + \left(\mu+\dfrac{\mu_t}{\sigma_k}\right)\dfrac{\partial k}{\partial x_j} \end{array} \right\} .
$$

(2.4)

The total energy is $\widetilde{E} = \widetilde{e} + 1/2 \widetilde{u}_j\, \widetilde{u}_j + k$, where the turbulent kinetic energy is defined as $k = 1/2 \overline{u_j u_j}$.

The averaged strain tensor, $\widetilde{S}_{ij}$, is given by

$$
\widetilde{S}_{ij} = \frac{1}{2}\left(\frac{\partial \widetilde{u}_i}{\partial x_j} + \frac{\partial \widetilde{u}_j}{\partial x_i}\right) - \frac{2}{3}\frac{\partial \widetilde{u}_k}{\partial x_k}\delta_{ij} .
$$

(2.5)

Here $\bar{\rho}$, $\widetilde{u}_i$, $\bar{p}$, $\widetilde{e}$ and $\widetilde{T}$ denotes averaged density, velocity, pressure internal energy and temperature.

The molecular viscosity for air μ is expressed as a function of temperature according to Sutherland's law

$$
\frac{\mu}{\mu_0} = \left(\frac{T}{T_0}\right)^{3/2}\frac{T_0 + S_1}{T + S_1} ,
$$

(2.6)

where $\mu_0 = 1.716 \cdot 10^{-5}$ kg/ms is the viscosity at the reference temperature $T_0 = 273.1$ K and the constant $S_1 = 110.6$ K.

For calorically perfect gases, the specific heats at constant volume and constant pressure are $C_v = R/(\gamma - 1)$ and $C_p = \gamma C_v$. For air, the ratio of specific heats $\gamma = 1.4$ and the gas constant $R = 287$. The Prandtl number, $Pr = 0.72$. The pressure must be related to the state vector and for calorically perfect gases, $\bar{p} = \bar{\rho}\, \widetilde{e}\, (\gamma - 1)$.

So far the Reynolds stresses have not been modelled while the other correlations have been modelled in a standard way using the gradient approximation, see Vandromme (1991).

2.2 The baseline k-ε turbulence model

The baseline two-equation turbulence model is the low Reynolds number k-ε model proposed by Chien (1982). The transport equations for the turbulent kinetic energy, k , and the turbulent dissipation rate, ε , are given by

$$
\frac{\partial}{\partial t} W_{k\text{-}\varepsilon} + \frac{\partial}{\partial x_j}\left(Q_{k\text{-}\varepsilon j} - V_{k\text{-}\varepsilon j}\right) = S_{k\text{-}\varepsilon} ,
$$

(2.7)

with the state vector

$$W_{k\text{-}\varepsilon} = \left\{ \begin{matrix} \bar{\rho}\, k \\ \bar{\rho}\, \varepsilon \end{matrix} \right\}, \tag{2.8}$$

the convective and viscous fluxes

$$(Q_{k\text{-}\varepsilon\, j} - V_{k\text{-}\varepsilon\, j}) = \left\{ \begin{matrix} \bar{\rho}\, k\, \tilde{u}_j - \left(\mu + \dfrac{\mu_t}{\sigma_k}\right) \dfrac{\partial k}{\partial x_j} \\[2mm] \bar{\rho}\, \varepsilon\, \tilde{u}_j - \left(\mu + \dfrac{\mu_t}{\sigma_\varepsilon}\right) \dfrac{\partial \varepsilon}{\partial x_j} \end{matrix} \right\}, \tag{2.9}$$

and the source term

$$S_{k\text{-}\varepsilon} = \left\{ \begin{matrix} P_k - \bar{\rho}\, \varepsilon \left(1 + \alpha_1\, M_t^2\right) - 2 \dfrac{\mu\, k}{n^2} \\[2mm] \left(C_{\varepsilon 1}\, P_k - C_{\varepsilon 2} f_2\, \bar{\rho}\, \varepsilon\right) \dfrac{\varepsilon}{k} - 2 \dfrac{\mu\, \varepsilon}{n^2} e^{-\frac{1}{2} n^+} \end{matrix} \right\}, \tag{2.10}$$

where the production of turbulent kinetic energy, P_k, is given by

$$P_k = -\bar{\rho}\, \widetilde{u''_i u''_j} \frac{\partial \tilde{u}_i}{\partial x_j}. \tag{2.11}$$

The term $\alpha_1\, M_t^2$ in the k source term is a compressibility correction of the dissipation as proposed by Sarkar, et al (1991) where $\alpha_1 = 1.0$ and the turbulent Mach number is defined as $M_t^2 = k/a^2$ where a is the speed of sound.

The function f_2 in the source term for the dissipation is

$$f_2 = 1 - 0.22\, e^{-\left(\frac{R_t}{6}\right)^2}, \tag{2.12}$$

where the turbulent Reynolds number is defined as $R_t = \bar{\rho}\, k^2 / \mu\, \varepsilon$.

The turbulent viscosity is derived from the turbulent quantities as

$$\mu_t = C_\mu f_\mu\, \bar{\rho} \frac{k^2}{\varepsilon}, \tag{2.13}$$

where $f_\mu = 1 - e^{-0.0115\, n^+}$. The constants involved are $C_\mu = 0.09$, $\sigma_k = \sigma_\varepsilon = 1.0$, $C_{\varepsilon 1} = 1.35$, $C_{\varepsilon 2} = 1.8$ and $Pr_t = 1.0$.

2.3 The Reynolds stresses

The Reynolds stresses appear in the momentum and energy equations and in the expression for the turbulent production and has to be modelled. The most used model is the Boussinesq hypothesis which is a linear relation between the turbulent stress and the mean flow strain. We will here form a more general relation between the turbulent stress and the mean flow and turbulence fields.

353

It is more convenient to formulate an expression for the Reynolds stress anisotropy tensor rather for the Reynolds stresses. The anisotropy tensor is here defined for compressible flow as

$$a_{ij} = \frac{\overline{u''_i u''_j}}{k} - \frac{2}{3}\delta_{ij} \quad . \tag{2.14}$$

The most general expression for the Reynolds stress anisotropy in terms of the mean flow strain and vorticity tensors and the turbulent kinetic energy and dissipation rate is

$$a_{ij} = \sum_{\lambda=1}^{11} \alpha_\lambda T_{ij}^{(\lambda)} \quad , \tag{2.15}$$

where the coefficients may be functions of the mean flow strain and vorticity tensor invariants. Fortunately, owing to the Cayley-Hamilton theorem, the number of independent invariants and linearly independent second-order tensors is finite. The first four terms are

$$\begin{aligned}
T_{ij}^1 &= S_{ij} \\
T_{ij}^2 &= S_{ik}\Omega_{kj} - \Omega_{ik}S_{kj} \\
T_{ij}^3 &= S_{ik}S_{kj} - \frac{1}{3}S_{mn}S_{nm}\delta_{ij} \\
T_{ij}^4 &= \Omega_{ik}\Omega_{kj} - \frac{1}{3}\Omega_{mn}\Omega_{nm}\delta_{ij}
\end{aligned} \tag{2.16}$$

where S and Ω are the mean flow strain and vorticity tensors normalised by the turbulent time scale k/ε. Keeping only the first term corresponds to the Boussinesq hypothesis with $\alpha_1 = -2\mu_t \varepsilon / \bar{\rho}k^2$. To make the model appliable for compressible flows, the strain and vorticity tensors are defined as

$$\begin{aligned}
S_{ij} &= \frac{1}{2}\frac{k}{\varepsilon}\left(\frac{\partial \tilde{u}_i}{\partial x_j} + \frac{\partial \tilde{u}_j}{\partial x_i} - \frac{1}{3}\frac{\partial \tilde{u}_k}{\partial x_k}\delta_{ij}\right) \\
\Omega_{ij} &= \frac{1}{2}\frac{k}{\varepsilon}\left(\frac{\partial \tilde{u}_i}{\partial x_j} - \frac{\partial \tilde{u}_j}{\partial x_i}\right)
\end{aligned} \quad . \tag{2.17}$$

Two different EARSM models have been implemented and tested.

The first model is proposed by Gatski & Speziale (1993) which gives coefficients as functions of the mean flow strain and vorticity tensor invariants ($\eta^2 = S_{ij}S_{ji}$ and $\zeta^2 = -\Omega_{ij}\Omega_{ji}$)

$$\begin{aligned}
\alpha_1 &= -0.227\,A \\
\alpha_2 &= -0.042\,A \\
\alpha_3 &= 0.040\,A \\
\alpha_4 &= 0
\end{aligned} \tag{2.18}$$

where

$$A = \frac{3\left(1 + C_1^2\eta^2\right)}{3 + C_1^2\eta^2 + 6\,C_1^2C_2^2\eta^2\zeta^2 + 6\,C_2^2\zeta^2} \tag{2.19}$$

and the constants $C_1 = 0.087375$ and $C_2 = 0.1864$.

The second model is proposed by Shih, Zhu & Lumley (1992) where the coefficients are:

$$\alpha_1 = -\frac{4}{3}\frac{1}{1.25 + \sqrt{2}\,\eta + 0.9\,\sqrt{2}\,\zeta}$$

$$\alpha_2 = \frac{-15}{1000 + 2\sqrt{2}\,\eta^3}$$

$$\alpha_3 = \frac{3}{1000 + 2\sqrt{2}\,\eta^3} \tag{2.20}$$

$$\alpha_4 = \frac{-19}{1000 + 2\sqrt{2}\,\eta^3} \;.$$

It is not obvious what to do in the near wall region. The EARSM is formulated as a correction to the Boussinesq hypothesis and this correction is damped out in the near wall region in the same way as the turbulent viscosity is damped according to Thangham, Abid & Speziale (1991). That means that $\alpha_1 = \alpha_1 f_\mu$ and $\alpha_{2,3,4} = \alpha_{2,3,4} f_\mu^2$.

2.4 Numerical method

The Explicit Algebraic Reynolds Stress Model, formulated as a correction to a k-ε eddy viscosity model, has been implemented in the CFD solver EURANUS developed by FFA together with VUB (Rizzi et al. 1993). EURANUS is a general Navier-Stokes solver for structured multiblock meshes with various discretizations in space and time. A multigrid method accelerates the convergence. The low Reynolds number k-ε model by Chien (1982) with compressibility corrections according to Sarkar et al. (1991) is implemented with central and upwind spatial discretization together with explicit time integration. For stability reasons the source terms are treated implicitly.

In this study the mean flow equations are discretized using a second order upwind scheme, symmetric TVD, with Van Leer limiters. The k-ε equations are discretized using first order upwind scheme. Care has been taken to have second order accuracy near the wall in the wall normal direction for the k-ε equations . The numerical dissipation is damped near the wall in the wall normal direction and the viscous terms are second order accurate also near the wall.

The equations are integrated to convergence using a five stage Runge-Kutta method.

3   RESULTS

3.1 Test case definition

The test case TC4 for the ETMA Work Shop was the classical compression ramp configuration for shock wave turbulent boundary layer interaction (Settles et. al. 1976, 1979). The free stream Mach number is 2.85 and two different ramp angles, 20° and 24°, have been considered for which the flow separates. The inflow conditions are tabulated in table 1.

Table 1   Inflow conditions for the compression ramp test case.

	TC4.C	TC4.D
Ramp angle	20°	24°
Unit Reynolds number	$63 \cdot 10^6$	$63 \cdot 10^6$
Total temperature [K]	258	262
Wall temperature [K]	274	276
Infinite temperature [K]	101	100
Infinite velocity [m/s]	562	569
Infinite static pressure [kPa]	26.0	24.0
Boundary layer thickness [m]	0.025	0.023
Displacement thickness [m]	0.0066	0.0061
Momentum thickness [m]	0.0013	0.0012

3.2 Computational grid and inflow boundary layer profile

Common grids with about 100x100 cells have been provided for the two cases, the 20° grid by the National Technical University of Athens and the 24° grid by the Institut de Mecanique de l' INSA, Rouen, France. The 20° grid is shown as the 'old' grid in Figure 1. It was however found that the grid was maybe not fine enough near the wall especially at the ramp. The nondimensional distance, $y+$, from the wall to the first grid line is shown in Figure 2 as the 'old' grid.

Two new grids were generated for the 20° and 24° cases with about the same number of grid points but with higher resolution near the wall, see the 'new' grid in Figures 1 and 2.

The inflow boundary layer profiles for the two cases were provided by the National Technical University of Athens with the velocity profile well corresponding to the measured inflow profile. The velocity profile together with the turbulence profile were, however, not quite consistent with our solver which gave disturbances near the inflow. To avoid that these disturbances would affect the results, a new inflow profile was generated by computing a flat plate with the k-ε turbulence model and use the profile where the skin friction corresponds to the measurements. The provided, 'old', and the 'new' boundary layer profile were compared with the measured profile in Figure 3. The new inflow profile does not fit the experiments as well as the old profile, but the new profile did not introduce any disturbances near the inflow.

Computations on the 20° case with the k-ε model with the provided grid and inflow profile (old) and the new grid and inflow profile were compared with the experiments in Figure 4. There surface pressure distribution and the separation length was not much affected by the new grid and inflow profile, but the overshoot in skin friction near the reattachment was much higher for the new case.

It was, however, not investigated if this difference was due to the new inflow profile or the new grid. Figure 2 shows the wall distance to the first grid line and in the region of the overshoot one finds that $y+\approx5$ for the old grid and $y+\approx2$ for the new grid. The wall distance should be less than one in low Reynolds number k-ε calculations to get a grid convergent solution and especially correct skin friction.

The conclusion is that the difference is mainly due to the grid and none of the solutions could be considered as grid converged. One has to be very cautious when drawing any conclusions from the following studies.

3.3 Computational results

The two test cases have been computed using the eddy viscosity k-ε turbulence model and two different EARSM turbulence models, Gatski & Speziale (1993), and Shih, Zhu & Lumley (1992). The notation in the figures are k-eps, GS and SZL. The wall pressure and skin friction for the 20° case are shown in Figure 5 and for the 24° case in Figure 7. Velocity profiles in different stations are shown in Figure 8 for the 20° case and in Figure 9 for the 24° case. In all cases the compressibility correction of the dissipation as proposed by Sarkar, et al (1991) has been used.

The results show a large influence of the EARSM models with a much larger separation compared to the k-ε model. The overshoot in skin friction near the reattachment was not much affected. The results with the k-ε model was much closer to the measurements compared to the EARSM models. The skin friction was, however, quite poor predicted also with the k-ε model.

The effect of the compressibility correction has been studied by computing the 20° case with the k-ε model and the Shih, Zhu & Lumley EARSM model. The results without compressibility correction and with compressibility correction of the dissipation as proposed by Sarkar, et al (1991), denoted '+ S' in the figures, have been compared in Figure 6. The compressibility correction gives a slightly larger separation.

3.4 Convergence and numerical problems

The solution was converged with two levels of multigrid cycling and the steady state was reached after about 20000 time steps. The EARSM models did not introduce any further numerical problems compared to the k-ε turbulence and the difference in computational effort to reach steady state was negligible.

4 CONCLUSIONS

To get a grid convergent solution and especially correct skin friction in low Reynolds number k-ε calculations, the nondimensionalised wall distance to the first grid line should be less than one. That is not fulfilled by the grids that are used in this study. The different grids gave also quite large differences. The conclusion is that the solutions are not grid independent and one has to be very cautious when drawing any further conclusions from this study.

For that reason one should not draw any conclusions of the comparison between computations and experiments although the k-ε model seams to be much better than the EARSM models. The trends are, however, quite clear and probably not very grid dependent and shows that the EARSM models gave a much larger separated region and also that the Sarkar compressibility correction of the dissipation gave a slightly larger separation.

The EARSM extension of the k-ε turbulence model did not introduce any further numerical problems and the difference in computational effort to reach steady state was negligible.

5   ACKNOWLEDGEMENT

This study has been carried out at FFA (the Aeronautical Research Institute of Sweden) within the ETMA project funded by FFA and NUTEK (the Swedish National Board for Industrial & Technical Development).

6   REFERENCES

[1]   Pope, S. B., "A more general effective-viscosity hypothesis", J. Fluid Mech., vol. 72, part 2, 1975.

[2]   Launder, B. E., Reece, G. J., Rodi, W., "Progress in the development of a Reynolds-stress turbulence closure", J. Fluid Mech., vol. 41, 1975.

[3]   Rodi, W., "A new algebraic relation for calculating the Reynolds stresses", Z. angew. Math. Mech. 56, T219-221, 1976.

[4]   Settles, G. S., Vas, I. E., Bogdonoff, S. M., "Details of a Shock Separated Turbulent Boundary Layer at a Compression Corner", AIAA Journal, Vol. 14, 1976.

[5]   Settles, G. S., Fitzpatrick, T. J., Bogdonoff, S. M., "Detailed Study of Attached and Separated Compression Corner Flow Fields in High Reynolds Number Supersonic Flow", AIAA Journal, Vol. 17, 1979.

[6]   Chien, K. Y., "Predictions of Channel and Boundary-Layer Flows with a Low-Reynolds-Number Turbulence Model", AIAA Journal, Vol. 20, No. 1, January, 1982.

[7]   Speziale, C. G., "On nonlinear K-l and K-ε models of turbulence", J. Fluid Mech., vol. 178, 1987.

[8]   Thangam, S., Abid, R., Speziale, C. G., "Application of a new k-t Model to Near Wall Turbulent Flows", ICASE Report No. 91-16, February, 1991.

[9]   Sarkar, S., Erlebacher, G., Hussaini, M. Y., Kreiss, H. O., "The analysis and modelling of dilatational terms in compressible turbulence", J. Fluid Mech., vol. 227, 1991.

[10] Vandromme, D., "Introduction to the Modeling of Turbulence", von Karman Institute for Fluid Dynamics, Lecture Series 1991-02, March, 1991.

[11] Shih, T. H., Zhu, J., Lumley, J. L., "A Realizable Reynolds Stress Algebraic Equation Model", NASA TM 105993, ICOMP-92-27, CMOTT-92-14, 1992.

[12] Gatski, T. B., Speziale, C. G., "On explicit algebraic stress models for complex turbulent flows", J. Fluid Mech., vol. 254, 1993.

[13] Rizzi, A., Eliasson, P., Lindblad, I., Hirsch, C., Lacor, C., Haeuser, J., "The engineering of multiblock multigrid software for Navier-Stokes flows on structured meshes", Computers and Fluids, Vol. 22, 1993.

7 FIGURES

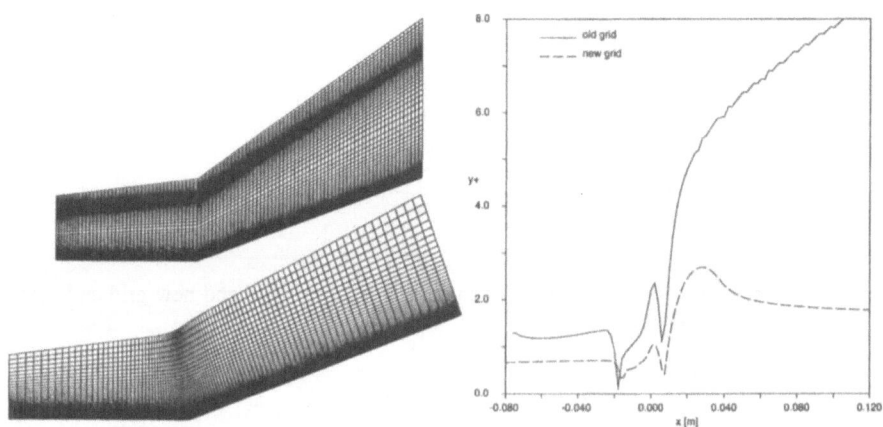

Figure 1. Computational grid for the 20° Figure 2. The y+ distance to the first grid
compression corner. Old grid   line over the surface for the old
(above) and new grid.   and new grid. The 20° case.

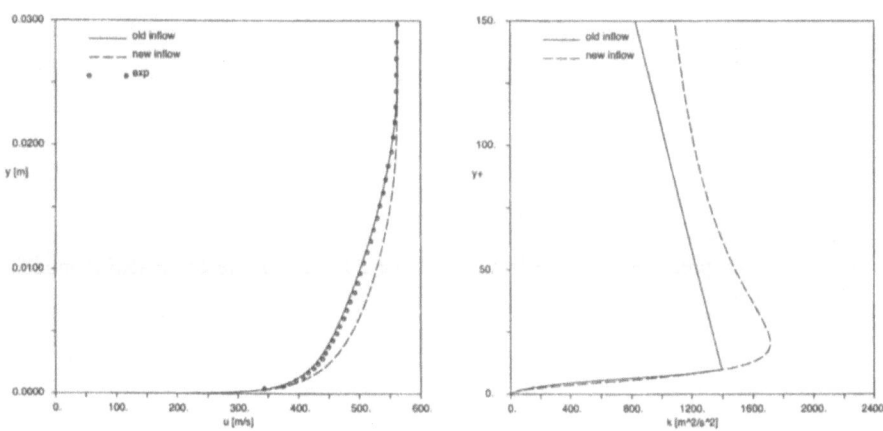

Figure 3. Inflow velocity and turbulent kinetic energy profiles for the 20° case. Old and new
profiles compared.

359

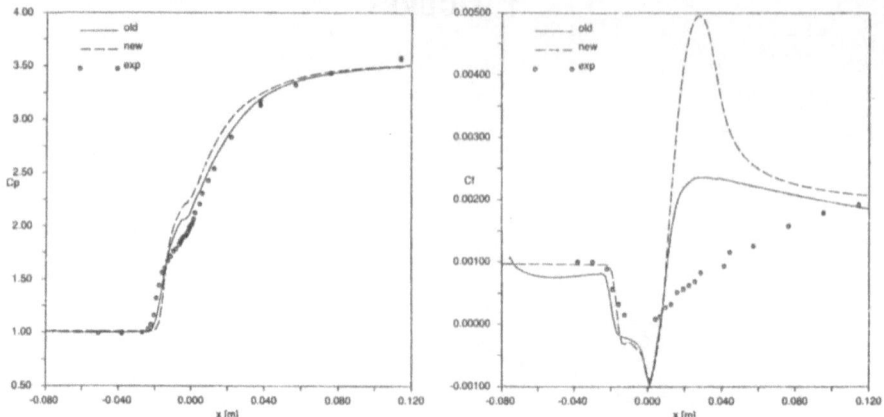

Figure 4. Surface pressure and skin friction for the 20° case. Old and new grid and inflow profiles compared.

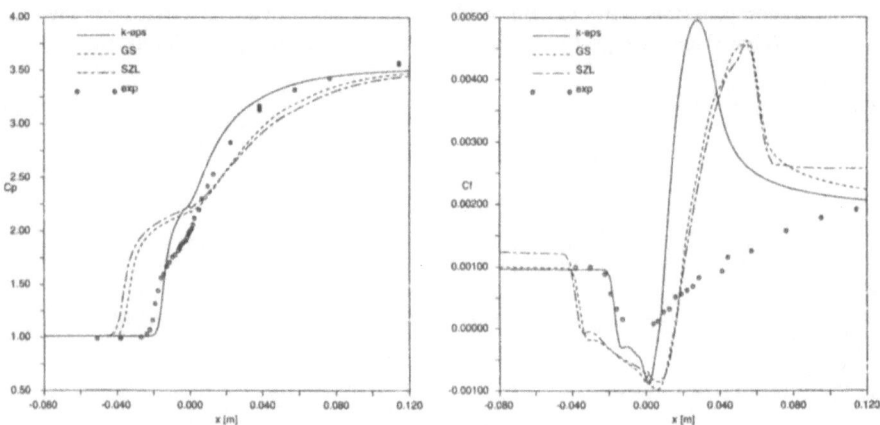

Figure 5. Surface pressure and skin friction for the 20° case. Different EARSM models compared.

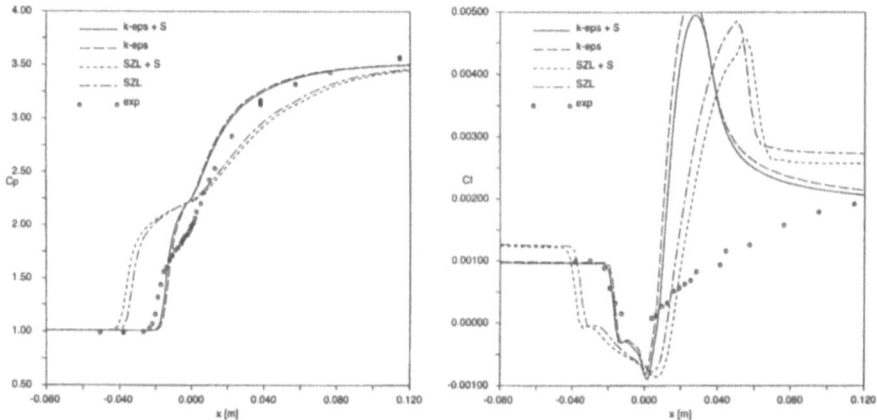

Figure 6.   Surface pressure and skin friction for the 20° case. Influence of Sarkar (1991) compressibility correction on the dissipation.

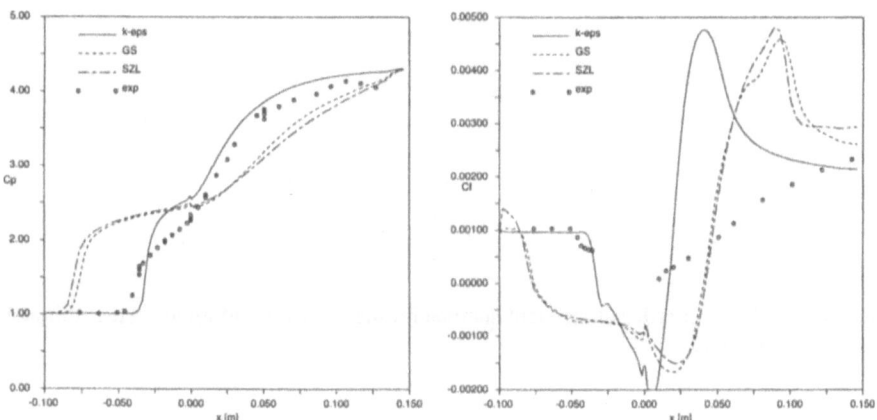

Figure 7.   Surface pressure and skin friction for the 24° case. Different EARSM models compared.

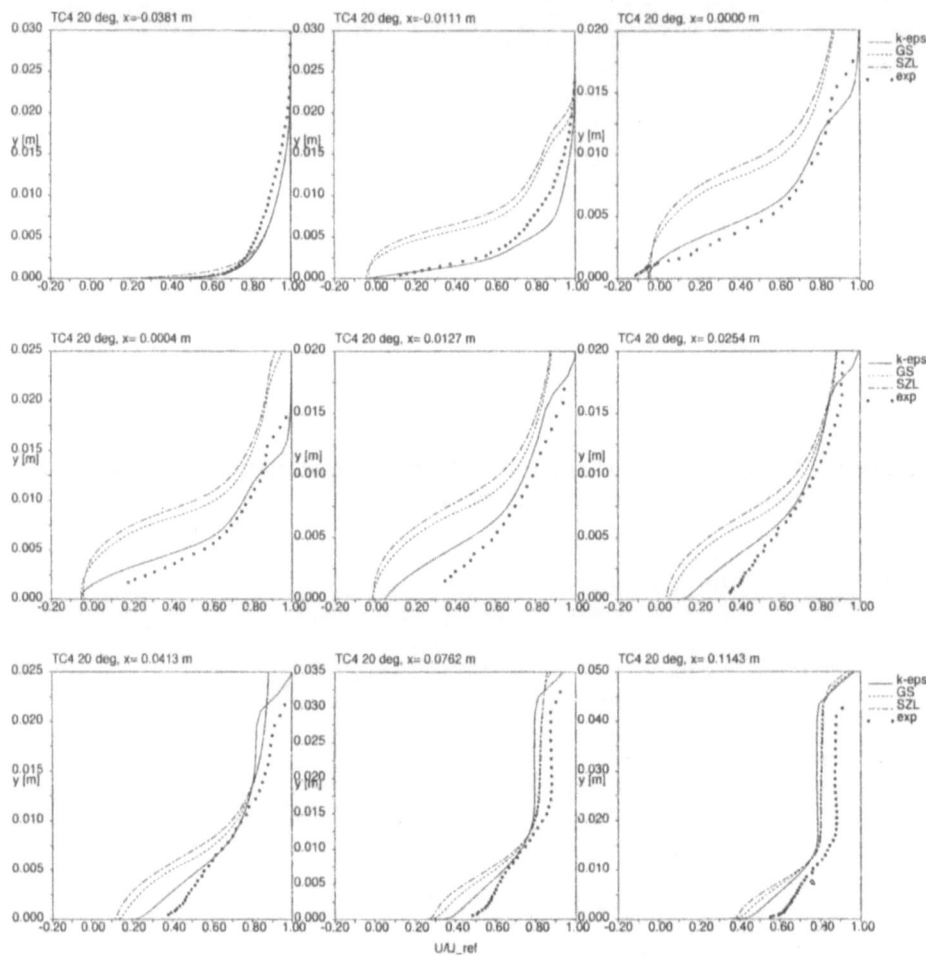

Figure 8.   Velocity profiles at different stations for the 20° case. Different EARSM models compared.

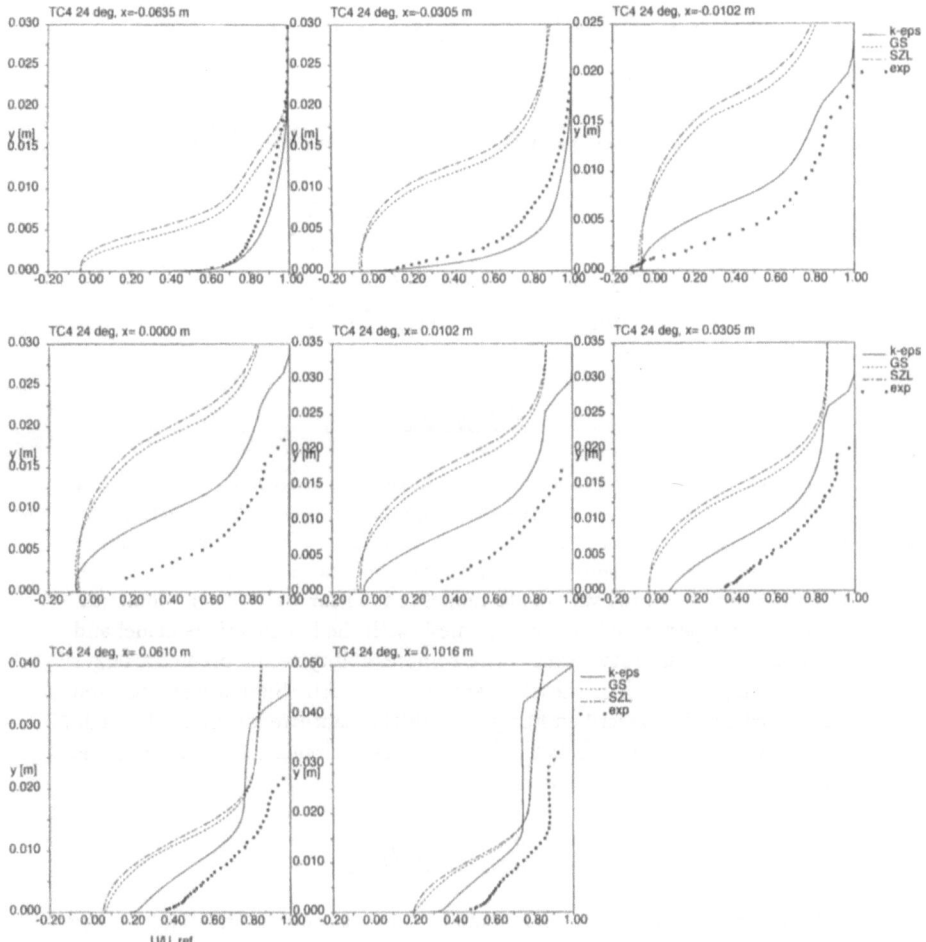

Figure 9.   Velocity profiles at different stations for the 24° case. Different EARSM models compared.

Numerical simulation of a 24° compression ramp by a $k - \epsilon$ model with compressibility terms

C. Le Ribault, M. Buffat

LMFA, CNRS URA 263, ECL, UCB Lyon I

36, avenue Guy de Collongue

69131 Ecully Cedex France

Abstract

Simulation of a 24° compression ramp have been performed with a $k - \epsilon$ model. Additional terms to take into account compressibility effects (compressible dissipation, pressure-dilatation correlation) have been introduced in the kinetic energy equation. Wall functions and a low Reynolds model of Lam-Bremhorst are available to modelise the near wall region. Computations are first performed on a coarse mesh with the law functions model and different compressibility terms. The compressibility terms predict a larger recirculating buble than the standard $k - \epsilon$ model. Simulations are then performed on the prescribed fine grid with the Lam-Bremhorst model. The standard $k - \epsilon$ underpredicts the lenght of the recirculation bubble and the compressibility terms improve the results.

Introduction

The compression ramp is a complex test case which embodies all the difficulties of compressible turbulent flows, with a shock/boundary layer interaction and compressibility effects. The experiments have been done by Settles[5] and Dolling[2] for a two dimensional interaction generated by compression ramps with different angles. The most severe test case, corresponding to the 24 degree angle, has been computed. For this configuration, the interaction between the shock and the boundary layer induces a large recirculation bubble, which modifies the shock system, depending on the extend of the recirculation zone, which experimentally is slightly more than twice the incoming boundary layer thickness. This test case is difficult because different factors influence the predicted length of the recirculation bubble (mesh, inlet boundary conditions, compressibility terms, near wall modeling...). To minimize numerical dependencies between the contributors, the prescribed structured mesh and boundary conditions have been used. Different simulations have been performed in order to compare different turbulence modelings.

Numerical method

The numerical method uses a mixed finite volume/finite element approximation on non structured grid initially develop by partner 1 (INRIA).

The convection fluxes are integrated by a Roe solver with a finite volume approximation and a 2nd order MUSCL interpolation, whereas the viscous and source terms are integrated using a finite element method. The time integration uses an implicit scheme, with performant preconditioned linear solvers. The resulting code, "NaturNG", developed at the LMFA, is a powerfull tool to simulate 2D and 3D compressible turbulent flows in complex geometries[3]. This code has been used to compute the test case using a classical $k - \epsilon$ model, with wall laws and different compressible terms, on a coarse finite element mesh (1556 nodes).

From this sequential code, a new project has started at the LMFA to develop a parallel code, using a new object oriented paradigm (C^{++}) and a standard message passing library (PVM). Efficient parallel numerical techniques have been developed, such as an automatic domain partitioning technique and a parallel implicit solver[1]. This new code, "NadiaNG", has been used to compute the test case with the low Reynolds turbulence model on the prescribed fine grid (99*99 nodes).

Turbulence models

The standard $k - \epsilon$ model does not include compressibility effects. To take into account an increase of the dissipation with the turbulent Mach number $Ma_t = \frac{\sqrt{2k}}{c}$, an additional compressible dissipation ϵ_c is added into the k equation. Different models, based on direct simulations results, have been proposed for ϵ_c (Sarkar [4], Zeman [7], Wilcox [6]) in term of the incompressible dissipation ϵ_s and the turbulent Mach number Ma_t:

$$\epsilon_c = F(Ma_t)\epsilon_s . \tag{1}$$

The difference between the models are the expression of the dependence on the turbulent Mach number $F(Ma_t)$:

$$F(Ma_t) = 0.5 * Ma_t^2 \text{ (Sarkar)} \tag{2}$$

$$F(Ma_t) = \begin{cases} 1.5 * (Ma_t^2 - 0.25^2) & \text{if } Ma_t \geq 0.25 \\ 0 & \text{if } Ma_t \leq 0.25 \end{cases} \quad \text{(Wilcox)} \quad (3)$$

$$F(Ma_t) = \begin{cases} 1 - e^{-\frac{(Ma_t - 0.25)^2}{0.66^2}} & \text{if } Ma_t \geq 0.25 \\ 0 & \text{if } Ma_t \leq 0.25 \ . \end{cases} \quad \text{(Zeman)} \quad (4)$$

$$(5)$$

For compressible flow, the pressure fluctuation induces a decrease of the production and the dissipation through the pressure-dilatation correlation. Sarkar proposed the following model for the pressure-dilatation term in the turbulent kinetic energy equation:

$$\overline{p'\frac{\partial u_i'}{\partial x_i}} = \alpha_1 Ma_t^2 \overline{\rho u_i' u_j'} \frac{\partial \tilde{u}_i}{\partial x_j} + \alpha_2 Ma_t^2 C_\mu \bar{\rho}\epsilon \tag{6}$$

with $\alpha_1 = 0.4$ and $\alpha_2 = 0.2$. $\qquad (7)$

Near wall model

To modelise the near region with the standard $k - \epsilon$ model, wall functions are used to impose boundary conditions on the velocity. By introducing the dimensionless wall distance $y^+ = \frac{\bar{\rho} u_f \delta}{\mu}$ and velocity $u^+ = \frac{\tilde{u}}{u_f}$ (u_f is the friction velocity), the wall laws functions can be written at a small distance δ from the wall as:

$$\begin{cases} u^+ = y^+ & \text{if } y^+ \leq 11.6 \\ u^+ = \frac{1}{K}log(y^+) + C & \text{if } y^+ \geq 11.6. \end{cases} \tag{8}$$

These relations are used to impose the turbulent shear stress inside the variational formulation of the momentum equation. For the turbulent quantities k and ϵ, equilibrium between production and dissipation rate gives the proper boundary conditions:

$$k_\delta = \frac{u_f^2}{\sqrt{C_\mu}} \qquad \epsilon_\delta = \frac{u_f^3}{\delta K} \ . \tag{9}$$

Whereas those wall functions have been established for incompressible flows, experiments with the compressible boundary layer show that they remain valid if there is no strong density variation in the boundary layer.

Low Reynolds turbulence model

To improve the modeling near the wall, the low Reynolds models of Shih-Lumley and Lam-Bremhost have also been tested. A first serie of computations has been done with the Shih Lumley model. Whereas good results have been obtained for incompressible flows (channel flow), the application to compressible flows was not satisfactory. Computations of compressible boundary layers have shown that the Lam Bremhorst model performs better. Thus the Lam Bremhorst model has been used to perform the test case computations.

To take into account the decrease of the turbulence near the wall, damping function are introduced in the turbulent viscosity μ_t:

$$\mu_t = C_\mu \rho f_\mu \frac{k^2}{\epsilon} \tag{10}$$

$$\text{with } f_\mu = (1 - e^{-0.0165R_y})^2 (1 + 20.5/Re_t). \tag{11}$$

Damping functions are also introduced in the turbulence dissipation equation to ensure the right behavior of ϵ at the wall:

$$\frac{\partial \tilde{\epsilon}}{\partial t} + div(\overline{\rho}\tilde{u}\tilde{\epsilon}) = \frac{\partial}{\partial x_i}(\frac{1}{Re} + \frac{\mu_t}{\sigma_\epsilon})\frac{\partial \epsilon}{\partial x_i} + C_{\epsilon 1}f_1\rho\frac{\epsilon}{k}Prod - C_{\epsilon 2}f_2\rho\frac{\epsilon^2}{k} \tag{12}$$

$$f_1 = 1 + (0.05/f_\mu)^3 \tag{13}$$

$$f_2 = 1 - e^{-Re_T^2}. \tag{14}$$

Boundary conditions for k and ϵ at the wall are :

$$k = 0 \qquad \epsilon = 2 * \mu \frac{\partial \sqrt{k}}{\partial y}\bigg|_{y=0}. \tag{15}$$

As low Reynolds models need very refined meshes near the walls (typically the first point is located at 10^{-4} time the height of the boundary layer), performant implicit solvers have been developed to compute the solution with those models.

Computational results

Two different grids are used for the simulations. A coarse grid (1556 nodes) (figure 1) is first used to perform computations with wall functions, in order to test the influence of the compressibility terms. The predicted pressure distributions with different models are plotted on figure 2, together with the experimental measurements of Settle and Dolling. The results clearly show the importance of the compressibility terms.

Using the parallel code, more precise simulations have been done on a prescribed structured grid built at INSA Rouen. Whereas this grid does not take advantage of the non-structured features of the numerical code, it ensures a more precise comparison between different models and their implementations. We have also used the inlet profiles prescribed by the coordinator to minimize the discrepancies due to different inlet conditions. Adiabatic boundary conditions are used at the wall, and a Sutherland law for the molecular viscosity is introduced.

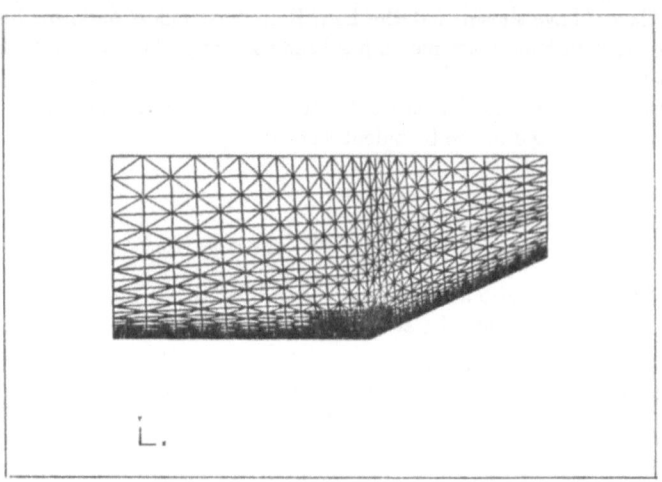

Figure 1: Finite Element Mesh 1556 nodes

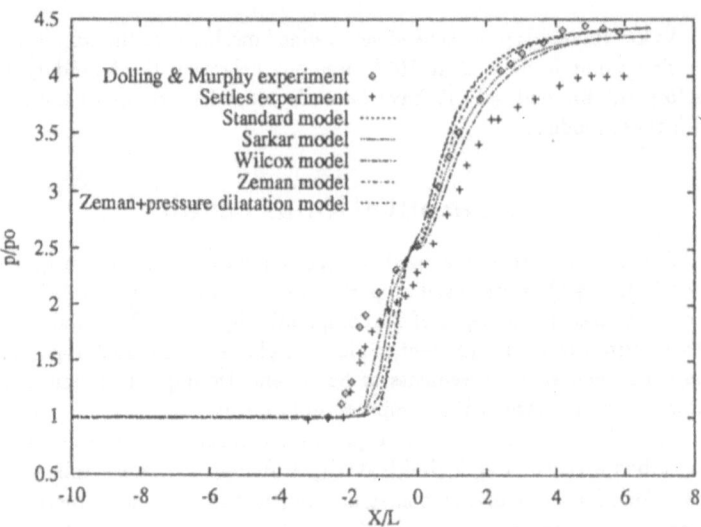

Figure 2: Wall pressure repartition with wall function

Iso Pressure (24 deg. ramp)

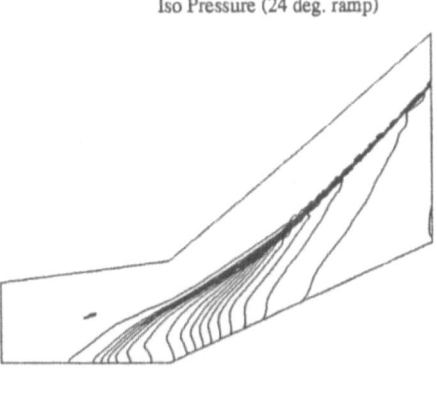

Figure 3: Pressure distribution with the Lam Bremhorst model

Two calculations have been done, with and without additional compressibility terms (the Wilcox model for the compressible dissipation and the Sarkar model for the pressure dilatation). The pressure distribution inside the computational domain are shown on figures 3 and 4. The results, with the compressibility terms (figure 4), show the strong modification of the shock system due to the large recirculation bubble, whereas in the first result the recirculation bubble is twice shorter, and the detached shock is less marked.

The predicted pressure distribution for the two simulations are plotted on figure 5 against the experimental results of Settle and Dolling. The best prediction is given by the second calculation with compressibility terms, which predicts a larger recirculation length. The computation runs on a cluster of DEC alpha workstations with an FFDI network. Using 4 nodes, the parallel computations take about $18sec.$ elapsed time per iteration, with a local time step and a local CFL of 5. Higher CFL number has not been used because, the boundary condition for ϵ is explicit and induces, in that test case, oscillations. All the residuals have decreased up to a plateau of 5.10^{-5} in normalized L_2 norm, except for ϵ, which has a higher value (4.10^{-2}). A converge solution is obtained after 2000 to 3000 iterations, starting from a different solution.

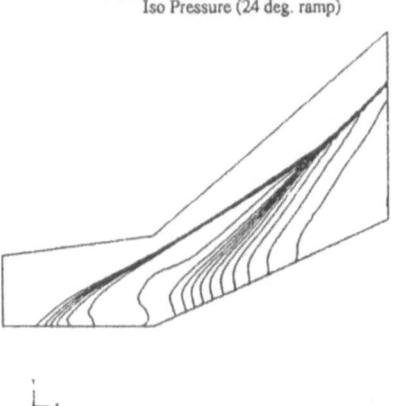

Iso Pressure (24 deg. ramp)

L.

Figure 4: Pressure distribution with the Lam Bremhorst model and the Wilcox and Sarkar models

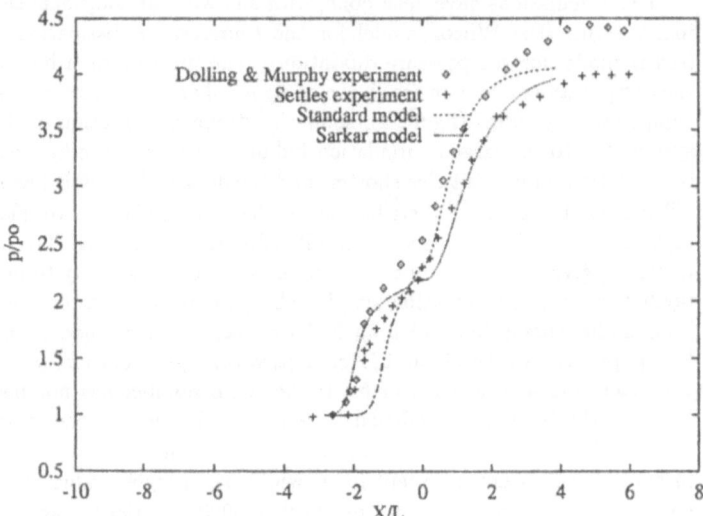

Figure 5: Wall pressure repartition with low Reynolds model

Conclusion

Numerical simulations have been performed in the test case of a 24 deg. ramp. Two different turbulent models have been used: a classical $k - \epsilon$ model with wall functions which allow computations on coarse meshes and a low Reynolds model of Lam Bremhorst which need refined meshes but gives more accurate results. Different compressibility terms have also been tested. Whereas the standard $k - \epsilon$ model underpredicts the recirculation bubble, the compressibility models improve the results.

With the developed numerical method and its parallel implementation, we have been able to obtain good results for this severe test case, without using all the possibilities of the non-structured approach. In the future, adapted finite element meshes with local grid refinement will be used to decrease the computational cost of Low Reynolds turbulence model.

References

[1] M. Buffat "Parallel simulation of compressible turbulent flows using non-structured domains partitioning and object oriented programming.", In Notes on numerical fluid mechanics:CFD on parallel systems. Vieweg-Verlag, (1994).

[2] D.S. Dolling, M.T. Murphy "Unsteadiness of the separation shock wavestructure in a supersonic compression ramp flowfield", AIAA Journal, 21(12), (dec. 1983).

[3] C.Le Ribault, L.Hallo, M. Buffat "An implicit mixed finite volume-finite element method for solving 3d turbulent compressible flows", In 8th international Conference on Numerical Methods in Laminar and Turbulent Flows, Swansea (July 1993).

[4] S. Sarkar, G. Erlabacher, M.Y. Hussaini "Compressible homogeneous shear: simulation and modeling", Eight Symposium on turbulent shear flows, Munich sept. 9-11 (1991).

[5] S.M. Bogdonoff, G.S. Settles, T.J. Fitzpatrick "Detailed study of attached and separated compression corner flow fields in high Reynolds number supersonic flow", AIAA Journal, 17:579-585, (1979).

[6] D.C. Wilcox "Turbulence modeling for CFD", Griffin Printing, Griffin Printing, Glendale, California, (1993).

[7] O. Zeman "Dilatational dissipation: the concept and application in modeling compressible mixing layers" Physics of Fluids A,2, 178-188 (1990).

CHAPTER 7 : FLOW OVER A BUMP

Test Case TC5: TWO DIMENSIONAL TRANSONIC BUMP

H. Loyau,D. Vandromme

LMFN - CORIA URA 230 - INSA de ROUEN
Place E. Blondel, 76130 Mont-Saint-Aignan

Introduction

This flow is one of the classical cases for shock/turbulent boundary layer interactions. In the incoming boundary layer, effects of compressible turbulence are probably very weak. Therefore, the observed evolutions are relevant of incompressible turbulence subjected to a shock wave which causes the plate to separate, inducing a lamdba shock structure at the trailing edge of the profile as seen fig. 1.

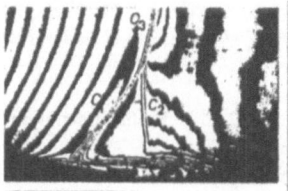

FIG. 1 - Interferogram in the interaction region

Experimentation

The flow configuration is the two-dimensional transonic turbulent flow corresponding to the asymetric geometry, studied by Délery and his co-workers and refered to as case C [1]. The fig. 2 depicts the nozzle whose inlet high and width are respectively 99,5 mm and 120 mm. The flow accelerates when passing over the bump which acts as a throat to reach supersonic conditions. The upper wall is a straight wall. At the nozzle exit, an adjustable second throat allows to control the downstream pressure and to choke the nozzle by maintaining a normal shock wave at the trailing edge of the bump.

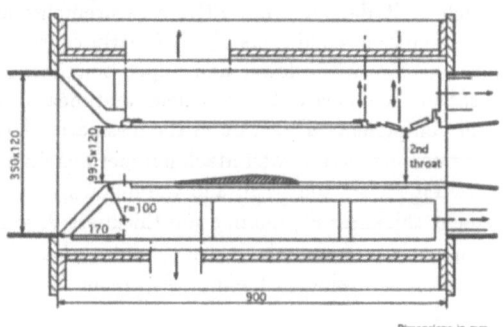

FIG. 2 - Experimental arrangement.

The reservoir conditions are for stagnation pressure and temperature: P_o=96000 Pa, T_o=300 K.

375

Geometry description

The fig. 3 shows the geometry of the bump. The width of the profile is 12 mm and the length 286 mm. The profile is made of a straight upstream part of 4° angle, and is followed by a circle arc up to the trailing edge. Radius of 20 mm ensure the junction between the profile and the upstream and downstream horizontal straight lines.

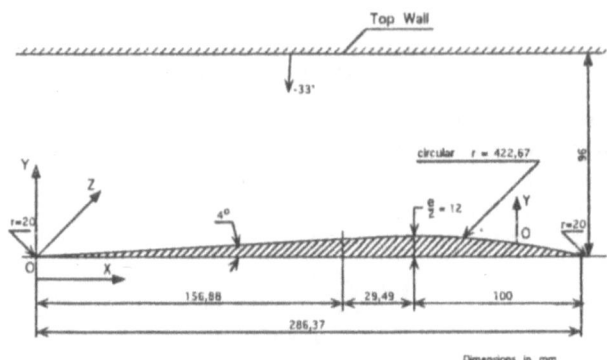

FIG. 3 - Profile description.

The figure depicts also the set of axis used:
—→the longitudinal axis OX is taken on the straight part of the lower wall
—→the vertical axis OY is perpendicular to OX
—→the transverse axis OZ corresponds to the width of the nozzle.
The origin of axis is taken at the leading edge of the profile. However, concerning the dissipative boundary layer parameters, the origin of the vertical axis is taken from the profile boundary.

Concerning the computation of this case, at the exit boundary, the static pressure will be adjusted assuming isentropic state, in order to have the shock location close to its experimental one.

Experimental data

Detailed data, much of it derived from LDA measurements are available [2]. The measurements include stagnation conditions P_o, T_o (nearly constant during all the test), wall pressure distributions P_w on the lower and upper boundaries, axial Mach number distributions M_{ax} in the inviscid part of the flow field, and mean-velocity components U, V and Reynolds tensor components u^2, v^2, uv in the dissipative layer. The data inferred from the previous mesurements include wall Mach number distribution M_w (assuming an isentropic relation between P_w/P_o and M_w), and the following boundary layer features: thickness δ, displacement thickness δ^*, momentum thickness θ, kinetic energy thickness Δ, and the incompressible shape parameter H_i in the interaction region. This exeptional level of resolution (summarized table 1) makes the present flow an interesting case for assessing turbulence models.

376

TABLE 1 : Experimental data set.

Exp. data	Location	Uncertaincies
$P_w/P_o, M_w$	*upper and lower wall*	P_w : ± 0.5 %
$U, U/U_e$	$X =$ 232 mm	U : ± 1. % *external flow* : ± 2m/s *elsewhere*
M_{ax}	$Y =$ 15, 20, 25, 30, 35, 40, 45, 50, 55 mm	
$U, V, u^2,$ v^2, uv	$X =$ 270, 275, 280, 285, 290, 295, 300, 305, 310, 315, 320, 325, 330, 340, 350, 360, 370, 380, 400, 410, 420, 430, 450, 470, 490, 510, 540 mm	V : ± 2m/s u^2 : ± 0.05 *max.prof.value* v^2 : ± 0.05 *max.prof.value* uv : ± 0.05 *max.prof.value*
$U_e, M_e, \delta,$ δ^*, θ, Δ	250 mm ≤ X ≤ 540 mm	

References

[1] DELERY J., COPY C. and REISZ J., (1980): "Analyse au vélocimètre laser bidirectionnel d'une interaction choc-couche l imite avec décollement étendu" ONERA Raport technique, No 37:7078 AY014

[2] KLINE et al. Ed., (1981): Proceedings of the 1980 HTTM Standford Conference on Complex Turbulent Flows, Stanford university.

ASSESSMENT OF A ONE-EQUATION POINTWISE TURBULENCE MODEL FOR COMPRESSIBLE FLOW

Nicola Ceresola

ALENIA - Ingegneria Velivoli Difesa
C.so Marche,41 - 10146 Torino Italy

Summary.

A pointwise version of the one-equation Baldwin-Barth model for the turbulence Reynolds number R_T is proposed. The y^* dependence of the wall damping terms is replaced with a dependence on R_T, that was obtained by comparison with the results of a direct simulation on a flat plate. The model is applied to the ONERA transonic channel testcase. The computed results are equivalent to those that can be obtained applying $\kappa - \epsilon$ models.

1 INTRODUCTION.

The present needs of modern industrial R&D require to make CFD simulations past more and more complex 3D geometries and in a variety of physical conditions. To fulfill the requests of the development engineers, it is therefore mandatory to look for numerical methods ensuring the maximum possible flexibility and generality in practical applications. The quick turnaround times that are required for a practical usage are, in fact, incompatible with setting up adhoc modifications when one has to do with different geometries or with changed physical conditions.

In that sense, the main difficulty that is associated with classical first order, two equations methods is the necessity to explicitly determine lengthscales, as the distance from solid walls, and velocity profiles. As an example, if multiple shear layers are present, such an evaluation may be very cumbersome, often compromising a correct prediction of the turbulence quantities. The presence of wall functions, moreover, requires the knowledge of the distance from the solid surfaces, precluding any general-purpose treatment of flows past arbitrary geometries, expecially when we have to do with structured multiblock or unstructured meshes.

In a recent work, Baldwin and Barth [1] proposed a self-consistent one-equation model for the turbulence Reynolds number Re_t avoiding the need to supply the algebraic length scales. A fully pointwise implementation of the model was later proposed by Goldberg [2], who substituted the y^* dependence of the damping factors with a dependence on Re_t itself. In the present paper, a form of the damping function is tried, which is mutuated both from [1] and from direct simulation data.

The scope of the present work is to carry on an evaluation of the model, as a possible candidate to be applied to geometrically complex industrial problems.

2 THE TURBULENCE MODEL.

2.1 The field equation.

The Baldwin-Barth model consists in a field equation for the turbulence Reynolds number

$$R_t = \frac{\kappa^2}{\nu \epsilon} .$$

The governing equation was derived from the $\kappa - \epsilon$ equations under the main hypothesis that the production equals the dissipation in the outer region of the boundary layer. It is given by

$$\frac{\partial \Phi}{\partial t} + \mathbf{V} \cdot \nabla \Phi = (c_{\epsilon_2} f_2 - c_{\epsilon_1})\sqrt{(\Phi P)} + \left(\nu + \frac{\nu_t}{\sigma_\epsilon}\right)\nabla^2 \Phi - (\nabla \nu_t) \cdot \nabla(\Phi)/\sigma_\epsilon \qquad (1)$$

where $\Phi = \nu R_t$, f_2 is a damping function and

$$P = \nu_t \left[\left(\frac{\partial u_i}{\partial x_j} + \frac{\partial u_j}{\partial x_i}\right)\frac{\partial u_i}{\partial x_j} - \frac{2}{3}\left(\frac{\partial u_k}{\partial x_k}\right)^2\right]$$

is the turbulence production.
The eddy viscosity $\nu_t = c_\mu \nu k^2/\epsilon$ is given, from (1), by

$$\nu_t = c_\mu f_\mu (\nu R_T)$$

where f_μ is a wall damping function.

During the preliminary testing on flat plate flow with zero pressure gradient, it was found that correct velocity profiles could only be obtained by halving the value of the dissipation coefficient σ_ϵ with respect to the one suggested in [1]. As it will be pointed out in the following, this is probably due to the fact that the model predicts a lack of dissipation with respect to the production in the outer part of the boundary layer.

2.2 Near-wall treatment.

In [1], the damping function f_μ is calibrated by comparison with the Cebeci-Smith model for an incompressible flat-plate flow, supposing that a linear relation exists between R_t and y^* in the near-wall region:

$$y_* = \frac{c_\mu}{k} R_t . \qquad (2)$$

The damping factor is given by

$$f_\mu = \left(1 - e^{-\frac{y^*}{A^*}}\right)\left(1 - e^{-\frac{y^*}{A_2^*}}\right) .$$

In the present work, a fully pointwise model was adopted, simply substituting () in (), to eliminate the explicit dependence of f_μ on the wall distance. In addition, correcting factors were introduced, to empirically fit the behavior of f_μ according with the prediction of direct simulations on a flat plate [3]. The final formula is

$$f_\mu = \left(1 - 0.5e^{-\frac{R_t c_\mu}{kA^*}}\right)\left(1 - 0.5e^{-\frac{R_t c_\mu}{kA_2^*}}\right) + 0.07\left(1 - e^{-(R_t/25)^2}\right) .$$

The standard Launder's formula was kept for f_2:

$$f_2 = \left(1 - 0.22e^{-0.1R_t^2}\right) .$$

379

3 BASIC NUMERICAL ALGORITHM.

The thin-layer Reynolds-averaged Navier-Stokes equations are solved with a finite differences, space-centered, implicit algorithm, with second and fourth-order numerical dissipation. A different treatment of the nonlinear fourth-order dissipation term was needed in solving Eq.() with respect to the Navier-Stokes equations. In the latter case, the scaling factor is taken to be proportional to the Euler spectral radius $\Phi = (|u| + a)/\Delta x$, while, in solving the R_t equation, it was redefined as $\Phi = a/\Delta x$. Besides this difference, the resolution schemes were identical, and the same numerical damping coefficients and CFL number were used.

Fully converged solution were obtained in all cases, with a five orders of magnitude drop in the residual in about 5000 iterations.

An implicit treatment of the turbulence production term was also attempted, following the identity

$$\sqrt{(\nu R_t P)} = \sqrt{\left(c_\mu f_\mu \frac{P}{\nu_t} \right)}(\nu R_t)$$

but no particular improvements in the rate of convergence were evidenced.

4 RESULTS.

The TC5 testcase, namely the ONERA transonic channel [4], constitutes a representative workbench to assess the performance of the above model. In spite of its geometrical simplicity, it includes very complex physical effects (strong shock-induced separation) which are very hard to be simulated by most turbulence models.

The computations were made on a 161x100 grid (fig.1). The distance of the first grid line from both walls was fixed so that y^+ is $o(1)$ at the beginning of the interaction. At inflow, all quantities except pressure were fixed assuming isentropic expansion from stagnation; extrapolation was done for pressure. At outflow, the pressure was fixed and adjusted until the shock wave was placed at the correct location. No slip conditions were enforced at the upper and lower walls.

No information was given to the turbulence model on the presence and location of the solid walls. The main purpose of the present investigation was, in effect, to assess if the model would have correctly "felt" the presence of the shear layers, only using mean-flow informations.

In Fig.2 the Mach number contours are depicted, that were computed using the Johnson-King model [5]. Due to their good agreement with the experiment, the Johnson-King results are here taken as a reference solution to be compared with the newly implemented model.

The most evident feature of the flow is the presence of a lambda shock, due to the streamline displacement above the shock-induced separation bubble. Such a flowfield is very hard to be reproduced numerically, the turbulence and the mean flow being out of equilibrium in the interaction region.

Fig.3 shows the Mach number contours, as they are computed with the Baldwin-Barth model. The qualitative features of the flowfield are in agreement both with the experiment and with the computation employing the Johnson-King model. A closer view of the interaction zone is shown in fig.4, where the complex structure of the predicted separation is well depicted.

The eddy viscosity contours in the interaction region are depicted in fig.5. The maximum $\frac{\mu_t}{\mu}$ ratio is

about 2400 and takes place in the shear layer above the separation bubble.

A more quantitative comparison is made in fig.6, where the pressure values on the lower wall are compared, showing a good overall agreement with the experiment results. However, in the reattachment region the flow deceleration is overestimated. This is a typical effect of an underestimation of the boundary layer thickness in the reattached zone. As it was shown by Délery, the equilibrium between the turbulence and the flowfield is here progressively recovered; this effect is probably the hardest to be reproduced by a conventional first-order model.

As it may be expected, the pressure distribution on the upper wall (fig.7) agrees with the measurements.

The velocity profiles at a station immediately upstream of the interaction (X=232 mm) are shown in fig.8. This ensures that the boundary layer behavior is quite correctly predicted until the interaction takes place, which is a necessary condition for the significance of any further comparison.

The profiles in the recirculation region (X=270, 280, 310 and 330 mm) are depicted in figs.9, 10, 11, 12. The computed profiles approximately follow the measured ones, except in that a somewhat thinner bubble is predicted. The extent of the recirculating bubble seems to be well predicted.

In the reattached flow (X=360, 380 mm, in fig.13, 14) the computation is not able to follow the rapid recovery of a stable attached boundary layer flow. In addition, as it was observed above, the boundary layer thickness is underestimated. As it was already noted, the flow is here characterized by a gradual return to the equilibrium between the mean flow and the turbulence, that cannot be simulated by equilibrium models.

A global view of the boundary layer features is given in figs.15 ,16 and 17, where the displacement thickness, momentum thickness and incompressible shape factor distributions are reported.

5 CONCLUSIONS.

The pointwise version of the one-equation Baldwin-Barth model was applied to the ONERA channel testcase. A damping function was tried, which seems to predict well enough the near-wall behaviour of the velocity profiles. The results seemed to be at least comparable whith those obtained with the standard versions of $\kappa - \epsilon$ In our view, this makes the model to be potentially attractive for engineering applications on complex geometries and in multiblock environments.

In addition, its operation count is about half than the one of two-equations models, and it can be implemented in a fully implicit way, enhancing the well conditioning of the matrices and the robustness of the code.

The main problem that came out from the present validation seemed to be a substantial overprediction of the $\frac{P}{\epsilon}$ ratio in the wake region of the profile. Further theoretical and numerical work is thought to be required to overcome this difficulty.

6 BIBLIOGRAPHY.

[1]   Baldwin, B.S., and Barth, T.: *A one-equation turbulence transport model for high Reynolds number wall-bounded flows.*
       NASA TM-102847, 1990.

[2]   Goldberg, U.C., and Ramakrishnan, S.V.: *A pointwise version of the Baldwin-Barth turbulence model.*
       Rockwell International Science Center, 1993.

[3]   Gilbert, N., and Kleiser, L.: *Turbulence model data derived from direct numerical simulations.*
Advances in Turbulence 2, H.H. Fernholz and H.E.Fiedler, Editors, Springer-Verlag, 1989.

[4]   Délery, J., and Marvin, J.G.: *Shock-wave boundary layer interactions.*
AGARDograph No.280, 1986.

[5]   Ceresola, N.: *Turbulence modelling in strong shock-boundary layer interaction.*
Brite-Euram 1051 Final Report, 1992.

[6]   Vandromme, D., and Solakoglu, E.: *Shock wave-turbulent boundary layer interaction.*
Brite-Euram 1051 Workshop Report, february 11-12, 1992.

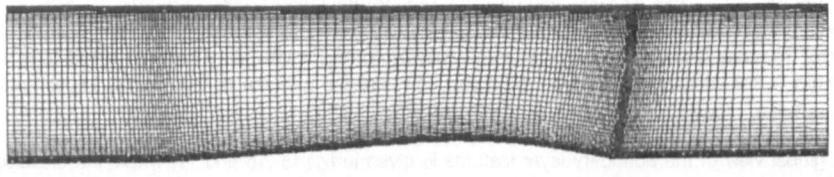

Fig.1 - Computational grid.

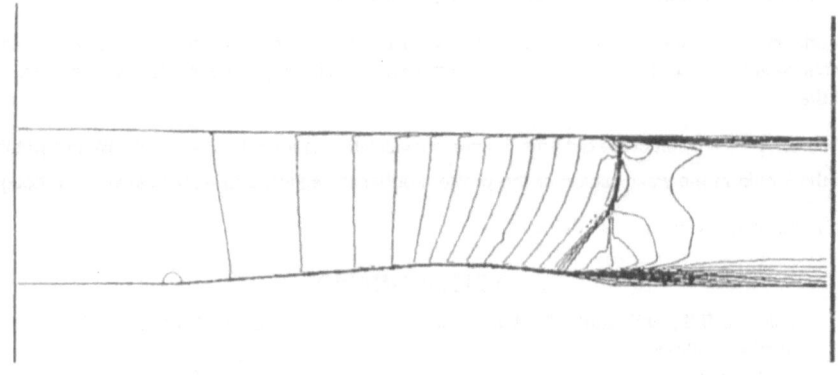

Fig.2 - Mach contours as computed with the Johnson-King model.

382

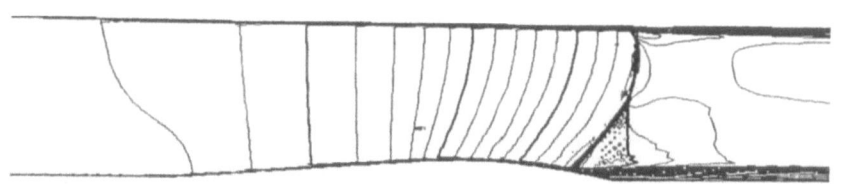

Fig.3 - Mach contours (present calculation).

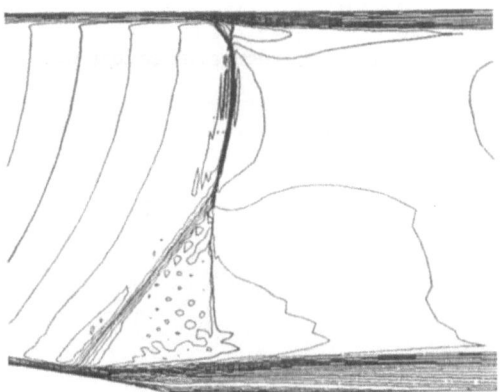

Fig.4 - Enlarged view of the separation region.

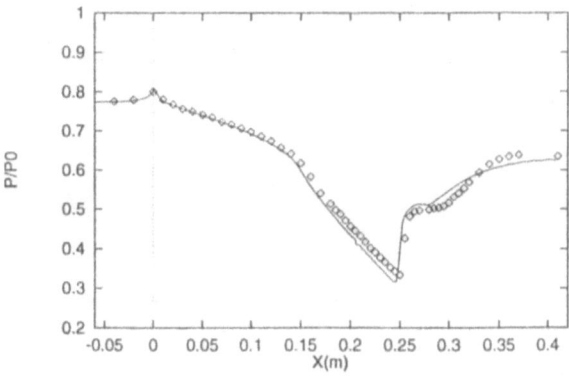

Fig.5 - Static to total pressure ratio on lower wall

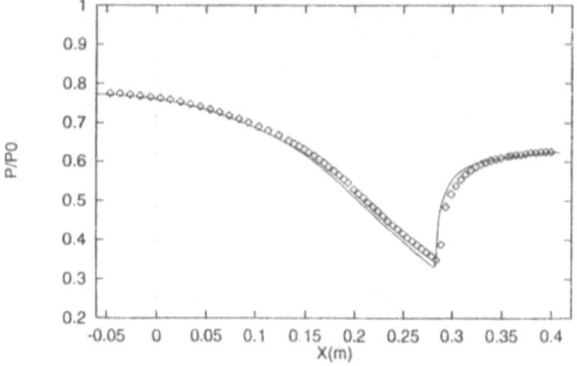

Fig.6 - Static to total pressure ratio on upper wall

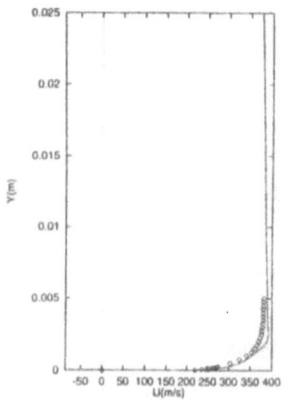

Fig.7 - Velocity profiles at x = 232 mm

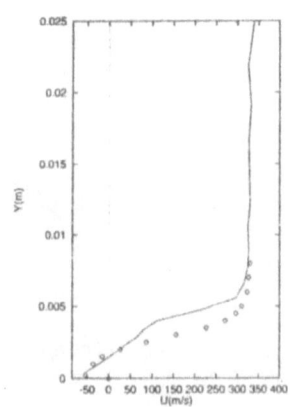

Fig.8 - Velocity profiles at x = 270 mm

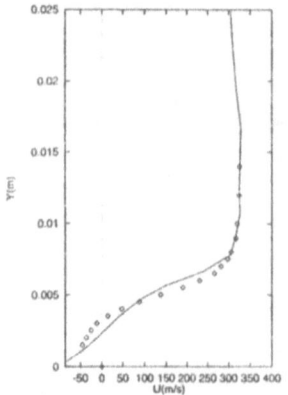

Fig.9 - Velocity profiles at x = 280 mm

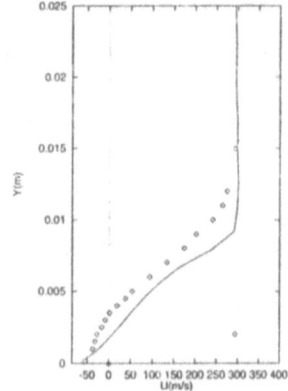

Fig.10 - Velocity profiles at x = 310 mm

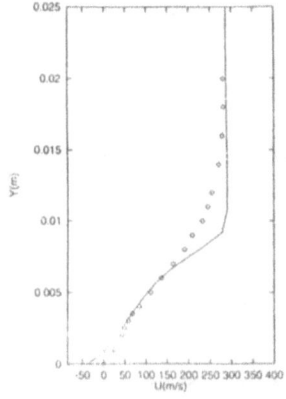

Fig. 11 - Velocity profiles at x = 330 mm

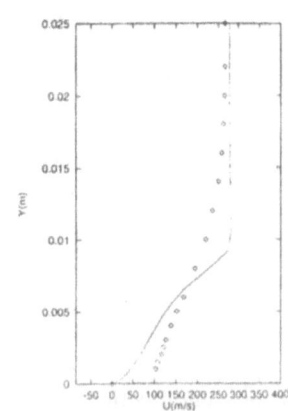

Fig. 12 - Velocity profiles at x = 360 mm

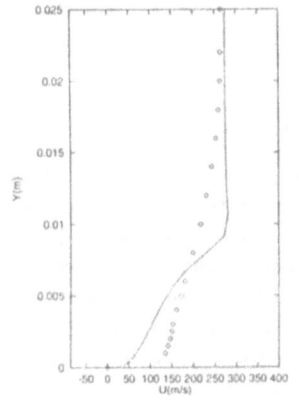

Fig. 13 - Velocity profiles at x = 380 mm

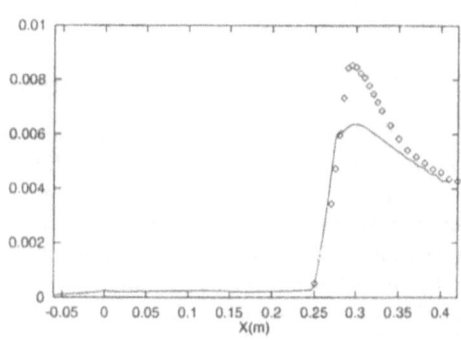

Fig. 14 - Displacement thickness

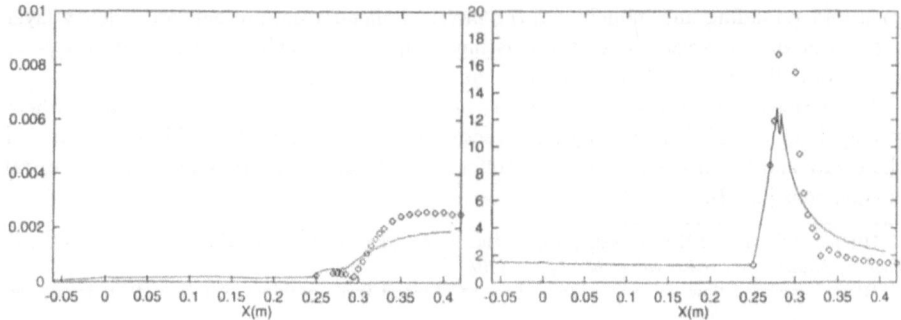

Fig. 15 - Momentum thickness

Fig. 16 - Shape parameter

Numerical simulation of the flow over a 2D transonic bump with extended separation *

G. Freskos [†]

Institute for Jet Propulsion and Turbomachinery, RWTH Aachen
Templergraben 55, 52056 Aachen, GERMANY

SUMMARY

This paper concerns the calculation of test case TC5 of ETMA project, corresponding to the two-dimensional transonic bump. The computation is performed with two numerical schemes developped for the mass-averaged Navier-Stokes equations in association with a low-Reynolds number $k - \varepsilon$ turbulence model. Furthermore, a modification, improving the numerical results on such geometry, is proposed for the two-equation turbulence model used.

Introduction

Shock-wave/boundary layer interaction phenomena are present in all aerodynamic systems in transonic regime and above. If uncontrolled, they can lead to severe problems regarding the overall efficiency of the flying system (flow separation and loss of pressure recovery at the engine inflow section). The character of this interaction is mainly three dimensional. A great part of the research devoted to this subject is up to now mostly restricted in two dimensional simulations. In this -2D- context an important simplification is achieved and valuable information concerning various phenomena can be obtained. Nevertheless the problem cannot be considered as resolved.

The complete -numerical- simulation of aerodynamic flows needs 3D codes, which should reply simultaneously to a number of important prerogatives. An efficient treatment of the convective terms of the governing equations should allow good resolution of shock waves, without penalizing the very sensitive boundary layer regions. Moreover, good understanding and modeling of the physical phenomena, present in boundary layer regions, is also necessary, since the existing computing capacities are still several years away from allowing direct simulation of complicated flows.

In [1] Délery presented measurements of three transonic flow configurations, where strong shock wave/boundary layer interaction is predominant. The wide availability of these data and the physical insight they offer, made these cases very popular for numerical comparisons [2], [3].

---

*Prepared for the ETMA Workshop on *Turbulence Modeling for Compressible Flows arising in Aeronautics*, Manchester, U.K., November 14-17, 1994*

[†]Post-doctorate fellow - Present address: Lab. of Thermal Turbomachines, N.T.U.A., P.O. Box 64069, 157 10 Athens, GREECE

Complete description of the code can be found in [4], where numerical schemes, turbulence models etc. are presented.

The governing equations

In conservative form, the time-dependent Navier-Stokes equation, for 2D planar compressible flow are expressed as follows in a Cartesian coordinate system (x, y):

$$\frac{\partial \vec{U}}{\partial t} + \frac{\partial \vec{F}}{\partial x} + \frac{\partial \vec{G}}{\partial y} = \frac{\partial \vec{F}_v}{\partial x} + \frac{\partial \vec{G}_v}{\partial y} \tag{1}$$

where $\vec{U}$ is the conservative variables vector:

$$\vec{U} = [\rho, \rho u, \rho v, \rho e_0]^{\mathrm{T}}. \tag{2}$$

Above, ρ is the density, u and v are the velocity components along x, and y directions, and e_0 is the total internal energy.

The convective and diffusive flux vector $\vec{F}$ and $\vec{F}_v$ in x direction are respectively (with p the static pressure) (analogous expressions hold for y direction):

$$\vec{F} = \begin{bmatrix} \rho u \\ \rho u^2 + p \\ \rho u v \\ u(\rho e_0 + p) \end{bmatrix}, \quad \vec{F}_v = - \begin{bmatrix} 0 \\ \sigma'_x \\ \tau_{xy} \\ \sigma'_x u + \tau_{yx} v - q_x \end{bmatrix}.$$

where σ'_i is the deviatoric normal stress ($\sigma'_i = \sigma_i - p$), $\tau_{ij}, i \neq j$ are the shear stresses and q_i the heat flux along the i direction.

If isotropic, Newtonian fluids, i.e., fluids for which a linear relation between the stress components and those of the rate of strain holds in all directions are considered, the stress components are expressed as follows:

$$\sigma_i = p - \lambda \frac{\partial u_j}{\partial j} - 2\mu \frac{\partial u_i}{\partial i}.$$

$$\tau_{ij} = -\mu \left(\frac{\partial u_i}{\partial j} + \frac{\partial u_j}{\partial i} \right).$$

where λ is the coefficient of bulk viscosity and μ is the ordinary viscosity coefficient. Assuming the mean value of the normal stresses to be equal to the thermodynamic pressure, leads to Stokes' hypothesis: $\lambda + \frac{2}{3}\mu = 0$. The viscosity coefficient μ is expressed as a function of the temperature through Sutherland's law.

The state equation for a perfect gas is

$$p = \rho R T.$$

where T is the temperature and R the air constant ($R = c_p(T) - c_v(T) = 287.04 \mathrm{m}^2\mathrm{s}^{-2}\mathrm{K}^{-1}$). For even low temperature levels ($T \leq 600\mathrm{K}$) the ratio of specific heats can be assumed constant (calorically perfect gas). Then, the kinetic theory of gases gives for air at standard conditions: $\gamma = 1.4$.

The heat flux in each direction is expressed according to Fourier's law, which is written as :

$$\vec{q} = -\chi \vec{\nabla} T.$$

where χ is the coefficient of thermal conductivity. It is computed by assuming the Prandtl number to be equal to 0.72.

The numerical schemes

Two numerical schemes are available. The first is the well-known Mac Cormack scheme combined with the Steger Warming flux-splitting. The other one is a TVD scheme, based on Roe's approximate Rieman solver. An entropy function, proposed by Harten and Hyman and modified by Lafon enforces the entropy condition. An approximate factorization (AF) implicit technique is used in both schemes for convergence acceleration.

The first scheme

MacCormack's predictor-corrector scheme [5] in two-dimensions is given by the following expressions:

$$\text{predictor} \; : \; \Delta U_{i,j}^n = -\Delta t \left(\frac{D_+ F_{i,j}^n}{\Delta x} + \frac{D_+ G_{i,j}^n}{\Delta y} \right),$$

$$U_{i,j}^{\overline{n+1}} = U_{i,j}^n + \Delta U_{i,j}^n. \tag{3}$$

$$\text{corrector} \; : \; \Delta U_{i,j}^{\overline{n+1}} = -\Delta t \left(\frac{D_- F_{i,j}^{\overline{n+1}}}{\Delta x} + \frac{D_- G_{i,j}^{\overline{n+1}}}{\Delta y} \right),$$

$$U_{i,j}^{n+1} = \tfrac{1}{2}(U_{i,j}^n + U_{i,j}^{\overline{n+1}} + \Delta U_{i,j}^{\overline{n+1}}). \tag{4}$$

Forward and backward discretizations are alternated during the iterative procedure. For supersonic flows a shock capturing technique is also needed. Steger-Warming's flux vector splitting [6] helps in resolving the shock waves.

The upwind scheme

The numerical treatment of the convective part of eq.(1) is based on the so-called Total Variation Diminishing approach (TVD). The above equation can take the following discretized form:

$$\frac{U_i^{n+1} - U_i^n}{\Delta t} + \frac{\tilde{F}_{i+\frac{1}{2},j} - \tilde{F}_{i-\frac{1}{2},j}}{\Delta x} + \frac{\tilde{G}_{i,j+\frac{1}{2}} - \tilde{G}_{i,j-\frac{1}{2}}}{\Delta y} = 0. \tag{5}$$

The numerical convective flux $\tilde{F}$ at $i + \frac{1}{2}, j$ will be expressed according to Yee's unified formulation (flux difference splitting):

$$\tilde{F}_{i+\frac{1}{2},j} = \frac{1}{2}(F(U^R_{i+\frac{1}{2},j}) + F(U^L_{i+\frac{1}{2},j}) + \hat{R}_{i+\frac{1}{2},j}\Phi_{i+\frac{1}{2},j}). \tag{6}$$

where MUSCL interpolation is assumed. The same analysis also applies to the $i, j + \frac{1}{2}$ interfaces for approximating the $\tilde{G}$ fluxes.

Based on scalar considerations and for a fully upwind scheme, the left and right states for the conservative variables are approximated as follows:

$$U^R_{i+\frac{1}{2},j} = U_{i+1,j} - \frac{1}{2}g_{i+1,j}.$$

$$U^L_{i+\frac{1}{2},j} = U_{i,j} + \frac{1}{2}g_{i,j}. \tag{7}$$

where g is the slope limiter function. The behavior of the solution as a function of various limiters can be studied. For the present calculations Roe's minmod limiter was retained. The expression for latter is:

$$g^l_i = \frac{\Delta^l_{i+\frac{1}{2},j}\Delta^l_{i-\frac{1}{2},j} + |\Delta^l_{i+\frac{1}{2},j}\Delta^l_{i-\frac{1}{2},j}|}{\Delta^l_{i+\frac{1}{2},j} - \Delta^l_{i-\frac{1}{2},j}}. \tag{8}$$

where

$$\Delta_{i+\frac{1}{2},j} = U_{i+1,j} - U_{i,j} .$$

Many authors prefer to impose the limiters on the characteristic variables. In our approach limiting was exclusively applied to the conservative variables.

The components of the dissipation term $\Phi_{i+\frac{1}{2},j}$ in (6) are expressed as follows:

$$\hat{\phi}^l_{i+\frac{1}{2},j} = -\psi(\hat{a}^l_{i+\frac{1}{2},j})\hat{\alpha}^l_{i+\frac{1}{2},j},$$

$$\hat{\alpha}_{i+\frac{1}{2},j} = \hat{R}^{-1}_{i+\frac{1}{2},j}(U^R_{i+\frac{1}{2},j} - U^L_{i+\frac{1}{2},j}). \tag{9}$$

where $\hat{a}^l_{i+\frac{1}{2},j}$ are the eigenvalues and $\hat{R}$ is the right eigenvectors matrix of the jacobian A^x of the flux vector F. They are computed at some symmetric average between $U^R_{i+\frac{1}{2},j}$ and $U^L_{i+\frac{1}{2},j}$. Here Roe's average state between $U^R_{i+\frac{1}{2},j}$ and $U^L_{i+\frac{1}{2},j}$ has been chosen [7]. ψ is an entropy function introduced in order to enforce the entropy condition. It is destined for regions where eigenvalues of the jacobian A^x tend to zero (shock waves or contact discontinuities). It eliminates expansion shocks and ensures a smooth transition from supersonic to subsonic flow. The choice of ψ is very important for viscous flow computations. Indeed, low numerical viscosity in boundary layer regions and stabilization of convective terms are simultaneously required. Harten [8] proposed the following expression for ψ, where a parameter δ adjusts the correction introduced by ψ:

$$\psi(\hat{a}^l_{i+\frac{1}{2},j}) = \begin{cases} |\hat{a}^l_{i+\frac{1}{2},j}|, & |\hat{a}^l_{i+\frac{1}{2},j}| > \delta \\ \dfrac{\hat{a}^{l2}_{i+\frac{1}{2},j} + \delta^2}{2\delta}, & |\hat{a}^l_{i+\frac{1}{2},j}| \le \delta . \end{cases} \tag{10}$$

Furthermore δ can be evaluated locally as [9]:

$$\delta = \delta_1(|u| + |v| + |c|). \tag{11}$$

In this way ψ can be switched on by choosing δ_1 large enough in regions where this is necessary (shock capturing), whereas it can be set equal to some small value in zones where it can provoke unwanted effects (boundary layers) [10].

The $k - \varepsilon$ turbulence model

The Jones-Launder $k - \varepsilon$ model is used. It has the advantage of not requiring computation of any wall distances, which can be of primary importance for the simplicity in which a turbulence model can be implemented and its extension in three dimensions.
The turbulent viscosity for the Jones-Launder low Reynolds number turbulence model [12], is expressed as follows:

$$\mu_t = c_\mu f_\mu \bar{\rho} \frac{\tilde{k}^2}{\tilde{\varepsilon}}.$$

Two supplementary equations expressing the transport of the turbulent kinetic energy and its dissipation rate are used:

$$\frac{\partial}{\partial t}\left(\bar{\rho}\tilde{k}\right) + \frac{\partial}{\partial x_i}\left(\bar{\rho}\tilde{k}\tilde{U}_i - (\mu + \mu_k)\frac{\partial \tilde{k}}{\partial x_k}\right) = S_k. \tag{12}$$

$$\frac{\partial}{\partial t}\left(\bar{\rho}\tilde{\varepsilon}^*\right) + \frac{\partial}{\partial x_i}\left(\bar{\rho}\tilde{\varepsilon}^*\tilde{U}_i - (\mu + \mu_\varepsilon)\frac{\partial \tilde{\varepsilon}^*}{\partial x_k}\right) = S_\varepsilon. \tag{13}$$

with:

$$\tilde{\varepsilon}^* = \tilde{\varepsilon} - 2\nu\left(\frac{\partial\sqrt{\tilde{k}}}{\partial x_n}\right)^2.$$

S_k, S_ε are source terms. In the present model they are expressed as follows:

$$S_k = P_k - \bar{\rho}\tilde{\varepsilon}^* - 2\mu\left(\frac{\partial\sqrt{\tilde{k}}}{\partial x_n}\right)^2.$$

$$S_\varepsilon = C_{\varepsilon_1}\frac{\tilde{\varepsilon}}{\tilde{k}}P_k - C_{\varepsilon_2}f_2\frac{\tilde{\varepsilon}}{\tilde{k}}\bar{\rho}\tilde{\varepsilon}^* + \frac{2\mu\mu_t}{\bar{\rho}}\left(\frac{\partial^2\tilde{U}_t}{\partial x_n^2}\right)^2.$$

where, $\tilde{U}_t$ is the tangential velocity component to the solid boundary(-ies) of the flow, x_n is the normal direction to these boundaries and P_k is the production term:

$$P_k = -\bar{\rho}\widetilde{u_i''u_j''}\frac{\partial\tilde{U}_i}{\partial x_j}.$$

The wall boundary conditions are $\tilde{k} = \tilde{\varepsilon}^* = 0$, since only the isotropic part of the turbulent dissipation is considered. The constants c_μ, C_{ε_1}, C_{ε_2} are equal to 0.09, 1.45,

390

1.92 respectively. The function f_μ allows to damp out the turbulent diffusion mechanisms in wall vicinity for the case of boundary layers:

$$f_\mu = \exp[-\frac{2.5}{1 + \frac{R_t}{50}}], \text{ where } : R_t = \frac{\rho \tilde{k}^2}{\mu \tilde{\varepsilon}}.$$

The f_2 is given by the relation:

$$f_2 = 1 - 0.3 \exp(-R_t{}^2).$$

The diffusivity coefficients are expressed as:

$$\mu_k = \frac{\mu_t}{Pr_k} \; ; \; \mu_\varepsilon = \frac{\mu_t}{Pr_\varepsilon}.$$

with $Pr_k = 1.0, Pr_\varepsilon = 1.3$. The turbulent Prandtl number is $Pr_t = 0.90$.

The numerical experiments

For the calculations presented in this work initialization profiles provided by INSA-Rouen were used. But since previous experience has showed that mainly the upwind model is sensitive to the initialization profiles of the turbulent quantities, the procedure described below was also used.

For the computations based on the Roe-TVD numerical scheme the minmod limiter is used, and the cutoff constants in the entropy function δ_{1x}, δ_{1y} were given the values 10^{-4}. An implicit treatment is combined with both numerical schemes. Approximate factorization technique is used for the resolution of the tri-diagonal linear system. Due to the AF technique an error is introduced in the time relaxation. For elevated CFL numbers, the results may be CFL dependent. (a detailed presentation can be found in [4])

A number of turbulence models can be used: the Baldwin-Lomax algebraic model, the Jones-Launder low Reynolds number $k - \varepsilon$ model, a wall functions model, taking into account compressibility, and finally an Algebraic Stress Model (A.S.M.). We will only use here the first model. Especially for this model, an implicit treatment of the source terms may further enhance the robustness of the numerical approach. An important parameter, which is often neglected in turbulent flow simulations, is the free-stream values of the turbulent quantities. We have noticed, that the flow under consideration, is very sensitive to the profiles and free-stream values of k and ε.

Initialization of the k and ε profiles

For the computations presented here, we used the following initialization process. The k-profile is defined as follows:

$$k = \min(c_\mu^{-\frac{1}{2}} \frac{U_\tau^2 y^+}{10}, 0.035 U_\tau^2 y_+^2), y^+ \leq 10.$$

$$k = \max(c_\mu^{-\frac{1}{2}} U_\tau^2 (1 - \frac{y^+ - 10}{400 - 10}), k_\infty), y^+ > 10$$

$$\left(\text{or, } k = \max\left(c_\mu^{-\frac{1}{2}} U_\tau^2 \left(1 - \frac{y^+ - 10}{\frac{\delta_0}{y_\tau} - 10}\right), k_\infty\right), y^+ > 10 \right).$$

where $c_\mu = 0.09$, $y^+ = \dfrac{y}{y_\tau}$, $y_\tau = \dfrac{\mu_w}{\rho_w U_\tau}$, $U_\tau = \sqrt{\dfrac{\tau_w}{\rho_w}}$ and k_∞ is the imposed free-stream

value. It was given the following value: $k_\infty = \left(\dfrac{U_\infty}{100}\right)^2$.

The turbulent dissipation is defined by mixing length model considerations: $\varepsilon = c_\mu^{\frac{3}{4}} k^{\frac{3}{2}} \dfrac{1}{l_t}$.

with: $l_t = \min(200 y^+ \, ; 0.4 y (1 - e^{-\frac{y^+}{27}}))$.

Flow over a Bump (Extended Separation)

The third configuration studied in [1] is characterized by the presence of a sufficiently strong shock wave to produce an important recirculation zone. Good description of the separation bubble by a two equation turbulence model remains up to date a non trivial subject.

In this case, the flow field is no more symmetric. A flat wall constitutes the upper boundary of the duct. The bump length is 0.286m and its height 0.096m. Stagnation inflow conditions are the same as in the previous case ($P_{t0} = 96$kPa, $T_{t0} = 300$K) (see also [13]).

A grid of (120x218) cells was used for the numerical computation (figure 1). Downstream pressure was fixed to $P_{out} = 59.5$kPa. In previous investigations a downstream pressure of 62kPa was used instead. This influences the shock wave position (no experimental value is available, since in this case a convergent-divergent nozzle is used to define the shock position).

Two computations were performed. In the first one the initialization provided by INSA, Rouen was used. This resulted to different shock positions when using the two numrical schemes. Therefore we have used the initialization profile presented above which resulted in better prediction of the shock wave position.

The iso-Mach contours and the velocity vector profiles (figures 2, 3, 4) show the existence of a separation bubble close to the end of the bump. But, the flow reattaches in a rather short distance. This is verified by the static pressure distribution along the bump wall (figure 5). The model fails to predict the flat region of this profile corresponding to the separation zone. No particular problems seem to exist at the upper boundary (figure 5).

Velocity profiles (figure 6) confirm the insufficiency of the model in the separation region. However, the rather good prediction of the downstream velocity profiles can be noticed.

In order to better understand the different mechanisms governing turbulent flows, we computed the bump flow using the same $k - \varepsilon$ model where the value of the constant C_{ε_1} was changed from 1.45 to 1.57. This modification should, in principle, increase the level of the turbulent dissipation. The turbulent kinetic energy is then reduced, and consequently, the flow separates more easily.

A well defined λ shock is apparent in figure 4. The separation region is much more extended now. The plateau region of the Pw profile was as well significantly improved. Velocity profiles in the separation zone were well described. In figures 5, 6 the results using the modified turbulence model seem to match quite well the experimental results with the exception of the downstream region where some differences appear. They are nevertheless considerably improved compared with analogous computations presented earlier in [4]. Further analysis of these results are needed before conclusions are drawn.

If nevertheless some comments are to be made, the imposed initial $k - \varepsilon$ profiles seem to affect considerably the numerical results. Furthermore, an important factor in low Reynolds number two-equation turbulence modeling is the near wall treatment of the turbulent quantities. Consequently, a more thorough examination of this aspect seems promising.

Second order modeling provides an alternative, which can improve the quality of the results for such flows. The Reynolds stress models are complex and their use is away from being generalized in practical flow computations. A simplified model using algebraic equations for the Reynolds stresses (the Algebraic Stress Model (ASM)) provides an interesting research domain. Nevertheless, it should be noted that according to some results reported in [15], 'there are strong indications that this approach, which does not account for turbulence transport, does not offer a sufficiently secure general modeling framework for shock / boundary layer interaction'.

Acknowledgements

The development of the two-dimensional code was supported by SNECMA, where the code is already installed. The present research is sponsored by the Commission of the European Communities in the framework of the Human Capital and Mobility Program.

References

[1] J.M. Délery. Experimental investigation of turbulence properties in transonic shock / boundary-layer interactions. *AIAA Journal*, 21:180-185, 1983.

[2] R. Benay, M.C. Coet, and J. Délery. Validation of turbulence models applied to transonic shock wave / boundary-layer interaction. *La Recherche Aérospatiale*, 3:1-16, 1987.

[3] J. Cousteix and B. Aupoix. Modèles de turbulence en écoulement compressible. Marseille, Octobre 1990. 27ème Colloque d'Aérodynamique Appliquée.

[4] G.O. Freskos. *Physical aspects and numerical simulation of the flows in supersonic air-intakes.* PhD thesis, INP, Toulouse, France, October 1992.

[5] R.W. Mac Cormack. The effect of viscosity in hypervelocity impact cratering. AIAA paper 69-0354, 1969.

[6] J.L. Steger and R.F. Warming. Flux vector splitting of the inviscid gadynamic equations with applications to finite difference methods. *Journal of Computational Physics*, 40(2):263-293,April 1981.

[7] P.L. Roe. Approximate Riemann solvers, parameter vectors and difference schemes. *Journal of Computational Physics*, 43:357-372, 1981.

[8] A. Harten. On a class of high resolution total-variation-stable finite-difference schemes. *SIAM J. Num. Anal.*, 21:1-23, 1984.

[9] J.L. Montagne, H.C. Yee, G.H. Klopfer, and M. Vinokur. Hypersonic blunt body computations including real gas effects. Technical Report TM100074, NASA, 1988.

[10] A. Lafon. Calcul d'écoulements visqueux hypersoniques. Raport Tecnique OA 32/5005.22, CERT-ONERA, Toulouse, France, Mars 1990.

[11] B.S. Baldwin and H. Lomax. Thin layer approximation and algebraic model for separated turbulent flows. AIAA paper 78-257, 1978.

[12] W.P. Jones and B.E. Launder. The calculation of low Reynolds number phenomena with a two-equation model of turbulence. *Int. J. Heat Mass Transfer*, 16:1119-1130, 1973.

[13] Analyse au vélocimètre laser bidirectionnel d'une interaction choc-couche limite turbulente avec décollement étendu. Rapport Technique No 37/7078 AY 014, ONERA, Chatillon, Août 1980.

[14] W. Rodi. Turbulence models and their application in hydraulics: A state of art review. The Netherlands, February 1984. International Association for hydraulic Research, 2nd edition.

[15] M.A. Leschziner. Validation of 2D Navier-Stokes codes with respect to turbulence modelling; ONERA bump flows: Case A and Case C. Composite midterm report, EUROVAL, February 1991.

Figures

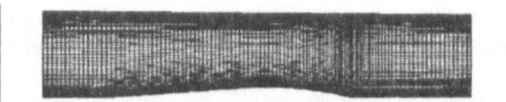

FIG. 1 - Computational grid.

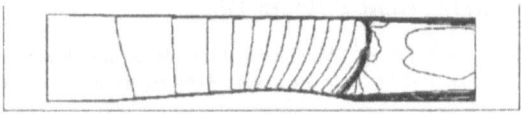

FIG. 2 - Mach contour lines ($k - \epsilon$ model/RWTH init.)

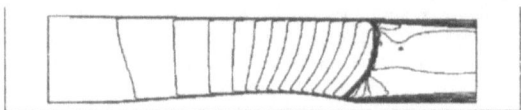

FIG. 3 - Mach contour lines ($k - \epsilon$ model/INSA init.)

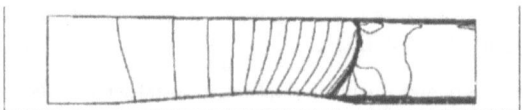

FIG. 4 - Mach contour lines (modified $k - \epsilon$)

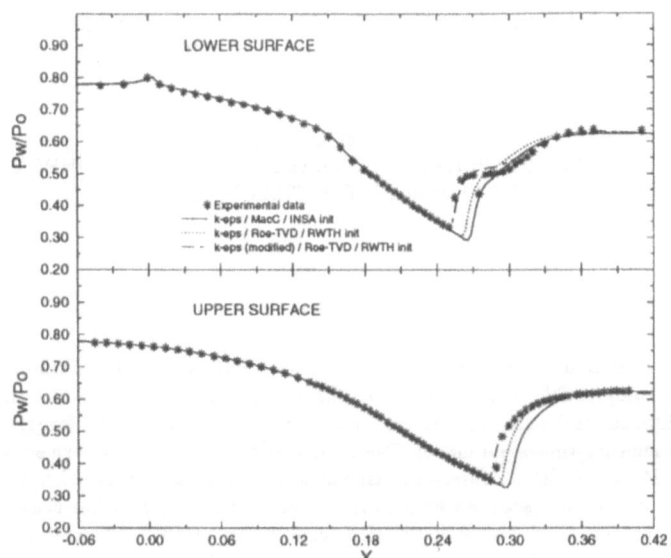

FIG. 5 - Upper and lower wall pressure distributions.

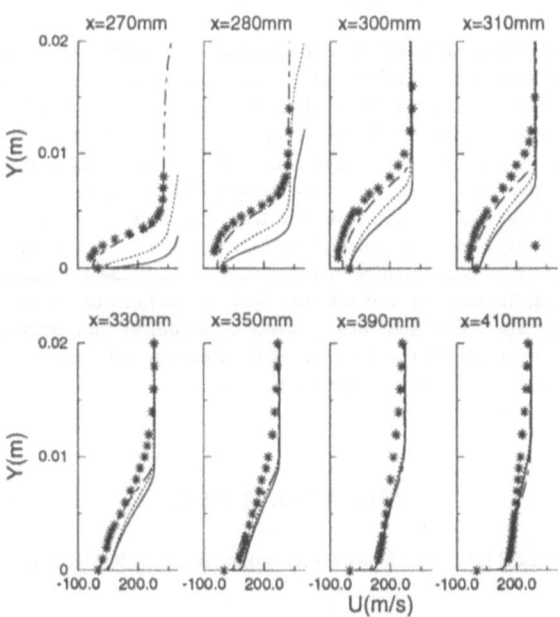

FIG. 6 - Velocity profiles

CALCULATION OF A TWO DIMENSIONAL TRANSONIC BUMP WITH A MULTIPLE-TIME-SCALE TURBULENCE MODEL

Vincent GLEIZE

Office National d'Etudes et de Recherches Aérospatiales (ONERA)

BP 72 - F-92322 CHATILLON CEDEX

SUMMARY

This paper concerns the calculation of one test case of ETMA workshop on turbulence modeling for compressible flow arising in aeronautics (TC5: two dimensional transonic bump [3]). Three different turbulence models are compared : an algebraic mixing-length model, a $k - \epsilon$ model and a multiple-time-scale model. These calculations have been performed with the solver CANARI [2, 15]. CANARI is a three-dimensional solver, developed at ONERA, for the averaged Navier-Stokes equations associated with a turbulence model. This code has been validated for a large number of complex practical applications concerning internal and external configurations for subsonic, transonic and supersonic flows, and it is already used by industrial companies.

INTRODUCTION

In this paper, three turbulent models are used and compared for a shock / boudary layer interaction problem. The two first models are part of usual turbulent models based on one-point closures which use a single time scale to describe the whole turbulent phenomenon. In the mixing-length algebraic model of Michel et al. [11], the eddy viscosity is expressed directly from the characteristic quantities of the boundary layer and the kinetic energy is assumed to be equal to zero. In the $k - \epsilon$ model of Jones-Launder [7], the eddy viscosity is related to the turbulent kinetic energy and to a characteristic length scale of turbulence. These models are based on an implicit hypothesis of spectral equilibrium due to this single time scale. Thus, they neglect the spectral characteristics of turbulence. Indeed the turbulence is characterised by a wide spectrum of fluctuations and the turbulent interactions are associated with different parts of this evolving spectrum. The idea to describe turbulent flow with a single time scale being simplistic [5], we present a third model which uses different scales to describe turbulent interaction process involving eddies of different sizes [8, 12, 13].

PHYSICAL MODEL

The physical model is the compressible Reynolds averaged Navier-Stokes equation system, written as follows :

$$\frac{\partial f}{\partial t} + div (F - F_v) = 0 , \tag{1}$$

$$f = [\rho, \rho\overline{V}, \rho E], \quad F = [\rho\overline{V}, \rho\overline{V} \otimes \overline{V} + p\overline{\overline{I}}, \rho E\overline{V} + p\overline{V}],$$

$$F_v = [0, \overline{\overline{\tau}} + \overline{\overline{\tau_R}}, (\overline{\overline{\tau}} + \overline{\overline{\tau_R}}).\overline{V} - \overline{q} - \overline{q_t}]].$$

In these expressions, averaged values are taken with the classical meaning for ρ and p, and with mass-averaged values for the velocity $\overline{V}$ and the specific internal energy e. The total energy E is the summation of the total energy $e + V^2 / 2$ of the mean flow and of the kinetic energy of turbulence K. $\overline{\overline{\tau}}$ is the shear stress tensor, $\overline{\overline{\tau_R}}$ is the Reynolds stress tensor, $\overline{q}$ is the heat flux and $\overline{q_t}$ is the turbulent heat flux.

Mixing-length algebraic model of Michel and al.

The turbulent viscosity μ_t is obtained as follows:

$$\mu_t = \rho\, l^2\, F^2\, |\, \overline{\omega}\, |, \quad l = 0.085\, \delta\, th\, (\frac{\kappa}{0.085} \cdot \frac{d}{\delta}\,), \tag{2}$$

where $\kappa = 0.41$ is the Von Karman constant. F is the Van Driest viscous sublayer damping function given by:

$$F\, (\, \xi\,) = 1 - exp\, (\, -\frac{\sqrt{\xi}}{26\, K}\,) \quad \xi = \rho\, l^2\, \frac{\mu + \mu_t}{\mu^2}\, |\, \overline{\omega}\, |. \tag{3}$$

In equation (2), d is Buleev's [1] "modified distance" which is used to account for the influence of several walls in corner flows, and δ is a "modified boundary layer thickness". The equations (2) and (3) give an implicit equation for μ_t, which is solved by Newton's method.

$K - \epsilon$ model of Jones-Launder

The $K - \epsilon$ transport equation model proposed by Jones and Launder , including low-Reynolds source terms, is written as follows:

$$\frac{\partial g}{\partial t} + div(G - G_v) = S, \tag{4}$$

where :

$$g = [\rho K; \rho\epsilon], \quad G = [\rho\overline{V}K; \rho\overline{V}\epsilon], \quad G_v = [(\mu + \frac{\mu_t}{\alpha_K})\nabla K; (\mu + \frac{\mu_t}{\alpha_\epsilon})\nabla\epsilon)],$$

$$S = [\overline{\overline{\tau_R}} : \nabla\overline{V} - \rho\epsilon - 2\mu\nabla\sqrt{K}; C_1 f_1 \frac{\epsilon}{K}\overline{\overline{\tau_R}} : \nabla\overline{V} - C_2 f_2 \rho\frac{\epsilon_2}{K} + 2\mu\mu_t(E)^2], \tag{5}$$

the source terms including generation terms, dissipation terms and the low-number Reynolds terms. In general case, $(E)^2$ is written like the sum of the double products of all the second derivative velocity. These equations are available until the wall, including low Reynolds turbulent number $R_t = \rho K^2/\mu\epsilon(\leq 100)$. Constants of the model and damping functions appearing in equations (5) are:

$$\alpha_K = 1, \quad \alpha_\epsilon = 1.3, \quad C_1 = 1.55, \quad C_2 = 2, \quad f_1 = 1, \quad f_2 = [1 - 0.3 exp(-R_t^2)].$$

The turbulent viscosity is evaluated by the relation :

$$\mu_t = C_\mu \rho \frac{K^2}{\epsilon} exp\left[\frac{-2.5}{(1 + \frac{R_t}{50})}\right], \quad C_\mu = 0.09. \tag{6}$$

397

Multi-time-scale model

The turbulent energy spectrum is decomposed into zones according to a wave number partition. Therefore, the generation, cascade, and dissipation of turbulent kinetic energy are described in the multiple-time-scale turbulence model by considering the energy flux rate between adjacent wave number regions. This description of the turbulent kinetic energy spectrum is realized as opposite to the classical turbulence models such as the $k - \epsilon$ turbulence models and the Reynolds stress turbulent models. The energy flux equations are performed by derivating the wave number partition law. Then, partial energy transport equation can be obtained by using two different methods. The first one, is based on an integration into Fourier space. The equation of the partitioned energy spectrum is integrated to obtain energy evolution on each region of the wave number partition. This method is particulary used in incompressible flows [12]. On an other hand, a second method consists in performing a wave number filtering operation in physical space [4]. This method seems to be more simple than the first one for compressible flows. Here, the energy flux equations are obtained from analogy with the dissipation rate equation. The present model is based on a two-scale scheme with algebraic modeling of partial Reynolds stresses with a turbulent viscosity μ_t. The first scale is related to the large eddies in the production range (index (1)) and the second scale is related to the fine eddies in the transfert and dissipation ranges (index (2)).

μ_t is evaluated with the following relation :

$$\mu_t = C_\mu^{(1)}\rho\frac{K^2}{F^{(1)}} + C_\mu^{(2)}\rho\frac{K^2}{F^{(2)}}, \quad C_\mu^{(1)} = \frac{0.09K^{(1)}}{K^{(1)} + 0.1K^{(2)}}, \quad C_\mu^{(2)} = \frac{0.1\ 0.09K^{(1)}}{K^{(1)} + 0.1K^{(2)}}. \tag{7}$$

K is the total kinetic energy $(K = K^{(1)} + K^{(2)})$. $K^{(1)}$ and $K^{(2)}$ are the partial kinetic energies. $F^{(1)}$ is the energy flux rate between the first and the second spectral regions. $F^{(2)}$ is the energy flux rate between the second region and the region corresponding to the Kolmogorov scales. We write also the multi-time-scale system with the compact form (4)

where : $g = [g^{(1)}; g^{(2)}]$, $\quad G = [G^{(1)}; G^{(2)}]$, $\quad G_v = [G_v^{(1)}; G_v^{(2)}]$ and $\quad S = [S^{(1)}; S^{(2)}]$,

$$g^{(m)} = [\rho K^{(m)}; \rho F^{(m)}], \quad G^{(m)} = [\rho\overline{V}K^{(m)}; \rho\overline{V}F^{(m)}],$$
$$G_v^{(m)} = [(\mu + \frac{\mu_t}{\alpha_K})\nabla K^{(m)}; (\mu + \frac{\mu_t}{\alpha_F})\nabla F^{(m)})],$$
$$S = [\overline{\overline{\tau_R^{(m)}}} : \nabla\overline{V} - \rho\epsilon^{(m)} - 2\mu_t\sqrt{\frac{K^{(m)}}{K}}(\nabla\sqrt{K^{(m)}})(\nabla\sqrt{K}); \tag{8}$$
$$C_1^{(m)}\frac{F^{(m)}}{K^{(m)}}\overline{\overline{\tau_R^{(m)}}} : \nabla\overline{V}C_1^{(m)}\rho\frac{F^{(m)}F^{(m-1)}}{K^{(m)}} - C_2^{(m)}\rho\frac{F^{(m)2}}{K^{(m)}} - C_2^{(m)}\rho\frac{F^{(m)}\epsilon^{(m)}}{K^{(m)}} + 2\mu C_\mu\frac{KK^{(m)}}{F^{(1)}}(E)^2],$$

with $m = 1$ and 2 and the energy flux $F^{(0)} = 0$. The constants are given by the following relations[4]:

$$\alpha_K = 1.1, \quad \alpha_F = 1.55, \quad C_1^{(1)} = 1.57, \quad C_2^{(1)} = 1.85, \quad C_1^{(2)} = 1.66, \quad C_2^{(2)} = 1.75.$$

NUMERICAL METHOD

The solver CANARI is built from the following numerical tools: multidomain approach with structured grids overlapping or not, centered schemes with implicit acceleration techniques, use of the characteristic relation method for boundary condition treatment. The numerical scheme used in this paper is a cell centered Runge-Kutta ("RK") scheme using space centered discretization operator. This type of scheme requires some artificial dissipation due to its non dissipative property in the sens of Kreiss. Then, a fourth order linear dissipation is added to ensure this property, associated with a second order non linear viscosity to capture the flow discontinuities correctly [6]. For stationary and unsteady problems, this scheme is coupled with an implicit acceleration stage to reduce the cost of computation. Concerning turbulent calculations using the turbulent transport equation systems presented above, we have chosen, in order to keep the modular feature of the solver, to treat the turbulent equation system apart from the mean flow system. At each iteration, we solve alternatively the system (1) assuming that the coefficient μ_t given by the relations ((6) and (7)) is frozen, and then we solve the system (4) assuming that the average values ρ and $\overline{V}$ are frozen. We now describe the time discretizations we use for the mean flow Navier-Stokes system (1) and for the turbulent transport equation systems, the space discretization being describe in [2]. The method "RK" which has been implemented is analogous to the classical method introduced by Jameson and al. [6]. It is a four step Runge-Kutta method defined as follows:

$$h^{(0)} = h^n,$$

For $k = 1$ to $k = 4$:

$$h^{(k)} = h^{(k-1)} - \alpha^{(k)} \Delta t [div(H^{(k-1)} - H_v^{(0)}) - T^{(0)}] + D(h^{(0)}),$$

$$h^{n+1} = h^{(4)}, \tag{9}$$

where $D(h^{(0)})$ is the artificial viscosity term and $\alpha^{(1)} = 1/4$, $\alpha^{(2)} = 1/3$, $\alpha^{(3)} = 1/2$ and $\alpha^{(4)} = 1$. For the Navier-Stokes system: $h \equiv f$, $\quad H \equiv F$, $\quad H_v \equiv F_v$ and $T = 0$. For the Turbulence equation system: $h \equiv g$, $\quad H \equiv G$, $\quad H_v \equiv G_v$ and $T = S$. and at each step only a fourth-order linear dissipation term is added. The turbulent variables are defined and updated at the cell centers by using an integral formula of the divergence terms. The gradients of K, ϵ, ∇K, $K^{(m)}$, $F^{(m)}$, $\nabla K^{(m)}$ and the second order derivatives of the velocity are also evaluated at the cell centers with integral formulae. The time step Δt used to integrate the solution in time with the above scheme is a local time step determined from the following formula which takes into account the convection and diffusion limitations :

$$\Delta t = \eta \ min[\frac{h}{V + c}, \frac{\rho h^2}{2}/\gamma(\frac{\mu}{Pr} + \frac{\mu_t}{Pr_t})], \tag{10}$$

where h is a characteristic length of the mesh size, representing an evaluation of the dimension of the numerical dependence domain, c the local sound speed. η is a numerical coefficient which has to reflect the stability conditions for the convection part (CFL criterion) and for the diffusion part. We use an implicit stage proposed by Lerat et al. [9] and implanted in CANARI with the "RK" scheme [10]. The treatment of the boundary conditions is based on the use of the characteristic relations as proposed by Viviand and Veuillot [14].

APPLICATION: TWO-DIMENSIONAL TRANSONIC BUMP (TC5)

Test Case TC5 is a classical case of flow configuration characterized by flow fields resulting from the interaction of fully turbulent boundary layers with shock or expension waves generated by the shape of the test surfaces. The selected flow configuration is the two-dimensional transonic turbulent flow studied at ONERA by Délery and his co-workers. That specific test case is commonly referred to as case C [3]. The nozzle is made of a lower wall with a bump and a straight upper wall (figure 1). The inlet Mach number is approximately equal to 0.63. The flow accelerates when passing over the bump which acts as a throat, to reach a Mach number peak value of 1.36, upstream of the shock. At the nozzle exit, an adjustable second throat allows to control the downstream pressure and to choke the nozzle by maintaining a normal shock wave at the end of the bump. The curvature of the lower wall at the trailing edge is combined with the pressure gradients in this region to induce the boundary layer separation with an extended recirculation zone. A standing normal shock wave C3 is produced in the nozzle. The boundary layer separates upstream of the shock feet, producing an oblique compression wave C1. The flow behind C1 being still supersonic is decelerated to subsonic conditions through the downstream leg C2 of the λ-shock structure (figure 2).

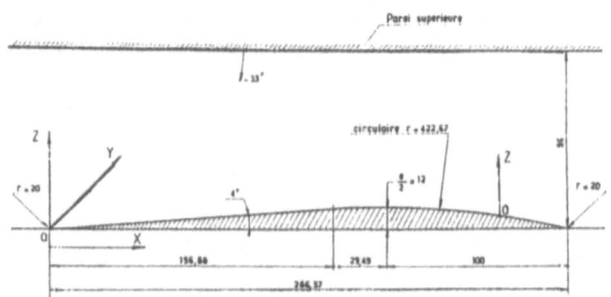

FIG. 1 - *Nozzle Geometry*

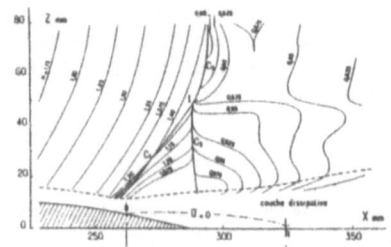

FIG. 2 - *Experimental Iso-Mach number Curves*

The Reynolds number is $Re_{io} = \rho_{io}V_{io}d/\mu_{io} = 2.078\ 10^6$ (d is the nozzle height), the reservoir conditions are $T_{io} = 300K$ and $p_{io} = 96000Pa$. The initial conditions correspond to the uniform flow defined by the inlet flow values. A no-slip condition and an isothermal condition are imposed on the lower and upper surfaces of the nozzle. At the inlet boundary, the velocity direction, the stagnation pressure and the stagnation temperature are prescribed. At the outlet boundary, static pressure is imposed and it is adjusted in order to have shock wave location close to the experimental one. The grid includes 153x65x2 = 19,890 points (although the case is two-dimensional, two mesh sections are requeried in the transverse direction by the three-dimensional code CANARI). The height of the first cell at the upper and lower surfaces is constant and equal to $3.5 10^{-6}m$, corresponding to values of $y^+ = \rho u_\tau y / \mu$ less than 3 on the whole length of the surfaces. For the computation made with the Michel turbulence model, the CFL number is equal to 6 and for the two other computations it is equal to 5. The iso-Mach number lines are plotted in figure 3 for the three turbulence model simulations. Before the shock wave, the fields are similar. On the other hand, shock structure and separation zone depend on the turbulence model.

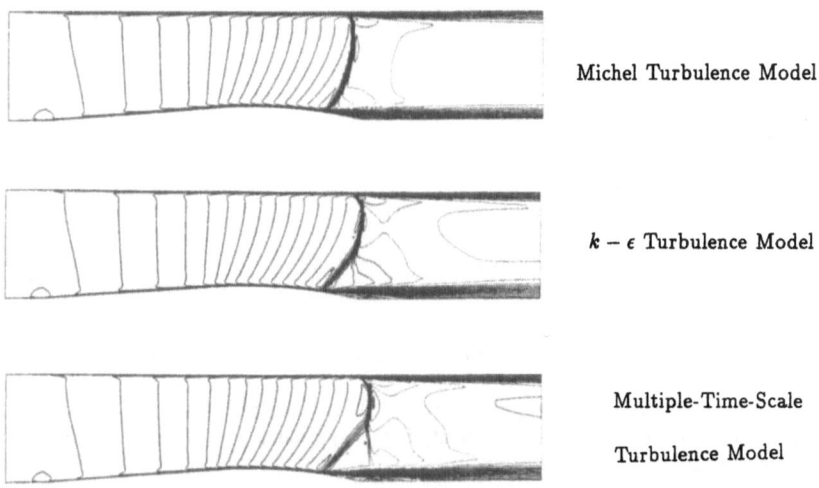

Michel Turbulence Model

$k - \epsilon$ Turbulence Model

Multiple-Time-Scale

Turbulence Model

FIG. 3 - *Iso-Mach Number Curves*

Figure 4 shows pressure distributions on the walls. Before the interaction, the curves coincide well with the experiment. The good position of C1 shock wave foot on the lower surface is obtained for each turbulence model taking the following outlet pressure values.

	Experiment	Multiple-Time-Scale	$k - \epsilon$	Michel et al.
P/P_{i0}	0.627	0.641	0.650	0.658

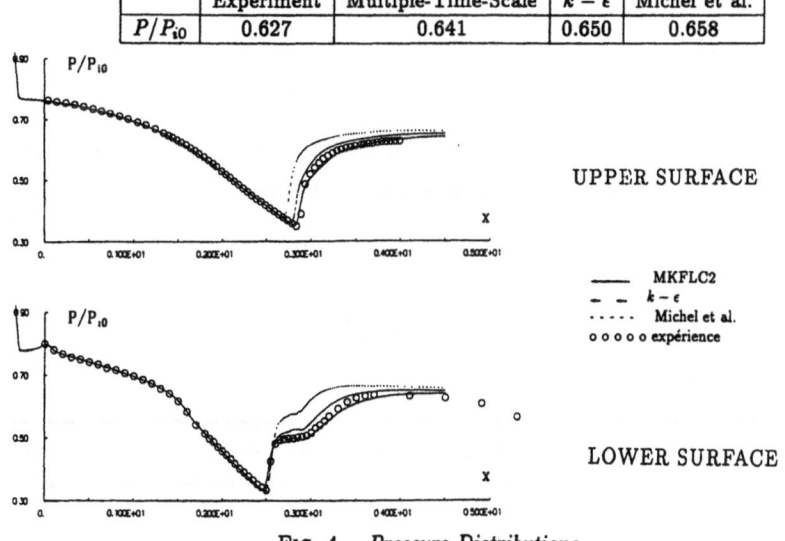

UPPER SURFACE

———— MKFLC2
– – $k - \epsilon$
· · · · · Michel et al.
o o o o o expérience

LOWER SURFACE

FIG. 4 - *Pressure Distributions*

With these values, only the multiple-time-scale model predicts a correct shock wave position on the upper surface. The normal velocity profiles presented on figure 5 show that the multiple-time-scale model gives a better agreement with experiment than the $k - \epsilon$ model. Behind reattachment point, all the computations underestimate normal velocities.

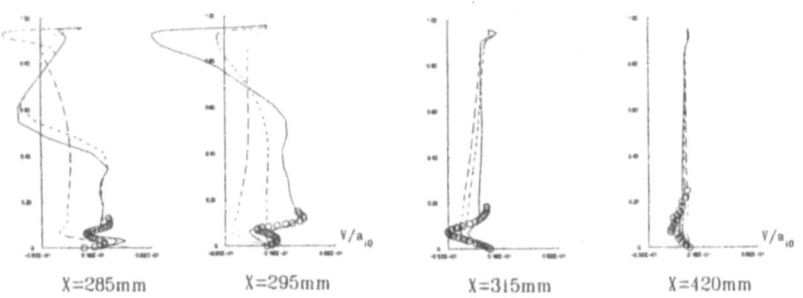

FIG. 5 - *Normal Velocity Profiles*

Figure 6 represent turbulent shear stress profiles. The multiple-time-scale model and the $k - \epsilon$ model give results rather close to experiment, but they predict slightly a too weak turbulent shear stress compared to the experiment.

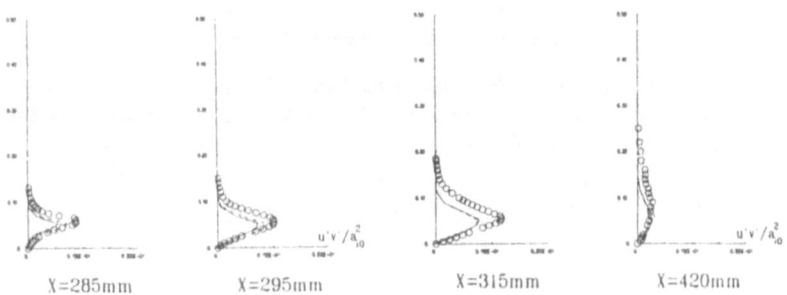

FIG. 6 - *Turbulent Shear stress Profiles*

CONCLUSION

A multiple-time-scale turbulence model and its application to complex flows have been presented in this paper. It was observed that the multiple-time-scale turbulence model could resolve details of this complex turbulent flow better than the standard two-equation model. The advantage of multiple-time-scale turbulence model lies in its capability to model the cascade process of turbulent kinetic energy and its potentiality to include more experimental observations in the future. Compared to the single-time-scale turbulence models like the $k - \epsilon$ model, the multi-time-scale model is more expensive in CPU time term, but it has a better capability to predict complex turbulent flows, as it has been proved for the application presented here.

References

[1] N.I.BULEEV - Theorical Model of Mechanism of Turbulent Exchange in Fluid Flows. (AERE Translation 957, Atomic Energy Research Establishment, Harwell, England, 1963.)

[2] V.COUAILLIER - Recent developments performed at ONERA for the simulation of 3D inviscid and viscous flows in turbomachinery by the solution of Euler and Navier-Stokes equations. (Proceedings of the 11th ISABE Symposium, edited by F.S.Billig, AIAA Washington D.C., Sept 93.)

[3] J.DELERY, C.COPY et J.REISZ - Analyse au vélocimètre laser bidirectionnel d'une interaction choc/couche limite turbulente avec décollement étendu. (O.N.E.R.A. RT, n° 37/7078 AY 014, Août 1980.)

[4] V.GLEIZE - Simulation numérique d'écoulements compressibles hors équilibre à l'aide de schémas à échelles multiples. (Thèse de Doctorat, Univ. Aix-Marseille II, 21 December 1994)

[5] K.HANJALIC, B.E.LAUNDER and R.SCHIESTEL - Multiple Time Scale Concepts in Turulence Transport Modelling. (2nd International Symposium on Turbulent Shear Flows, Imperial College, London, 2/4 July 1979.)

[6] A.JAMESON, W.SCHMIDT and E.TURKEL - Numerical solution of the Euler equations by finite-volume methods using Runge-Kutta time stepping schemes. (AIAA Paper, n° 8-1259, June 1981.)

[7] W.P.JONES and B.E.LAUNDER - The calculation of low-Reynolds-number phenomena with a two-equation model of turbulence. (Journal Heat Mass Trans., vol. 16, 1973.)

[8] B.E.LAUNDER et R.SCHIESTEL - Application d'un modèle de turbulence à échelles multiples au calcul d'écoulements libres turbulents. (C.R. Acad. Sci. Paris, t.288, Serie B, 1979)

[9] A.LERAT, J.SIDES and V.DARU - An implicit Finite-Volume Method for Solving the Euler Equations. (Lecture Notes in Physics, 170, 1982.)

[10] N.LIAMIS and V.COUAILLIER - Unsteady Euler and Navier-Stokes Flow Simulations with an Implicit Runge-Kutta Method. (Proceedings of the Second European Computational Fluid Dynamics Conference, 5-8 September 1994, Stuttgart, Germany.)

[11] R.MICHEL, C.QUEMARD et R.DURANT - Application d'un schéma de longueur de mélange à l'étude de couches limites turbulentes d'équilibre. (ONERA NT, n° 154, 1969.)

[12] R.SCHIESTEL - Multiple-time-scale modeling of turbulent flows in one-point closure. (Physics of Fluids, 30(3), March 1987)

[13] R.SCHIESTEL - Modélisation et simulation des écoulements turbulents. (Traités des Nouvelles Technologies, Série mécanique, Editions HERMES, 1993)

[14] H.VIVIAND et J.P.VEUILLOT - Méthodes pseudo-instationnaires pour le calcul d'écoulements turbulents transsoniques. (ONERA Publication, n° 1978-4)

[15] A.M.VUILLOT, V.COUAILLIER et N.LIAMIS - 3-D Turbomachinery Euler and Navier-Stokes calculations with a multidomain cell-centered approach. (AIAA/SAE/ASME/ASEE, 29th Joint Propulsion Conference and Exhibit, Monterey, CA (USA), June 28-30, 1993.)

COMPUTATION OF A TWO DIMENSIONAL TRANSONIC BUMP FLOW USING A POINTWISE $k - k^2/\epsilon$ TURBULENCE MODEL

Uriel C. Goldberg
CFD Department, Rockwell Science Center
Thousand Oaks, California 91360, USA

SUMMARY

A recently introduced $k - k^2/\epsilon$ turbulence model is used to compute the Délery transonic internal flow in a two dimensional channel including a bump on one surface. This case involves a $\lambda-$shock interacting with the boundary layer, producing a large separated flow region, and is considered a challenging test for turbulence models. The model, in its present form, predicts well the $\lambda-$shock and the pressure distributions on both walls, performs fairly well in predicting detailed flow profiles but overpredicts the extent of the separated flow region on the lower wall.

INTRODUCTION

The standard high Reynolds number $k - \epsilon$ model is a good performer in free shear flows but its extension to low Reynolds number wall-bounded flows proved over the years a rather difficult task with only limited success.

One approach follows the method of Jones-Launder [1] and Launder-Sharma [2] in which the dissipation rate of turbulence kinetic energy, ϵ, is replaced by $\tilde{\epsilon} = \epsilon - 2\nu(\partial\sqrt{k}/\partial y)^2$ which, unlike ϵ itself, vanishes at solid surfaces. This necessitates adding the term $S = 2\nu\nu_t(\partial^2 U/\partial y^2)^2$ to the $\tilde{\epsilon}$ transport equation. The presence of the normal-to-wall first and second derivatives in $\tilde{\epsilon}$ and in S, respectively, introduces two disadvantages; a numerical and a conceptual one: these derivatives create sometimes severe numerical "stiffness", limiting the usable size of the time step and increasing the chance of numerical transients severe enough to terminate a computation; they also prevent the model from being pointwise (local) because of the need to find normal-to-wall unit vectors.

Another approach is to use the ϵ transport equation unaltered, such as in the Lam-Bremhorst [3] model, but then the presence of the unnatural wall boundary condition $\epsilon = \nu\partial^2 k/\partial y^2$ becomes the source of numerical difficulties and inability to render the model pointwise.

In both the above approaches there is a need for two or three near-wall functions to convey the effect of viscous damping and to ensure proper asymptotic behaviour in the viscous sublayer.

This state of affairs regarding low Reynolds number extensions to the standard $k - \epsilon$ model has been a constant source of difficulties and frustration to CFD users attempting to predict real life flow problems.

Since the source of trouble is evidently the dissipation rate equation, some researchers replaced it with alternative transport equations to determine the length scale. A significant example is the Wilcox $k-\omega$ model [4] in which ϵ is replaced by $\omega = \epsilon/(C_\mu k)$. The model performs better that the $k - \epsilon$ model in adverse pressure gradient flows and is compatible with near-wall flows without the need to use damping functions; it is, however, a poor performer in free shear flows because of its sensitivity to the freestream condition of ω. Another source of difficulty is the singularity of ω at solid surfaces. This

404

requires setting it to its value at the first mesh point off walls, $\omega_1 = C(\nu/y^2)_1$ which is, again, a non-pointwise attribute and also introduces an extra measure of sensitivity to normal-to-wall mesh distribution.

Baldwin and Barth [5] and later Goldberg [6] introduced two equation models in which the ϵ equation was replaced by one for $R \equiv k^2/\epsilon$. With both k and R vanishing at walls the latter model is completely pointwise and involves only a single near-wall damping function. The model proved to be a good performer for wall-bounded flows including those involving separated flow regions. It does, however, exhibit some sensitivity to the freestream value of R which impairs its performance in predicting free shear flows. The reason for this behaviour is the action of the extra difusion term present in that model while it enhances turbulence propagation from solid surfaces into the flow, it inhibits such propagation in defect portions of boundary layers and in free shear flows. The next section explores this aspect of the model in detail and suggests a possible remedy to be explored in future work.

MODEL FORMULATION

We start with the standard high Reynolds number $k - \epsilon$ model.

$$\frac{Dk}{Dt} = \nabla \cdot \left(\frac{\nu_t}{\sigma_k} \nabla k \right) + P - \epsilon , \tag{1}$$

$$\frac{D\epsilon}{Dt} = \nabla \cdot \left(\frac{\nu_t}{\sigma_\epsilon} \nabla \epsilon \right) + C_{\epsilon 1} \frac{\epsilon}{k} P - C_{\epsilon 2} \frac{\epsilon^2}{k} . \tag{2}$$

The turbulence production term is given in terms of the Boussinesq concept

$$P = \left[\nu_t \left(\frac{\partial U_i}{\partial x_j} + \frac{\partial U_j}{\partial x_i} - \frac{2}{3} \frac{\partial U_k}{\partial x_k} \delta_{ij} \right) - \frac{2}{3} k \delta_{ij} \right] \frac{\partial U_i}{\partial x_j} . \tag{3}$$

Here k is the turbulence kinetic energy; ϵ is its dissipation rate; U_i are the Cartesian mean velocity components; x_i are the corresponding coordinates; ν and ν_t are the molecular and eddy kinematic viscosities, respectively; and D/Dt is the material derivative. The constants appearing in these transport equations will be discussed later.

A transport equation for the undamped eddy viscosity, $R \equiv k^2/\epsilon$, may be derived from these equations. For example, $\nabla \epsilon = (2k/R)\nabla k - (k/R)^2 \nabla R$. Carrying out the algebra and assuming that $\sigma_k = \sigma_\epsilon = \sigma$ results in the following equation, written in conservation form:

$$
\begin{aligned}
\frac{D\rho R}{Dt} = \nabla \cdot \left(\frac{\mu_t}{\sigma} \nabla R \right) &+ (2 - C_{\epsilon 1}) \frac{R}{k} \rho P - (2 - C_{\epsilon 2}) \rho k \\
&- \frac{2}{\sigma} C_\mu \rho \nabla R \cdot \nabla R \quad (I) \\
&- \frac{2}{\sigma} \frac{\mu_t R}{k^2} \nabla k \cdot \nabla k \quad (II) \\
&+ \frac{4}{\sigma} \frac{\mu_t}{k} \nabla k \cdot \nabla R. \quad (III)
\end{aligned}
\tag{4}
$$

Here ρ is the density, and $\mu_t = \rho \nu_t = C_\mu \rho R$.

In the process of deriving their one-equation model, Baldwin and Barth [5] neglected the last two diffusion terms, retaining only the first in the slightly different form $-2/\sigma \rho \nabla \nu_t \cdot \nabla R$. Goldberg [6] proposed a pointwise version of the Baldwin-Barth model, still using this term only. It is of interest, therefore, to reexamine the effect of the three terms on the behaviour of the model. Experience indicates that the original models, using term (I) only, enable good prediction of near-wall flows but may fail in the defect

405

layer and in free shear flows. The reason and possible remedy may be understood from evaluating the normal-to-wall entrainment velocity. By Eq.(4) this velocity is given as

$$\tilde{v} = v + 2\frac{C_\mu}{\sigma}\frac{\partial R}{\partial y} + v_{II} - \frac{4}{\sigma}\frac{\nu_t}{k}\frac{\partial k}{\partial y} \ , \tag{5}$$

where v_{II} is the contribution of term (II) which is not readily translatable into a normal-to-wall velocity but behaves similar to v_I. Eq.(5) provides a rather clear picture: in the near-wall region $\partial R/\partial y$ and $\partial k/\partial y$ are both positive, so that v_I enhances while v_{III} reduces diffusion away from the wall, making the former desirable and the latter undesirable; in defect layers the situation is reversed: now both $\partial R/\partial y$ and $\partial k/\partial y$ are negative, hence v_{III} becomes instrumental in diffusing the turbulence into the non-turbulent zone. A possible criterion for incorporating term (III) is to invoke it only if

$$[(\mathbf{V} \cdot \nabla)R]\,(\nabla k \cdot \nabla R) < 0 \tag{6}$$

where $\mathbf{V}$ is the velocity vector. This constitutes a pointwise generalisation of the one-dimensional criterion $v(\partial k/\partial y) < 0$.

A simple yet useful initial test case is fully developed flow in a channel, since it includes both near-wall and outer flow regimes. Using Wilcox's [7] PIPE flow solver the two velocity terms v_I and v_{III} are plotted against wall distance in Fig. (1). The regions of positive and negative entrainment velocities from the two diffusion terms, (I) and (III), are clearly seen, in correspondence with Eq.(5). In particular, the negative portion of v_{III} adjacent to the walls is observed; this is the portion that must be avoided to circumvent the corresponding reduction in entrainment velocity.

The proposed transport equation for R, written in low Reynolds number form, thus reads

$$\frac{D\rho R}{Dt} = \nabla \cdot [(\mu + \mu_t/\sigma)\nabla R] + (2 - C_{\epsilon 1})\frac{R}{k}\rho P - (2 - C_{\epsilon 2})\rho k$$
$$- 2\frac{C_\mu}{\sigma}\rho\nabla R \cdot \nabla R + \mathcal{D} \tag{7a}$$

where the cross-diffusion term is given by

$$\mathcal{D} = \frac{4}{\sigma}\frac{\mu_t}{k}\nabla k \cdot \nabla R \tag{7b}$$

if the criterion in Eq.(6) is met, otherwise $\mathcal{D} = 0$.

Evaluating Eq.(7) at the logarithmic overlap, the following relation results:

$$\sigma = \frac{\kappa^2}{\sqrt{C_\mu}(C_{\epsilon 2} - C_{\epsilon 1})} \tag{8}$$

where $\kappa = 0.41$ and $C_\mu = 0.09$. The other model constants are chosen to be those suggested by Myong and Kasagi [8], namely, $C_{\epsilon 1}=1.42$, $C_{\epsilon 2}=1.83$, and from Eq.(8) it follows that $\sigma=1.367$.

The eddy viscosity is now given by

$$\mu_t = C_\mu f_\mu \rho R \tag{9}$$

where the near-wall damping function is chosen as

$$f_\mu = \frac{1 - e^{-A_\mu R_T}}{1 - e^{-A_\epsilon R_T}} \tag{10}$$

using the turbulence Reynolds nubmer $R_T \equiv R/\nu$. The constants here are determined by optimisation for flat plate near-wall flow, resulting in $A_\mu = 0.017$ with $A_\epsilon = C_\mu^{3/4}/2\kappa = 0.2$.

The two transport equations, Eqs.(1) and (7) (the former in low Reynolds number conservation form with ϵ replaced by its equivalent, k^2/R), are subject to the following boundary conditions:

(i) Solid Walls

$$k = 0 , R = 0 , \tag{11}$$

(ii) freestream (and initial) conditions

$$k/U_{\text{ref}}^2 \ll 1 , R/\nu_{\text{ref}} < 1 . \tag{12}$$

In general k_∞ is determined by the level of freestream turbulence, otherwise a value of $k/U_{\text{ref}}^2 = 10^{-6}$ is used. R_∞ is kept below the level of the molecular viscosity.

The $k - R$ turbulence model, with $\mathcal{D} \equiv 0$ in Eq.(7), was included in the USA Reynolds-averaged Navier-Stokes multi-block structured grid flow solver, which is an up to 3rd order accurate solver based on an upwind TVD scheme for the convection terms within a finite volume framework [9].

Délery Transonic Bump Flow (ETMA TC5)

In this case a Mach 0.65 flow enters a 2-dimensional channel comprising of a flat upper wall and a lower surface which includes a bump-like profile protruding from the otherwise flat wall (see Fig. 2). A transonic shock-wave forms upstream of the bump trailing edge and its interaction with the boundary layer induces a separated flow region. Experimental data for this case were taken by Délery and his associates [10]. In this experiment the shock location was controlled by an adjustable throat downstream of the bump. Since no geometrical details of this throat are provided, computors are instructed to adjust the downstream pressure in order to replicate numerically the experimental shock location. Figures 3(a,b) show comparisons of predictions with experimental data of lower and upper wall pressure profiles. The pressure plateau in the separated flow region on the lower wall is well predicted by the model. Velocity, turbulence kinetic energy and Reynolds stress profiles over the lower wall at five streamwise locations are shown in Figs 4(a-e). The extent of the separation bubble is overpredicted by the model, as seen in Figs. 4(d,e). Fig. 5 shows Mach contours; the experimentally observed λ-shock is well predicted by the model.

The computations were done on a 150×120 grid with 14 points inside the viscous sublayers, first point being located at $y^+ = 0.1$ (see Fig. 2.) The computational domain extended from $x = -90$ to $x = 610$ mm. Results similar to those shown here were reported by Lien and Leschziner [11] using two versions of the $k - \epsilon$ turbulence model as well as a full Reynolds stress model.

CONCLUSIONS

This paper analysed the effect of the extra diffusion terms which arise from deriving a transport equation for the undamped eddy viscosity, k^2/ϵ, from the standard $k - \epsilon$ turbulence model. Based on evaluation of the diffusion velocities it was shown that a selective utilisation of the cross-diffusion term, in addition to the diffusion term already used in the original model, may improve predictive capability of boundary layer outer regions and free shear flows. While this potential improvement remains subject to future investigation, the original model was used here to compute the Délery transonic flow over a bump, involving shock/boundary layer interactions and separated flow zones. The model's performance was good in predicting the λ-shock and the pressure distributions on both lower and upper walls, and fair in predicting detailed flow profiles.

REFERENCES

[1] Jones, W. P., Launder, B. E.: "The Prediction of Laminarization with a Two-Equation Model of Turbulence", *International Journal of Heat and Mass Transfer*, **15** (1972), pp. 301–314.

[2] Launder, B. E., Sharma, B. I.: "Application of the Energy-Dissipation Model of Turbulence to the Calculation of Flow Near a Spinning Disc", *letters in Heat and Mass Transfer*, **1** (1974), pp. 131–138.

[3] Lam, C. K. G., Bremhorst, K. A.: "Modified Form of $k - \epsilon$ Model for Predicting Wall Turbulence", *ASME Journal of Fluids Engineering*, **103** (1981), pp. 456–460.

[4] Wilcox, D. C.: "Reassessment of the Scale Determining Equation for Advanced Turbulence Models", *AIAA Journal*, **26** (1988), pp. 1299–1310.

[5] Baldwin, B. S., Barth, T. J.: "A One-Equation Turbulence Transport Model for High Reynolds Number Wall-Bounded Flows", NASA TM 102847 (1990).

[6] Goldberg, U. C.: "Toward a Pointwise Turbulence Model for Wall-Bounded and Free Shear Flows", *ASME J. Fluids Eng.*, **116** (1994), pp. 72–76.

[7] Wilcox, D. C.: *Turbulence Modeling for CFD*, DCW Industries, Inc., La Cañada, California (1993).

[8] Myong, H. K., Kasagi, N.: "Prediction of Anisotropy of the Near-Wall Turbulence With an Anisotropic Low-Reynolds-Number $k - \epsilon$ Turbulence Model", *ASME Journal of Fluids Engineering*, **112** (1990), pp. 521–524.

[9] Chakravarthy, S. R., Szema, K.-Y., Haney, J. W.: "Unified 'Nose-to-Tail' Computational Method for Hypersonic Vehicle Applications", AIAA Paper 88-2564 (1988).

[10] Délery, J. M.: "Experimental Investigation of Turbulence Properties in Transonic Shock-wave/Bo-undary-Layer Interactions", *AIAA Journal*, **21** (1983), pp. 180–185.

[11] Lien, F.-S., Leschziner, M. A.: "A Pressure-Velocity Solution Strategy for Compressible Flow and Its Application to Shock/Boundary-Layer Interaction Using Second-Moment Turbulence Closure", *ASME Journal of Fluids Engineering*, **115**, No. 4 (1993), pp. 717–725.

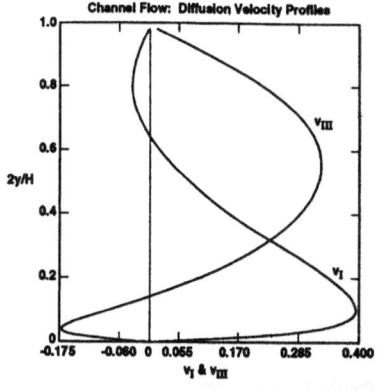

Channel Flow: Diffusion Velocity Profiles

2y/H

v_{III}

v_I

v_I & v_{III}

1. Channel flow: diffusion velocity profiles

SC.1435E.011095

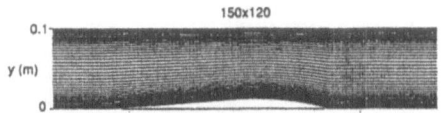

150x120

y (m)

2. Transonic 2D bump flow: computational grid

SC.54333.011195

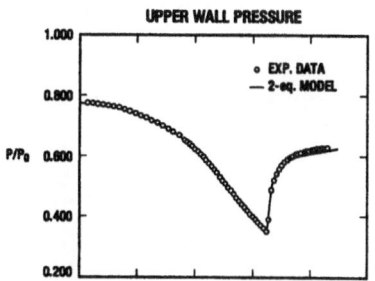

UPPER WALL PRESSURE

o EXP. DATA
— 2-eq. MODEL

P/P_0

3 (a) Mach 0.65 2D bump flow: upper wall pressure profile

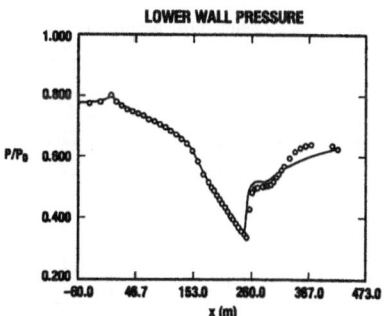

LOWER WALL PRESSURE

P/P_0

x (m)

3 (b) Mach 0.65 2D bump flow: lower wall pressure profile

SC.1436E.011195

409

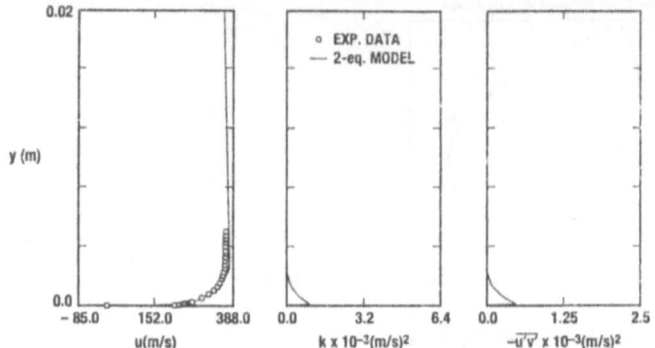

4 (a) Mach 0.65 2D bump flow: profiles at $x = 232$ mm

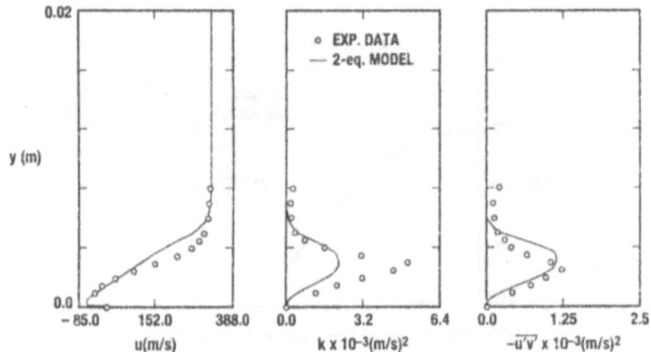

4 (b) Mach 0.65 2D bump flow: profiles at $x = 270$ mm

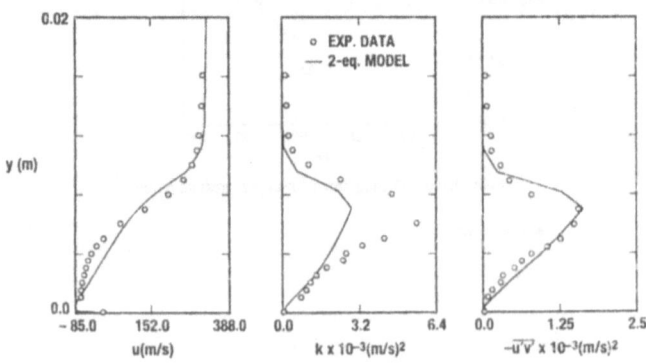

4 (c) Mach 0.65 2D bump flow: profiles at $x = 290$ mm

410

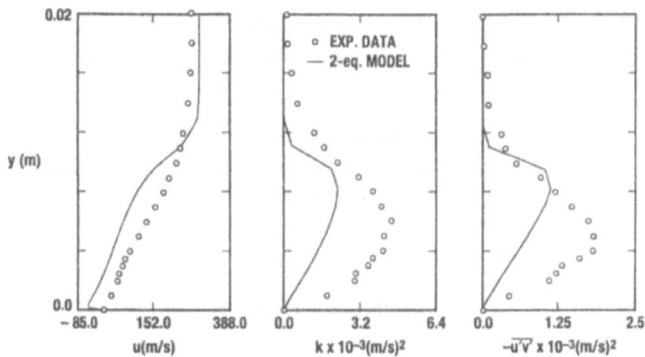

4 (d) Mach 0.65 2D bump flow: profiles at $x = 330$ mm

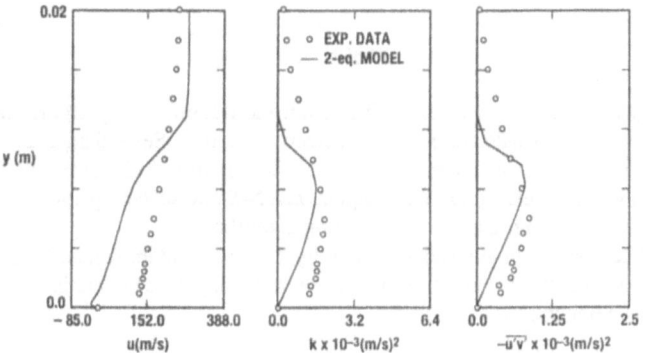

4 (e) Mach 0.65 2D bump flow: profiles at $x = 380$ mm

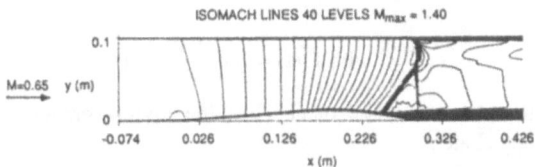

ISOMACH LINES 40 LEVELS $M_{max} = 1.40$

5. Mach 0.65 2D bump flow: Mach contours predicted by the 2-Eq. model

411

EXTENSION OF AN INCOMPRESSIBLE ALGORITHM FOR COMPRESSIBLE FLOW CALCULATIONS : VALIDATION ON A TRANSSONIC FLOW IN A BUMP.

L. Léal De Sousa, J. Duplex, A. Caruso.
Electricité De France, Direction Des Etudes Et Recherches,
6 Quai Watier, 78401 Chatou Cedex, France.

1 - INTRODUCTION

The computation of a transsonic flow in a Bump has been carried out with the Finite Element Code N3S [1] (developped at EDF). The particularitie of this problem is the presence of a shock structure so called Lambda shock. The shock is curved due to viscous sublayer effects. Here we try to assess the capability of the standard k-ε model to capture such interaction.

2- DESCRIPTION OF THE MODELS

The following numerical procedure derives from the fractional step algorithm proposed by Chorin [2] which already takes into account density variations. Several adjustements have been implemented in order to account for the specific compressibily terms in Navier-Stokes equations, and will be presented in this chapter. The N-S and scalar equations are solved in the same way, using a quasi-unsteady time marching algorithm.

Turbulence effects are taken into account either by a coupled and implicited k-ε model written in terms of time increments, or during the advection step along the characteristic curves. The later will be described beyond.

2 - 1 BASIC EQUATIONS

Instantaneous equations are ensemble averaged, written in terms of Favre average defined as (The ~ symbol is relative to the Favre averaged quantities while the - symbol is relative to the Reynolds average) :

$$\tilde{G} = \frac{\overline{\rho\,G}}{\overline{\rho}} \text{ with } \overline{G}(t) = \lim_{N \to \infty} \frac{1}{N} \sum_{n=1}^{N} G_n(t) \tag{1}$$

where G(t) is the instantaneous value of G. Thus, the Favre-averaged equations for velocity vector $\tilde{u}$ and enthalpy $\tilde{H}$ are :

Mass conservation :
$$\frac{\partial \overline{\rho}}{\partial t} + \text{div}(\overline{\rho}\tilde{u}) = 0. \tag{2}$$

Mean velocity equation :
$$\overline{\rho}(\frac{\partial \tilde{u}}{\partial t} + \tilde{u}\nabla\tilde{u}) = -\nabla\overline{p} - \text{div}(\overline{\tau}). \tag{3}$$

412

where

$$\bar{\tau} = (\mu + \mu_t).(\nabla\tilde{u} + \nabla'\tilde{u} - \frac{2}{3}\nabla\tilde{u}.I) - \frac{2}{3}\bar{\rho}\tilde{k} .$$ (4)

Enthalpy equation :

$$\bar{\rho}(\frac{\partial\tilde{H}}{\partial t} + \tilde{u}\nabla\tilde{H}) = div((\frac{\lambda}{C_p} + \frac{\mu_t}{C_p\sigma_t})\tilde{H}) + \frac{\partial\bar{p}}{\partial t} + \tilde{u}\nabla\bar{p} + S_H$$ (5)

where μ_t and σ_t are respectively the turbulent viscosity and turbulent Prandtl number ; H is the enthalpy per unit of mass.

2 - 2 TURBULENCE MODELS

The standard k-ε model is used for turbulence modelling. The transport equations of $\tilde{k}$ (mean turbulent kinetic energy) and $\tilde{\varepsilon}$ (its mean dissipation rate) are :

Kinetic energy equation :

$$\bar{\rho}(\frac{\partial\tilde{k}}{\partial t} + \tilde{u}\nabla\tilde{k}) = div((\mu + \frac{\mu_t}{\sigma_k})\nabla\tilde{k}) + \tilde{P} + \tilde{G} - \bar{\rho}\tilde{\varepsilon} .$$ (6)

Dissipation rate equation :

$$\bar{\rho}\frac{\partial\tilde{\varepsilon}}{\partial t} + \tilde{u}\nabla\tilde{\varepsilon} = div((\mu + \frac{\mu_t}{\sigma_\varepsilon})\nabla\tilde{\varepsilon}) + c_{\varepsilon_1}\frac{\tilde{\varepsilon}}{k}(\tilde{P} + \tilde{G}) + c_{\varepsilon_2}\bar{\rho}\frac{\tilde{\varepsilon}^2}{k}$$ (7)

where $\tilde{P}$ is the production rate defined as :

$$\tilde{P} = 2\mu_t.Tr\left[(\nabla\tilde{u} + \nabla'\tilde{u})^2\right] - \frac{2}{3}\bar{\rho}\tilde{k}.div(\tilde{u}) - \frac{2}{3}\mu_t.\left[div(\tilde{u})\right]^2$$ (8)

and $\tilde{G}$ the contribution of buoyancy :

and the buoyancy term

$$\tilde{G} = \frac{\mu_t}{\sigma_t}\nabla\bar{p}.\nabla\bar{\rho} .$$ (9)

Turbulent viscosity :

$$\mu_t = C_\mu.\bar{\rho}\frac{\tilde{k}^2}{\tilde{\varepsilon}} .$$ (10)

The constants of the model are given below (Launder and Spalding) :

σ_t	σ_ε	σ_k	C_μ	$C_{\varepsilon 1}$	$C_{\varepsilon 2}$
0.7	1.3	1.	0.09	1.44	1.92

For boundary conditions we use the so called wall functions that modelize the viscous sublayer near the wall. Consequently the mesh boundary is supposed to be at distance y from the wall where the velocity profile is known. In practice, we use the well known Reichardt law :

$$u^+ = 2.5\ln(1 + 0.4y^+) + 7.8(1 - \exp(-\frac{y^+}{11}) - \frac{y^+}{11}\exp(-0.33))$$ (11)

where :

$$u^+ = \frac{u_{wall}}{u_*}$$ (12)

413

$$y^+ = y \frac{u_*}{v} \tag{13}$$

u* is the friction velocity linked to the friction stress σ* given by :

$$\sigma_* = -\rho(u_*)^2 \tag{14}$$

y^+ and u* are determined iteratively at each wall point for each time step. The corresponding boundary conditions for $\tilde{k}$ and $\tilde{\varepsilon}$ are :

$$k = \frac{u_*^2}{\sqrt{C_\mu}} \text{ and } \varepsilon = \frac{|u_*|^3}{Ky} \,. \tag{15}$$

3 - NUMERICAL TECHNIQUES

Transport-diffusion operator of equations (3) to (7) are solved using a quasi-unsteady algorithm where time discretization uses a first order scheme [3],[4] :

$$\overline{\rho}(\frac{G^{n+1} - G^n}{\delta t} + \tilde{u}\nabla G^n) = \text{div}(K_G.G^{n+1}) + S_G \tag{16}$$

where δt is the time step, G^n and G^{n+1} are the values of G at the n th and (n+1) th time step. The rates of change for Favre averaged velocity, k, ε and scalar variables, after the advection step, are completed explicitly (explicit pressure gradient, explicit diffusion and source terms). In order to avoid instability, this explicit treatment is completed by an implicit integration of diffusion and source terms for the increments for all variables ; for steady flows computed as limit of a transient, all increments tend towards zero with convergence in time ; thus splitting approximations have little influence and solutions are fairly independent of the chosen time step.
The solution of (16) is divided into two parts :
An advection step, solving for the velocity components U_x, U_y, the enthalpy H, k and ε :

$$\overline{\rho}(\frac{\hat{G} - G^n}{\delta t} + \tilde{u}^n\nabla\hat{G}) = 0 \,. \tag{17}$$

The solution $\hat{G}$ is obtained using the characteristics methods with quadratic interpolation allowing to minimise numerical diffusion : each particle trajectory passing by a node at time t_{n+1} is computed backwards in time, using a Runge-Kutta method, to time t_n where the transported quantities (U,H,k,ε) are interpolated on the finite element mesh:

$$\hat{G}(\vec{x}) = G^n(\vec{x}_c) \text{ and } \vec{x}_c = \vec{x} + \int_{dt} \tilde{u}^n(\chi(t),t)dt \,. \tag{18}$$

$\chi(t)$ is the characteristic curve.
To improve the stability of this step, the sources terms of (6) (7) can be taken into account on the right hand side of equation (16). For these two variables the advection contribution is :

$$\hat{G}(\vec{x}) = G^n(\vec{x}_c) + \int_{\delta t} S_G^n(\chi(t),t) \, dt \,. \tag{19}$$

A diffusion step (implicit scheme) :

414

$$\overline{\rho}\frac{G^{n+1} - \hat{G}}{\delta t} = div(K_G \cdot \mathbb{G}^{n+1}) + S_G^n .$$ (20)

Once the advection step done, the set of equations (3) to (6) becomes :

$$\overline{\rho}\frac{\tilde{u}^{n+1} - \hat{u}}{\delta t} = -\nabla\overline{p}^{n+1} + div((\mu + \mu_t) . \nabla\tilde{u}^{n+1}) + S_u$$ (21)

$$\overline{\rho}\frac{\tilde{k}^{n+1} - \hat{k}}{\delta t} = div((\mu + \frac{\mu_t}{\sigma_k})\nabla\tilde{k}^{n+1}) + S_k$$ (22)

$$\overline{\rho}\frac{\tilde{\epsilon}^{n+1} - \hat{\epsilon}}{\delta t} = div((\mu + \frac{\mu_t}{\sigma_k})\nabla\tilde{\epsilon}^{n+1}) + S_\epsilon$$ (23)

$$\overline{\rho}\frac{\tilde{H}^{n+1} - \hat{H}}{\delta t} = div((\frac{\lambda}{C_p} + \frac{\mu_t}{C_p\sigma_t})\tilde{H}^{n+1}) + \theta.\tilde{u}^n\nabla\overline{p}^n + S_H .$$ (24)

Stokes problem (coupling velocity and pressure) :
For the computation of the velocity a third step is required in order to satisfy the continuity condition (2). For this step the density will remain explicit at time t_n but density time variations are taken into account by linearisation :

$$\frac{\partial\rho}{\partial t} = \left[\frac{\partial\rho}{\partial p}\right]_h . \left[\frac{\partial p}{\partial t}\right] + \left[\frac{\partial\rho}{\partial H}\right]_p . \left[\frac{\partial H}{\partial t}\right] .$$ (25)

The time derivative of the enthalpy is given on equation (16) in which $\theta=0$ (prediction step) and will be updated after the computation of P and U at time t_{n+1} with $\theta=1$. Moreover, a state law (for instance the perfect gas law) enables us to close the system and to compute analytically the partial derivatives (density through the pressure and through the enthalpy). Finally the continuity equation reads as :

$$\frac{1}{c^2}\frac{\delta p}{\delta t} + div(\overline{\rho}\tilde{u}^{n+1}) = f_H^n$$ (26)

where f_h is an explicit contribution of the enthalpy after diffusion step and "c" the sound speed :

$$\frac{1}{c^2} = \left[\left(\frac{\partial\rho}{\partial P}\right)_H^n + \frac{1}{\rho}\left(\frac{\partial\rho}{\partial H}\right)_P^n\right] .$$ (27)

4 - NUMERICAL RESULTS

At the inlet we imposed a reservoir condition : $P_0 = 96000$ Pa , $T_0 = 300$ K
At the outlet we imposed a constant pressure level : $P = 59500$ Pa .
The mesh is composed with 17 000 elements and 34000 nodes P2. The smallest mesh size (near the wall) is about 0.001 m (transversal direction). The P1-Iso P2 finite element has been used for space discretization. In order to accelerate the temporal convergence, we have employed the local time step technique.

415

The results show a shock pattern whose foot is located at x = 0.27. The shock is slightly curved but not enough. The lower wall pressure profile shows that the turbulent model, may be the modelisation of the viscous sublayer, is not capable of capturing the Lambda shape.

Turbulent kinetic energy profiles as well as Reynolds stress figures indicate an underestimation of the turbulent production near the wall.

The streamwise velocity level seems too high. On the vicinity of the shock location, its maximum value is about 400 m/s, whereas it should worth 300 m/s. We question the outlet boundary condition that may be too weak in comparison with the experimental data.

Though, the overall behaviour of all experimental profiles are quietly well captured even if they are underestimated. Perhaps it would be interesting to perform a low Reynolds computation on a more refined mesh in order to account for the viscous sublayer effects in a better way.

5 - REFERENCES

[1]   Pot G., Gregoire J.-P., Léal De Sousa L., Souffez Y. : "Improvement of Finite Element Algorithms For Industrial CFD Code N3S", Proceedings of VIII International Conference on Finite Elements in Fluids. New Trends and applications, Barcelone, Spain 1993.

[2]   Chorin A.J : "Numerical solutions of the Navier-Stokes equations", Math. Comp. 22, 745, 1968.

[3]   A. Caruso, N. Mechitoua : "Computations of Turbulent Reactive or Compressible flows (using Finite Element Method)", Proceedings of VIII International Conference on Finite Elements in Fluids", Barcelone, Spain 1993.

[4]   Baron F., Caruso A., Duplex J., Lefevre L. : "Extension of Incompressible Algorithm to Compressible Flows : Validation on a governing valve Mock-up", Proceedings of IFAIF Conference, Prague 1993.

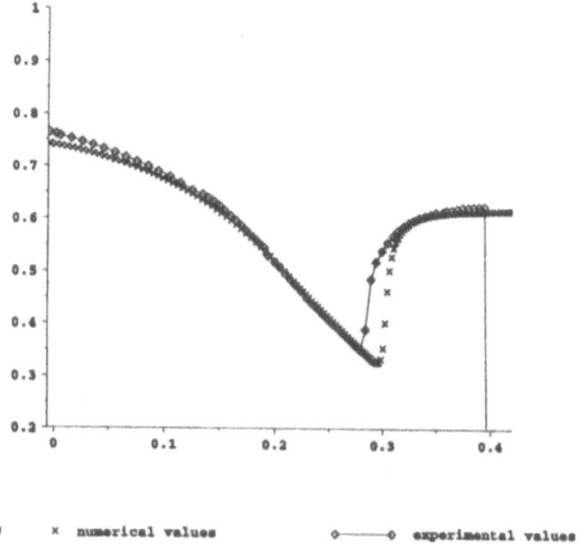

FIG. 1   Lower wall pressure

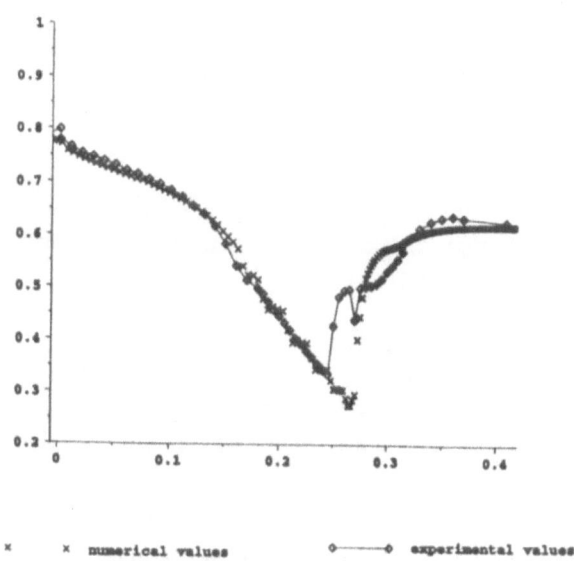

FIG. 2   Upper wall pressure

417

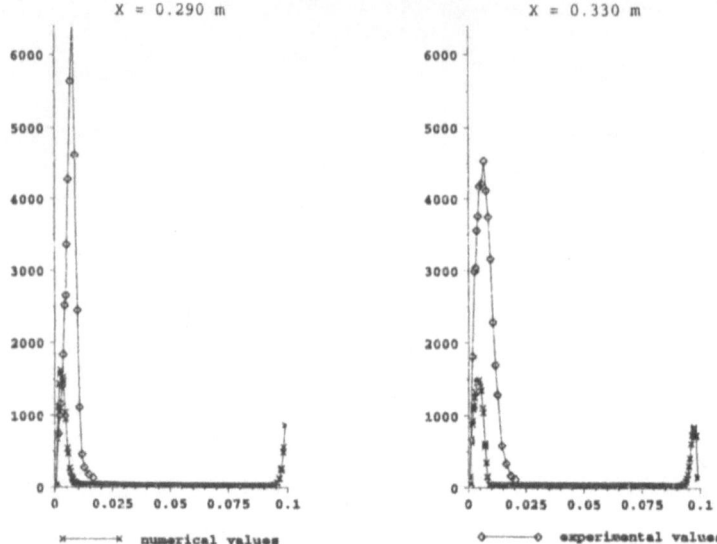

FIG. 3   Turbulent Kinetic Energy

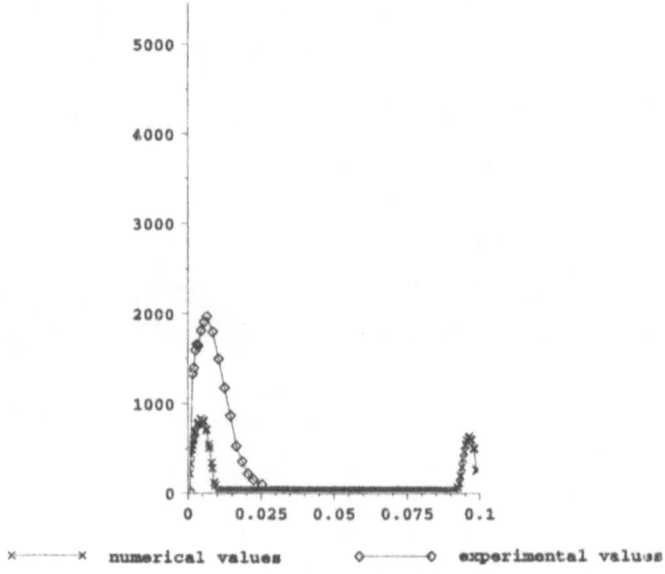

FIG. 4   Turbulent Kinetic Energy : x = 0.380

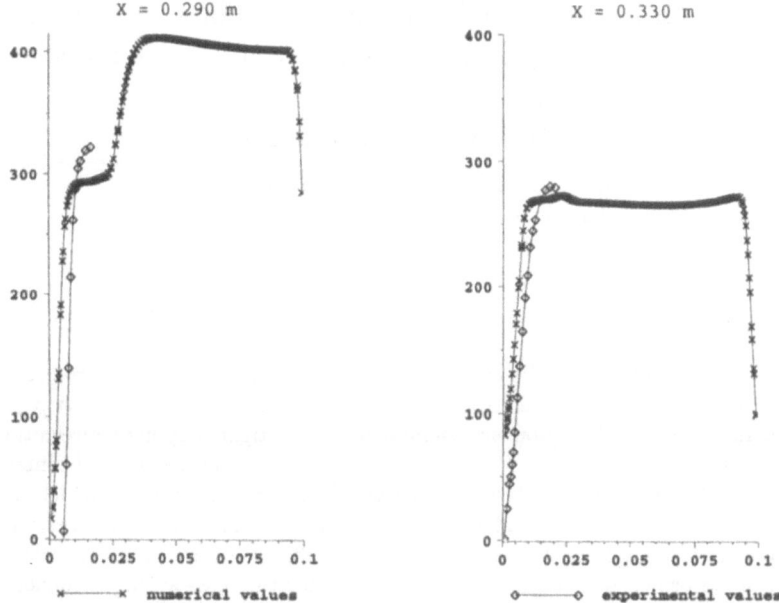

FIG. 5   Streamwise Velocity

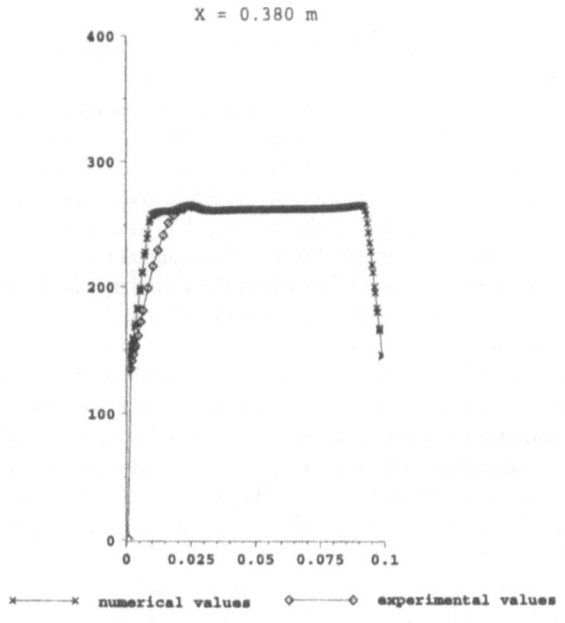

FIG. 6   Streamwise Velocity

Computation of a shock wave boundary layer interaction in a nozzle with different anisotropic tubulence models

H. Loyau,D. Vandromme

LMFN - CORIA URA 230 - INSA de ROUEN
Place E. Blondel, 76130 Mont-Saint-Aignan

SUMMARY

This paper is devoted to the calculation of test case TC5 of ETMA project, corresponding to the two-dimensional transonic bump geometry. For this case, strong shock wave/boundary layer interaction is predominant and the tight coupling between the incoming boundary layer and the sudden change of anisotropy of turbulence in the interaction region makes this experimental test case a severe challenge for turbulence models. For this purpose, three different models allowing to take into account anisotropic beaviour near wall are compared together with Low-Reynolds number $k - \epsilon$ model. These calculations have been performed with a 2D solver developped for the mass-averaged Navier-Stokes equations in association with the turbulence models under study.

Introduction

The eddy viscosity concept, originally due to Boussinesq, is based on the assumption that the turbulent stresses depend linearly on the strain rate of the mean motion. In such formulation, the eddy viscosity μ_t is assumed proportional to the mean density, to a turbulent characteristic velocity scale and to a turbulent lenght scale. Depending on the choice of these scales, various types of eddy viscosity models can be found. Such models whether based on purely local, algebraic prescription or involving transport equations for scalar turbulence quantities like $k - \epsilon$ model, are in fact tuned to yield the required level of the shear stress (the only stress of importance) in two-dimensional thin shear flow. Therefore the interactions governing the regions of flow where the turbulence is strongly anisotropic, due to the highly unequal level of generation rate of different Reynolds stresses, cannot be described by turbulence models based on eddy viscosity concept. On the other hand, more general modelling offered by second order closure in which transport equations are solved for each Reynolds stress is difficult to implement. These purposes have motivated the present suggestion of a anisotropic correction to the eddy viscosity, analogous to the Algebraic Reynolds Stress Model developped by Rodi [7], but without the solution of the whole non-linear set of equations, in which the eddy viscosity is not scalar but assimilated to a tensor quantity. In the present work, the proposed anisotropic correction and two existing non-linear $k - \epsilon$ models of Pope [6] and Saffman [8] are compared together with the baseline Low-Reynolds number $k - \epsilon$ model of Jones and Launder [2]. The calculations are performed on a flow configuration including shock wave/boundary layer interaction with extended recirculation region studied by Delery et al. [1].

The governing equations

The formulation of conservation of mass, momentum and energy, together with the physical properties for 2D compressible viscous fluid is expressed in the following form:

$$\frac{\partial U}{\partial t} + \frac{\partial F}{\partial x} + \frac{\partial G}{\partial y} = S_o.$$

where U the conservative variable vector is given by:

$$U = [\rho, \rho u, \rho v, \rho E].$$

and F respectively G, the convective/diffusive fluxes in x, respectively y direction by:

$$F = \begin{bmatrix} \rho u \\ \rho u^2 + p^* + \sigma_x \\ \rho u v + \tau_{xy} \\ (\rho E + p^* + \sigma_x)u + \tau_{xy}v - q_x \end{bmatrix} \quad G = \begin{bmatrix} \rho v \\ \rho u v + \tau_{xy} \\ \rho v^2 + p^* + \tau_{xy} \\ (\rho E + p^* + \sigma_y)v + \tau_{xy}u - q_y \end{bmatrix}.$$

In this formulation, ρ denotes the averaged density, u and v are the velocity components along x, and y directions, E is the mass-averaged total energy, p^* the effective pressure, σ_x and σ_y the deviatoric normal stresses and τ_{xy} the shear stress being given, according to the Stokes hypothesis, by:

$$\sigma_x = 2\mu\frac{\partial u}{\partial x} - \frac{2}{3}\mu\left(\frac{\partial u}{\partial x} + \frac{\partial v}{\partial y}\right), \quad \sigma_y = 2\mu\frac{\partial v}{\partial y} - \frac{2}{3}\mu\left(\frac{\partial u}{\partial x} + \frac{\partial v}{\partial y}\right), \quad \tau_{xy} = \mu\left(\frac{\partial u}{\partial y} + \frac{\partial v}{\partial x}\right).$$

where μ the molecular viscosity, is expressed as a function of temperature according to the Sutherland's law. The heat flux is expressed following the Fourier's law

$$q = -\lambda \nabla T = -\frac{C_p \mu}{p_r}\nabla T.$$

where P_r the Prandtl number is assumed to be equal to 0.72.

Turbulence modeling

● Baseline model

The baseline model of this study is the Low-Reynolds number $k - \epsilon$ model proposed by Jones and Launder [2]. This model base on the Boussinesq eddy viscosity concept, relates linearly the Reynolds stresses to the mean flow quantities:

$$-\rho \widetilde{u''_i u''_j} + \frac{2}{3}\rho k\, \delta_{ij} = \mu_t\left(\frac{\partial u_i}{\partial x_j} + \frac{\partial u_j}{\partial x_i} - \frac{2}{3}\frac{\partial u_l}{\partial x_l}\delta_{ij}\right).$$

and the two-supplementary equations of turbulent kinetic energy and its dissipation rate are given by:

$$\frac{\partial \rho k}{\partial t} + \nabla \rho k u + \nabla(\mu_k \nabla k) = S_k = P_k - \rho\epsilon - 2\mu\left(\frac{\partial \sqrt{k}}{\partial x_n}\right)^2.$$

$$\frac{\partial \rho \epsilon}{\partial t} + \nabla \rho e u + \nabla(\mu_\epsilon \nabla \epsilon) = S_\epsilon = C_{\epsilon 1}\frac{\epsilon}{k}P_k - C_{\epsilon 2}f_2\frac{\epsilon}{k}\rho\epsilon + \frac{2\mu\mu_t}{\rho}\left(\frac{\partial^2 U_t}{\partial x_n^2}\right)^2.$$

where

$$P_k = -\rho \widetilde{u''_i u''_j} \frac{\partial u_i}{\partial x_j}, \qquad \mu_k = \mu + \frac{\mu_t}{\sigma_k}, \qquad \mu_\epsilon = \mu + \frac{\mu_t}{\sigma_\epsilon}, \qquad \mu_t = C_\mu f_\mu \mu R_t.$$

$$R_t = \frac{\rho k^2}{\mu \epsilon}, \qquad f_2 = 1 - 0.3 \exp(-R_t^2), \qquad f_\mu = \exp\left(\frac{-2.5}{1 + \frac{R_t}{50}}\right).$$

with the following set of constants:

$$C_\mu = 0.09, \qquad C_{\epsilon 1} = 1.45, \qquad C_{\epsilon 2} = 1.92, \qquad \sigma_k = 1, \qquad \sigma_\epsilon = 1.3.$$

With this model, the singularity when approching the wall, since at the wall $k \longrightarrow 0$ while ϵ tends toward a constant value, is removed by splitting the dissipation into an isotropic and an anisotropic part as suggested by Jones and Launder together with the use of near wall corrective terms.

● Non linear $k - \epsilon$ model of Saffman

This model derived from the original one, initially promoted by P.G. Saffman [8] for $k - \omega^2$ model, proposed a correction on the Boussinesq equality assuming that the Reynolds stresses are not necessarily align with the mean rate of strain.

By considering that the mean rate of strain tensor and the Reynolds stress tensor do not require the same relaxation time, Saffman [8] developed at first a relaxation stress model. Afterwards, this model have been adapted by D.C. Wilcox and M.W. Rubesin [11] for compressible flows by neglecting the relaxation hypohtesis, and its adaptation to the Low-Reynolds number $k - \epsilon$ model, leads to the following closure equation:

$$-\rho \widetilde{u''_i u''_j} + \frac{2}{3} \rho k \, \delta_{ij} = \mu_t \left(\frac{\partial u_i}{\partial x_j} + \frac{\partial u_j}{\partial x_i} - \frac{2}{3} \frac{\partial u_l}{\partial x_l} \delta_{ij} \right) + A_\omega \left(S_{im} \Omega_{mj} + S_{jm} \Omega_{mi} \right).$$

$$A_\omega = \frac{8}{9} \frac{f_\mu^* \rho k}{\frac{\epsilon^2}{\beta^* k^2} + 2\left(\frac{\partial u}{\partial x}\right)^2 + \left(\frac{\partial u}{\partial y} + \frac{\partial v}{\partial x}\right)^2 + 2\left(\frac{\partial u}{\partial x}\right)^2}.$$

In which appears the incompressible form of strain tensor, together with the vorticity tensor, and a damping function issued from the Low-Reynolds $k - \epsilon$ model of Launder and Sharma [3] given by:

$$S_{ij} = \frac{1}{2}\left(\frac{\partial u_i}{\partial x_j} + \frac{\partial u_j}{\partial x_i}\right), \qquad \Omega_{ij} = \frac{1}{2}\left(\frac{\partial u_i}{\partial x_j} - \frac{\partial u_j}{\partial x_i}\right), \qquad f_\mu^* = \exp\left[\frac{-3.4}{(1 + Rt/50)^2}\right].$$

The consecutive relationship which is a rather drastic simplification of the original Reynolds stress equilibrium equation, conserves the idea of misalignment. The anisotropic corrections are afterwards introduced in the terms of momentum, energy and turbulence equations, in which the Reynolds stresses appear explicitly (ie: the turbulent fluxes are not modeled). The implementation of the model leads to the following supplementary source terms:

$$S_u = -\frac{1}{2} \frac{\partial}{\partial x}\left[A_\omega\left(\left(\frac{\partial u}{\partial y}\right)^2 - \left(\frac{\partial v}{\partial x}\right)^2\right)\right] - \frac{1}{2}\frac{\partial}{\partial y}\left[A_\omega\left(\frac{\partial u}{\partial x} - \frac{\partial v}{\partial y}\right)\left(\frac{\partial v}{\partial x} - \frac{\partial u}{\partial y}\right)\right].$$

$$S_v = -\frac{1}{2}\frac{\partial}{\partial x}\left[A_\omega\left(\frac{\partial u}{\partial x} - \frac{\partial v}{\partial y}\right)\left(\frac{\partial v}{\partial x} - \frac{\partial u}{\partial y}\right)\right] + \frac{1}{2}\frac{\partial}{\partial y}\left[A_\omega\left(\left(\frac{\partial u}{\partial y}\right)^2 - \left(\frac{\partial v}{\partial x}\right)^2\right)\right].$$

$$S_E = -\frac{1}{2}\frac{\partial}{\partial x}\left[A_\omega\left(\left(\frac{\partial u}{\partial y}\right)^2 - \left(\frac{\partial v}{\partial x}\right)^2\right)u + A_\omega\left(\frac{\partial u}{\partial x} - \frac{\partial v}{\partial y}\right)\left(\frac{\partial v}{\partial x} - \frac{\partial u}{\partial y}\right)v\right],$$

$$-\frac{1}{2}\frac{\partial}{\partial y}\left[A_\omega\left(\frac{\partial u}{\partial x} - \frac{\partial v}{\partial y}\right)\left(\frac{\partial v}{\partial x} - \frac{\partial u}{\partial y}\right)u - A_\omega\left(\left(\frac{\partial u}{\partial y}\right)^2 - \left(\frac{\partial v}{\partial x}\right)^2\right)v\right].$$

We can note that the anisotropic corrections vanished in the kinetic energy production term, letting the turbulent transport equations in the same form as the original one.

• Non linear $k - \epsilon$ model of Pope

This model developed by S.B. Pope [6] is a generalization of the effective viscosity hypothesis, in form of finite degree tensor polynomial.

By invoking the same equilibrium hypothesis as Rodi [7], the coefficients of the polynomial are related to the constants of the modeled High-Reynolds Launder, Reece and Rodi [4] model. The consecutive Reynolds stress tensor is expressed as a tensor polynomial of mean flow strain tensor and rotation tensor (incompressible form) invariants. In this study, following Pope's suggestion a simplified form of the consecutive closure equation is used in association with the baseline $k - \epsilon$ model as a non-linear $k - \epsilon$ model given by:

$$-\rho\widetilde{u''_i u''}_j + \frac{2}{3}\rho k\, \delta_{ij} = \rho C_\mu^* f_\mu \frac{k^2}{\epsilon}\left(\frac{\partial u_i}{\partial x_j} + \frac{\partial u_j}{\partial x_i} - \frac{2}{3}\frac{\partial u_l}{\partial x_l}\delta_{ij}\right) + 2gb_3\frac{k}{\epsilon}\left(S_{im}\Omega_{mj} + S_{jm}\Omega_{mi}\right).$$

$$C_\mu^* = \frac{1}{2}\frac{b_1 g}{1 - 2\left(b_3 g\frac{k}{\epsilon}\right)^2\{\Omega^2\} - \frac{2}{3}\left(b_2 g\frac{k}{\epsilon}\right)^2\{S^2\}}. \qquad g = \frac{1}{C1 + \frac{P_k}{\epsilon} - 1}.$$

$$b_1 = \frac{8}{15}. \qquad b_2 = \frac{5 - 9C_2}{11}. \qquad b_3 = \frac{7C_2 + 1}{11}. \qquad C_1 = 1.5. \qquad C_2 = 0.4.$$

Proceeding in this way, the model obtained is very closed to the one of P.G Saffman as remarked by Saffman [8] himself.

The implementation of the model is the same as the one done for the previous model in which only anisotropic corrections are introduced, the kinetic energy production term being unchanged and the corrections on each stresses given by:

$$A_u = \rho f_\mu \frac{k^2}{\epsilon}\left[(C_\mu^* - C_\mu)\left(\frac{4}{3}\frac{\partial u}{\partial x} - \frac{2}{3}\frac{\partial v}{\partial y}\right) - gb_3 C_\mu^*\frac{k}{\epsilon}\left(\left(\frac{\partial u}{\partial y}\right)^2 - \left(\frac{\partial v}{\partial x}\right)^2\right)\right].$$

$$A_v = \rho f_\mu \frac{k^2}{\epsilon}\left[(C_\mu^* - C_\mu)\left(\frac{4}{3}\frac{\partial v}{\partial y} - \frac{2}{3}\frac{\partial u}{\partial x}\right) + gb_3 C_\mu^*\frac{k}{\epsilon}\left(\left(\frac{\partial u}{\partial y}\right)^2 - \left(\frac{\partial v}{\partial x}\right)^2\right)\right].$$

$$A_{uv} = \rho f_\mu \frac{k^2}{\epsilon}\left[(C_\mu^* - C_\mu)\left(\frac{\partial u}{\partial y} + \frac{\partial v}{\partial x}\right) - gb_3 C_\mu^*\frac{k}{\epsilon}\left(\frac{\partial u}{\partial x} - \frac{\partial v}{\partial y}\right)\left(\frac{\partial v}{\partial x} - \frac{\partial u}{\partial y}\right)\right].$$

• Simplified Algebraic Reynolds Stress Model

In this original model, the correlations between fluctuations of scalar quantities and velocity fluctuations are modeled with a generalized gradient transport hypothesis:

$$-\widetilde{u''_i \Phi''} = C_\Phi \frac{k}{\epsilon} \widetilde{u''_i u''_k} \frac{\partial \tilde{\Phi}}{\partial x_k}.$$

this assumption applied to the Reynolds stresses leads, after several rearrangements, to the following closure equation:

$$-\rho \widetilde{u''_i u''_j} + \frac{2}{3} \rho k \, \delta_{ij} = \rho C^* f_\mu \frac{k}{2\epsilon} \left(\widetilde{u''_i u''_k} \frac{\partial \tilde{u}_j}{\partial x_k} + \widetilde{u''_j u''_k} \frac{\partial \tilde{u}_i}{\partial x_k} - \frac{2}{3} \widetilde{u''_l u''_n} \frac{1}{2} \left(\frac{\partial \tilde{u}_l}{\partial x_n} + \frac{\partial \tilde{u}_n}{\partial x_l} \right) \delta_{ij} \right).$$

We obtain then a system of 6 linear algebraic equation in which the turbulent viscosity is no longer scalar.

Furthermore, in case of two-dimensional flow, the set of equations can be reduced to the following form:

$$
\begin{bmatrix}
\left(1 + \frac{2}{3}\alpha\frac{\partial u}{\partial x}\right) & -\frac{\alpha}{3}\frac{\partial v}{\partial y} & \frac{\alpha}{3}\left(2\frac{\partial u}{\partial y} - \frac{\partial v}{\partial x}\right) \\
-\frac{\alpha}{3}\frac{\partial u}{\partial x} & \left(1 + \frac{2}{3}\alpha\frac{\partial v}{\partial y}\right) & \frac{\alpha}{3}\left(2\frac{\partial v}{\partial x} - \frac{\partial u}{\partial y}\right) \\
\frac{\alpha}{2}\frac{\partial v}{\partial x} & \frac{\alpha}{2}\frac{\partial u}{\partial y} & \frac{\alpha}{3}\left(2\frac{\partial v}{\partial x} - \frac{\partial \tilde{u}}{\partial y}\right)
\end{bmatrix}
\begin{bmatrix}
\widetilde{u''^2} \\[4pt]
\widetilde{v''^2} \\[4pt]
\widetilde{u''v''}
\end{bmatrix}
\begin{bmatrix}
\frac{2}{3}k \\[4pt]
\frac{2}{3}k \\[4pt]
0
\end{bmatrix}
$$

with:

$$\alpha = C^* f_\mu \frac{k}{\epsilon} = \frac{C_2}{C_1} f_\mu \frac{k}{\epsilon}.$$

the system being completed by:

$$\widetilde{w''^2} = 2k - \widetilde{u''^2} - \widetilde{v''^2}.$$

At this point it is necessary (in case of flow with shock-waves) to protect the system in order to avoid the presence of negative values for normal stresses during the computation. In this way, the Reynolds stresses have to verify realisability conditions expressed by Schumann [9]:

$$\widetilde{u''^2} \geq 0, \qquad \widetilde{v''^2} \geq 0, \qquad \widetilde{w''^2} \geq 0, \qquad (\widetilde{u''v''})^2 \leq \widetilde{u''^2}\widetilde{v''^2}.$$

The formulation of such conditions depending strongly on the linear form of the set, the system is triangulated and limitations are imposed on the new matrix coefficients. By considering the following form:

$$
\begin{pmatrix}
a_{11}^1 & a_{12}^1 & a_{13}^1 \\
 & a_{22}^2 & a_{23}^2 \\
\mathbf{O} & & a_{33}^3
\end{pmatrix}
\begin{pmatrix}
\widetilde{u''v''} \\
\widetilde{v''^2} \\
\widetilde{u''^2}
\end{pmatrix}
=
\begin{pmatrix}
0 \\
b_2^2 \\
b_3^3
\end{pmatrix}
$$

we obtained for each coefficient the following formulation:

$$a_{33}^3 = 1 + Max\left[0,\ \frac{\alpha}{3}\frac{a(2+(a+b)\alpha)(2+\alpha b)-\alpha d(2c-d)-\alpha^2 cd(a+b)}{2+\alpha(a+b)+\frac{\alpha}{3}\left(2b(2+(a+b)\alpha)-\alpha c(2d-c)\right)}\right].$$

$$b_3^3 = \frac{2}{3}k\left[1 + Max\left[0,\ \frac{\alpha}{3}\frac{b(2+(a+b)\alpha)+\alpha c(2c-d)}{2+\alpha(a+b)+\frac{\alpha}{3}\left(2b(2+(a+b)\alpha)-\alpha c(2d-c)\right)}\right]\right].$$

$$a_{22}^2 = 1 + Max\left[0,\ \frac{\alpha}{3}\frac{2b(2+(a+b)\alpha)-\alpha c(2d-c)}{2+\alpha(a+b)}\right].$$

$$a_{23}^2 = Min\left[\frac{a_{33}^3 b_2^2}{b_3^3},\ Max\left[0,\ -\frac{\alpha}{3}\frac{a(2+(a+b)\alpha)-\alpha d(2d-c)}{2+\alpha(a+b)}\right]\right].$$

$$a_{22}^2 = Max\left[a_{22}^2,\ \frac{b_2^2 a_{33}^3 - a_{23}^2 b_3^3}{2ka_{33}^3 - b_3^3}\right].\quad a_{33}^3 = Max\left[a_{33}^3,\ \frac{b_3^3(a_{22}^2 - a_{23}^2)}{2ka_{22}^2 - b_2^2}\right].\quad b_2^2 = Max\left[\frac{a_{23}^2 b_3^3}{a_{33}^3}\right].$$

$$a_{11}^1 = 1 + Max\left[0,\ \frac{\alpha}{2}(a+b)\right].\quad a_{12}^1 = \frac{\alpha}{2}c.$$

$$a_{13}^1 = Max\left[a_{13}^1,\ \left(a - \left[\frac{a_{12}^1 a_{33}^3}{a_{22}^2 b_3^3}\left(b_2^2 - \frac{b_3^3 a_{23}^2}{a_{33}^3}\right)\right] - \frac{a_{11}^1 a_{33}^3}{b_3^3}\sqrt{\frac{b_3^3}{a_{22}^2 a_{33}^3}\left(b_2^2 - \frac{b_3^3 a_{23}^2}{a_{33}^3}\right)}\right)\right].$$

$$a_{13}^1 = Min\left[a_{13}^1,\ \left(-\left[\frac{a_{12}^1 a_{33}^3}{a_{22}^2 b_3^3}\left(b_2^2 - \frac{b_3^3 a_{23}^2}{a_{33}^3}\right)\right] + \frac{a_{11}^1 a_{33}^3}{b_3^3}\sqrt{\frac{b_3^3}{a_{22}^2 a_{33}^3}\left(b_2^2 - \frac{b_3^3 a_{23}^2}{a_{33}^3}\right)}\right)\right].$$

the following notations being adopted:

$$a = \frac{\partial\tilde{u}}{\partial x}\quad ;\quad b = \frac{\partial\tilde{v}}{\partial y}\quad ;\quad c = \frac{\partial\tilde{u}}{\partial y}\quad ;\quad d = \frac{\partial\tilde{v}}{\partial x}.$$

The implementation of this model is the same as the previous ones concerning the velocity correlation except for the turbulent kinetic energy production term in wich the Reynolds stresses are now directly introduced. Furthermore the turbulent fluxes are splitted in an isotropic part modeled by the $k - \epsilon$ model and an anisotropic part modeled by a generalized gradient transport hypothesis (as in classical Algebraic Reynolds Stress Model) and treated in source terms, in which only the following anisotropic corrections are introduced:

$$A_u = \rho f_\mu\left[\left(\frac{2}{3}k - \widetilde{u''^2}\right) - C_\mu\frac{k^2}{\epsilon}\left(\frac{4}{3}\frac{\partial u}{\partial x} - \frac{2}{3}\frac{\partial v}{\partial y}\right)\right].$$

$$A_v = \rho f_\mu\left[\left(\frac{2}{3}k - \widetilde{v''^2}\right) - C_\mu\frac{k^2}{\epsilon}\left(\frac{4}{3}\frac{\partial v}{\partial y} - \frac{2}{3}\frac{\partial u}{\partial x}\right)\right],\quad A_{uv} = \rho f_\mu\left[-\widetilde{u''v''} - C_\mu\frac{k^2}{\epsilon}\left(\frac{\partial u}{\partial y} + \frac{\partial v}{\partial x}\right)\right].$$

Concerning the computation done with this model the diffusion coefficient for the kinetic energy and the dissipation rate take respectively the following values: 0.22 and 0.16.

425

Numerical method

The previous equations are solved on a computational domain of variables ξ and η (transformed coordinates of the physical domain), by the use of finite volume discretisation technique on structured mesh. The new system of equation is solved by using MacCormack[5]'s explicit-implicit finite volume method. This two-step predictor-corrector algorithm is of second-order accuracy in space and time, and the basic discretisation for the convective fluxes is modified in order to take into account the information propagation as done initially by Steger and Warming[10]. The flux splitting is made second order accurate, but is lowered to first order in shock regions. The viscous terms are centered and the source terms are integrated in the center of each control volumes in both ξ and η directional sweeps. The explicit discretisation is complemented with an implicit numerical approximation which is free from stability conditions but in order to minimize the numerical stiffness these terms are under-relaxed.

Numerical results

Concerning the 2D transonic bump studied at ONERA by Déléry and his co-workers [1], the incoming flow is a fully developped turbulent boundary layer and the reservoir conditions are $p_o = 96000 Pa$ and $T_o = 300K$ whereas the inflow Mach number is $M \approx 0.63$.

For the computation of this case, a 120×220 grid size was used, first point off walls being located at $y^+ \approx 1$. As inlet boundary conditions, the stagnation pressure and temperature are conserved, the velocity is taken parallel to the wall and a characteristic relation supplies downstream-upstream influence. At the exit boundary, the static pressure is adjusted in order to have the shock location close to its experimental one ($P_e = 60100 Pa$).

The results joined to this paper are isomach lines, wall pressure distributions, velocity, kinetic energy and Reynolds stresses profiles at $x = 0.232$, 0.270, 290, 330, 380m.

Qualitatively, as seen on the isomach lines the λ shock structure is poorly estimated by the models except for the simplified ASM model (LMFN). The comparision of prediction with experimental lower wall pressure distribution shows clearly that linear and non-linear $k - \epsilon$ models are not capable to well predict the pressure plateau in the separated region although they are in rather good accordance with the experimental data of upper wall pressure distribution. Concerning the simplified ASM model, a small pressure plateau is detected but the upstream and downstream shock feet are shiffted, and the shock location given by the model on the upper wall is not correctly predicted.

On the velocity profile, the models are in good agreement with the experimental data for the incoming Boundary layer ($x = 0.232m$)except for the simplified ASM model witch seems to be not sufficiently damped in the near wall region. Later, in the recirculation region ($x = 0.270, 0.290m$), all the models show some discrepancies with the experimental profiles and tend to under-estimate this zone. However concerning the relaxation region ($x = 0.330, 0.380m$) all models are in rather good agreement with the experimental profiles.

For the turbulent profiles, the linear and non-linear $k-\epsilon$ models do not predict correctly the peak locations at the begining of the recirculation region and only the simplified ASM model corrects this tendancy. At the end of the separation region and the begining of the relaxation zone the ASM model shows a very good tendancy against the other non linear models by predicting corrctly peak locations on the turbulent profiles, but the prediction given by this model seems to be the worst latter in the relaxation region, in wich the $k - \epsilon$ model gives the more accurate predictions.

Conclusion

Anisotropic corrections and their applications to a complex flow geometry have been detailed in the present paper. The anisotropic corrections of the non-linear $k - \epsilon$ studied seems not to be sufficient to radically improved the computation on this test case. Concerning the simplified ASM model a supplementary effort on the modeling has to be done, in order to correct the behavior of the model in the near wall region, and to improved the stability of the model during the convergence.

References

[1] DELERY J., COPY C. and REISZ J., (1980): "Analyse au vélocimètre laser bidirectionnel d'une interaction choc-couche l imite avec décollement étendu".
ONERA Raport technique, No 37:7078 AY014.

[2] JONES W.P. and LAUNDER B.E., (1972): "The calculation of Low-Reynolds-Number phenomena with a two-equation model of turbulence".
J. of Heat and Mass Transfer, vol. 15, No 3, pp. 301-314.

[3] LAUNDER B.E., SHARMA B.I., (1974): "Application of the energy-dissipation model of turbulence to the calculation of flow near a spinning disc".
Letter in Heat and Mass Transfer, vol. 1, pp. 131-138.

[4] LAUNDER B.E., REECE G.J. and RODI W., (1975): "Progress in the development of a Reynolds stress turbulence closure".
J.F.M., vol. 68, No 3, pp. 537-566.

[5] MAC CORMACK R.C., (1985): "Current status of numerical solution of the Navier-Stokes equations".
AIAA paper 85-0032.

[6] POPE S.B., (1975): "A more general effective-viscosity hypothesis".
J.F.M., vol. 72, part 2, pp. 331-340.

[7] RODI W., (1987): "Turbulence models for practical applications".
von Karman Institute for fluid dynamic, Lecture Serie 1987-06.

[8] SAFFMAN P.G., (1976): "Development of a complete model for the calculation of turbulent shear flows".
Duke Turbulence Conf., Duke University, Durham.

[9] SCHUMANN V., (1977): "Realizability of Reynolds stress turbulence model".
Phys. Fluids. vol. 20, No 5, pp. 721-725.

[10] STEGER J. and WARMING R.F., (1979): "Flux vector splitting of the inviscid gas dynamics equations with application to finite difference methods".
NASA TM-78605.

[11] WILCOX D.C. and RUBESIN M.W. (1980): "Progress in turbulence modeling for complex flow fields including effects of compressibility".
NASA Technical paper, 1517.

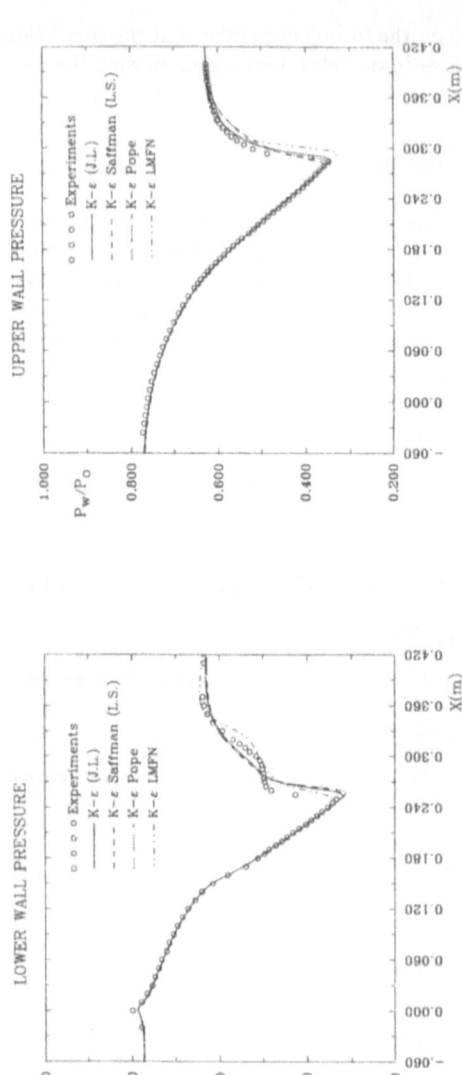

FIG. 1 - Wall pressure distributions

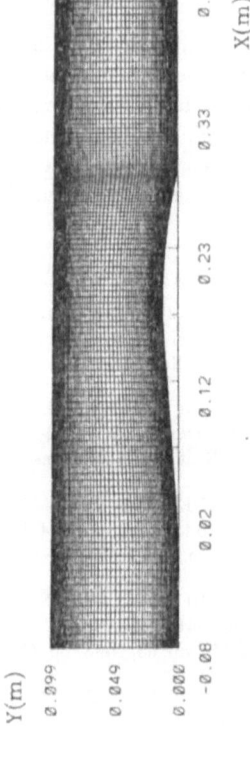

FIG. 2 - Computational grid

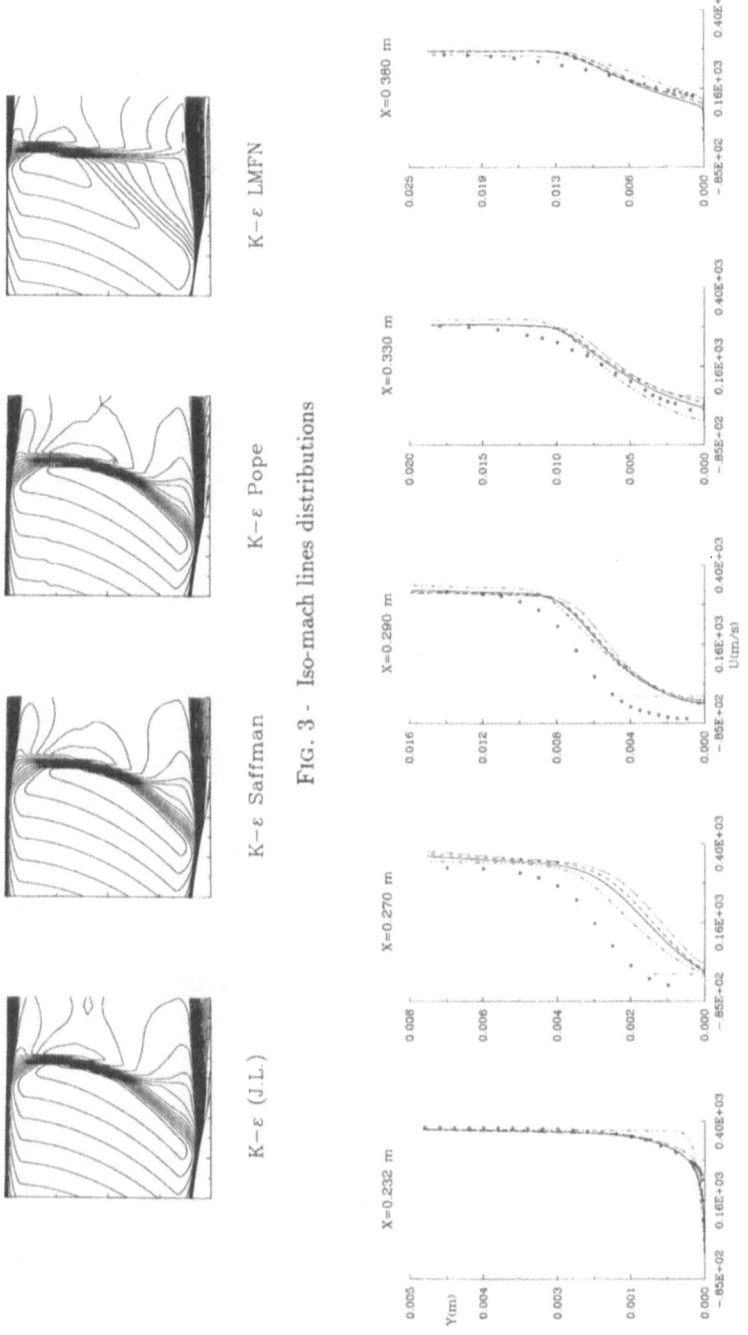

FIG. 3 - Iso-mach lines distributions

FIG. 4 - Longitudinal velocity profiles. Symbols as in figure 1.

429

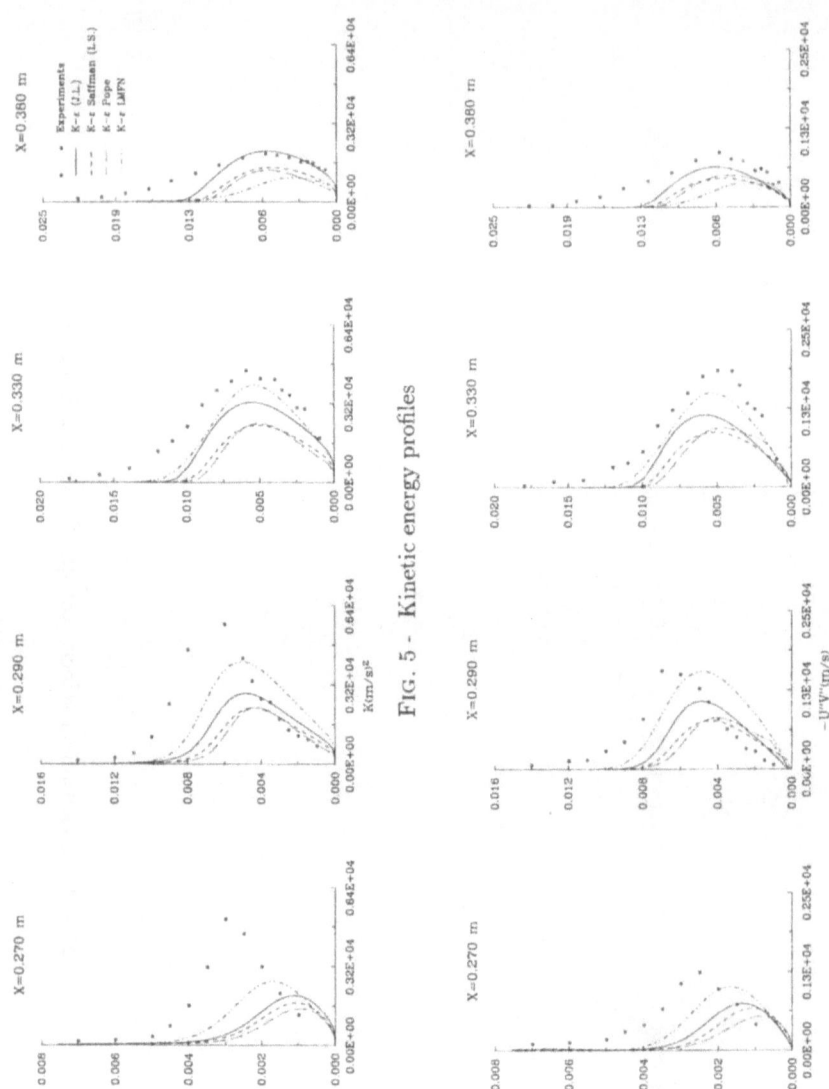

FIG. 5 - Kinetic energy profiles

FIG. 6 - Shear stress profiles. Symbols as in figure 5.

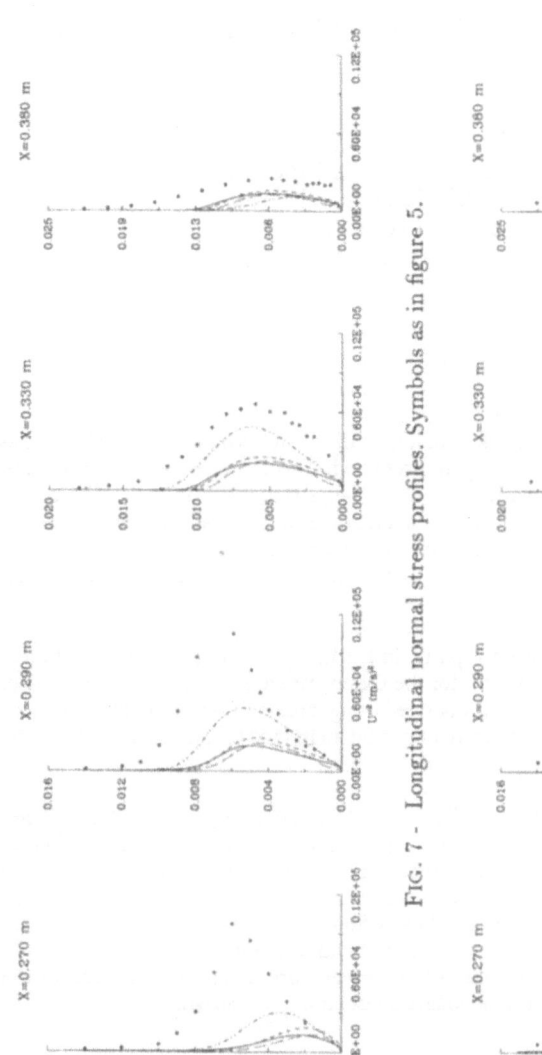

FIG. 7 - Longitudinal normal stress profiles. Symbols as in figure 5.

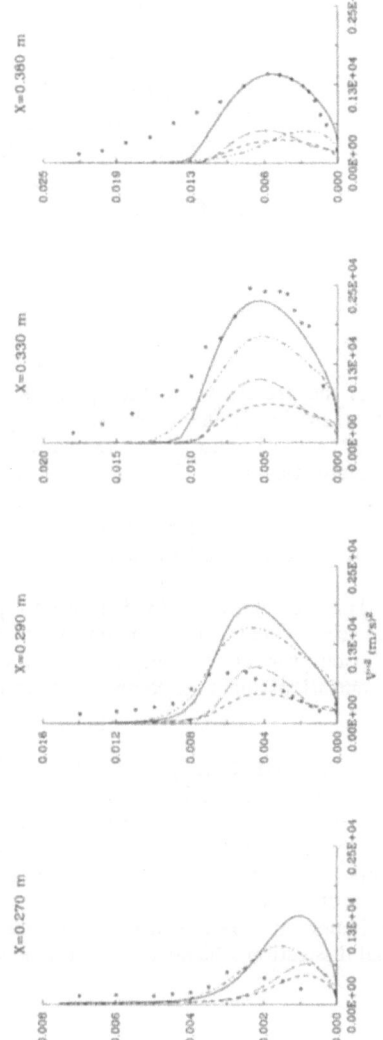

FIG. 8 - Transversal normal stress profiles. Symbols as in figure 5.

431

2-D TRANSONIC BUMP FLOW CALCULATIONS USING AN EXPLICIT FRACTIONAL-STEP METHOD

G. Simandirakis, C. Vassilopoulos,
K.C.Giannakoglou and K.D.Papailiou
Lab. of Thermal Turbomachines
National Technical University of Athens
P.O. Box 64069, 157 10 Athens, Greece.

INTRODUCTION

In the present work, the application of three different turbulence models, in the transonic flow inside the Delery bump, [1], is presented. The three models used are the algebraic Baldwin-Lomax [2] and two variants of the two-equation k-ε model; the first one is the standard k-ε model [3] and the second is a modified version where the turbulence dissipation is externally controlled in the separation region. By adjusting the balance between the production and the dissipation part in the turbulence equations inside the separation region, better results can be obtained.

The presentation is divided into two parts. In the first part, the numerical method is given in brief, while in the second one, results for the Delery case are presented and discussed. By performing runs and by comparing results obtained using three different turbulence models, useful conclusions about the comparative behaviour of the examined turbulence models can be drawn.

Code ATHENA (A Turbulent Hyperbolic Explicit Navier-Stokes Algorithm), developed in the Lab. of Thermal Turbomachines of NTUA, [4], [5], constitutes the basic numerical kernel for the present study. This is an explicit, time-marching fractional-step method which solves the unsteady Navier-Stokes equations by means of a successive application of one-dimensional operators. The application of this method in turbulent flow calculations, requires a particular treatment of the source terms in the k-ε model, which could create numerical instabilities. A second important aspect, which is addressed in the present work, is the modification of the artificial dissipation scheme, through the introduction of an ε-based sensor.

GOVERNING EQUATIONS

The conservative form of the two-dimensional, unsteady Favre-averaged Navier-Stokes equations, coupled with the standard Jones and Launder k-ε turbulence model [3], can be written in the curvilinear coordinates (ξ, η) as

$$\frac{\partial Q}{\partial t} + \frac{\partial F}{\partial \xi} + \frac{\partial G}{\partial \eta} = \frac{\partial F_\nu}{\partial \xi} + \frac{\partial G_\nu}{\partial \eta} + S . \tag{1}$$

The unknown vector Q is given by

$$Q = J \left[\rho , \rho u , \rho v , E_t , \rho k , \rho \varepsilon \right]^T$$

where

$$E_t = \rho (c_v T + \frac{1}{2} \vec{V} \cdot \vec{V} + k).$$

The Jacobian of the transformation reads

$$J = \frac{\partial(x,y)}{\partial(\xi,\eta)} = x_\xi y_\eta - x_\eta y_\xi$$

where $x_\xi, x_\eta, x_\xi, y_\eta$ are the metrics of the transformation.

The inviscid flux F is given by

$$F = J \left[\rho V^1 , \rho u V^1 + \xi_x p^* , \rho v V^1 + \xi_y p^* , [E_t + p^*] V^1 , \rho k V^1 , \rho \varepsilon V^1 \right]^T.$$

A similar expression holds for the inviscid flux G, in the η-direction. V^1 and V^2 stand for the contravariant velocity components and $p^* = p + 2/3 \rho k$ is the so-called effective pressure. On the other hand, the viscous flux F_V, in the ξ-direction, may be expressed in the form

$$F_V = J \begin{bmatrix} 0 \\ \xi_x \tau_{xx} + \xi_y \tau_{xy} \\ \xi_x \tau_{xy} + \xi_y \tau_{yy} \\ \xi_x \left[\tau_{xx} u + \tau_{xy} v + \frac{\mu_t}{Pr_k} k_x + KT_x \right] + \xi_y \left[\tau_{xy} u + \tau_{yy} v + \frac{\mu_t}{Pr_k} k_y + KT_y \right] \\ \mu_k (\xi_x k_x + \xi_y k_y) \\ \mu_\varepsilon (\xi_x \varepsilon_x + \xi_y \varepsilon_y) \end{bmatrix}$$

where the cartesian viscous stress tensor components are given by

$$\tau_{xx} = 2\mu u_x - \frac{2}{3} \mu(u_x + v_y) \quad , \quad \tau_{yy} = 2\mu v_y - \frac{2}{3} \mu(u_x + v_y) \quad , \quad \tau_{xy} = \mu(u_y + v_x).$$

When the eddy viscosity concept is applied, the following effective viscosity coefficients are defined

$$\mu = \mu_l + \mu_t \quad , \quad \mu_k = \mu_l + \frac{\mu_t}{Pr_k} \quad , \quad \mu_\varepsilon = \mu_l + \frac{\mu_t}{Pr_\varepsilon}$$

with $Pr_k = 1.0$ and $Pr_\varepsilon = 1.3$. The molecular viscosity μ_l is given by the Sutherland's law. K is the effective thermal conductivity

$$K = C_p \left[\frac{\mu_l}{Pr} + \frac{\mu_t}{Pr_t} \right] \quad , \quad Pr = 0.72 \quad , \quad Pr_t = 0.9.$$

The turbulent viscosity μ_t is obtained from the Prandtl-Kolmogorov relation

$$\mu_t = c_\mu f_\mu \frac{\rho k^2}{\varepsilon} .$$

(2)

Non-zero entries in the source term array S, exist only for the turbulence equations and these will be denoted by S_k and S_ε. Thus,

$$S = [\, 0\, ,\, 0\, ,\, 0\, ,\, 0\, ,\, S_k\, ,\, S_\varepsilon\,]^T$$

$$S_k = P_k - \rho\varepsilon - 2\mu\left(\frac{\partial\sqrt{k}}{\partial n}\right)^2$$

(3)

$$S_\varepsilon = C_{\varepsilon_1} P_k \frac{\varepsilon}{k} - C_\varepsilon f_2 \frac{\rho\varepsilon^2}{k} + \frac{2\mu_l\mu_t}{\rho}\left(\frac{\partial^2 V_t}{\partial n^2}\right)^2$$

where V_t denotes the velocity component tangential to solid walls and n is the normal to the wall direction.

The turbulence production term P_k is defined as

$$P_k = \mu_t\left[(u_y+v_x)^2 + 2(u_x^2+v_y^2) - \frac{2}{3}(u_x+v_y)^2\right] - \frac{2}{3}\rho k(u_x+v_y) .$$

The constants C_μ and $C_{\varepsilon2}$ are set equal to 0.09 and 1.92 respectively and the functions f_μ and f_2 are given by

$$f_\mu = \exp\left[\frac{-3.4}{(1+R_t/50)^2}\right] \quad , \quad f_2 = 1 - 0.3\exp(-R_t^2) \quad , \quad R_t = \frac{\rho k^2}{\mu_l\varepsilon} .$$

The $C_{\varepsilon1}$ constant bears the burden of the adjustment of the level of turbulent energy dissipation in turbulent flows. Higher $C_{\varepsilon1}$ values lead to increased dissipation and consequently to smaller local values of the turbulent kinetic energy. Therefore, by using a higher $C_{\varepsilon1}$ value, easier flow separation and a more extended separation zone occur. In the present study, the standard $C_{\varepsilon1} = 1.44$ value shifts to the higher value $C_{\varepsilon1} = 1.57$, locally within the separation region. This considerably improves the calculated flowfield, as it could be seen in the Results section.

The inlet and exit boundary conditions are imposed according to the theory of characteristics. In this particular case of a transonic flow with subsonic inlet conditions the total flow conditions and a zero flow angle are imposed at the inlet, while static pressure is imposed at the exit. All other exit flow quantities are extrapolated from the interior. Both velocity components, k and ε are set to zero along solid walls and the pressure is calculated through the normal momentum equation.

NUMERICAL PROCEDURE

The finite-difference form of the discretized equation (1) is decomposed into a sequence of multiple single-directional operators, according to the time-splitting method, [6].

434

The involved one-dimensional operators will be denoted by L. Each of them corresponds to a specific physical component of the equations. They will be superscribed by H, P or ST in order to distinguish the type of the equation part which is solved. Superscripts H, P and ST represent the inviscid part (H = Hyperbolic), the viscous part (P = Parabolic) and the source terms in the equations, respectively. Subscripts ξ and η, will be used to denote operators written along the η = const. or ξ = const. grid lines respectively.

A symbolic representation of the fractional-step algorithm, is given below,

$$Q^{n+2} = L_\xi^H L_\eta^H L_\xi^P L_\eta^P L^{ST} L^{ST} L_\eta^P L_\xi^P L_\eta^H L_\xi^H Q^n. \tag{4}$$

In this equation, the time evolution of the dependent variables array Q, is numerically obtained by sequentially applying the ensemble of single-directional operators. The second order accuracy in time is achieved by means of a double and inverse sequence of the 1-D operators. According to this scheme, the solution array is updated and recorded at the expiration of a $2\Delta t$ time interval (i.e. from n to n+2 time-level). It is worth noting that all flow quantities obtained at the end of any intermediate step have no physical meaning. The second order accuracy in space is ensured by employing the predictor-corrector MacCormack scheme [7], for the L^H and L^P operators.

Particular attention must be paid to the treatment of the L^{ST} operators, since numerical problems could appear, due to the stiffness of source-terms in the turbulence equations. Source-terms operators, representing 1-D equations of the form

$$\frac{\partial Q}{\partial t} = S \tag{5}$$

are solved using a semi-implicit scheme. Since all quantities involved in the source-term expressions (3) appear with predefined signs (for instance the turbulent energy dissipation and the low-Reynolds terms are always positive), it is possible to handle separately the positive and negative entries in the RHS of (5). If

$$S = S^+ + S^-$$

where S^+ and S^- contain non-negative and non-positive terms respectively, numerical stability is ensured by enhancing the diagonal dominality of the discretized operator. Using a Newton linearization for S^-, a delta formulation scheme for the solution of (5) is established, which reads

$$\left[I - \Delta t \left(\frac{\partial S^-}{\partial Q} \right)^{old} \right] \Delta Q = \Delta t \, (S^+ + S^-)^{old}$$

where

$$Q^{new} = Q^{old} + \Delta Q .$$

For steady state calculations, the local time-stepping technique is employed to speed up convergence. Local Δt at each grid node, is obtained by

$$\Delta t = CFL \cdot \min[\Delta t_\xi , \Delta t_\eta]$$

where CFL stands for the Courant-Friedriechs-Lewis number and

$$\Delta t_\xi = \min \left[\frac{1}{|V^1| + c\sqrt{g^{11}}} \ , \ \frac{2\rho Pr}{\gamma\mu(4g^{11} + g^{12})} \right]$$

$$\Delta t_\eta = \min \left[\frac{1}{|V^2| + c\sqrt{g^{22}}} \ , \ \frac{2\rho Pr}{\gamma\mu(4g^{22} + g^{12})} \right]$$

are time-steps corresponding to the ξ and η directions.

ARTIFICIAL DISSIPATION

Due to the centered discretization scheme used, a blend of second- and fourth-order dissipation, [8], must be explicitly added to the solution array Q, to overcome decoupling effects. At the end of a complete calculation period, the solution array takes the form

$$Q^{n+2} = Q^{n+2} + D_\xi^2 - D_\xi^4 + D_\eta^2 - D_\eta^4 \tag{6}$$

where the D terms are superscribed by either 2 or 4 to denote second or fourth order terms. As an example, the D_ξ terms read

$$D_\xi^2 = \nabla_\xi \left[\Sigma_\xi(\sigma \, J)\epsilon_\xi^{(2)} \Delta_\xi \left(\frac{Q^{n+2}}{J} \right) \right]$$

$$D_\xi^4 = \nabla_\xi \left[\Sigma_\xi(\sigma \, J)\epsilon_\xi^{(4)} \Delta_\xi \nabla_\xi \Delta_\xi \left(\frac{Q^{n+2}}{J} \right) \right] \tag{7}$$

where ∇_ξ, Δ_ξ are the backward and forward difference operators in the ξ-direction and Σ_ξ is the forward averaging operator in the same direction.

The amount of the numerical dissipation added, is controlled by means of the two coefficients

$$\epsilon_\xi^{(2)} = k_2 \Delta t \ \max_\xi(Y) \qquad , \qquad k_2 = 0.25$$

$$\epsilon_\xi^{(4)} = \max(0, k_4 \Delta t - \epsilon_\xi^{(2)}) \qquad , \qquad k_4 = 0.01 \tag{8}$$

where $\max_\xi$ expresses maximum value in the ξ-direction. Different expressions are proposed for the sensor Y appearing in (8) for the mean flow and the turbulence equations. Thus, for the mean flow equations, Y is based on local static pressure variations

$$Y = \frac{|\nabla_\xi \Delta_\xi P|}{|4\bar{P}|} \ . \tag{9}$$

On the other hand, the sensor Y used in the k-ε equations is defined as

$$Y = \frac{|\nabla_\xi \Delta_\xi \varepsilon|}{|4\bar{\varepsilon}|}$$ (10)

where overbar denotes local average in the ξ-direction.

The dual definition of Y, has been proved very advantageous, since it protects the evolution of the k-ε field from any abnormal transient behaviour of k and ε. More specifically, the system of differential equations of k and ε is prone to oscillations in regions where rapid changes in turbulence occur; this is where the sensor defined in (10) is activated.

RESULTS AND DISCUSSION

The problem examined herein, concerns the transonic flow in a 2-D channel with a characteristic lower wall configuration [1]. The computational grid, having 121x219 nodes, is shown in figure (1). The inlet total pressure was 96 kPa and the inlet total temperature was 300 K. At the exit, the static pressure was kept constant and equal to 62 kPa.

In figures (2), (3) and (4) the calculated iso-Mach contours using the Baldwin-Lomax, the standard k-ε and the version of the k-ε model with the dual $C_{\varepsilon 1}$ definition are presented. The Baldwin-Lomax model fails to provide the λ-shock patterns, which, on the contrary are well captured by means of the standard k-ε model, as shown in figure (3). Better results are obtained by increasing the $C_{\varepsilon 1}$ parameter in the separation zone, allowing thus the shock induced separation to occur easier. The use of a higher $C_{\varepsilon 1}$ value leads to a separation region with the correct length. Figure (5) shows a blow-up of the flow-field in the region of interest, i.e. close to the shock location.

Figure (6) shows the non-dimensional static pressure distribution along the lower wall of the channel, close to the shock wave- boundary layer interaction region. This figure shows more clearly the difference between results obtained through the Baldwin-Lomax model and the k-ε ones.

Figures (7) and (8) present the two Cartesian velocity components at various axial positions along the channel. Figure (9) shows the Reynolds stress profiles at these positions. At the same figures measurements [10] are also presented.

CONCLUSIONS

The transonic asymmetric Delery bump has been numerically analyzed, using three different turbulence models. The standard k-ε model with a modified constant $C_{\varepsilon 1}$ value, provides results which are in good agreement with the available measurements. It ensures an accurate representation of the shock induced separation and a correct extent of the separation zone.

REFERENCES

[1]   Delery, J.M.: "Experimental investigation of turbulence properties in transonic shock/boundary-layer interactions", AIAA Journal 21, No. 2 (1983).

[2]   Baldwin, B.S., Lomax, H.: "Thin layer approximation and algebraic model for separated turbulent flows", AIAA Paper 78-257 (1978).

[3]   Jones, W.P., Launder, B.E.: "The prediction of laminarization with a two-equation model of turbulence", Int. Journal Heat Mass Transfer 15 (1972), pp. 301-314.

[4]   Simandirakis, G.: "Numerical solutions of Navier-Stokes equations for transonic flows inside turbine bladings", Ph.D. Thesis, Athens, February 1992.

[5]   Dejean, F., Vassilopoulos, C., Simandirakis, G., Giannakoglou, K., Papailiou, K.D.: "Analysis of transonic turbomachinery flows using a 2-D explicit low-Reynolds k-ε Navier-Stokes solver', ASME Paper 94-GT-63 (1994).

[6]   Laval, P.: "Nouveaux schémas de désintégration pour la résolution des problèmes hyperboliques et paraboliques non linéar: Application aux equations d'Euler et de Navier-Stokes", Recherche Aérospatiale 4 (1983).

[7]   MacCormack, R.W.: "On the development of efficient algorithms for three-dimensional fluid flow", Recent Developments in Computational Fluid Dynamics, T.E.Tezchigar et al., ASME AMD 95 (1988), pp. 117-138.

[8]   Jameson, A., Schmidt, W., Turkel, E.: "Numerical solutions of the Euler equations by finite volume methods using Runge-Kutta time stepping schemes', AIAA Paper 81-1259 (1981).

[9]   Simandirakis, G., Vassilopoulos, C., Giannakoglou, K., Papailiou, K.D.: "Steady and unsteady two-dimensional flow calculations using an explicit fractional step algorithm", Proceedings of the Second European Computational Fluid Dynamics Conference, Stuttgart, September, 5-8 (1994).

[10]   Delery, J., Copy, C., Reisz, J..: "Analyse au vélocimètre laser bidirectionnel d'une interaction choc-couche limite turbulente avec décollement étendu", Rapport Technique n° 37/7078 AY 014, August (1980).

Acknowledgements

Part of the work was funded by the ETMA (Efficient Turbulence Models for Aeronautics) Project (BRITE-EURAM-2076/2032).

438

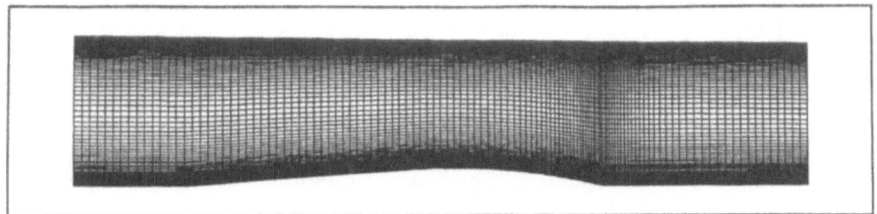

Figure 1 :   Computational grid 121x219

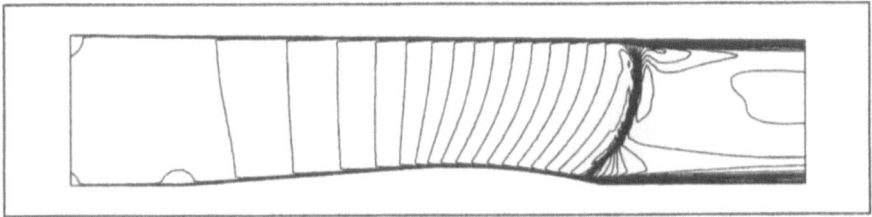

Figure 2 :   Calculated iso-Mach contours using Baldwin-
                 Lomax model

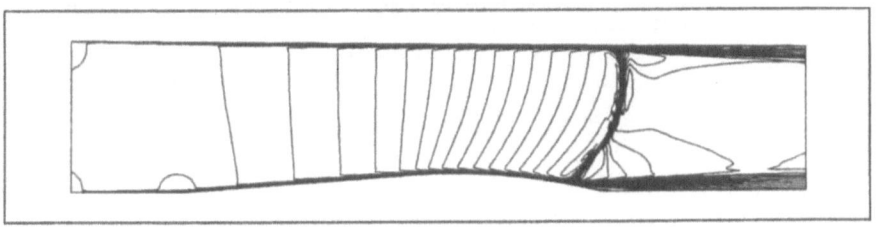

Figure 3 :   Calculated iso-Mach contours using the
                 standart k-ε model (ce1=1.44)

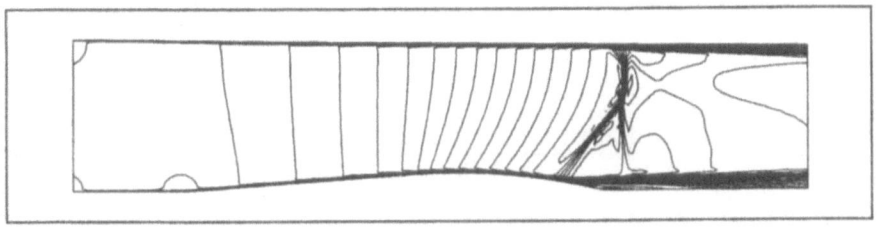

Figure 4 :   Calculated iso-Mach contours using the
                 standart k-ε model (ce1=1.44 or ce1=1.57)

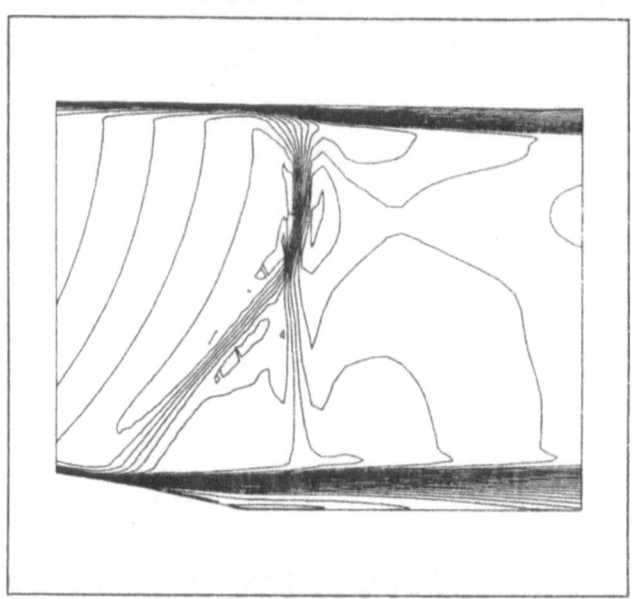

Figure 5: Blow-up of the calculated Iso-Mach contours using the modified k-ε model (ce1=1.44 or ce1=1.57), near the shock.

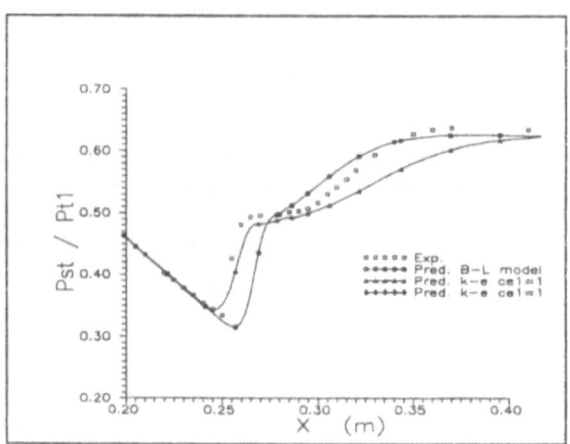

Figure 6: Non-dimentional pressure distribution along the lower wall.

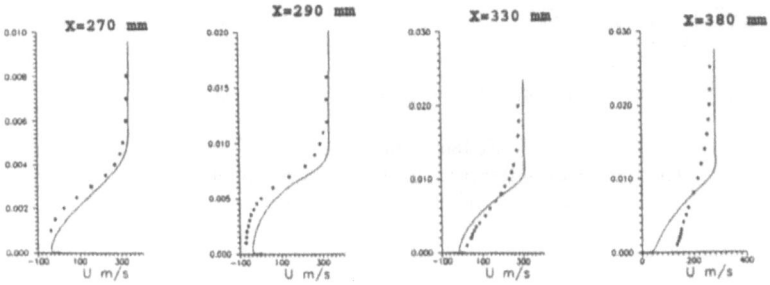

Figure 7 :   Ux velocity component at four axial
              positions. (ooo=exper., ---=calc.)

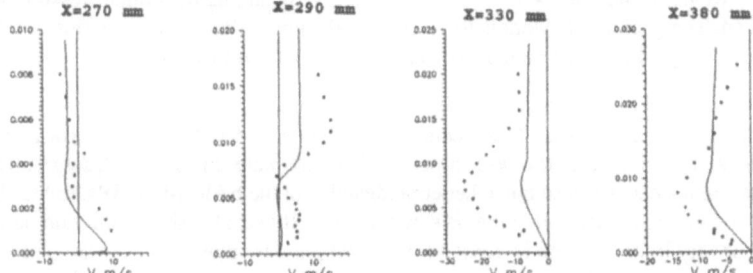

Figure 8 :   Uy velocity component at four axial
              positions.   (ooo=exper., ---=calc.)

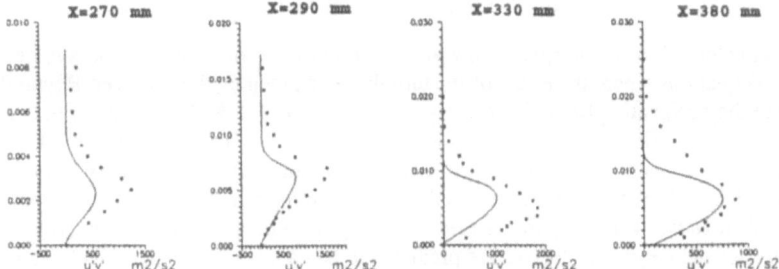

Figure 9 :   Reynolds stress profiles at four axial
              positions. (ooo=exper., ---=calc.)

COMPUTATIONS ON THE TRANSONIC BUMP USING EXPLICIT ALGEBRAIC REYNOLDS STRESS MODELS

Final report for the ETMA Workshop, Test Case 5, February 1995.

Stefan Wallin

The Aeronautical Research Institute of Sweden, FFA

Box 110 21, S-161 11 Bromma, Sweden

ABSTRACT

The transonic channel with a bump, test cases TC5 in the ETMA Workshop, have been computed with two different Explicit Algebraic Reynolds Stress Models, EARSM, together with Chien low Reynolds number k-ε model. The effect of changing the outlet pressure has also been studied. The goal of this work is to investigate the possibilities and limitations of the Explicit Algebraic Reynolds Stress Models for this kind of separated flow.

The EARSM models, especially the Shih, Zhu & Lumley model, gave a clear improvement compared to the k-ε model. The iso Mach contours for the two EARSM models show a well formed λ shock structure while the k-ε model fails to estimate this shock structure. The lower wall pressure plateau was also much better predicted with the EARSM models, especially the Shih, Zhu & Lumley model, while the k-ε model fails. It was also shown that the outlet pressure was not very important. An increase of the outlet pressure moved the shock upstream but did not affect the overall flow structure, as the λ shock structure and the wall pressure plateau.

The EARSM extension of the k-ε turbulence model did not introduce any further numerical problems and the difference in computational effort to reach steady state was negligible.

1 INTRODUCTION

Most practical calculations of turbulent flow at the present time are based on the averaged Navier-Stokes equations where the effect of the turbulence is given by the averaged Reynolds stresses, hence the designation Reynolds Averaged Navier-Stokes (RANS). The most popular models used in practical flow calculations are the zero, one and two equation models where the Reynolds stresses are related to the mean flow quantities by the generalised Boussinesq eddy viscosity hypothesis. It is well known that eddy viscosity models are unable to properly describe turbulent flows with body forces effects arising from curvature or system rotation. Moreover, the normal Reynolds stresses are predicted to be equal, which is incorrect even in simple cases such as homogeneous shear flow. In situations other than thin boundary layers this may cause significant changes to the mean flow. It is thus a generally accepted fact that the popular k-ε model does not predict the correct size of the separated region behind a backward facing step.

During the 1970s, second-order closure models became popular in which closure was achieved based on the Reynolds stress transport equation (RSTM). History and nonlocal effects are then included in the modelling as well as a more correct description of the turbulent anisotropy. Algebraic Reynolds Stress Models (ARSM) were obtained from the Reynolds stress transport equation where the transport of the Reynolds stress anisotropy was neglected (see Rodi 1976). The Reynolds stresses are here related to the mean flow field by an implicit relation. However, this model has been found to be cumbersome to implement in complex flow, and numerical stiffness problems can result from the need for successive matrix inversion at each iteration.

Pope (1975) developed an algebraic Reynolds stress model (EARSM) by invoking the same equilibrium hypothesis as Rodi (1976) but presented a methodology for obtaining an explicit relation for the Reynolds stresses. The Reynolds stress anisotropy tensor is in his model developed as a tensor polynomial of the mean flow strain and rotation tensors where the coefficients may be functions of the mean flow strain and rotation tensor invariants, and the turbulent Reynolds number.

The functional form of the coefficients can be derived formally from an RST model with the aid of an anisotropy equilibrium hypothesis, Pope presented the solution for a two-dimensional mean flow derived from the RST model by Launder, Reece and Rodi (1975). Gatski & Speziale (1993) extended the results of Pope to three-dimensional turbulent flow and demonstrated a way to remove singular behaviour of the model for large strain rates. It is also possible to find the coefficients from general conditions such as requiring the kinetic energy in each velocity component to remain nonnegative (realizability) and then fit the functions to fundamental turbulence experiments as shown by, e.g. Speziale (1987), Thangham, Abid & Speziale (1991) and Shih, Zhu & Lumley (1992).

These Explicit Algebraic Reynolds Stress Models (EARSM) are used together with a two-equation turbulence model and are simple to implement in an existing Navier-Stokes solver avoiding the numerical stiffness of the implicit stress equation. The number of boundary conditions is also unchanged from the two-equation models. History and nonlocal effects of the anisotropy can not be accounted for in these models, but they have the potential to correctly describe the Reynolds stresses for flows with effects arising from curvature, system rotation and nonequal normal Reynolds stresses.

2   GOVERNING EQUATIONS

2.1   The Favre averaged Navier-Stokes equations

The classical way to decompose a variable into a constant and a fluctuating part is the Reynolds decomposition where the variable $q = \bar{q} + q'$ is split into a time average $\bar{q}$ and the fluctuating part q' where $\overline{q'} = 0$. A favourable decomposition for compressible flow is a mass weighted averaging according to Favre where the variable $q = \tilde{q} + q''$ is split into a mass weighted time average $\tilde{q}$ and the fluctuating part q'' where $\overline{q''} \neq 0$. The time averaged Navier-Stokes equations can be expressed in the following form,

$$\frac{\partial}{\partial t} W + \frac{\partial}{\partial x_j} (Q_j - V_j) = 0 , \tag{2.1}$$

where the state vector, W, is given by

$$W = \left\{ \begin{matrix} \bar{\rho} \\ \bar{\rho}\,\tilde{u}_i \\ \bar{\rho}\,\tilde{E} \end{matrix} \right\} \tag{2.2}$$

and the convective, Q, and viscous, V, fluxes are

$$Q_j = \left\{ \begin{matrix} \overline{\rho}\tilde{u}_j \\ \overline{\rho\tilde{u}_i\tilde{u}_j} + \overline{\rho u''_i u''_j} + \bar{p}\delta_{ij} \\ \overline{\rho \tilde{E}\tilde{u}_j} + \overline{\rho u''_i u''_j}\tilde{u}_i + \bar{p}\,\tilde{u}_j \end{matrix} \right\}, \tag{2.3}$$

$$V_j = \left\{ \begin{matrix} 0 \\ 2\mu\widetilde{S}_{ij} \\ 2\mu\widetilde{S}_{ij}\tilde{u}_i + C_p\left(\dfrac{\mu}{Pr}+\dfrac{\mu_t}{Pr_t}\right)\dfrac{\partial \widetilde{T}}{\partial x_j} + \left(\mu+\dfrac{\mu_t}{\sigma_k}\right)\dfrac{\partial k}{\partial x_j} \end{matrix} \right\}. \tag{2.4}$$

The total energy is $\widetilde{E} = \tilde{e} + 1/2\tilde{u}_j\,\tilde{u}_j + k$, where the turbulent kinetic energy is defined as $k = 1/2\widetilde{u_j u_j}$.

The averaged strain tensor, $\widetilde{S}_{ij}$, is given by

$$\widetilde{S}_{ij} = \frac{1}{2}\left(\frac{\partial \tilde{u}_i}{\partial x_j} + \frac{\partial \tilde{u}_j}{\partial x_i}\right) - \frac{2}{3}\frac{\partial \tilde{u}_k}{\partial x_k}\delta_{ij}. \tag{2.5}$$

Here $\bar{\rho}$, $\tilde{u}_i$, $\bar{p}$, $\tilde{e}$ and $\widetilde{T}$ denotes averaged density, velocity, pressure internal energy and temperature.

The molecular viscosity for air μ is expressed as a function of temperature according to Sutherland's law

$$\frac{\mu}{\mu_0} = \left(\frac{T}{T_0}\right)^{3/2}\frac{T_0 + S_1}{T + S_1}, \tag{2.6}$$

where $\mu_0 = 1.716 \cdot 10^{-5}$ kg/ms is the viscosity at the reference temperature $T_0 = 273.1$ K and the constant $S_1 = 110.6$ K.

For calorically perfect gases, the specific heats at constant volume and constant pressure are $C_v = R/(\gamma - 1)$ and $C_p = \gamma\,C_v$. For air, the ratio of specific heats $\gamma = 1.4$ and the gas constant $R = 287$. The Prandtl number, $Pr = 0.72$. The pressure must be related to the state vector and for calorically perfect gases, $\bar{p} = \bar{\rho}\,\tilde{e}\,(\gamma - 1)$.

So far the Reynolds stresses have not been modelled while the other correlations have been modelled in a standard way using the gradient approximation, see Vandromme (1991).

2.2 The baseline k-ε turbulence model

The baseline two-equation turbulence model is the low Reynolds number k-ε model proposed by Chien (1982). The transport equations for the turbulent kinetic energy, k, and the turbulent dissipation rate, ε, are given by

$$\frac{\partial}{\partial t} W_{k\text{-}\varepsilon} + \frac{\partial}{\partial x_j}(Q_{k\text{-}\varepsilon j} - V_{k\text{-}\varepsilon j}) = S_{k\text{-}\varepsilon} , \qquad (2.7)$$

with the state vector

$$W_{k\text{-}\varepsilon} = \left\{ \begin{array}{c} \overline{\rho}\, k \\ \overline{\rho}\, \varepsilon \end{array} \right\} , \qquad (2.8)$$

the convective and viscous fluxes

$$(Q_{k\text{-}\varepsilon j} - V_{k\text{-}\varepsilon j}) = \left\{ \begin{array}{c} \overline{\rho}\, k\, \widetilde{u}_j - \left(\mu + \dfrac{\mu_t}{\sigma_k}\right)\dfrac{\partial k}{\partial x_j} \\[3mm] \overline{\rho}\, \varepsilon\, \widetilde{u}_j - \left(\mu + \dfrac{\mu_t}{\sigma_\varepsilon}\right)\dfrac{\partial \varepsilon}{\partial x_j} \end{array} \right\} , \qquad (2.9)$$

and the source term

$$S_{k\text{-}\varepsilon} = \left\{ \begin{array}{c} P_k - \overline{\rho}\,\varepsilon\left(1 + \alpha_1\, M_t^2\right) - 2\dfrac{\mu\, k}{n^2} \\[3mm] \left(C_{\varepsilon 1}\, P_k - C_{\varepsilon 2} f_2\, \overline{\rho}\,\varepsilon\right)\dfrac{\varepsilon}{k} - 2\dfrac{\mu\,\varepsilon}{n^2} e^{-\frac{1}{2}n^+} \end{array} \right\} , \qquad (2.10)$$

where the production of turbulent kinetic energy, P_k, is given by

$$P_k = -\overline{\rho\, u''_i u''_j}\,\frac{\partial \widetilde{u}_i}{\partial x_j} . \qquad (2.11)$$

The term $\alpha_1\, M_t^2$ in the k source term is a compressibility correction of the dissipation as proposed by Sarkar, et al (1991) where $\alpha_1 = 1.0$ and the turbulent Mach number is defined as $M_t^2 = k/a^2$ where a is the speed of sound.

The function f_2 in the source term for the dissipation is

$$f_2 = 1 - 0.22\, e^{-\left(\frac{R_t}{6}\right)^2} , \qquad (2.12)$$

where the turbulent Reynolds number is defined as $R_t = \overline{\rho}\, k^2/\mu\,\varepsilon$.

The turbulent viscosity is derived from the turbulent quantities as

$$\mu_t = C_\mu f_\mu \overline{\rho}\,\frac{k^2}{\varepsilon} , \qquad (2.13)$$

where $f_\mu = 1 - e^{-0.0115\, n^+}$. The constants involved are $C_\mu = 0.09$, $\sigma_k = \sigma_\varepsilon = 1.0$, $C_{\varepsilon 1} = 1.35$, $C_{\varepsilon 2} = 1.8$ and $Pr_t = 1.0$.

445

2.3 The Reynolds stresses

The Reynolds stresses appear in the momentum and energy equations and in the expression for the turbulent production and has to be modelled. The most used model is the Boussinesq hypothesis which is a linear relation between the turbulent stress and the mean flow strain. We will here form a more general relation between the turbulent stress and the mean flow and turbulence fields.

It is more convenient to formulate an expression for the Reynolds stress anisotropy tensor rather for the Reynolds stresses. The anisotropy tensor is here defined for compressible flow as

$$a_{ij} = \frac{\overline{u''_i u''_j}}{k} - \frac{2}{3}\delta_{ij} \ . \tag{2.14}$$

The most general expression for the Reynolds stress anisotropy in terms of the mean flow strain and vorticity tensors and the turbulent kinetic energy and dissipation rate is

$$a_{ij} = \sum_{\lambda=1}^{11} \alpha_\lambda \, T_{ij}^{(\lambda)} \ , \tag{2.15}$$

where the coefficients may be functions of the mean flow strain and vorticity tensor invariants. Fortunately, owing to the Cayley-Hamilton theorem, the number of independent invariants and linearly independent second-order tensors is finite. The first four terms are

$$
\begin{aligned}
T_{ij}^1 &= S_{ij} \\
T_{ij}^2 &= S_{ik}\,\Omega_{kj} - \Omega_{ik}\,S_{kj} \\
T_{ij}^3 &= S_{ik}\,S_{kj} - \frac{1}{3}\,S_{mn}\,S_{nm}\,\delta_{ij} \\
T_{ij}^4 &= \Omega_{ik}\,\Omega_{kj} - \frac{1}{3}\,\Omega_{mn}\,\Omega_{nm}\,\delta_{ij}
\end{aligned}
\tag{2.16}
$$

where S and Ω are the mean flow strain and vorticity tensors normalised by the turbulent time scale k/ε. Keeping only the first term corresponds to the Boussinesq hypothesis with $\alpha_1 = -2\mu_t\,\varepsilon/\bar{\rho}k^2$. To make the model appliable for compressible flows, the strain and vorticity tensors are defined as

$$
\begin{aligned}
S_{ij} &= \frac{1}{2}\frac{k}{\varepsilon}\left(\frac{\partial \tilde{u}_i}{\partial x_j} + \frac{\partial \tilde{u}_j}{\partial x_i} - \frac{1}{3}\frac{\partial \tilde{u}_k}{\partial x_k}\delta_{ij}\right) \\
\Omega_{ij} &= \frac{1}{2}\frac{k}{\varepsilon}\left(\frac{\partial \tilde{u}_i}{\partial x_j} - \frac{\partial \tilde{u}_j}{\partial x_i}\right)
\end{aligned}
\tag{2.17}
$$

Two different EARSM models have been implemented and tested.

The first model is proposed by Gatski & Speziale (1993) which gives coefficients as functions of the mean flow strain and vorticity tensor invariants ($\eta^2 = S_{ij}S_{ji}$ and $\zeta^2 = -\Omega_{ij}\Omega_{ji}$)

$$
\begin{aligned}
\alpha_1 &= -0.227\,A \\
\alpha_2 &= -0.042\,A \\
\alpha_3 &= 0.040\,A \\
\alpha_4 &= 0
\end{aligned}
\tag{2.18}
$$

where

$$A = \frac{3\left(1 + C_1{}^2\eta^2\right)}{3 + C_1{}^2\eta^2 + 6\,C_1{}^2C_2{}^2\eta^2\zeta^2 + 6\,C_2{}^2\zeta^2} \tag{2.19}$$

and the constants C_1=0.087375 and C_2=0.1864 .

The second model is proposed by Shih, Zhu & Lumley (1992) where the coefficients are:

$$\begin{aligned}
\alpha_1 &= -\frac{4}{3}\frac{1}{1.25 + \sqrt{2}\,\eta + 0.9\,\sqrt{2}\,\zeta} \\
\alpha_2 &= \frac{-15}{1000 + 2\sqrt{2}\,\eta^3} \\
\alpha_3 &= \frac{3}{1000 + 2\sqrt{2}\,\eta^3} \\
\alpha_4 &= \frac{-19}{1000 + 2\sqrt{2}\,\eta^3}
\end{aligned} \tag{2.20}$$

It is not obvious what to do in the near wall region. The EARSM is formulated as a correction to the Boussinesq hypothesis and this correction is damped out in the near wall region in the same way as the turbulent viscosity is damped according to Thangham, Abid & Speziale (1991). That means that $\alpha_1 = \alpha_1 f_\mu$ and $\alpha_{2,3,4} = \alpha_{2,3,4} f_\mu{}^2$.

2.4 Numerical method

The Explicit Algebraic Reynolds Stress Model, formulated as a correction to a $k\text{-}\varepsilon$ eddy viscosity model, has been implemented in the CFD solver EURANUS developed by FFA together with VUB (Rizzi et al. 1993). EURANUS is a general Navier-Stokes solver for structured multiblock meshes with various discretizations in space and time. A multigrid method accelerates the convergence. The low Reynolds number $k\text{-}\varepsilon$ model by Chien (1982) with compressibility corrections according to Sarkar et al. (1991) is implemented with central and upwind spatial discretization together with explicit time integration. For stability reasons the source terms are treated implicitly.

In this study the mean flow equations are discretized using a second order upwind scheme, symmetric TVD, with Van Leer limiters. The $k\text{-}\varepsilon$ equations are discretized using first order upwind scheme. Care has been taken to have second order accuracy near the wall in the wall normal direction for the $k\text{-}\varepsilon$ equations . The numerical dissipation is damped near the wall in the wall normal direction and the viscous terms are second order accurate also near the wall.

The equations are integrated to convergence using a five stage Runge-Kutta method.

3 RESULTS

3.1 Test case definition and computational grid

The test case TC5 for the ETMA Work Shop was the two-dimensional transonic channel flow with a bump, case C, studied by Délery (1980). The inflow conditions are tabulated in table 1.

The walls were assumed adiabatic and the outlet pressure was adjusted to the measured pressure at the outlet of the computational domain.

Table 1   Inflow conditions for the transonic bump flow.

Total pressure [kPa]	96.0
Total temperature [K]	300
Mach number	0.63

The distributed grid with 120x220 cells has been used in this study (see Figure 1). The nondimensional distance, $y+$, from the wall to the first grid line is well below one (see Figure 2). Figure 3 shows the lower wall surface pressure where the solution on the fine grid has been compared to the solution on the medium grid where every second grid point have been removed. The difference between the grids was negligible and the solution could be regarded as grid independent.

3.2 Computational results

The test case has been computed using the eddy viscosity k-ε turbulence model and two different EARSM turbulence models, Gatski & Speziale (1993), and Shih, Zhu & Lumley (1992). The notation in the figures are k-eps, GS and SZL. Computations for two different outlet pressures have been made.

First, the influence of the EARSM modelling was studied with a constant outlet pressure of 60.5 kPa. The wall pressure is shown in Figure 4, iso Mach contours in Figure 6, velocity profiles in Figure 7 and turbulence profiles in Figure 9.

The iso Mach contours for the two EARSM models show a well formed λ shock structure while the k-ε model fails to estimate this shock structure. The lower wall pressure plateau was also much better predicted with the EARSM models, especially the Shih, Zhu & Lumley model, while the k-ε model fails. The velocity and turbulence profiles also show that the EARSM models performs better than the k-ε model. The Gatski & Speziale model shows however some strange behaviour downstream of the separation.

Then, the influence of different outlet pressures was studied for the k-ε model and the Shih, Zhu & Lumley model. Two different outlet pressure were set. 'pa' corresponds to 60.5 kPa, the same as in the previous case, and 'pb' corresponds to 61.0 kPa. The wall pressure is shown in Figure 5, velocity profiles in Figure 8 and turbulence profiles in Figure 10. When the outlet pressure was increased, the shock moved slightly upstream and improved the results with the Shih, Zhu & Lumley model slightly.

3.4 Convergence and numerical problems

Steady state was reached after about 20000 time steps. Multigrid cycling was however not possible due to stability reasons for this particular case. The EARSM models did not introduce any further numerical problems compared to the k-ε turbulence and the difference in computational effort to reach steady state was negligible.

4 CONCLUSIONS

To get a grid convergent solution and especially correct skin friction in low Reynolds number k-ε calculations, the nondimensionalised wall distance to the first grid line should be less than one. This is fulfilled and there are also negligible differences in wall pressure between the fine and medium grids. The conclusion is that the solutions are grid independent.

The EARSM models, especially the Shih, Zhu & Lumley model, gave a clear improvement compared to the k-ε model. It was also shown that the outlet pressure was not very important. An increase of the outlet pressure moved the shock upstream but did not affect the overall flow structure, as the λ shock structure and the wall pressure plateau.

There are still differences between the computations and the experiments, especially for the velocity and turbulence profiles, and the models have to be further developed. The differences could also be due to nonlocal transport effects of the anisotropy where a local algebraic Reynolds stress model fails.

The EARSM extension of the k-ε turbulence model did not introduce any further numerical problems and the difference in computational effort to reach steady state was negligible.

5 ACKNOWLEDGEMENT

This study has been carried out at FFA (the Aeronautical Research Institute of Sweden) within the ETMA project funded by FFA and NUTEK (the Swedish National Board for Industrial & Technical Development).

6 REFERENCES

[1]   Pope, S. B., "A more general effective-viscosity hypothesis", J. Fluid Mech., vol. 72, part 2, 1975.

[2]   Launder, B. E., Reece, G. J., Rodi, W., "Progress in the development of a Reynolds-stress turbulence closure", J. Fluid Mech., vol. 41, 1975.

[3]   Rodi, W., "A new algebraic relation for calculating the Reynolds stresses", Z. angew. Math. Mech. 56, T219-221, 1976.

[4]   Delery, J., Copy, C., Reisz, J., "Analyse au vélocimètre laser bidirectionnel d'une interaction choc-couche limite avec décollement étendu", ONERA Raport technique, n°. 37:7078 AY014, 1980.

[5]   Chien, K. Y., "Predictions of Channel and Boundary-Layer Flows with a Low-Reynolds-Number Turbulence Model", AIAA Journal, Vol. 20, No. 1, January, 1982.

[6]   Speziale, C. G., "On nonlinear K-l and K-e models of turbulence", J. Fluid Mech., vol. 178, 1987.

[7]   Thangam, S., Abid, R., Speziale, C. G., "Application of a new k-t Model to Near Wall Turbulent Flows", ICASE Report No. 91-16, February, 1991.

[8]   Sarkar, S., Erlebacher, G., Hussaini, M. Y., Kreiss, H. O., "The analysis and modelling of dilatational terms in compressible turbulence", J. Fluid Mech., vol. 227, 1991.

[9]   Vandromme, D., "Introduction to the Modeling of Turbulence", von Karman Institute for Fluid Dynamics, Lecture Series 1991-02, March, 1991.

[10] Shih, T. H., Zhu, J., Lumley, J. L., "A Realizable Reynolds Stress Algebraic Equation Model", NASA TM 105993, ICOMP-92-27, CMOTT-92-14, 1992.

[11] Gatski, T. B., Speziale, C. G., "On explicit algebraic stress models for complex turbulent flows", J. Fluid Mech., vol. 254, 1993.

[12] Rizzi, A., Eliasson, P., Lindblad, I., Hirsch, C., Lacor, C., Haeuser, J., "The engineering of multiblock multigrid software for Navier-Stokes flows on structured meshes", Computers and Fluids, Vol. 22, 1993.

7 FIGURES

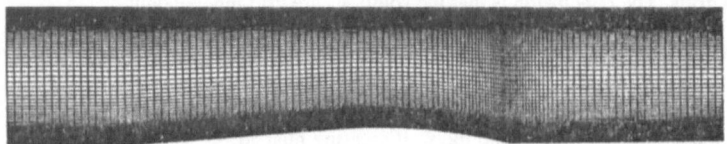

Figure 1.   Computational grid for the transonic bump.

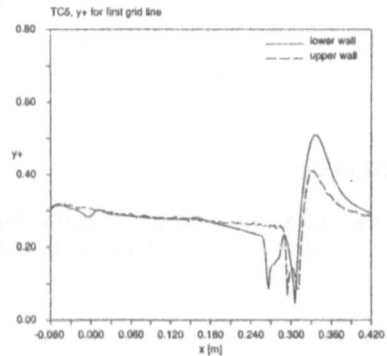

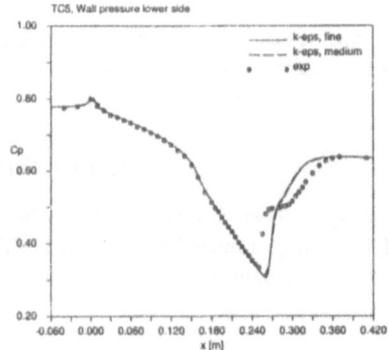

Figure 2.   The y+ distance to the first grid   Figure 3.   Wall pressure lower and upper
            line over the upper and lower   side, outlet pressure 60.5 kPa.
            walls.   Fine and medium grid compared.

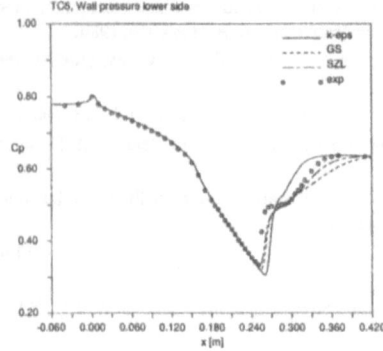

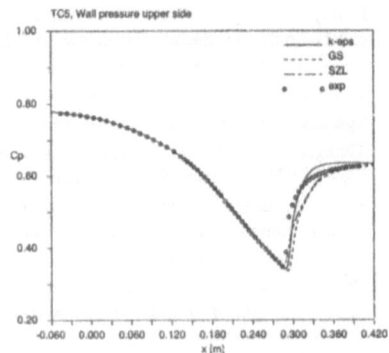

Figure 4.   Wall pressure lower and upper side, outlet pressure 60.5 kPa. Different EARSM
            models compared.

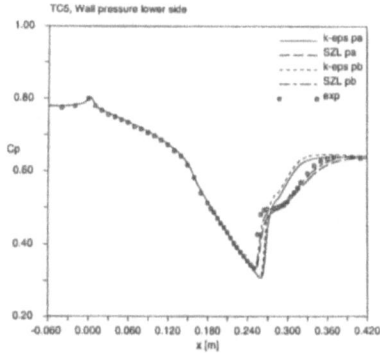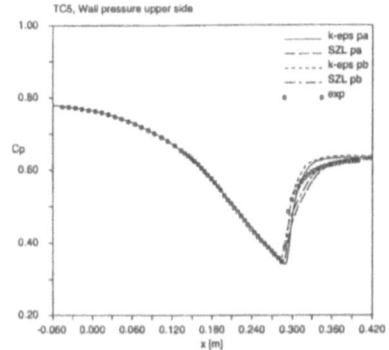

Figure 5.   Wall pressure lower and upper side. Different EARSM models and outlet pressure compared. Outlet pressure is for pa - 60.5 kPa and for pb - 61.0 kPa.

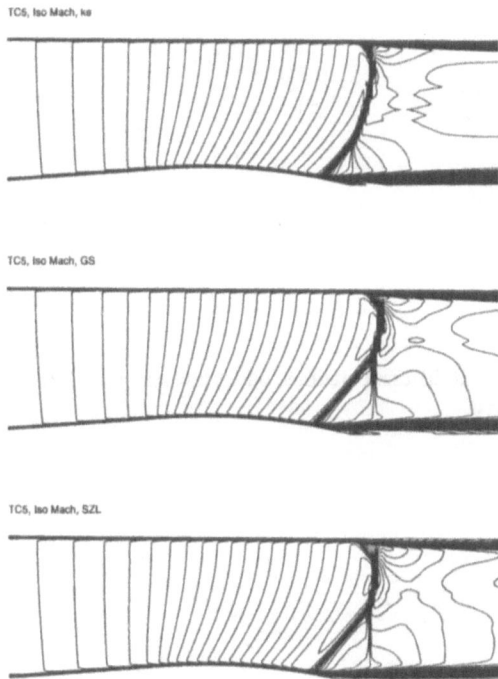

Figure 6.   Iso Mach number lines, outlet pressure 60.5 kPa. Different EARSM models compared.

451

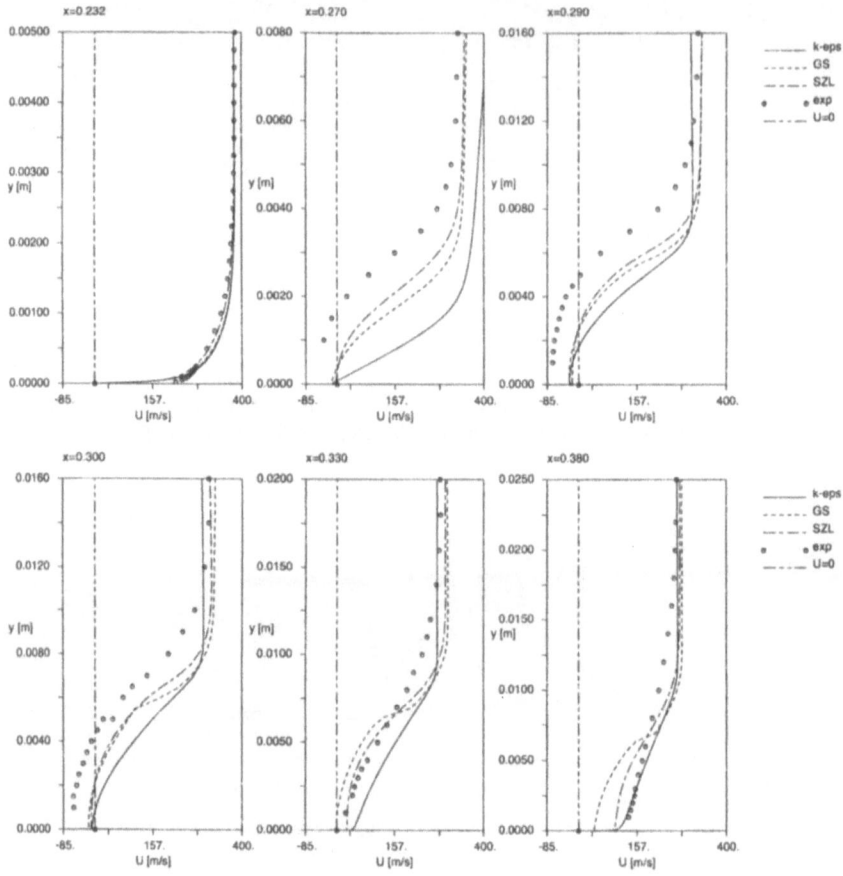

Figure 7. Velocity profiles at different stations, outlet pressure 60.5 kPa. Different EARSM models compared.

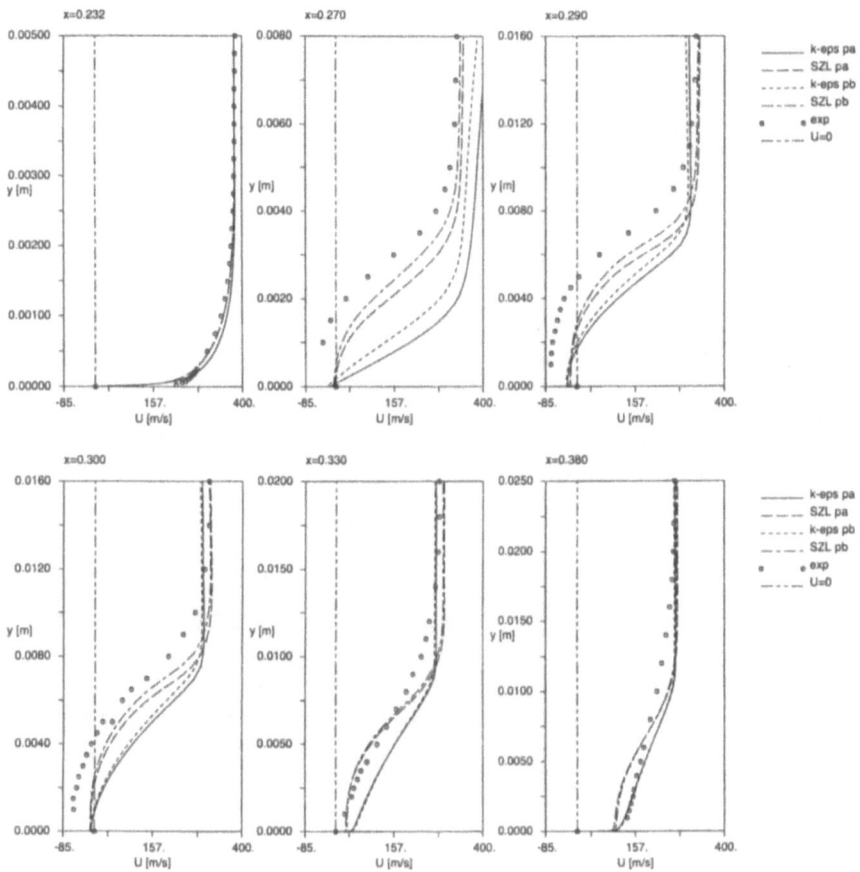

Figure 8.   Velocity profiles at different stations. Different EARSM models and outlet pressure
compared. Outlet pressure is for pa - 60.5 kPa and for pb - 61.0 kPa.

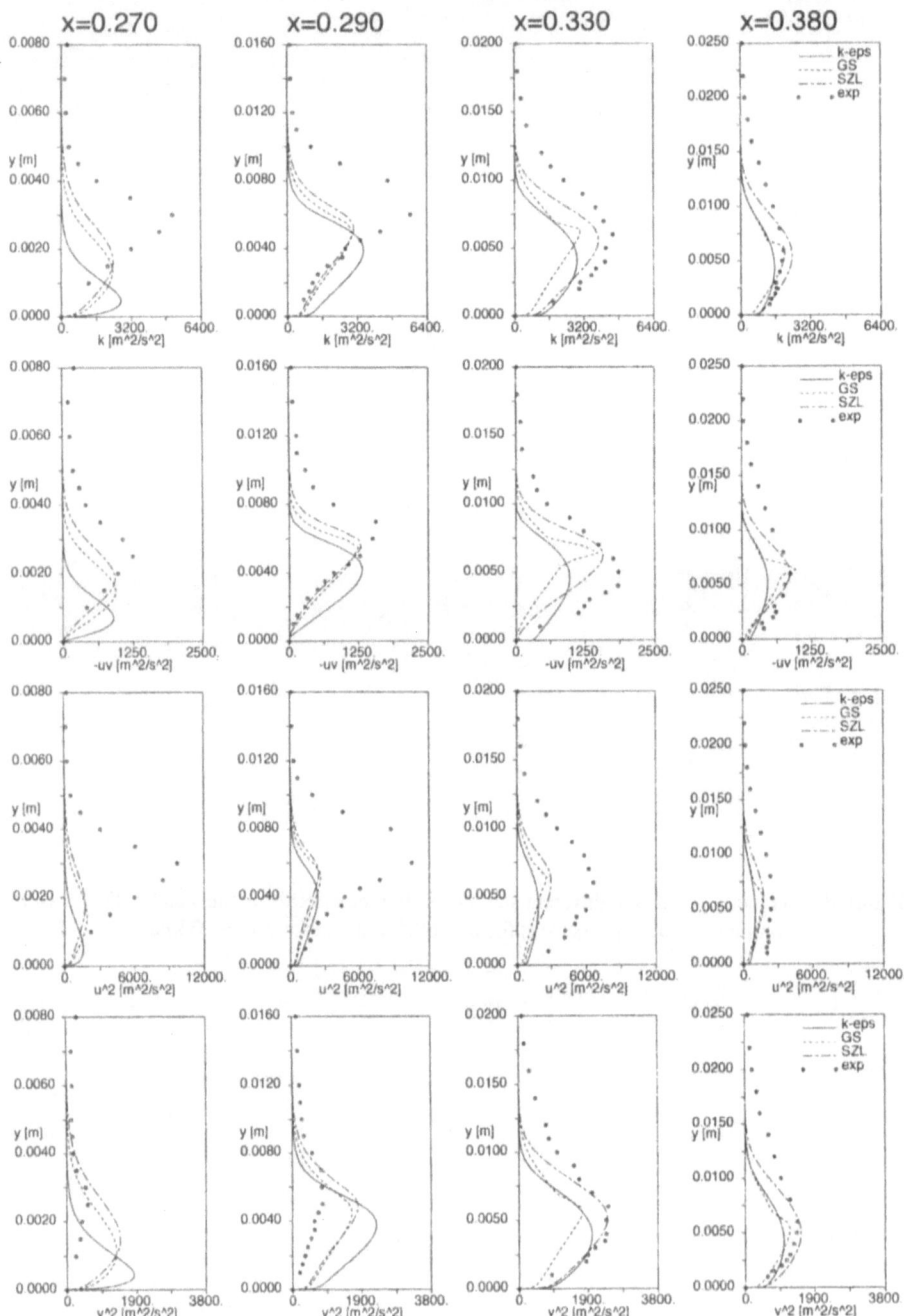

Figure 9.   Turbulence profiles at different stations, outlet pressure 60.5 kPa. Different
EARSM models compared.

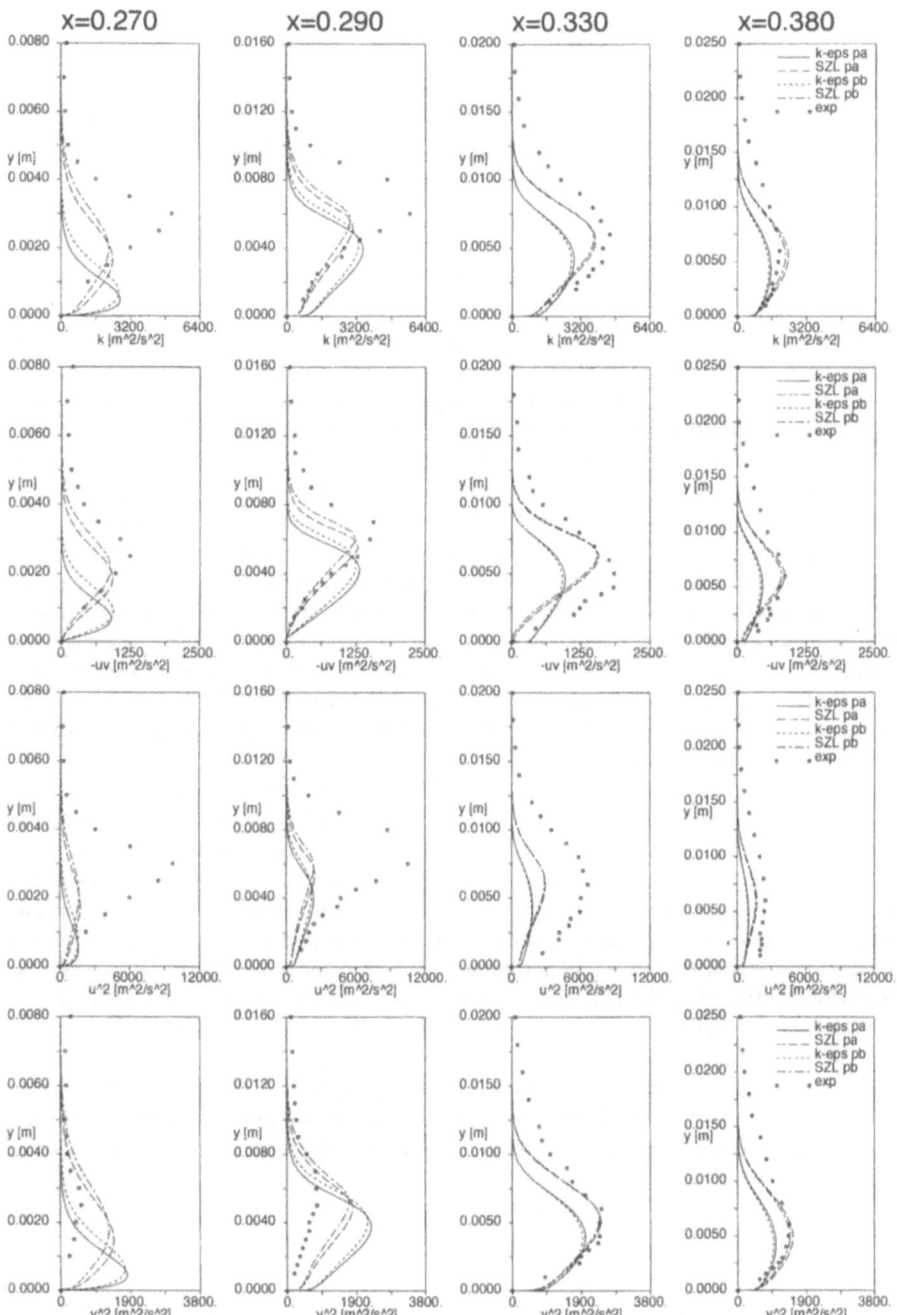

Figure 10. Turbulence profiles at different stations. Different EARSM models and outlet pressure compared. Outlet pressure is for pa - 60.5 kPa and for pb - 61.0 kPa.

TC5 Synthesis

H. Loyau, D. Vandromme

LMFN - CORIA URA 230 - INSA de ROUEN
Place E. Blondel, 76130 Mont-Saint-Aignan

Introduction

Shock wave boundary layer interactions represent an important problem in fluid mechanics because of their widespread occurence in transonic as well as in higher regimes. In particular in the transonic regime where a strong interaction causes the boundary layer thickening or separation largely responsible for airfoil and wing drag rises, and of air intakes performance losses. This feature being tightly coupled together with the state of the incoming boundary layer and the turbulent transport processes in the boundary layer during the interaction, it follows that the turbulence model used plays an important role in the prediction of such flows. This purpose motivated the present study of two-dimensional transonic channel referred as test case TC5 in the ETMA project.

The description of the geometry can be found in the previous test case specification section. In this experiment conducted by Delery and his co-workers, a bump accelerates an inlet subsonic flow to supersonic conditions. The downstream pressure level adopted allows to choke the nozzle by maintaining a normal shock wave at the trailing edge of the bump, sufficiently strong to provoke in combination with the curvature of the lower wall an extended area of recirculation.

For this case, a total of thirteen models of turbulence covered many variants or modifications of base models have been computed, some contributors testing two or three models. These calculations being described by the computors, this synthesis do not intend to detail the numerical solvers and turbulence models used, and the reader is invited to refer to the corresponding articles. In comparison with these articles some turbulence models, used as reference models by some contributors who prefer focusing the comparison on their most elaborate model, do not appear in the present study. Nevertheless they can be briefly mentioned in the following. The simplest closure used was the algebraic model of Michel et al., while the most sophisticated involved anisotropic corrections, close to algebraic Reynolds-stress formulation, on two-equation models.

The following 5 sections of this report are related respectively to a summary of contributions, a brief description of turbulence models, numerical practices (grids, boundary conditions, ...), comparison between numerical and experimental data, and conclusion.

Summary of Contributions

For this case a set of seven contributions was retained. The calculations have been conducted respectively by:

ALENIA (N. Ceresola)
FFA (S. Wallin)
LMFN (H. Loyau and D. Vandromme)
NTUA (G. Simandirakis, C. Vassilopoulos, K.C. Giannakoglou and K.D Papailiou)
ONERA (V. Gleize)
ROCKWELL Science Center (U.C. Goldberg)
RWTH Aachen (G. Freskos)

the following table giving for each contributor in alphabetical order, the list of exit pressures adjusted, turbulence models tested, and numerical choices adopted.

In the following, for most convenience, the models are identified by codes when they are directly compared together. Designations adopted will be specified in the next section.

Turbulence models

By considering the table 1, one can see that a wide range of turbulence models have been used, and that the overlap of models among contributors is focused on the Jones-Launder model. Obviously the turbulence closure levels adopted are mostly concentrated on two-equation model with several variants. One of the original features in the present work lies in the presence of several anisotropic two-equation models applied by FFA and LMFN which cover a wide range of anisotropic corrections including some of the most recent formulations like the Explicit Algebraic Reynolds Stress Models tested by FFA or the simplified Algebraic Stress Model developed by LMFN. Another interesting fact is the use of recent models like the Baldwin-Barth modified by ALENIA, the $k - k/\epsilon^2$ model of Goldberg (ROCKWELL), or the multi-time-scale $k - \epsilon$ model developed by ONERA on a such configuration flow.

The standard models used are essentially the algebraic model of Michel, the one-half equation model of Johnson-King, and the two-equation models of Goldberg, Chien and Jones-Launder, identified in the following by "M", "JK", "G", "C" and "JL1".

A particular attention must be paid on the Jones-Launder model for which different values for the two constant $C_{\epsilon 1}$ and $C_{\epsilon 2}$ have been used. Among the three variants tested, the standard JL1 model is related to the following standard values: $C_{\epsilon 1} = 1.45$ and $C_{\epsilon 2} = 1.92$. Concerning the model tested by both NTUA and RWTH the modification consists in shifting the standard value to the higher value $C_{\epsilon 1} = 1.57$ locally within the recirculation zone, whereas in the case of the ONERA version, the two constants are set in the following steady conditions: $C_{\epsilon 1} = 1.55$ and $C_{\epsilon 2} = 2$, these two versions being respectively identified by "JL2" and "JL3".

Concerning the modified turbulence model used by ALENIA the starting point is the standard Baldwin-Barth model in which the y^+ dependance of the damping functions is corrected by a dependance on the turbulent Reynolds number itself the consecutive formulation being identified by "BB-A".

Table1 Contributors and numerical practices adopted

Contributors	Pe (Pa)	Turbulence Models	Numerical methods	grids
ALENIA	61440.	Baldwin-Barth (BB-A) Johnson-King (JK)	Implicit Finite difference; Space centered 4th order artificial dissipation	161 × 100
FFA	60500. } 61000. }	$k - \epsilon$ Chien (C) EARSM: • Gatski-Speziale (GS) • Shih-Zhu-Lumley (SZL)	Upwind-TVD Structured Multiblock Scheme Multigrid acceleration	120 × 220
LMFN	60100.	$k - \epsilon$ Jones Launder (JL1) $k - \epsilon$ Saffman (S) $k - \epsilon$ Pope (P) Simpl. ASM (LMFN)	Implicit Predictor/Corrector Mac Cormack scheme Steger-Warning Flux splitting	120 × 220
NTUA	62000.	$k - \epsilon$ Jones Launder: • $C_{\epsilon 1}$: 1.45 (JL1) • $C_{\epsilon 1}$: 1.45 − 1.57 (JL2)	Semi-Implicit space-centered Predictor/Corrector Mac Cormack scheme $2^{nd}/4^{th}$ order artificial dissipation	120 × 220
ONERA	63170. 62400. 61540.	Michel algebraic model (M) $k - \epsilon$ Jones Launder (JL3) $k - \epsilon$ Multi-time-scale (MTS)	3D Cell-centered Runge-Kutta scheme 4th order artificial dissipation ADI Implicit stage	153 × 65 × 2
ROCKWELL	61790.	$k - \dfrac{k^2}{\epsilon}$ Goldberg (G)	Finite volume 3rd order accurate solver TVD Formulation	120 × 120
RWTH	62000.	$k - \epsilon$ Jones Launder: • $C_{\epsilon 1}$: 1.45 (JL1) • $C_{\epsilon 1}$: 1.45 − 1.57 (JL2)	Implicit; Finite volume Roe-TVD Numerical scheme	120 × 220

The Multi-time-scale model tested by ONERA (identified as MTS) is based on a two-scale scheme with algebraic modeling of partial Reynolds stress with an eddy viscosity in which the turbulent quantities are splitted in two part corresponding respectively to the large (production range) and fine.(transfert and dissipation ranges) eddies.

The remaining models concern anisotropic corrections added to standard $k - \epsilon$ models. In these models, the Reynolds Stress tensor is rewritten as a tensor polynomial of mean flow strain tensor and rotation tensor the coefficients being related to the mean flow strain and rotation invariants. Doing so, the Boussinesq equality appears as the first order part of the previous polynomial form, the remaining terms forming the anisotropic corrections. Among the present models the formulations adapted from the ones of Saffman and Pope to the JL1 model by LMFN and respectively identified as "S" and "P" represent the simpliest corrections, whereas those adapted from the ones of Gatski-Speziale and Shi-Zhu-Lumley (respectively "GSZ" and "SZL") to the C model by FFA are most elaborate with supplementary correction terms. The last model developed by LMFN appears rather as an anisotropic correction to the eddy viscosity analogous to the Standard Algebraic Reynolds Stress Model but in which the classical non-linear set of equations is reduced to a linear set. Nevertheless the implementation of this model identified as "LMFN" is done in similar way and it can be directly compared with the previous corrections.

As pointed out previously, the JL3, JK and M turbulence models do not appear in the comparative plots, but may be mentioned as references in following.

Numerical Strategies

Following the ETMA project philosophy turned towards data base production a whole set of detailed experimental and numerical data on this case was sent to each participant. This data set supplied a whole numerical field from which a non-uniform 'mandatory' grid (Fig. 1) could be easily extracted. For this grid of 120×220 cells, the first point off the wall is associated to a non-dimensional distance y^+ well below 1, the grid-independence of computations performed on this grid being confirmed and illustrated in the FFA communication. Nevertheless, some contributors preferred to use their own grid. ALENIA performed calculations with 161×100 grid ensuring for the distance of the first grid line from walls that y^+ is in order of magnitude of 1 at the beginning of the interaction, whereas in case of ROCKWELL computation a 120×120 grid size was used, first point off walls being located at $y^+ = 1$. For the computation done by ONERA, the use of a three-dimensional code imposed a supplementary section in the transverse direction, the corresponding $153 \times 65 \times 2$ grid providing a constant first cell height at both walls related to value of y^+ less than 3 on the whole length of the computational geometry.

Concerning the precribed boundary conditions, the main elements of the standard set consisted of experimental reservoir conditions at inflow, no-slip conditions at both upper and lower boundaries, the walls being assumed adiabatic. Additionally, since there are no informations concerning the downstream throat of the nozzle controlling experimentally the location of the shock, contributors were instructed to adjust the exit level pressure in order to replicate numerically the experimental shock conditions in order to be in good agreement with the experimental upper wall pressure distribution.

For inlet flow conditions, a wide range of numerical procedures was adopted for the evaluation of mean-flow quantities. Some computors conserved the stagnation pressure and temperature; characteristic relations supplying upstream-downstream influences

(ONERA,LMFN), others preferred to fix all quantities except pressure which is extrapolated (ALENIA). Concerning the turbulence quantities, most of contributors imposed for the two-equation models, profiles for k and ϵ. But concerning the BB-A model tested by ALENIA, one can notice that no particular initialisation was done.

At the exit of the geometry, an explicit prescription of the static pressure level has been adopted by all contributors. Table 1 gives an overview of the values adopted for which a certain level of conformity can be observed for similar computations.

Concerning the numerical solvers, table 1 gives also a general survey of the numerical codes used in the present study. Most contributors employed an implicit conservative finite volume approach execpt FFA using explicit time-marching scheme with multigrid acceleration. Few codes use high-resolution shock-capturing techniques for non-viscous terms discretization the viscous terms being treated with space-centered discretisation. Whereas the other part of the codes use centered schemes for non-viscous terms with 4th-order or combination of 4th and 2nd-order artificial dissipation.

Results

In the following section results are compared together for a set of flow parameters which are capable to underline the predictive abilities of the turbulence models under study. For this purpose, a first set of these plots provides a qualitative view of the predicted interaction region flow. Afterwards the wall pressure distributions, velocity profiles and shear stress profiles lead to more quantitative comparisons between the turbulence models studied. Concerning the velocity and the shear stress profiles, it was felt that the study of four stations - two in the interaction zone $X = 270, 290 \ mm$ and two in the recovery region $X = 330, 380 \ mm$ - would be sufficient to make a synthetic comparison. Other turbulent quantities profiles would be informative but depends here on the transport turbulence models or on the corrections used. Therefore in the interest of uniformity and conciseness, it was decided to focuse the present study on the most universal turbulence profiles. Readers interested in other quantities are invited to refer to the corresponding articles. In comparison presented below all quantities have been cross-plotted from numerical data supplied by participants except the shear stress in the case of ROCKWELL and RWTH, which had to be inferred from the available data, whereas in the case of ALENIA the shear stress profiles could not be obtained from the available data. These calculations have been done from the velocity components and the eddy viscosity values by using the Boussinesq equality, the derivatives being discretized with centered space formulation.

An overall view of the contributions is given (Fig. 2) in terms of predicted Mach number contours in the interaction region. The relevant feature of a λ-shock structure (reflecting the strength of the shock-boundary layer interaction) is poorly estimated by the standard JL1 and C two-equation models. Similarly the simpliest anisotropic corrections of the S and P models seem not to bring any significant improvements. Concerning the JL2 model the structure seems to be in better agreement with the the experiment with a more accurate prediction of the extent which seems to indicate that the correction on the $C_{\epsilon 1}$ value play a not insignificant role in the region of interest. But a more pronounced λ-structure is observed with the BB-A, G, MTS, LMFN, GS and SZL, for which the normal-shock leg of the structure is most visible. One can underline, though it is not directly compared with the others, the good prediction given by the JK model.

The wall pressure distributions are plotted in the two following figures corresponding to underestimate the boudary layer thickness. Only the MTS model still remains in good agreement with the experimental profiles in this zone. The standard JL1 and C models give basically identical predictions by overestimated the near-wall velocity level, which seems to indicate that both models underpredict the recirculation extend and tend prematurely towards recovery state, but recover later a rather good level of correspondance with the experimental data. At the opposite, the JL2 model predict correctly the near-reattachment profile but is a little behind the recovery behavior observed experimentally. The both BB-A and G models exhibit a similar behavior by predicting a too extended recirculation bubble, and show therefore some difficulties to recover in time the appropriate recovery profiles. This feature is also observed with the MTS model, but lesser than for the previous ones, with a good correspondance between numerical and experimental shapes in the boundary layer defect region. In the case of the anisotropic corrections, the S and P models show some similarity with the JL1 standard model but tend slightly to correct the discrepancies observed between the JL1 profiles and the experimental ones. On the same plots, the GS model overpredict the recirculation extend and shows a pronounced shear-layer velocity profile at the edge of the boundary layer, which actually occurs experimentally but not at the same stations and particularly not at the same intensity level which underlines the difficulties of this model to recover after a such interaction the satisfactory equilibrium state. The SZL model predicts better than the GS model the velocity profiles on the two last stations, giving a rather good approximation of the recirculation extend which is overestimated in the case of the LMFN model, this model recovering more easily the equilibrium state after the interaction.

When looking at the shear stress profiles given in the following, it is clear that during the interaction the standard JL1 and C models underestimate the experimental peak location. The S and P models tend to correct this tendency but not as significantly as the JL2 model. Later, all these models give in the recovery part of the flow similar predictions which underestimate high shear stress level. Concerning the remaining models a wide range of behavior can be observed which leads to rather good predictions, in term of shear stress level, either only in the interaction region or only in the recovery part of the flow, except for the SZL model which shows a rather well correspondance with the whole set of experimental profiles. One can notice the profiles predicted by the GS model in accordance with the pronounced shear layer velocity profiles observed above.

Conclusion

Whilst the present study has been focused on short range of turbulence model, it has yielded some valuable indications which are likely to have a usefull influences in a more practical context. Thus, the standard two-equation models tested are not sufficiently accurate to well predict the present shock-wave/boundary-layer interaction, giving a reverse flow region which is confined to a thin layer close to the wall. The simpliest attempt like the corrections of Saffman or Pope, in order to take into account the anisotropic characteristics of the flow, suffer from the same level of inaccuracy, probably due to the level of correction and their implementation too close to the standard equilibrium model ones. Afterwards, the more sophisticated correction given by the models of Gatski-Speziale corrects rather well this tendency but shows some difficulties to reproduce the gradual return to equilibrium between the mean flow and the turbulence, whereas those given by the Shih-Zhu-Lumley model shows a more stable behavior by predicting rather well the whole respectively to the upper (fig. 3) and the lower bump (fig. 4) walls. The pressure distribution relating to the upper boudary confirms that most of the contributions match reasonably well with the experimental data points. Nevertheless one can remark a smaller defect on the minimum value given by the GS ($P_e = 60500\ Pa$) and the LMFN model ($P_e = 61000\ Pa$), which seems to indicate that the exit pressure level prescribed could be more accurately adjusted. But for the both LMFN and FFA contributions, due to the similarity between the models tested, the numerical stategy was rather at first to performed the computations with the same exit pressure level. Afterwards, FFA provided new results for the C and SZL models ($P_e = 61000\ Pa$), from which are extracted the present results, correcting this tendency.

From physical point of view, the pressure distribution along the lower bump wall is much interesting. It shows, as it was sensed previously on the iso-Mach lines, that the standard two-equation models (C and JL1) fail to predict correctly the actual interaction, except in the case of the JL2 model for which, the modification adopted contributes to enhance the sensitivity of the model in the interaction region, but without bringing a definite agreement with the experimental variations. The previous remark concerning the standard models can be extended to the simpliest anisotropic corrections of the S and P models, whereas other anisotropic corrections like the SZL or the LMFN models, show a greater sensitivity to the experimentally observed pressure plateau. Among these anisotropic models, the GS model yield to an insuficient response of the boundary layer to the 2nd leg of the shock structure. The both behaviours of the G and BB-A models are very similar showing a quantitative correspondance with the experiemental pressure plateau, but underestimate the return to the post-shock pressure level. Finally, the best result is given by the MTS model which predict rather well with the experimental values along the whole part of the geometry considered.

Longitudinal velocity profiles at four stations are next plotted in Fig. 5. Concerning the recirculation region ($X = 270, 290\ mm$), the reattachment point being approximately located at $X = 325\ mm$, these profiles confirm the insufficiency of the standard JL1 and C models to predict correctly the interaction. The modification adopted in the JL2 model which increase the level of the turbulent dissipation in this region improves the flow separation by predicting a much more developped recirculation bubble than those given by the standard formulations. The G and BB-A models give different predictions in this part of the flow, though their predicted level of reverse velocity are very close. Thus, in the beginning of the recirculation region the G model gives a rather good approximation of the velocity profile, whereas the BB-A model underestimates the boundary layer thickness by predicting a less developped reverse flow region. Afterwards, on the second velocity profiles the both models are very close in the reverse part of the profiles but diverge in the remaining part, where the G model recovers the experimental profile values faster than the BB-A model which still underestimates the boundary layer edge. On the same profiles, the predictions given by the MTS model is close to the experimental distributions, this model giving here again the most accurate prediction. Concerning the anisotropic corrections, the different models tested react with different levels of accuracy. Thus, the prediction given by both S and P models still remain close to the standard model ones. The GS and the SZL models predict with a better accuracy the recirculation bubble sizes by better taking into account the reverse part of the profiles, which is also the case of the LMFN models at a lesser level , but nevertheless underestimate the experimental ones.

Further in the recovery region ($X = 330, 380\ mm$), most of the models show profile shapes which do not agree with the experimental ones in the boundary layer defect region, leading flow characteristics. The simplified Algebraic Stress model shows encouraging results but nevertheless some precautions have to be taken with this model which is probably too recent and still requires some improvements (better calibration, damping terms, ...) which could correct its behavior in a such type of interactions. Concerning the multi-time-scale model developed by ONERA the predicted displacement and reverse-flow characteristics are in rather well accordance with the experimental ones, the interaction being generally well predicted, but the model suffers from some deffects in the post-shock region, predicting a too low rate of flow recovery. Finally the BB-A and the G models still remain an interesting alternative for a such type of flows, but predict a too extended recirculation region and are therefore far behind the recovery profiles observed experimentally.

Acknowledgements

The authors are indebted to A. Hadjadj and A. Stoukov (LMFN/CORIA URA 230) for their help during the Workshop preparation. The treatment of the result files has been supported by CRIHAN, Mt-St-Aignan.

Figures

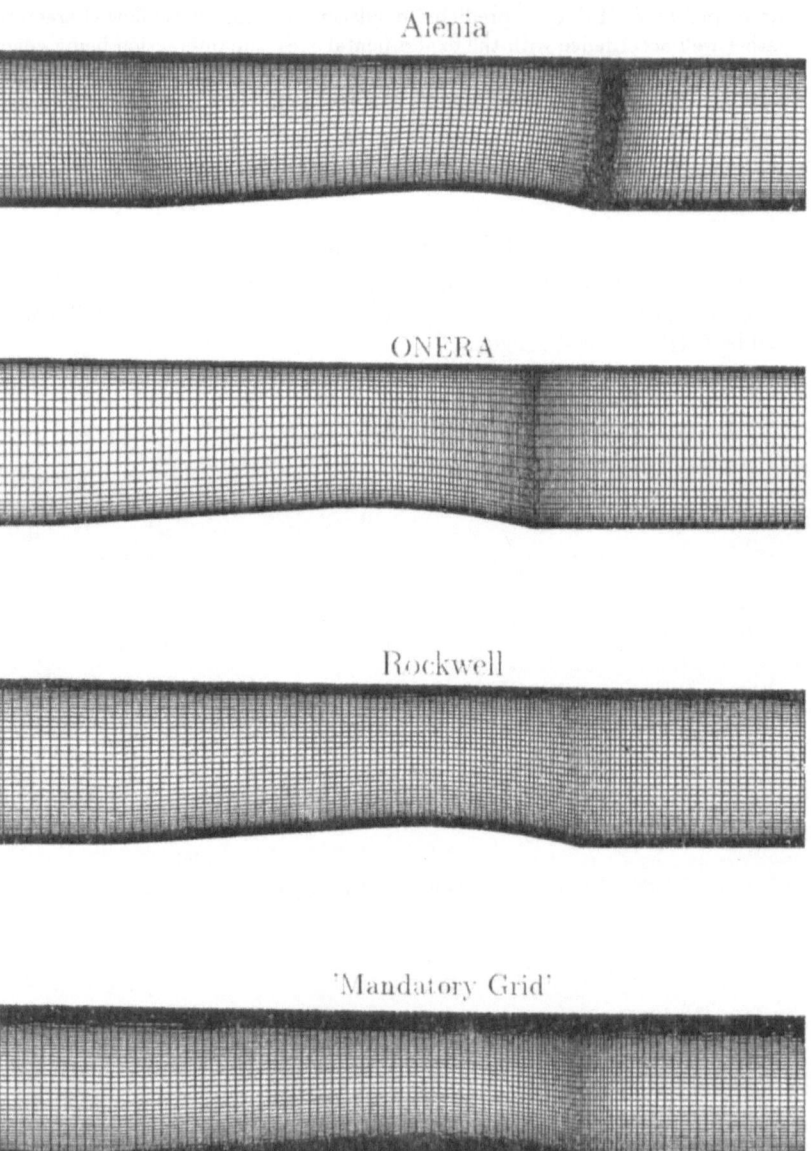

FIG. 0.1 - Computational grids.

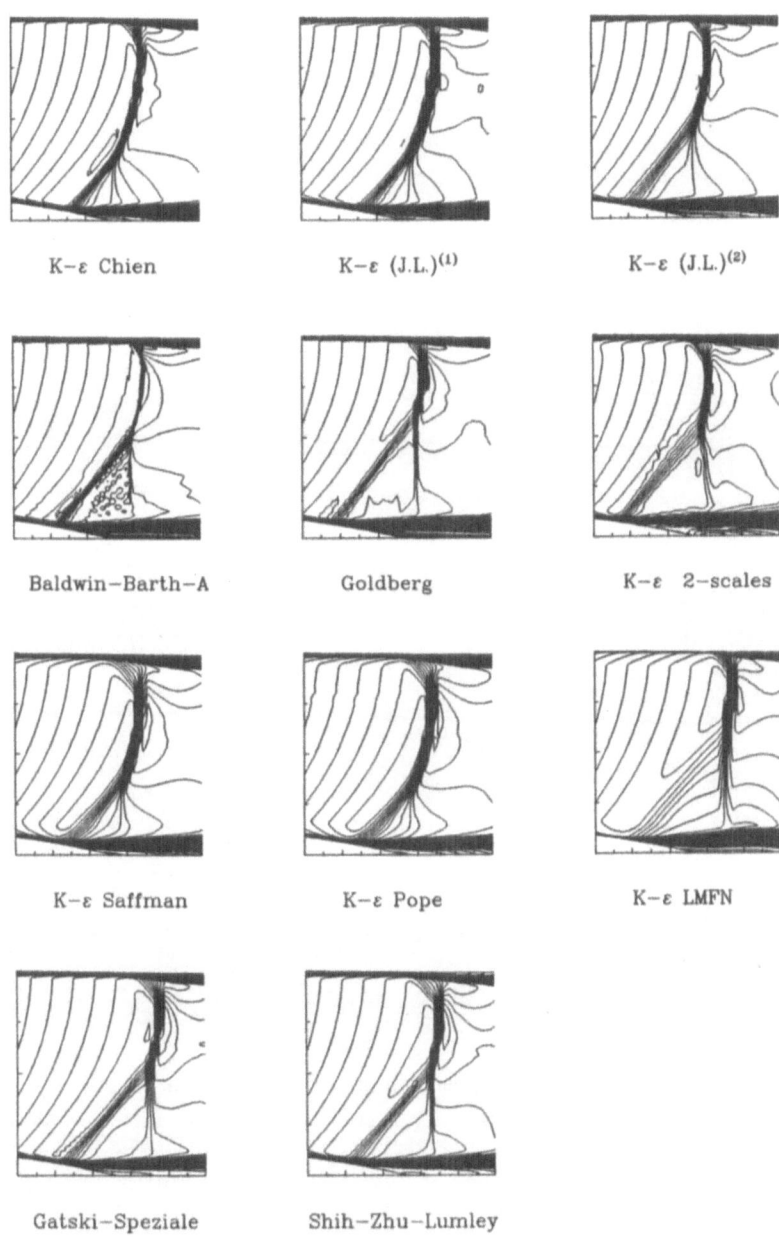

K–ε Chien   K–ε (J.L.)[1]   K–ε (J.L.)[2]

Baldwin–Barth–A   Goldberg   K–ε 2–scales

K–ε Saffman   K–ε Pope   K–ε LMFN

Gatski–Speziale   Shih–Zhu–Lumley

FIG. 0.2 - Isomach-line distributions in the interaction region, $M_{max} = 1.46$

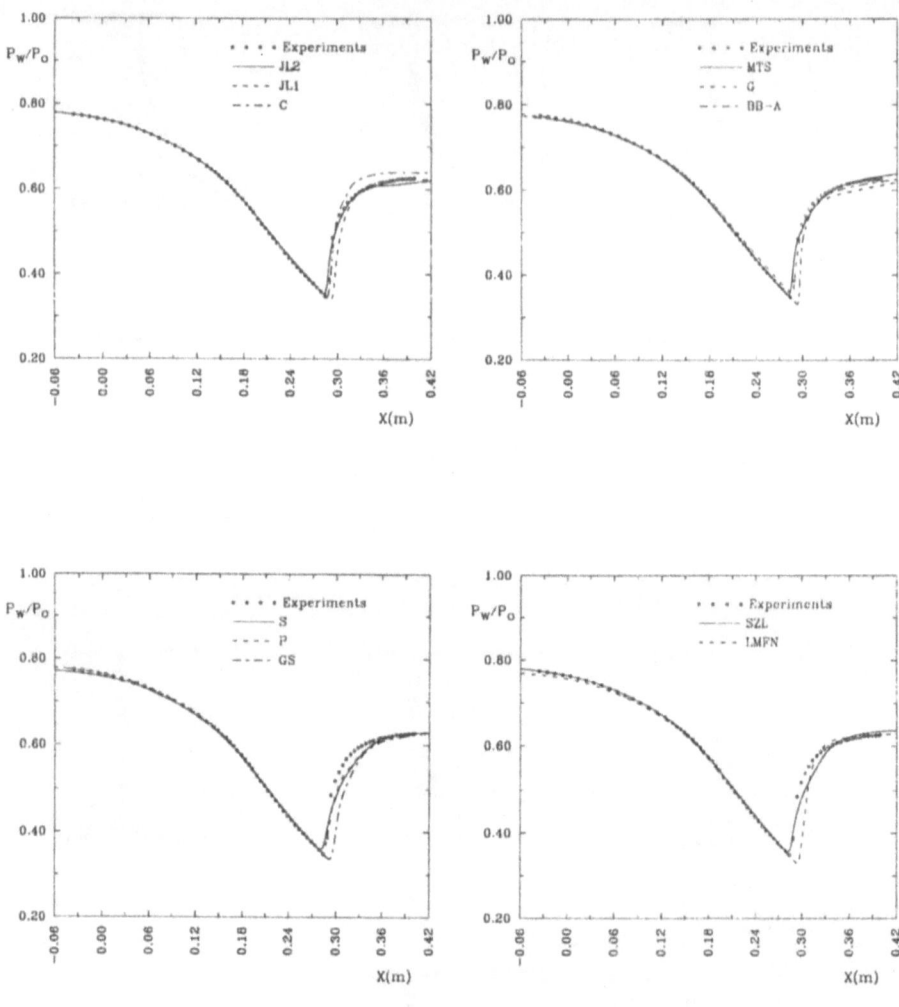

FIG. 0.3 - Upper wall pressure distribution.

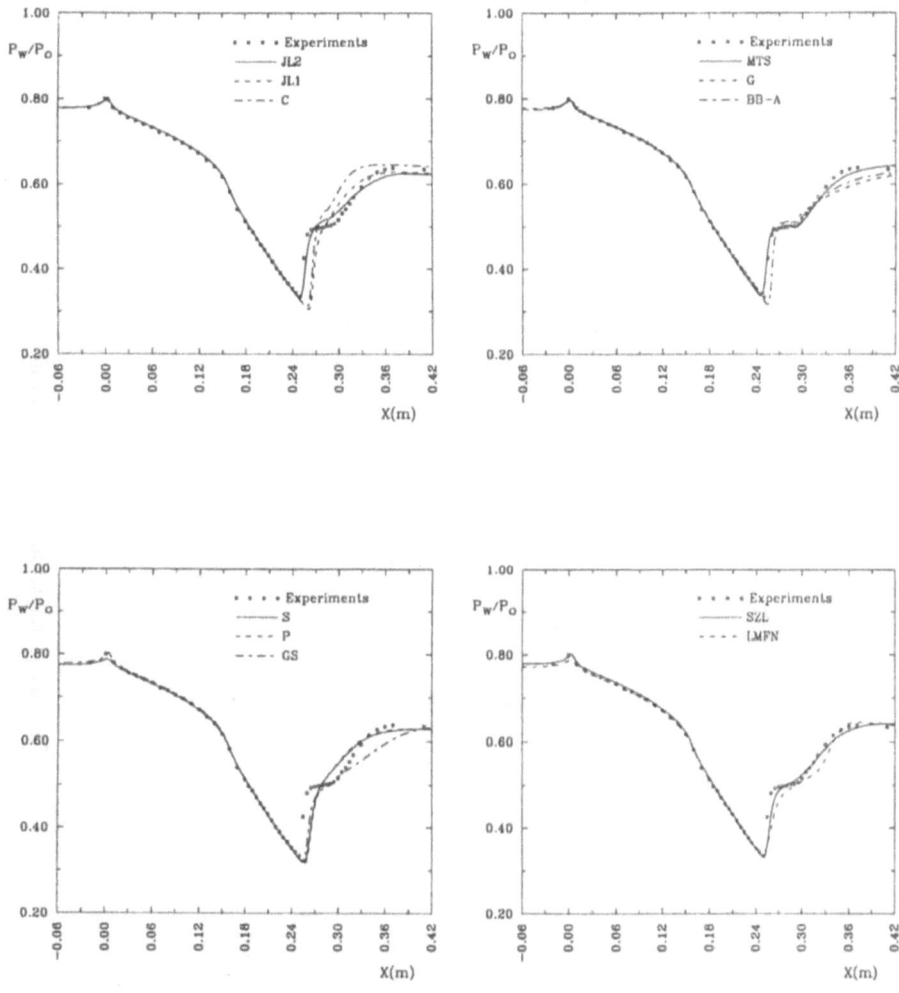

FIG. 0.4 - Lower wall pressure distribution.

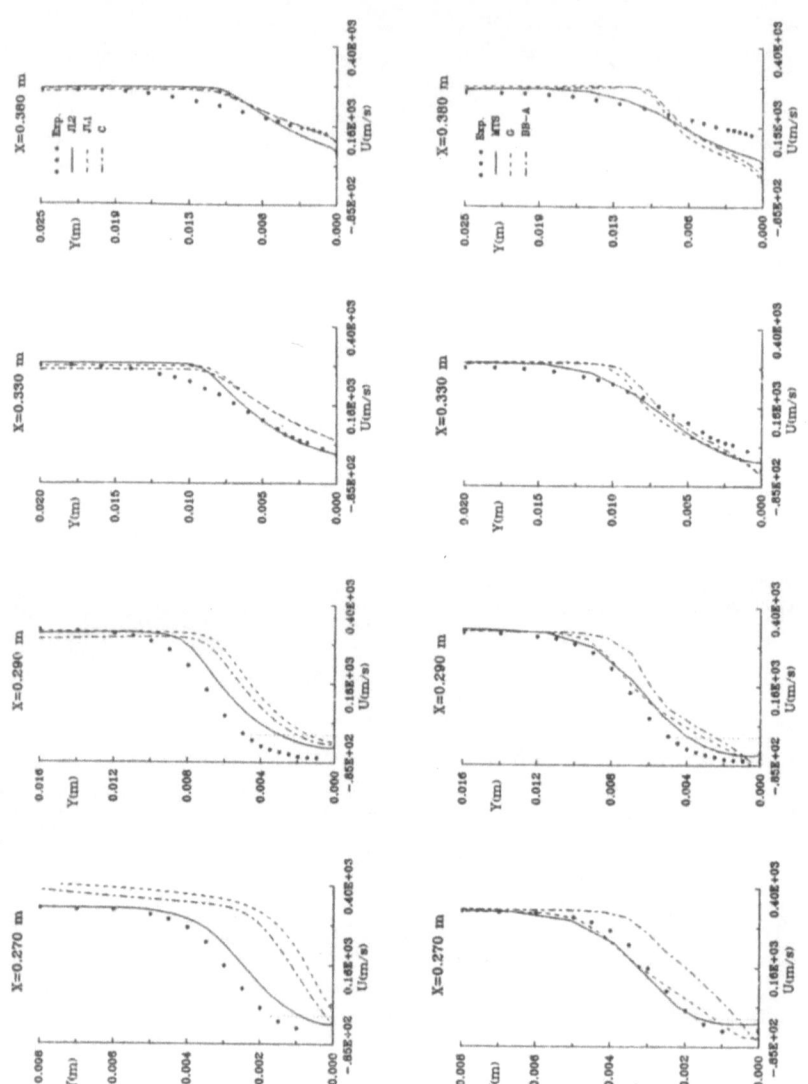

FIG. 0.5 - Longitudinal velocity profiles.

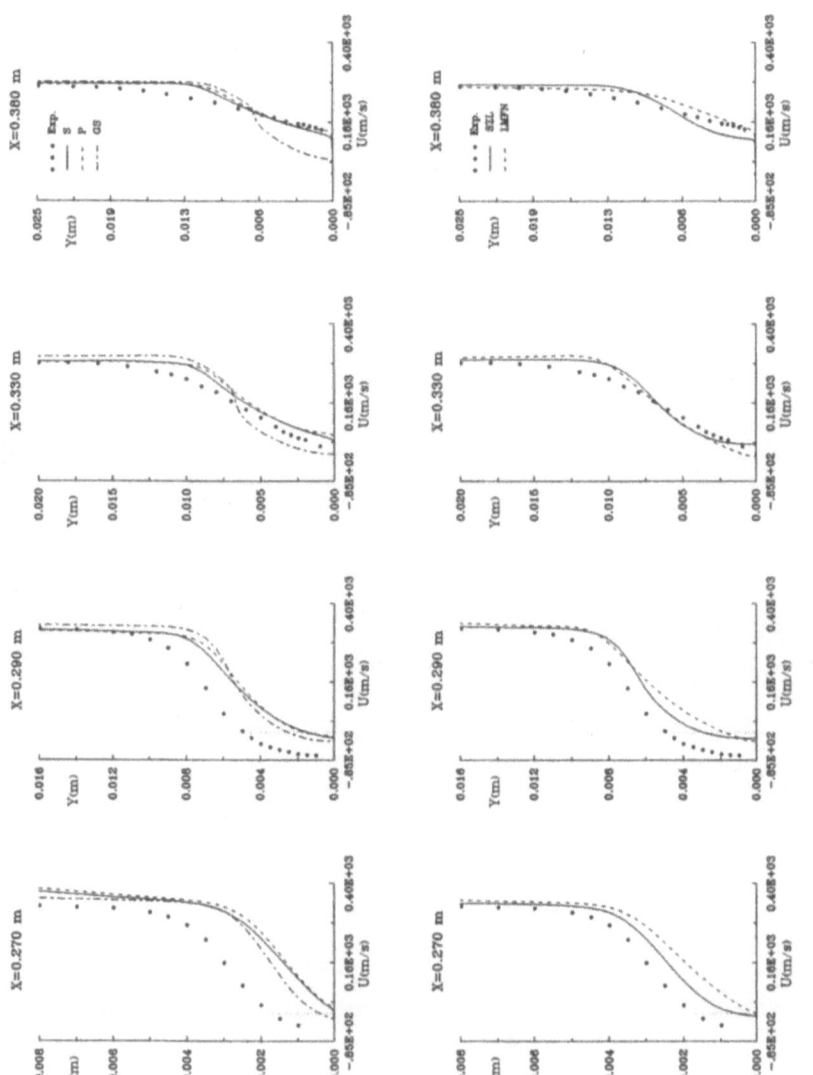

FIG. 0.5 - Longitudinal velocity profiles (cont.)

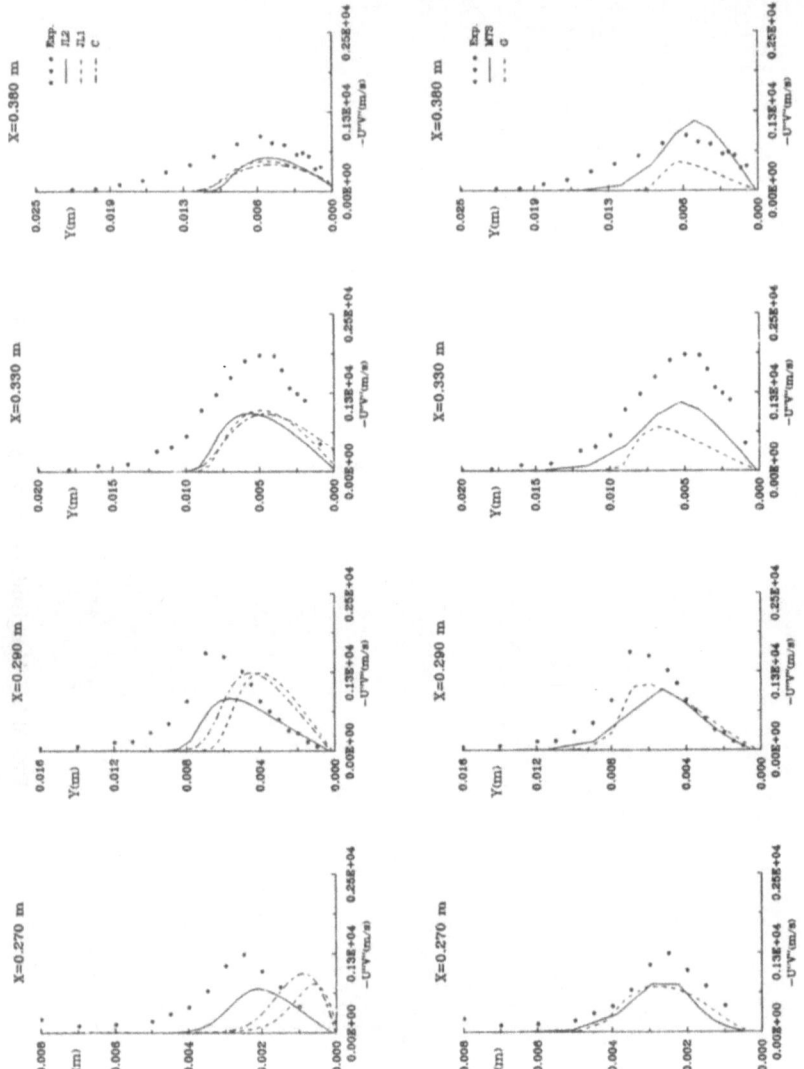

FIG. 0.6 - Shear stress profiles.

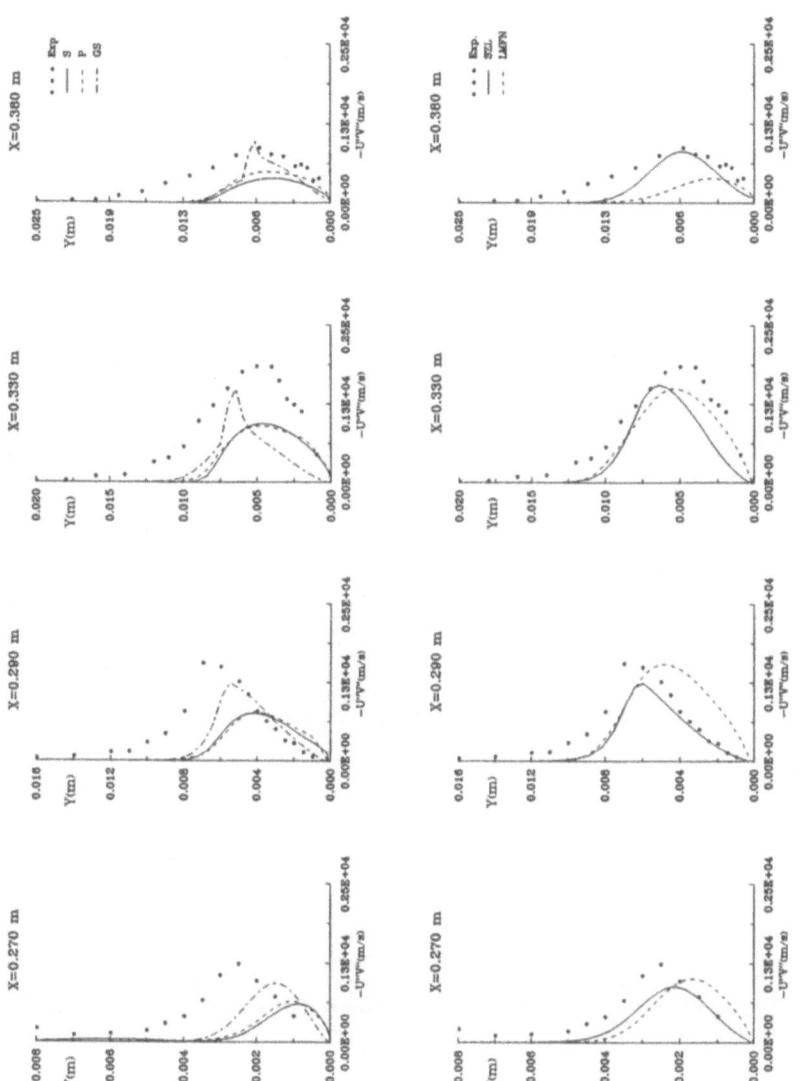

FIG. 0.6 - Shear stress profiles (cont.)

471

CHAPTER 8 : STEADY AIRFOIL FLOW

Presentation of TC8: Flow around airfoil (steady)

Leslie J. Johnston
Department of Mechanical Engineering
UMIST
PO BOX 88
MANCHESTER M60 1QD
U.K.

1. Motivation

It is proposed to use two series of test cases to evaluate the recent turbulence modelling developments for steady flow about aerofoils sections. The first series wil consider two isolated cases from the well-known RAE 2822 transonic aerofoil data set. These will enable "bench-mark" computations for comparison with existing results in the litterature. The second series of test cases will be for a different aerofoil section, and will be concerned with predictions for a range of increasingly severe transonic flow conditions.
The most severe flow condition will involve extensive shock-induced separation. The aim will be to investigate if the new developments in turbulence modelling are able to predict the experimentally-observed upstream movement and weakening of the shock wave with the onset of shock-induced separation.

2. Short Review

2.1. RAE2822 Aerofoil

Cook et al. [1] present experimental data from an extensive study of this aerofoil in the 8ft×6ft transonic wind tunnel at RAE Farnborough. The data-set includes surface pressure, skin friction and integral thickness distributions, as well as mean velocity profiles, at a range of transonic flow conditions. As such, it is one of the more complete transonic aerofoil data-sets and has been used extensively to evaluate numerical methods (see [2]).

The aerofoil model tested had an aspect ration of 3. and so the wind tunnel sidewalls should have had little influence on the flow developement. Similarly, the tunnel height/chord ratio of 4 was large enough for linearised theory to be used to determine corrections for the upper and lower walls of the wind tunnel.

The flow conditions for the two cases selected are as follows (where M is the freestream Mach number, R the freestream Reynolds number based on chord and α the incidence angle in degrees).

Case 9 : M=0.73, R=6.5 Millions, α=2.79.

Case 10 : M=0.75, R=6.2 Millions, α=2.81.

In the review presented at the Stanford Conference, some reservations were made on these measurements, essentially because of the possible blockage effects and of the probe/flow interferences. It seems however that this experiment remains a good reference for transonic flows around airfoils, in which measurements are particularly difficult.

Case 9 represents a fully-attached flow condition and so good agreement with experiments is expected.
Case 10 has proved to be a more difficult test case for computational methods, the experiment results indicating an extended region of shock-induced separation.

Computation should use the **manufactured geometry** (and not the design geometry) as tabulated in [1]. Transition was fixed in the experiments at 3% chord on both the upper and the lower surfaces. Finally, the computations should use farfield boundary conditions which include the influence of the aerofoil's circulation (farfield vortex correction).

Output required:

(1) tabulated surface pressure coefficient distributions.

(2) integrated loads:

C_l: lift coefficient

$C_d(P)$ pressure drag coefficient

$C_d(V)$ viscous drag coefficient (skin friction integration)

$C_d(T)$ total drag coefficient $= C_d(P) + C_d(V)$

$C_m(1/4)$ pitching moment coefficient about quarter chord point

(3) optional input:

(a) mean-velocity profiles at selected upper and lower surface stations

(b) distributions of displacement and momentum thickness, and local skin friction coefficient for the upper and lower surfaces.

2.2. MBB-A3 Aerofoil

Bucciantini et al. [3] present surface pressure distributions and integrated loads for an extended range of flow conditions in two wind tunnels. It is porposed to select a series of test cases at approximatively constant incidence angle but an increasing freestream Mach number. In this way, it will be possible to investigate, in a systematic fashion, the performance of the computational methods in increasingly adverse flow conditions. The more extreme flow conditions involve extensive shock-induced separation.

The test cases selected are for tests in the ARA Bedford 8inch×18inch transonic wind tunnel. The aerofoil model tested has an aspect ration of 1.6 and a tunnel height/chord ratio of 3.6. Again, linearised theory has been used to obtain corrections for the influence of the upper and lower wind tunnel walls. The low value of the model aspect ratio remains a concern, however.

The ARA Bedford wind tunnel tests involved free transition. Previous experience with computing these flows suggests fixed transition positions of 3% and 40% chord on the upper and lower surfaces respectively are reasonable for computation (see [4]).

The following table gives the flow conditions for which data are available; **recommanded cases** for computation are case 100 and 113.

ARA RUN # 84 : M=0.700; R=6.08 millions; α=1.82

ARA RUN # 13 : M=0.755; R=6.01 millions; α=1.69

ARA RUN # 18 : M=0.760; R=6.01 millions; α=1.68

ARA RUN # 16 : M=0.765; R=6.00 millions; α=1.68

ARA RUN # 44 : M=0.771; R=6.09 millions; α=1.68

ARA RUN # 100 : M=0.798; R=6.20 millions; α=1.75

ARA RUN # 113 : M=0.850; R=6.08 millions; α=1.78.

Output required:

(1) tabulated surface pressure coefficient distributions

(2) integrated loads:

C_l: lift coefficient

$C_d(P)$ pressure drag coefficient

$C_d(V)$ viscous drag coefficient (skin friction integration)

$C_d(T)$ total drag coefficient = $C_d(P) + C_d(V)$

$C_m(1/4)$ pitching moment coefficient about quarter chord point.

References

[1] COOK, P.H., MCDONALD, M.A. and FIRMIN, M.C.P., "Aerofoil RAE 2822. Pressure distributions and boundary layer measurements", AGARD AR 138, May 1979, A6-1 to A6-77.

[2] HOLST, T.L., " Viscous transonic airfoil workshop compendium of results", AIAA Paper 87-1460, 1987

[3] BUCCIANTINI, G., OGGIANO, M. and ONORATO, M., "Supercritical airfoil MBB-A3 surface pressure distributions, wake and boundary condition measurement", AGARD AR 138, May 1979, A8-1 to A8-25

[4] JOHNSTON, L.J., "Transonic aerofoil performance by solution of the Reynolds-averaged Navier-Stokes equations with a one-equation turbulence model", Notes on Numerical Fluid Mechanics, Volume 35, Vieweg, 1992, 427-436

TRANSONIC NAVIER-STOKES COMPUTATIONS ON UNSTRUCTURED GRIDS USING A DIFFERENTIAL REYNOLDS STRESS MODEL

Franck J-J CANTARITI and Leslie J JOHNSTON
Department of Mechanical Engineering, UMIST,
PO Box 88, Manchester, M60 1QD, ENGLAND

INTRODUCTION

Following the work of Stolcis [5], the present paper describes a computational method able to predict the viscous transonic flow development around single and multi-element aerofoils. The Reynolds-averaged Navier-Stokes equations applicable to compressible, two-dimensional turbulent flow are solved using a cell-centred, finite volume spatial discretisation. A multi-stage, explicit, time-marching scheme is used to advance the unsteady flow equations in time to a steady-state solution. Turbulence closure is achieved by using a differential Reynolds stress model (DRSM), which solves modelled transport equations for the Reynolds stress components themselves. Also, the method makes use of unstructured grids in order to be able to deal routinely with complex geometries such as multi-element aerofoil configurations. Results are presented for the RAE 2822 and MBB A3 transonic single-element aerofoils, comparing predictions using a differential Reynolds stress model and a two-equation k-ϵ turbulence model with experiment.

TURBULENCE MODEL

The turbulence-transport equations are modelled following the work of Launder, Reece and Rodi [3]. This DRSM is applicable only to the fully-turbulent and outer regions of the viscous layers and therefore, an appropriate turbulence model closure is required in the molecular viscosity-dominated near-wall region.

Wall Function Approach

Firstly, the so-called wall function approach is described here. In this method, the wall friction velocity U_τ in the fully-turbulent region is given by the law-of-the-wall:

$$\frac{U_T}{U_\tau} = \frac{1}{\kappa}\ln(y^+ E) \text{ with } y^+ = \frac{\rho_w U_\tau y_n}{\mu_w} \tag{1}$$

where U_T is the mean-velocity component parallel to the wall, y_n is the surface normal distance, $\kappa = 0.41$ is the von Karman constant, $E = 9$ is the appropriate value for smooth walls and subscript $_w$ indicates conditions at the wall. In the wall function approach, the turbulence transport-equations are solved only in the fully-turbulent and outer regions of the wall-bounded viscous layers. This is achieved by arranging that the centres of the first near-wall computational cells are in the range $30 < y^+ < 300$. The modelled equations for the Reynolds stresses and the turbulent kinetic energy dissipation rate ϵ are not solved in these near-wall cells, but have their values prescribed by assuming local equilibrium of the turbulence. Following Launder et al [3], the wall boundary conditions are given by:

$$\overline{uv_{s_w}} = -U_\tau^2 + y_n \frac{dP}{dX_s}, \quad \epsilon = -\overline{uv_{s_w}} \frac{dU_T}{dy_n}, \quad \overline{u_{s_w}^2} = 5.1U_\tau^2, \quad \overline{v_{s_w}^2} = 1.0U_\tau^2, \quad \overline{w_{s_w}^2} = 2.3U_\tau^2 \quad (2)$$

where subscript $_w$ indicates conditions at the wall and $_s$ refers to a coordinate system parallel to the wall. The expressions given in equation (2) must be transformed into the cartesian coordinate system used for the mean-flow and turbulence-transport equations.

Algebraic ϵ Near-Wall Formulation

A first low-Reynolds number wall boundary condition is described here. It is based on the definition of an algebraic length scale for the dissipation rate of turbulent kinetic energy ϵ. In regions adjacent to the surface, where the turbulent Reynolds number $R_y = \rho\sqrt{k}y/\mu$ is less than 250, the mean-flow and Reynolds stress transport equations are solved, whilst a characteristic length scale for ϵ is determined via an algebraic relation:

$$l_\epsilon = c_1 y_n [1 - \exp(-R_y/2c_1)] \text{ where } c_1 = \kappa c_\mu^{-\frac{3}{4}} \text{ and } c_\mu = 0.09. \quad (3)$$

The dissipation rate of turbulent kinetic energy is then defined as:

$$\epsilon = k^{\frac{3}{2}}/l_\epsilon. \quad (4)$$

Also, two low-Reynolds number versions of the pressure-strain terms are investigated. The first one follows the work of Hanjalic and Launder [1] whilst the second one was developed by Launder and Shima [4].

One-Equation Eddy-Viscosity Near-Wall Formulation

The second low-Reynolds number wall boundary condition investigated in this work is based on the one-equation model by Wolfshtein [6]. In this model, a transport equation for the turbulent kinetic energy ρk is solved in place of that for $\rho\overline{w^2}$. In the low-Reynolds number region, the Reynolds stresses are computed using the Boussinesq assumption, and the eddy viscosity μ_t and the dissipation rate of turbulent kinetic energy ϵ are defined in terms of two algebraic length scales.

NUMERICAL SCHEME

For the convection terms, the mean-flow and turbulence-transport equations are discretised using centred and first-order upwind schemes, respectively. For the diffusion terms, both sets of equations are discretised using a centred scheme. The fully centred formulation is non-dissipative and numerical dissipation is explicitly added to the mean-flow equations. The approach of Jameson et al [2] is adopted here to construct the dissipation terms. Also, the amount of artificial dissipation introduced is controlled in the boundary layers regions. Both sets of equations are integrated in time to a steady-state solution by using an explicit four-stage time-marching scheme and convergence is accelerated using local time stepping and implicit residual smoothing.

480

RESULTS

All results were computed using a fixed set of flow algorithm parameters, with the solution considered to be converged when the average density residual was reduced by four orders of magnitude. The computational grids were generated by direct triangulation of a 201x41 C-type structured grid, and contained 8220 nodes, 16000 cells and 24220 edges. The outer boundary was placed at about 15 chords from the leading edge. The normal distance of the first node varies from 0.0005 to 0.0015 for wall function calculations and from 0.00005 to 0.0001 for low-Reynolds number computations. Figures 1 and 2 show the inner regions of the grids for the RAE 2822 and MBB A3 aerofoils, respectively.

RAE 2822 Aerofoil

Figures 3 and 4 show the surface pressure distributions for Case 9, which is a fully-attached transonic flow condition. There is good agreement between all the predictions and with experiment, apart from the Hanjalic-Launder formulation which will be discarded for the rest of this work. The mean-velocity profiles are shown in Figure 5 and again, all models agree quite well with experiment. Finally, the experimental and computational lift, drag and pitching moment coefficients are compared in the table below.

Case 9	C_L	$C_D(P)$	$C_D(V)$	$C_D(T)$	$C_m(\frac{1}{4})$
Experiment	0.803	-	-	0.0168	-0.099
k-ϵ model + w.f.	0.8151	0.01373	0.00668	0.02041	-0.0947
DRSM + w.f.	0.8161	0.01390	0.00611	0.02001	-0.0953
k-ϵ model + one-eq.	0.8364	0.01469	0.00592	0.02061	-0.0997
DRSM + algebraic ϵ (LS)	0.8290	0.01430	0.00519	0.01949	-0.0994
DRSM + one-eq.	0.8302	0.01437	0.00584	0.02021	-0.0983

Figures 6 and 7 show the surface pressure distributions for Case 10, which is a more difficult case since shock-induced separation is present in the experiment. The agreement between the computations and with experiment is still good, apart for the shock-wave position, which is predicted too far downstream, and for the pressure recovery downstream of the shock-wave. Oil-flow visualisation in the experiment shows that the separation region extends from an X/c of 0.62 to 0.72. Figure 8 shows that separation is quite correctly predicted in position and extent by the DRSM when using the wall function approach and by both models when using the one-equation near-wall formulation. The mean-velocity profiles are shown in Figure 9 and indicate that only the DRSM with wall function boundary conditions gives satisfactory results in the shock-wave region. However, the flow is recovering to slowly further downstream, which is typical of DRSM calculations. Finally, the experimental and computational lift, drag and pitching moment coefficients are given in the table below.

Case 10	C_L	$C_D(P)$	$C_D(V)$	$C_D(T)$	$C_m(\frac{1}{4})$
Experiment	0.743	-	-	0.0242	-0.106
k-ϵ model + w.f.	0.7923	0.02394	0.00644	0.03038	-0.1081
DRSM + w.f.	0.7680	0.02329	0.00597	0.02926	-0.1041
k-ϵ model + one-eq.	0.8166	0.02522	0.00581	0.03103	-0.1142
DRSM + one-eq.	0.8014	0.02481	0.00578	0.03059	-0.1116

481

MBB A3 Aerofoil

The next set of calculations is a Mach number sweep at a corrected incidence angle of 1.7°. The predicted surface pressure distributions, presented in Figures 10 and 11, are only for Run 113 because there exist very little differences between the computations using the DRSM and k-ϵ turbulence models for the lower Mach number cases. The two turbulence models agree quite well with experiment, although the calculation using the DRSM predicts a better shock-wave position, especially when using wall function boundary conditions. However, neither model captures correctly the pressure plateau downstream of the shock-wave, where all computations predict flow separation from the foot of the shock-wave up to the trailing edge. The experimental and computational lift, drag and pitching moment coefficients for this case are compared in the table below.

Run 113	C_L	$C_D(P)$	$C_D(V)$	$C_D(T)$	$C_m(\frac{1}{4})$
Experiment	0.490	-	-	0.03236	-0.1103
k-ϵ model + w.f.	0.4963	0.04166	0.00556	0.04721	-0.1106
DRSM + w.f.	0.4319	0.03616	0.00502	0.04118	-0.0906
k-ϵ model + one-eq.	0.5436	0.04630	0.00525	0.05155	-0.1281
DRSM + algebraic ϵ (LS)	0.5109	0.04280	0.00370	0.04650	-0.1163
DRSM + one-eq.	0.5133	0.04338	0.00500	0.04837	-0.1175

CONCLUSIONS

A method solving the Reynolds-averaged Navier-Stokes equations applicable to compressible, two-dimensional flows has been presented. Turbulence closure was achieved by using a differential Reynolds stress model. Results were presented for the RAE 2822 and MBB A3 transonic single-element aerofoils, comparing the predictions using the DRSM and a k-ϵ turbulence model with experiment. These results indicate that improved modelling of the complex flow physics by the differential Reynolds stress model leads to better predictions, especially for the more extreme flow conditions. However, some further improvements are needed in the shock-wave and separated flow regions. Also, the low-Reynolds number versions of the DRSM were found to be less satisfactory than the wall function approach. Hence, further work should focus on improving the DRSM, in particular by looking at some of the more recent developments at UMIST in this area.

References

[1] HANJALIC, K. and LAUNDER, B.E., Contribution towards a Reynolds-stress closure for low-Reynolds-number turbulence, Journal of Fluid Mechanics, vol 74, pp 593-610, 1976.

[2] JAMESON, A., SCHMIDT, W. and TURKEL, E., 'Numerical Solution of the Euler Equations by Finite Volume Methods Using Runge-Kutta Time Stepping Schemes', AIAA Paper 81-1259, June 1981.

[3] LAUNDER, B.E., REECE, G.J. and RODI, W., 'Progress in the Development of a Reynolds-Stress Turbulence Closure', Journal of Fluid Mechanics, vol 68, pp 537-566, part 3, 1975.

[4] LAUNDER, B.E., SHIMA, N., 'Second Moment Closure for the Near-Wall Sublayer: Development and Application', AIAA Journal, vol 27, No 10, 1989.

[5] STOLCIS, L., 'Computation of the Turbulent Flow Development Around Single- and Multi-Element Aerofoils', PhD Thesis, University of Manchester, University of Science and Technology (UMIST), 1992.

[6] WOLFSHTEIN, M., The velocity and Temperature Distribution in One-Dimensional Flow with Turbulence Augmentation and Pressure Gradient, Int. J. Mass Heat Transfer, vol 12, pp 301-318, 1969.

FIGURES

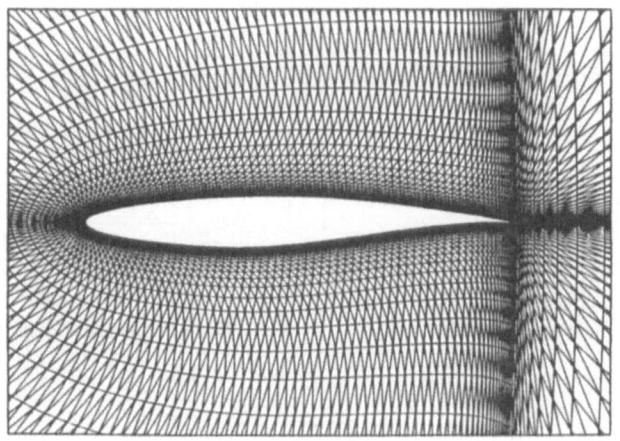

Figure 1
RAE 2822 Aerofoil - Computational grid

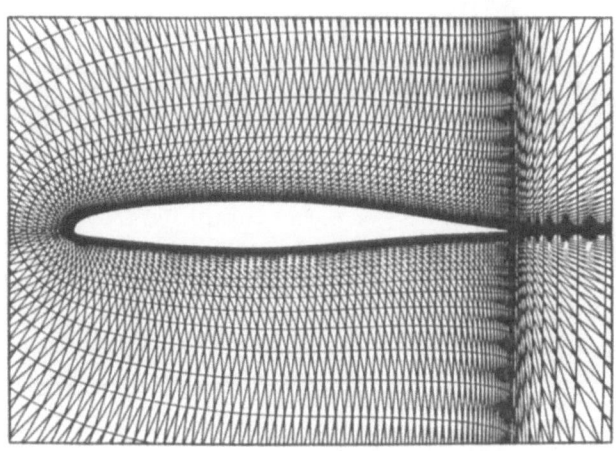

Figure 2
MBB A3 Aerofoil - Computational grid

483

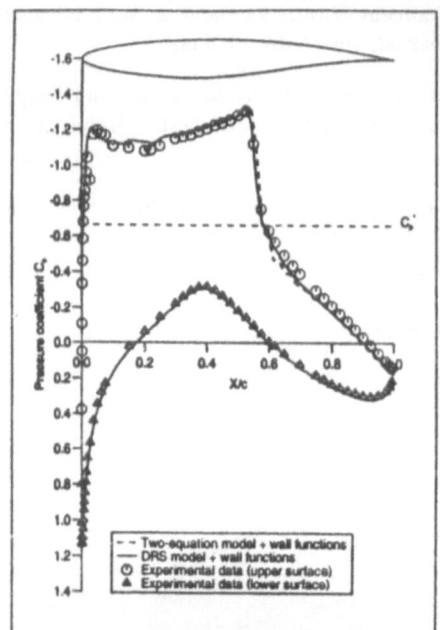

Figure 3
Surface pressure distributions

Figure 4
Surface pressure distributions

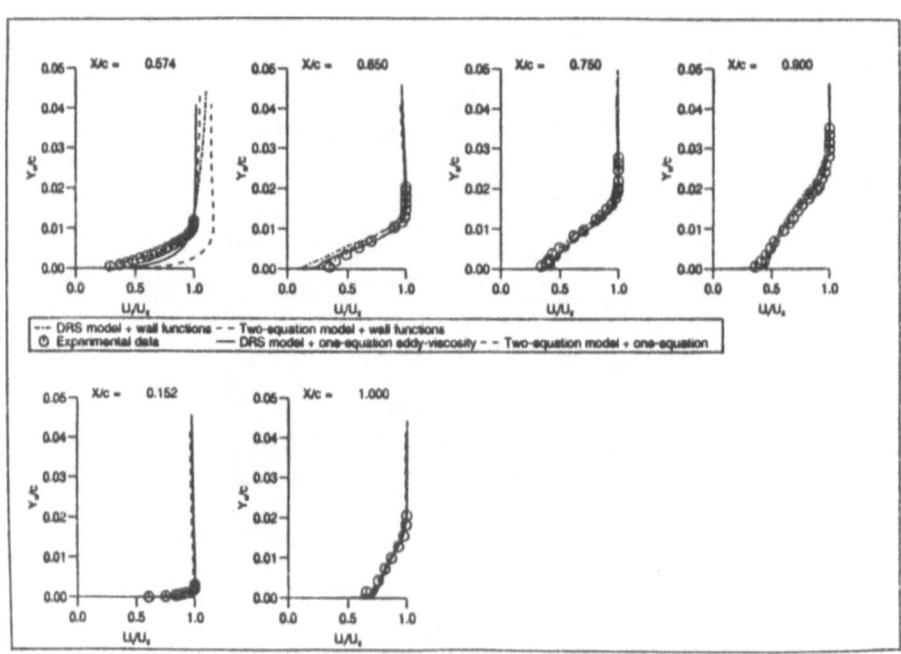

Figure 5
Upper and lower surface velocity profiles

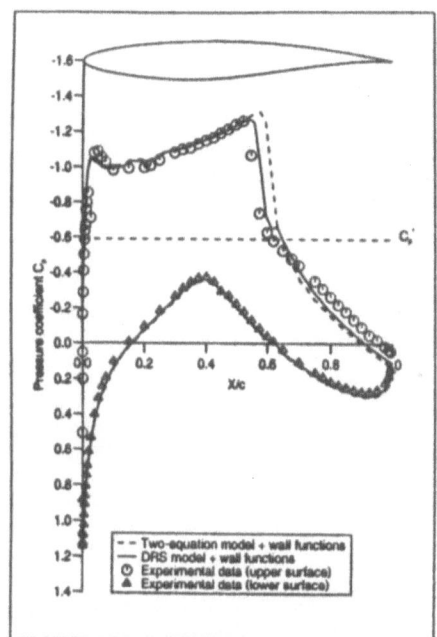

Figure 6
Surface pressure distributions

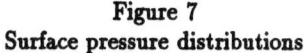

Figure 7
Surface pressure distributions

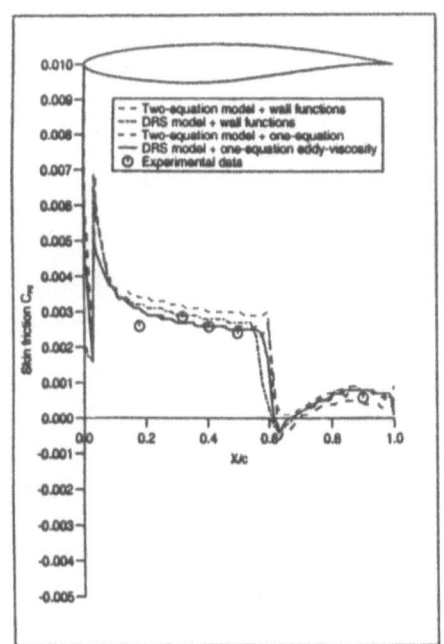

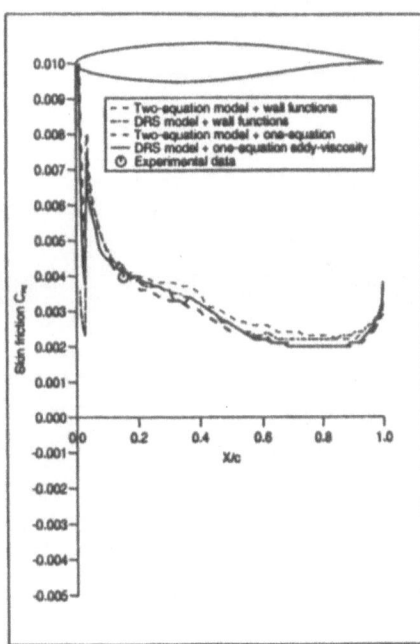

Figure 8
Upper and lower surface skin friction distributions

485

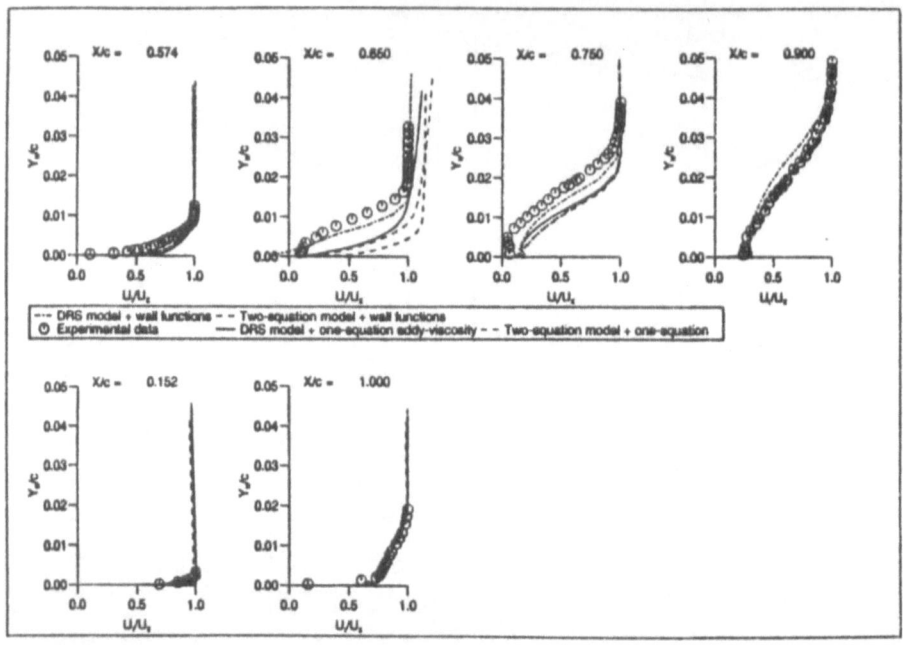

Figure 9
Upper and lower surface velocity profiles

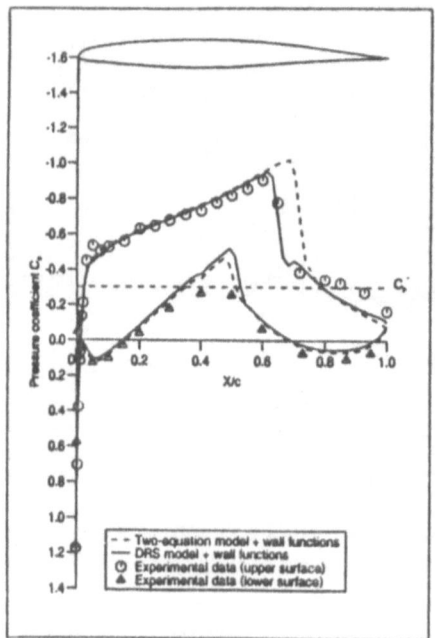

Figure 10
Surface pressure distributions

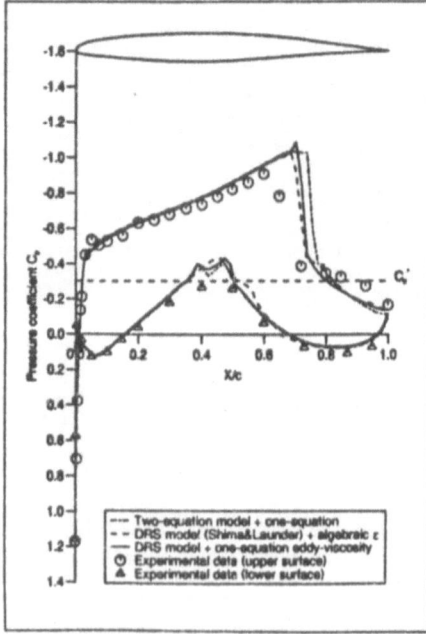

Figure 11
Surface pressure distributions

IMPLICIT MULTIGRID COMPUTATIONS OF A AIRFOIL FLOW

G. Carre

SNECMA Moissy Cramayel, France and

INRIA, BP 93, 06902 Sophia-Antipolis Cedex, France.

INTRODUCTION

We apply a numerical approach using a linear multigrid method by agglomeration technique to solve the Navier-Stokes equations, coupled with a first order turbulence model. Several numerical results, for different flows, on a single airfoil, are presented.

Both experiments and numerical results allow a better understanding of the physical processes of complex flows. The purpose of this work is to experiment a new computational method, involving a multigrid process [2], for solving turbulent flows around a single airfoil configuration. We solve different test cases, using the experimental data for comparison, in order to evaluate the influence of numerical viscosity and of the mesh convergence.

THE $K - \epsilon$ TURBULENCE MODEL

Governing equations : the full system of the compressible Navier-Stokes equations coupled with a two-equation turbulence model, is considered and we first recall in short some standard features of this model. Favre averages are used for the instantaneous variables except for the pressure and density. The Reynolds stresses which appear after the statistical treatment are modeled by the Boussinesq assumption. It involves an eddy viscosity term defined by a length-scale (related to the dissipation rate ϵ) and a velocity-scale (k) written as :

$$\mu_t = C_\mu \frac{\rho k^2}{\epsilon}$$

where C_μ is a empirical constant. To close the system of equations, two additionnal transport equations for turbulent quantities are solved.

In a conservative form, these equations can be written as :

$$\frac{\partial W}{\partial t} + \frac{\partial F(W)}{\partial x} + \frac{\partial G(W)}{\partial y} = \frac{1}{Re}\left(\frac{\partial R(W)}{\partial x} + \frac{\partial S(W)}{\partial y}\right) + \frac{\partial \frac{1}{Rt}\tilde{R}(W)}{\partial x} + \frac{\partial \frac{1}{Rt}\tilde{S}(W)}{\partial y} + \Omega(W)$$

where :

- $W(x, y, t)$ contains laminar conservative variables $\rho, \rho u, \rho v, E$ and turbulent conservative variables $\rho k, \rho \epsilon$

- $F(W)$, $G(W)$ are convectives fluxes

Boundary treatment : wall functions, deduced from the incompressible boundary layer equations, express that the mean and turbulent flows are solved up to a distance δ (assumed to be in the turbulent region) from the wall [7].

SPATIAL APPROXIMATION

The spatial approximation is defined on a triangulation; degrees of freedom are the values of conserved variables on the nodes, situated on the vertices of the mesh. The spatial discretization results from a Mixed finite Element, finite Volume (MEV) :

- For the convective terms a Finite Volume integral formulation is applied to the dual tessellation defined by medians. We assume that the computational domain Ω is bounded by a polygonal, denote by :

    - I_h a given triangulation of Ω

    - N_h the total number of vertices in I_h

    - ϕ_j the basis function associed with each node s_j.

We derive a new finite volume partition of Ω, called the dual mesh of I_h and built by the control volume C_i around each vertex s_j.

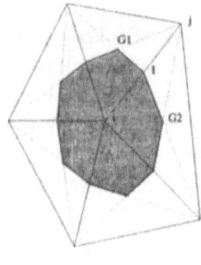

Figure 1: Cell C_i

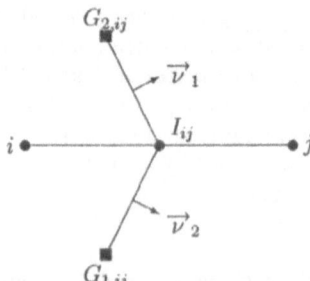

Figure 2: Interface ∂C_{ij} separating nodes i and j

The discretization of advection terms uses a MUSCL-like second-order Godunov upwinding, involving an approximated Riemann solver. For fluid variables, the flux throughout the ∂C_{ij} interface (see Figure 2) is solved with Roe 's numerical flux function whereas a positivity preserving multi-component Riemann flux developed in [6] is used for turbulent fluxes.

- For diffusive terms, a finite-element variational formulation is applied [3]

Theoretical proofs of the compatibility between the two formulations have been built in [1].

It is well known that Newton iteration cannot be applied to compressible flows starting from arbitrary initializations. In order to approach the convergence domain of a (modified) Newton iteration, an Implicit, Euler Backward time advancing is applied [4]:
The following TVD limitations options can be applied:
- no limitation.
- van Albada limiter applied to k and ϵ variables
- van Albada limiter applied to all primitive variables
In this paper, except when it is explicitly mentioned, computations are made without limiters on fluid variables and with van Albada limiters on turbulent variables. That gives a global second-order accurate. The space accurate approximation can be modified throughout a upwinding parameter ($\beta \in [0,1]$) :
- $\beta = 0$ gives a central differencing
- $\beta = 1$ gives a fully upwind differencing
- $\beta = \frac{1}{2}$: gives a half-upwind second- order approximation for the convective flux and corresponds to the effect of van Albada limiters on a smooth function
- $\beta = \frac{1}{6}$: results in an upwind biased scheme involving a numerical viscosity that is only 33% of the previous option.

NUMERICAL EXPERIMENTS

The advantage of an implicit scheme is to obtain solutions rather independantly on the spatial dissipation. We present comparaisons for different levels of upwinding and for two nested meshes. Test cases are RAE2822 and A3MBB3, also considered in [9].

TEST CASES CONFIGURATIONS FOR RAE2822
TC9 :
The flow is composed by a freestream Mach number of 0.734, a Reynolds number of 6.5×10^6 and a flow incidence of 2.79 degrees. The transition, on the upper and lower surfaces, is fixed at 3% of the chord length in accordance with experiments.
TC10 :
The flow is composed by a freestream Mach number of 0.754, a Reynolds number of 6.2×10^6, and a flow incidence of 2.81 degrees. Initialization and transition are the same as previously.

WALL LAW THICKNESS
In order to avoid a computation with an analytic thickness dependant of the shear velocity at each time step, the analytic thickness δ will be considered as a constant. This assumption decreases the complexity of the computation, but we must verify that the y^+ condition is respected in every case. In fact, a boundary layer can be divided in different parts.

- $0 \leq y^+ \leq 3$ represents the viscous sublayer

- $3 - 5 \leq y^+ \leq 30 - 40$ represents the buffer zone

- $30 - 40 \leq y^+ \leq 100 - 300$ represents the logarithmic zone .

489

Several expressions for the logarithmic law in the buffer zone and the logarithmic zone can be found (see [8]). But in our case, we suppose that the behaviour is similar in these two regions. Futhermore, the logarithmic law can be assumed to be valid up to $y^+ = 1000$ (depending on the Reynolds number) [10]. In order to respect this last condition, we choose a thickness in accordance to the previous assumption :

- $\delta = 5 \times 10^{-4}$ involving $y^+_{max} \simeq 100$ and the mean value is equal to $y^+ \simeq 60$ at the beginning of the computation.

TC9 : Figures 3, 4 depict the pressure and the skin friction coefficients. The shock position, except the bottom, is close to the experiment, and the pressure profile on the lower surface approach correctly the experimental data.

TC10 : Results with the same value of the analytic thickness than previously are given on Figure 8. In this case, the turbulence model is not in accordance with experimental data. The computation gives a shock located at the rear part of the airfoil. furthermore, the experience gives a separated region at the foot of the shock, but the computation shows that the flow is attached along the profile as depicted on Figure 7, The large discrepancy between the experimental data and the computation can be explained by the wall law function employed. In our case, the simple wall law does not predict correctly the interaction between the shock wave and the boundary layer. In fact the assumption that production is equal to the dissipation is not true in this region.

NUMERICAL VISCOSITY DEPENDANCY

In order to show the influence of the space accurate approximation, we studied several approximation of convective terms. We tried two values of upwinding parameter:
- $\beta = \frac{1}{2}$ (b1)
- $\beta = \frac{1}{6}$ (b2)

TC9 - TC10

The computations of TC9 and TC10, with the previous values of β give the results sketched in Figures 4 and 8. The pressure coefficient profiles show that the different spacial accuracies give similar results; this seems to indicates that these results are not much dependant of the numerical approximation.

MESH CONVERGENCE FOR TC9 AND TC10

We have refined the initial mesh, containing 8220 vertices, to show the mesh independancy solution. A computation on a finest grid, obtained by subdividing the initial mesh, contains 32440 vertices. Figures 5 and 9 illustrate the solution independancy throughout the pressure coefficient profiles for the coarse and the fine meshes.

TEST CASE CONFIGURATION FOR A3MBB3 RUN100-RUN113

RUN113 :

The flow is composed by a freestream Mach number of 0.85, a Reynolds number of 6.08×10^6 and a flow incidence of 1.78 degrees. The transition, on the upper surface, is fixed at 3% of the chord length in accordance with experiments. The computational grid has been generated by using the structured mesh given by UMIST, and it contains 8220 nodes. The pressure coefficient profile exhibited in Figure 18 shows many differences with the experimental data. The shock is located at the rear part of the airfoil. As for test-case 10, the turbulence model is not in accordance with experiment in the boundary layer

490

detachment region.

MESH CONVERGENCE FOR RUN113

We have refined the initial mesh, containing 8220 vertices, as the previous case, in order
to show the mesh convergence. A computation on a finest grid, Figure 18, illustrates the
solution independancy throughout the pressure coefficient profiles for the coarse and the
fine meshes. For this test-case, we note some differences between the fine and the coarse
grid at the rear part of the airfoil, behind the shock.

RUN100 :

The flow is composed by a freestream Mach number of 0.798, a Reynolds number of
6.2×10^6 and a flow incidence of 1.75 degrees. The transition, is fixed at 3% of the
chord length , on the upper surface, and 40% on the lower surface, in accordance with
experiments. The pressure coefficient profile exhibited in Figure 14 shows many differences
with the experimental data. The shock is located at the rear part of the airfoil. As for
test-case 10, the turbulence model is not well in accordance with experiment in the shock-
wave/boundary layer interation region.

CONCLUDING REMARKS

The high Reynolds turbulence model, initially developed to solve internal flows, has been
applied to two airfoil problems. Iterative convergence is obtained in 300-400 time steps.
However, comparaison to experiments is not completely satisfactory. To be more precise,
the approximation of flow for Case 9 gives good agreement with experiment data. Unlike the previous test case, Case 10 shows a rather large discrepancy with experimental
data. The mesh convergence independance and the numerical viscosity variation studies
show that the several problems encountered with case 10 are independant of numerical approach. Futhermore, the two others test-cases (RUN100-RUN113) shows the same
discrepancy between the experimental data and computation as Case 10 for the shock
position. We believe that the bad approximation of these flows is caused by the logarithmic law applied for this transonic application. In fact, these wall functions, deduced from
the incompressible boundary layer equations, do not take in account the adverse pressure
gradient term, which is very important in the interaction of shock-wave / boundary layer.
In order to show the validity of the last remark, the results are depicted on the Figures 11,
15 and 19 where we modified the standard wall functions. The first approach consist to
applied a Mellor model (see Figure 10), in order to take in account the pressure gradient
term, coupled to a Menter model which is a non-equilibrium assumption ([11] and [12]).

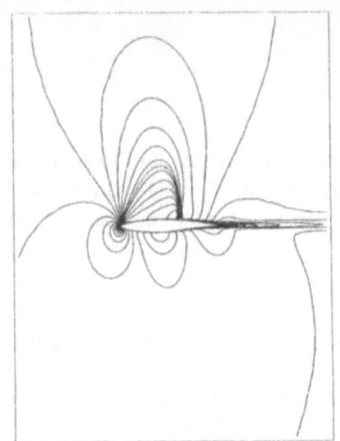

Figure 3: Isomach lines for TC9

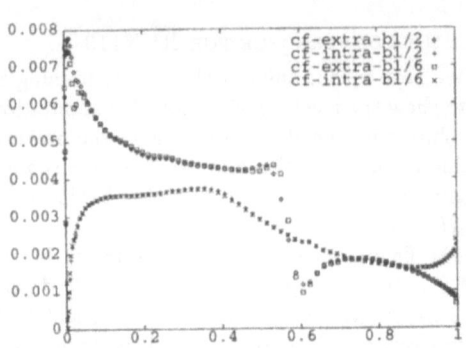

Figure 4: Skin friction (cf)

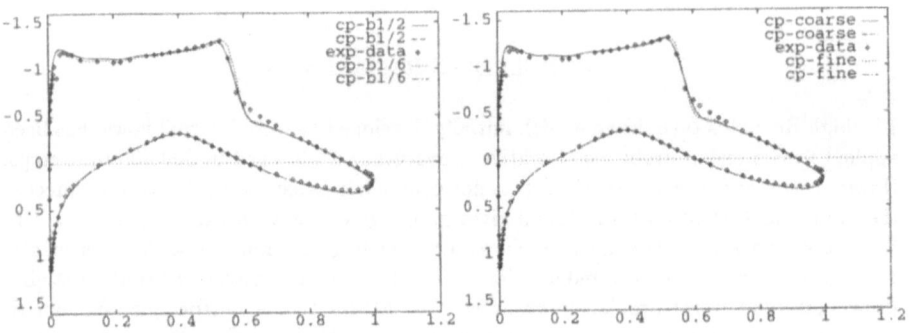

Figure 5: Numerical viscosity variation on pressure coefficient (cp)

Figure 6: Mesh dependance

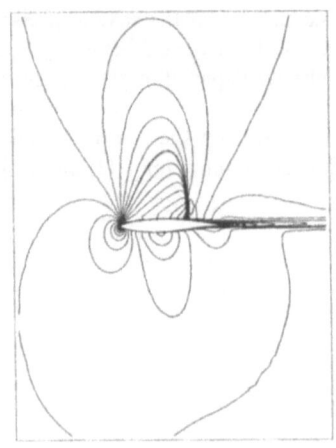

Figure 7: Isomach lines for TC10

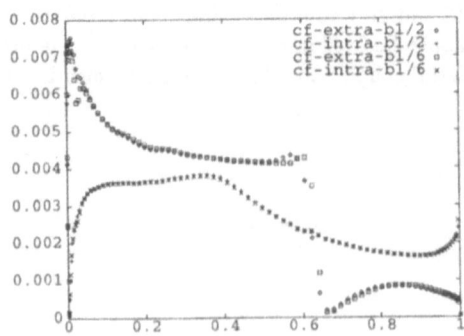

Figure 8: Skin friction (cf)

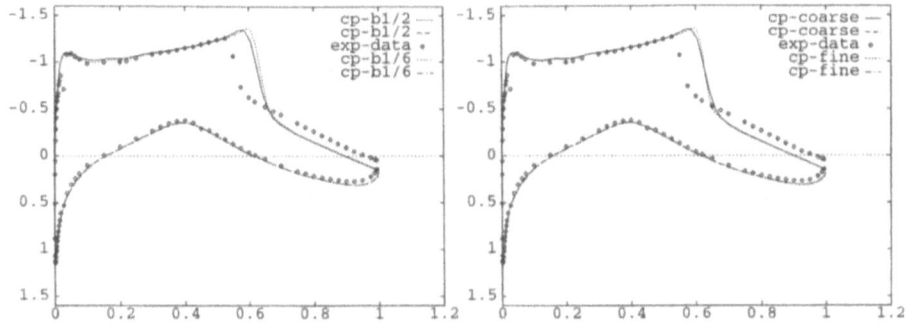

Figure 9: Numerical viscosity variation on pressure coefficient (cp)

Figure 10: Mesh dependance

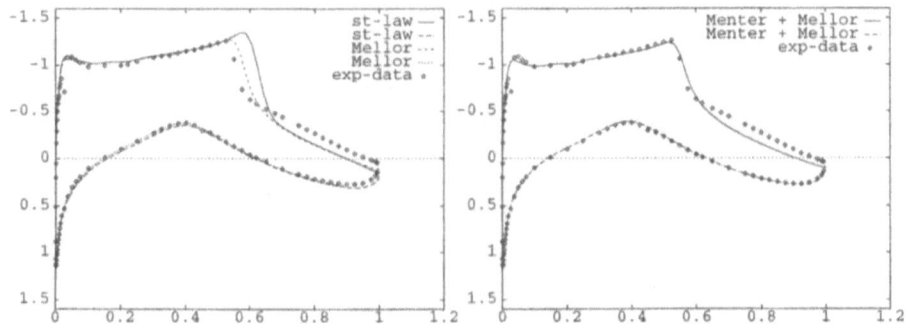

Figure 11: Wall function comparison for TC10

Figure 12: Mellor-Menter model

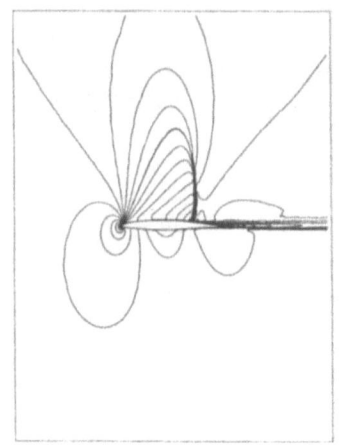

Figure 13: Isomach lines for RUN100

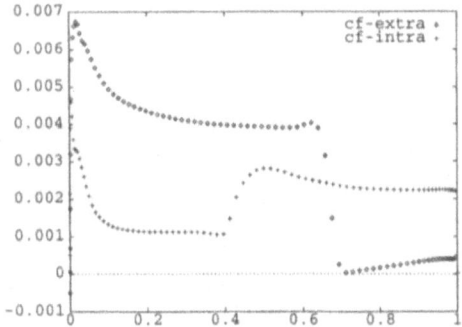

Figure 14: Skin friction (cf)

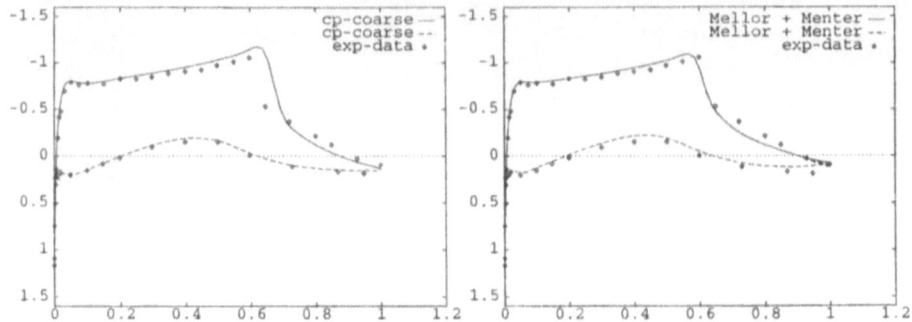

Figure 15: Pressure coefficient (cp)

Figure 16: Mellor-Menter model

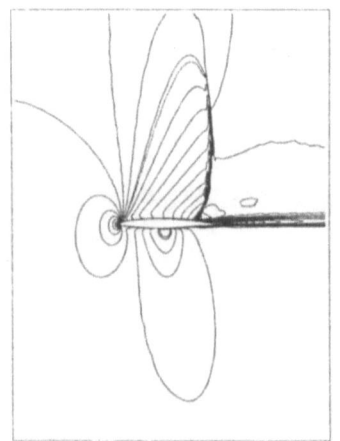

Figure 17: Isomach lines for RUN113

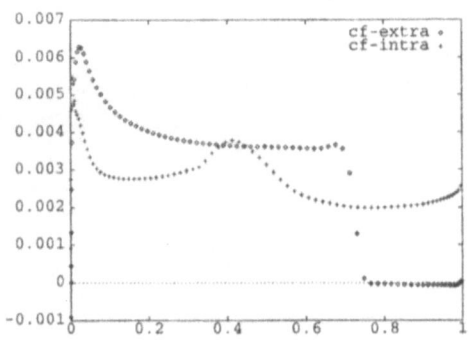

Figure 18: Skin friction (cf)

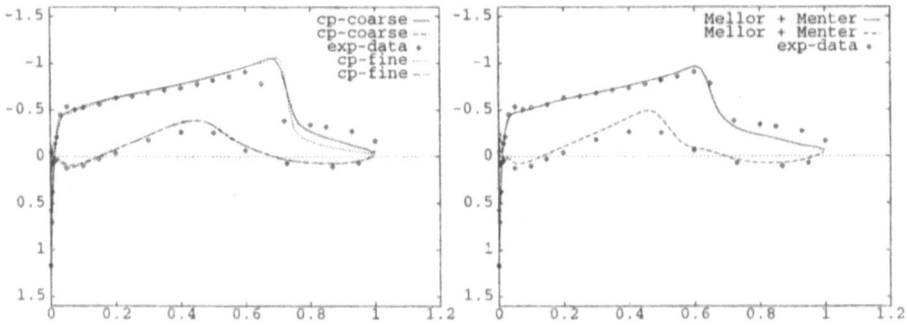

Figure 19: Mesh dependance

Figure 20: Mellor-Menter model

Acknowledgements
We thank C. OLIVIER for the generation of the grids, M. RAVACHOL and F. CAN-
TARITI for fruitfull discussions

REFERENCES

[1] MER K., "Variational Analysis of a Mixed Finite Element / Finite Volume Scheme on General Triangulations"- *Rapport de Recherche INRIA N° 2213 (1994)*

[2] KOOBUS B., LALLEMAND M.H., DERVIEUX A. -"Unstructured Volume-Agglomeration MG : Solution of Poisson Equation"- *Rapport de Recherche INRIA N° 1946, (1993)*

[3] FEZOUI L. and al. - "Résolution numérique des équations de Navier-Stokes pour un fluide compressible en maillage triangulaire" - *Rapport de recherche INRIA N° 1033, (1989)*

[4] STEVE H. - "Schémas implicites linéarisés décentrés pour la résolution des équations d'Euler en plusieurs dimensions" - *Thèse, Université de Provence, (1988)*

[5] Le RIBAULT C. - "Simulation des écoulements turbulents compressibles par une méthodes mixtes EF/VF" - *Thèse, Ecole Centrale de Lyon (1991)*

[6] LARROUTUROU B., "How to preserve the mass fraction positivity when computing compressible multi-component flows"- *Rapport de recherche INRIA N° 1080, (1989)*

[7] LARROUTUROU B., OLIVIER C. "On the numerical approximation of the $K - \epsilon$ model for two dimensionnal compressible flow"-*Rapport de recherche INRIA N° 1526, (1991)*

[8] JAEGER M. and DHATT G. - "An extended $K - \epsilon$ finite element model - *Int. J. Numer. Methods Fluidf, Vol.14, pp.1325-1345 (1992)*

[9] HAASE W. and al. EUROVAL -An european initiative on validation of CFD codes- *Notes on Numerical Fluid Mechanics- Vol 42*

[10] STOLCIS L. "Computation of the turbulent flows development around single and multi-element aerofoils"- *PhD Thesis, Univ. of Manchester (1992)*

[11] MELLOR G.L. "The effects of pressure gradients on turbulent flow near a smooth wall" - *J. Fluid Mech., vol. 24, pp 255-274, (1966)*

[12] MENTER F. "zonal two equations K-w Turbulence Models for Aerodynamic Flows" - *AIAA 93-2906 24 th F.D. Conference Orlando, (1993)*

Unstructured Grid Solutions using k-ε with Wall Functions

F. Fortin and D.J.Jones
Institute for Aerospace Research
National Research Council
Montreal Road
Ottawa, Canada
e mail: denis.jones@nrc.ca

INTRODUCTION

The method used for solution is very similar to that presented in Ref 1 and is based on Jameson's type of algorithm. Improvements have been made to the grid generation technique and for the present exercise a structured triangular grid is used very close to the airfoil surface and an unstructured triangular grid is generated outside of this layer to a distance of 1.5 chords; outside of this grid is a structured triangular grid that extends the flow region to about 12 chords from the centre of the airfoil. An unstructured grid Navier Stokes code [1] that will handle laminar and turbulent flows is used as the flow solver. For turbulence closure a k-ε model with wall functions is used.

GRIDS

The grids for both the RAE2822 and MBB-A3 airfoils were generated using IAR grid generation codes. First a structured grid is generated very close to the surface for about 6 layers, each of thickness 0.001 (for a chord of 1), and extending a small distance of about 0.01 downstream of the trailing edge. This structured layer is based on a C-grid normal to the surface that generates quadrilaterals which are then split diagonally to give triangles. Outside this layer the grid is unstructured and is based on a Delaunay algorithm which also forces triangles to be quite elongated near the join to the structured layer so that there is not usually a too drastic change in triangle areas across the join especially for the fine grid (see Fig 1). Another point to note is that, beyond a radius of 1.5 from the centre of the airfoil, a structured grid is used that extends the grid to about 12 chords; in this region the triangles are nearly equilateral.

This strategy of using structured layers joining to an unstructured grid was used so that the present grids are very similar to IAR grids obtained for multi-element airfoil computations. This is as opposed to using a structured quadrilateral grid and cutting across the diagonals to form triangles in the whole flow field; such grids cannot be used for complex geometries.

The results for each case are shown in Figs 2 (pressure coefficients) and in Figs 3 (skin friction). Also, tabulated below, are the force and moment coefficients.

In each case three grids (coarse, medium, fine) having 60, 120 and 240 points on the airfoil were used. The structured grid near the surface had 5 layers for RAE2822 and 7 layers for MBB-A3, each of thickness 0.001 (giving y+ of about 50 to 100). In the far field which is a circle at about 12 chords there were 36 points at 10 deg apart.

RAE2822.

		CL	CM	CD	CDvisc	CDpress
Case 9.	60 points	0.8359	−0.1009	0.0262	0.0077	0.0185
	120	0.8427	−0.1008	0.0234	0.0074	0.0160
	240	0.7862	−0.0879	0.0195	0.0065	0.0130

Comments: The Cp results change slightly for the 3 grids and the fine grid, GRID 3, shows a very good prediction except aft of the shock. The drop in lift for GRID 3 is surprising but is due to a small shift in the Cp level on the upper surface prior to the shock and also since the shock moves forward by about 2%. Drag appears to improve with grid size although one has to expect an overprediction with the wall function method (Ref 2).

		CL	CM	CD	CDvisc	CDpress
Case 10.	60 points	0.8298	−0.1173	0.0367	0.0077	0.0290
	120	0.8439	−0.1196	0.0349	0.0074	0.0275
	240	0.7415	−0.0948	0.0278	0.0064	0.0214

Comments: The agreement of the pressure distribution with the experimental data improves markedly as the grid is refined. The trailing edge pressure decrease and the negative skin friction just aft of the shock for the fine grid result are consistent with the experiment although because of using wall functions these results must be viewed with caution.

MBB-A3 Airfoil

General comments: For all cases at the lower Mach numbers (up to and including Case 44, M=0.771 - not shown due to lack of space), the pressure distributions for the three grids are in good agreement with each other but they are consistently different to the experimental results. In view of the good agreement for the RAE 2822 case it would appear that the experimental wall corrections are in doubt.

Case 100.

		CL	CM	CD	CDvisc	CDpress
	60 points	0.7026	−0.1089	0.0379	0.0062	0.0317
	120	0.5943	−0.0777	0.0284	0.0058	0.0226
	240	0.6110	−0.0858	0.0295	0.0053	0.0242

For this case the pressure distributions differ quite a lot for the three grids and the finest grid Cp appears to match the experiment very well; however this is fortuitous due to the comments on wall corrections above. The trailing edge pressure decrease (shown very clearly for the fine grid) is indicative of separated flow on the upper or lower surface. The skin friction also showed a negative value just aft of the shock on the upper surface.

Case 113.

	CL	CM	CD	CDvisc	CDpress
60 points	No convergence				
120	0.4416	−0.0925	0.0439	0.0057	0.0382
240	0.4559	−0.0976	0.0441	0.0054	0.0387

For this case, convergence could not be obtained for the coarse grid while the medium and fine grid results did not display the usual very low residuals. However it can be seen that the medium and fine grid Cp results show consistency everywhere except aft of the shock on the upper surface and here only the fine grid results look acceptable. Again the decrease in trailing edge Cp indicates separation and drag rise. The skin friction indicates negative values right from the shock to the trailing edge of the airfoil showing, as expected, that this is the most difficult case to predict. Since the assumptions of law of the wall are not valid for this large region of separated flow, it is doubtful whether these results can be viewed with any confidence.

REFERENCES

[1] FORTIN F. and JONES D.J. 'Solution of Compressible Inviscid and Viscous Flows around Single and Multi-element Airfoils on Unstructured Meshes'. Proceedings of the Second Annual Conference of the CFD Society of Canada. Toronto, June 1994.

[2] STOLCIS L. 'Computation of the Turbulent Flow Development around Single and Multi-Element Airfoils'. Doctoral Thesis, University of Manchester Institute of Science and Technology, 1992.

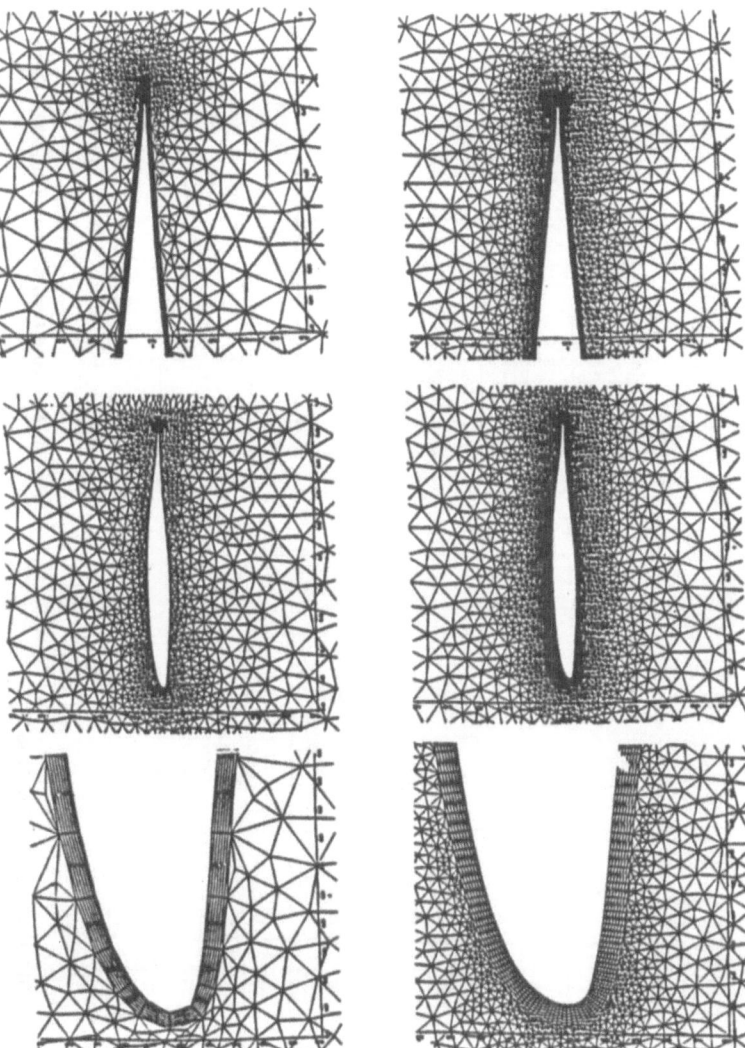

Fig 1.   MBB-A3 airfoil.
Details of Coarse Grid (60 points on foil)
and Fine Grid (240 points on foil)

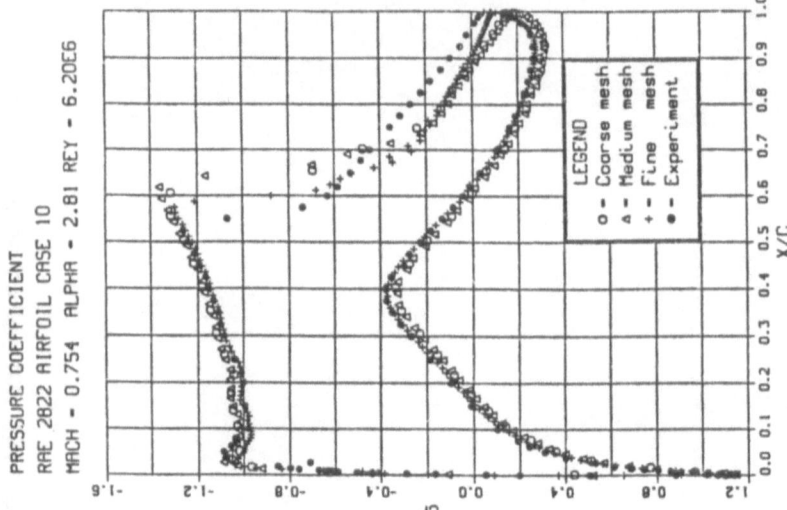

Fig 2b. Pressure Distributions. RAE 2822, Case 10

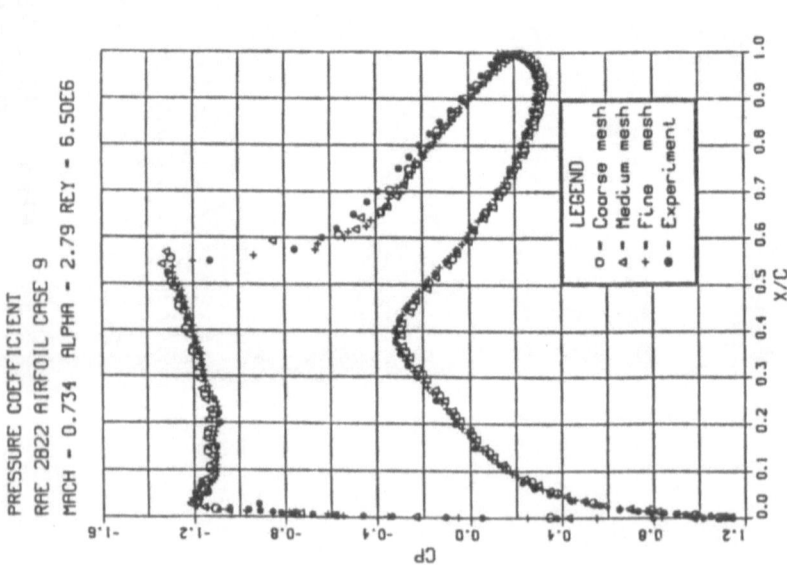

Fig 2a. Pressure Distributions. RAE 2822, Case 9

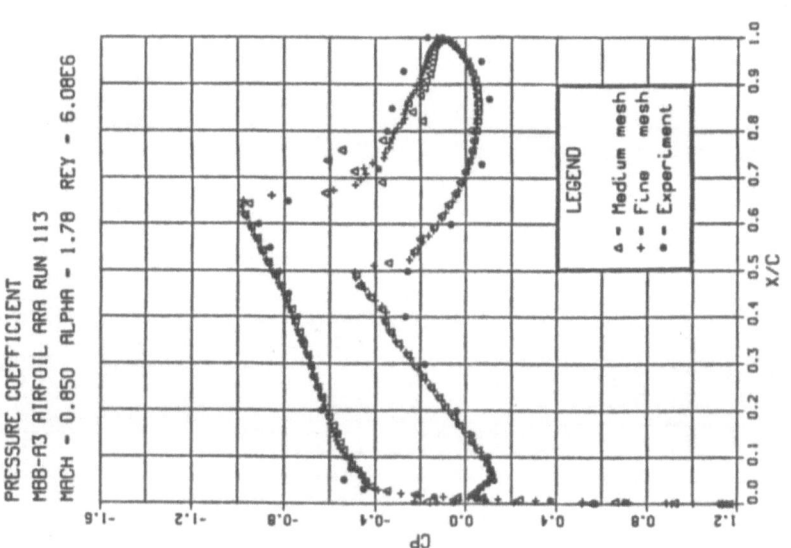

Fig 2d. Pressure Distributions. MBB-A3, Case 113

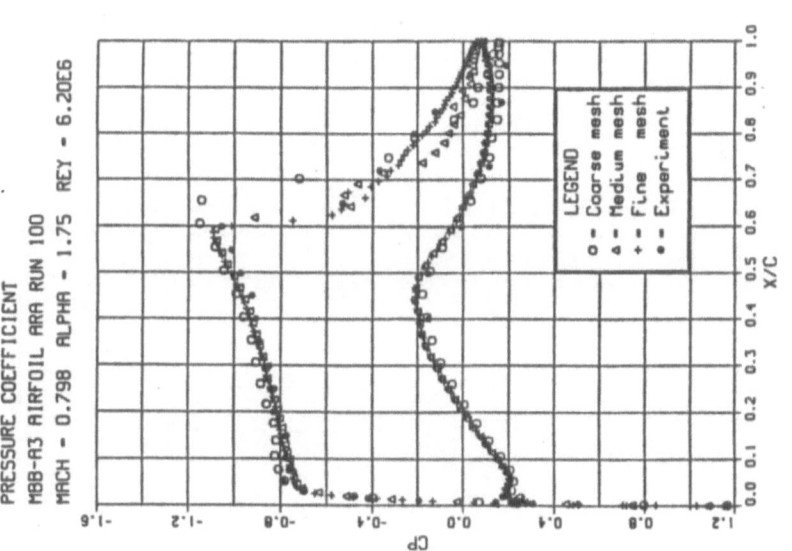

Fig 2c. Pressure Distributions. MBB-A3, Case 100

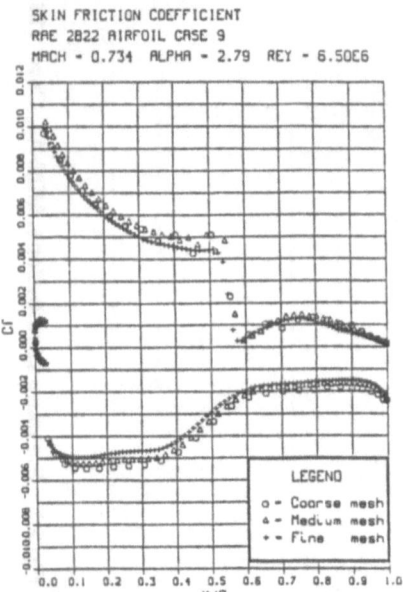

Fig 3a. Skin Friction. RAE 2822, Case 9

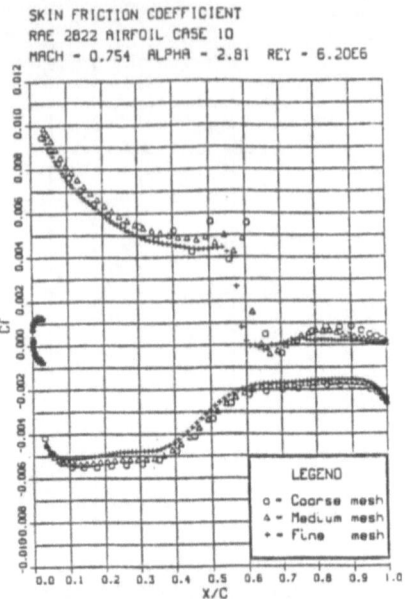

Fig 3b. Skin Friction. RAE 2822, Case 10

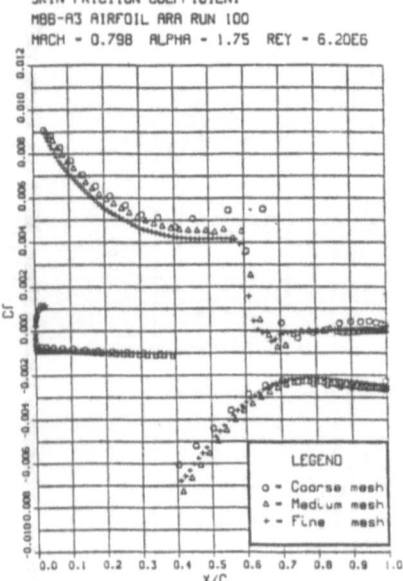

Fig 3c. Skin Friction. MBB-A3, Case 100

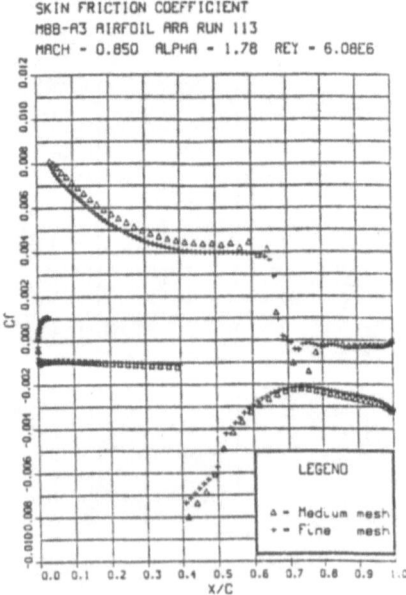

Fig 3d. Skin Friction. MBB-A3, Case 113

502

The Dornier 2D Navier-Stokes Approach Applied to Transonic Airfoil Flow

Werner Haase, Dornier Luftfahrt GmbH, D–88039 Friedrichshafen, Germany

Summary

The Dornier Navier–Stokes method is used to present results for both the RAE2822 and MBB–A3 airfoil using a variation of 0– and 1/2–equation turbulence models, in particular the models by Cebeci–Smith, Johnson–King and Johnson–Coakley. All calculations are based on a multi–level approach applying multigrid on all meshes of a desired mesh sequence. Results exhibit a reasonable grid–independent behaviour stepping from a medium mesh with 256x64 volumes to the fine mesh with 512x128 volumes.

1 The Method

The Navier–Stokes equations, describing two–dimensional, unsteady and compressible flows in conservation form, are solved by means of a finite volume approach using a Runge–Kutta time–stepping method with multigrid acceleration.

The system of ordinary differential equations in time is solved explicitly by a 3–stage Runge–Kutta time–stepping method using the coefficients 0.8/0.8/1.0 and being stable up to a Courant number of 1.5. An advantage of that scheme, compared to the more commonly used coefficients of 0.6/0.6/1.0, is an increased damping property in the low(er) frequency range. The disadvantage, i.e. the reduction of the maximum CFL–no. from 1.8 to 1.5, can be easily overcome by applying acceleration techniques.

Blended second and fourth order filtering is applied to prevent odd–even decoupling. Filtering is applied only once to provide the best damping properties. Numerical dissipation is minimized by taking the eigenvalues in x– and y–direction independently instead of using the sum of those. Moreover, filtering is switched off in the boundary layer region by scaling the filter value with the ratio of local–to–free–stream Mach number.

The introduction of the residual averaging approach permits stable calculations beyond the ordinary Courant number limit. In all present calculations, the Courant number is chosen to be 3.5. Since only the steady state is of interest, a variable timestep approach is used.

Moreover, for the grid–dependence study, a multi–level approach is applied running 3–mesh W–cycles on the coarsest (128x32) mesh, 4–mesh W–cycles on the medium (256x64) and 5–mesh W–cycles on the finest (512x128) mesh.

In all mesh levels of the multi–level approach, the steady state is defined to be reached if the force coefficients (c_D and c_L) and the sum of pressure along airfoil and wake do not vary by more than 0.01 % using a monitoring sequence of 10 iterative (multigrid) cycles, i.e. the mentioned limits are checked each 10th iteration. Typically, this leads to an error–norm (L2–norm) reduction of approximately 5 decades in the coarsest desired mesh and further 2.5 decades in the very fine one.

2 Boundary Conditions

The following boundary conditions are chosen for all present calculations:

At the solid–wall boundary no–slip conditions are implemeneted and the flow is assumed to be adiabatic with a zero normal–pressure gradient.

A wake boundary is defined by overlapping lower and upper wake boundaries.

At the outflow boundary, linear extrapolation is used for density and mass fluxes, the static pressure is fixed to the static pressure at infinity.

The farfield boundary condition is implemented using fixed and interpolated Riemann invariants in combination with a compressible farfield vortex, which seems to be essential for increased accuracy due to a finite farfield distance of the computational domain.

3 Mesh Generation

For all meshes, i.e. for the RAE2822 (the same mesh is used for case 09 and 10, respectively) and the MBB–A3 airfoil, 512x128 volumes are chosen for specifying the finest desired mesh. Moreover, it should be mentioned that all meshes have been slightly adapted to the measured shock location(s). However, in order to provide reliable results for different turbulence model with differently predicted shock locations, this adjustment has been applied very moderately as it is shown in Fig. 1 for the second mesh level of the RAE2822–airfoil mesh.

In the fine meshes, 512 volumes are located in the i– or wrap–around–direction and 128 volumes are used to discretise the j– or wall–normal–direction. 64 volumes have been taken from the 512 for wake representation, i.e. 384 volumes are distributed on the airfoil surface.

3.1 RAE2822 airfoil

The first volume height (adjacent to the airfoil surface) is selected to be between 3.5×10^{-7} in the apex region and about 6.0×10^{-6} in the trailing–edge part on lower and upper surface, respectively. Calculations in the finest 512x128 mesh exhibit 50–78 mesh points in the boundary layer, starting with 50 points (volumes) at the stagnation point and reaching 56 on the lower trailing edge and 71 (78) for case 09 (case10) on the upper trailing edge.

3.2 MBB–A3 airfoil

The first volume height (adjacent to the airfoil surface) is selected to be between 3.4×10^{-7} in the apex region and 7.0×10^{-6} in the trailing–edge part on lower and upper surface, respectively. Calculations for test case 113 exhibit in the finest 512x128 mesh 45–67 mesh points in the boundary layer, starting with 45 points (volumes) at the stagnation point and reaching 59 on the lower trailing edge and 67 on the upper trailing edge.

4 Grid Dependence Study

An attempt was made to reach something like a "gold standard" for both sets of applications, the RAE2822 and the MBB–A3 airfoil. Although in general – and from an engineering point of view – the results obtained when stepping from the 256x64 to the 512x128 mesh can be seen as grid independent, in each of the three test cases still some deviations are evident.

In all subsequent figures, thick lines indicate upper–surface distributions while thinner lines correspond to lower–surface distributions. All figures which depict pressure distributions contain additional information regarding the computed and measured force coefficients.

4.1 RAE2822 airfoil (case 09 and10)

Figures 2 and 3 exhibit the dependence of mesh fineness with respect to pressure for case 09 (Fig. 2) and case 10 (Fig. 3). The turbulence model in use for the study of grid (in)dependence is the Johnson–Coakley model. When switching from the 256x64 to the 512x128 mesh for both cases, some minor discrepancies in pressure occur on the upper surface in the areas of suction peak and shock. Additionally, Fig. 5 depicts for case 10 the y^+_{Wall} distribution for the different mesh levels. It can be easily seen that for the medium (256x64) and the fine (512x128) mesh the maximum values do not violate the "$y^+_{Wall}=1.0$ restrictions"; even in the coarsest (128x32) mesh, the y^+_{Wall}

values on the upper surface are still less than 1.3

4.2 MBB–A3 airfoil

For the MBB–A3, test case 113, Fig. 4 exhibits the dependence of mesh fineness with respect to the pressure coefficient by using (again) the Johnson–Coakley model. All distributions investigated, depict a globally mesh–independent behaviour when switching from the fine (256x64) to the very fine (512x128) mesh, apart from the lower surface distribution where a steep adverse pressure gradient is forming up in the fine mesh(es).

It should be mentioned at this point that the fine–mesh solutions for the MBB–A3, case 113, turned out to be difficult to obtain; after 1500 iterations still some (although rather small) oscillations of the force coefficients are present; e.g., a maximum relative "error" of 0.01 for the drag coefficient, c_D, can be detected. The y^+_{Wall} values for the medium and fine mesh are rather similar to those obtained for RAE2822 test cases.

5 Results

All computations presented hereafter are based on the flow parameters given in Table 1, while force coefficients and separation locations are presented in Table 2 for the RAE2822, case 09, in Table 3 for case 10 and in Table 4 for the MBB–A3, case 113.

Table 1 Flow parameters for presented test cases

Airfoil / Case	Ma	Re	α [°]	Transition lower surf.	Transition upper surf.
RAE2822 / 09	0.734	6.5×10^6	2.79	0.03	0.03
RAE2822 / 10	0.754	6.2×10^6	2.81	0.03	0.03
MBB–A3 / 113	0.85	6.08×10^6	1.78	0.4	0.03

It should be noted at this point, that – unfortunately – the defined flow parameters for the RAE2822 airfoil are slightly different to those selected for EUROVAL, making a direct comparison "a bit" difficult.

5.1 RAE2822 airfoil

For the three tested turbulence models, Cebeci–Smith (CS), Johnson–King (JK) and Johnson–Coakley (JC), the obtained computational results are given in Tabel 2 and 3, compared with the measured values.

Table 2 Summary of computational results for the RAE2822 airfoil, case 09

TurbulenceModel	Mesh	c_D	c_L	c_M	Separation [x/c]
CS	256x64	0.02005	0.83780	−0.09986	0.574
CS	512x128	0.01900	0.83741	−0.09782	0.573
JK	256x64	0.01828	0.79066	−0.08944	0.555
JK	512x128	0.01771	0.79918	−0.08945	0.557
JC	256x64	0.01993	0.82546	−0.09728	0.575 (incipient only)
JC	512x128	0.01939	0.83486	−0.09749	0.585 (incipient only)
Measurements		0.0168	0.8030	−0.099	

Table 3 Summary of computational results for the RAE2822 airfoil, case 10

TurbulenceModel	Mesh	c_D	c_L	c_M	Separation [x/c]
CS	256x64	0.02916	0.79746	−0.10915	0.614 (until t.e.)
CS	512x128	0.02836	0.80199	−0.10802	0.614 (until t.e.)
JK	256x64	0.02599	0.72591	−0.09481	0.560
JK	512x128	0.02580	0.73782	−0.96219	0.560
JC	256x64	0.02834	0.77091	−0.10433	0.595
JC	512x128	0.02827	0.78528	−0.10623	0.599
Measurements		0.0242	0.743	−0.106	

Apart to the shock location, pressure distributions for case 09 and case 10 look rather similar, Fig. 6 and 11. This holds also for the situation concerning the use of the turbulence models selected. For both test cases, the CS model as an equilibrium turbulence model predicts stronger shocks more downstream of the expected position, while the use of the JK model results in the most up-stream position of the upper–surface shock. The JC version, designed to overcome some weaknesses of the JK model, gives a somewhat "better" shock location for case 09, however, it is off in case 10. This might be correlated to the general (?) insufficiency of the JC model to predict separated flow in an accurate way. This finding seems to be in line with the problems of the Johnson–Coakley model to reasonably predict pressure induced separation in subsonic high–lift flow.

Figures 7 and 8 (case 09) as well as Figs. 12 and 14 (case 10) present skin friction and displacement thickness distributions for both test cases and all turbulence models employed. Additionally, for test case 09 velocity profiles are shown in Figs. 9 and 10 at x/c=0.65 and x/c=1.0, i.e. some percent of chord behind the shock (according to the turbulence model in use) and at the upper-surface trailing edge. The CS model provides reasonable results for the trailing edge, where the flow is again near to equilibrium, however, it fails aft of the shock in the non–equilibrium region. Astonishing to see is the overpredicted momentum loss in the near wall region for the 1/2–equation models, Fig. 9, and for the JK model, Fig. 10.

For test case 10, equivalent velocity profile plots are given in Figs. 14 and 15. The position x/c=0.65 is now just aft of the shock and the results depend still strongly on the shock position. Nevertheless, the shape of the separation region predicted by the algebraic CS model is well known to be too "kinky", the 1/2–equations models behave much better in predicting the separated part of the velocity profile. At the trailing edge, where the flow is still in non–equilibrium (in contrast to case 09), the CS model still shows separation. Again, the momentum loss in the near wall region for the 1/2–equation models is significant.

5.2 MBB–A3 airfoil
The following table, Table 5, summarizes the main results for the MBB–A3 airfoil:

Table 4 Summary of computational results for the MBB–A3 airfoil, case 113

TurbulenceModel	Mesh	c_D	c_L	c_M	Separation [x/c]
CS	256x64	0.04674	0.51038	−0.11489	0.716
CS	512x128	0.04763	0.52125	−0.11832	0.726
JK	256x64	0.03672	0.39152	−0.08022	0.609
JK	512x128	0.03574	0.38188	−0.07664	0.604
JC	256x64	0.04213	0.45793	−0.09964	0.666

JC	512x128	0.04183	0.45522	-0.09819	0.665
Measurements		0.03236	0.515	-0.1165	

The brief discussion concerning the influence of the selected turbulence models given above for the RAE2822 airfoil holds also for the chosen MBB–A3 test case. However, an additional problem can be detected related to the lower–surface–flow behaviour. Having already seen in Fig. 4 that the shock formation on the lower surface does rely on the fineness of the mesh, the results presented in Figs. 16 to 18 for pressure (Fig. 16), skin friction (Fig. 17) and displacement thickness (Fig. 18) indicate, moreover, that a strong turbulence–model dependence is evident not only for the upper surface properties. While the CS model does not predict a lower–surface shock (or steep adverse pressure gradient) at all, the JK model overpredicts that (possible) shock area by a lot although the history of the oncoming laminar boundary layer seems to be similar for all three turbulence models. A first idea that the transition position (according to the lower–surface suction peak near the leading edge) might be wrong did not hold. A check run with a lower–surface transition at x/c=0.03 did not really change the lower surface pressure distribution and it had an opposite effect on the upper–surface shock which moved a bit more upstream (!).

Comparable to the findings in subsonic flows, the CS model – and to some extent the JC model as well – cannot predict the pressure in the separation zone adequately. Although the JK model results in an upper–surface shock position located too much upstream, the pressure level in the separation zone exhibits a correct (measurement–like) behaviour; of course, due to the difference in the shock location the correct level is slightly different.

For all turbulence models involved, the predicted flow situation along the lower surface, cumulating in a shock–like structure and exhibiting a large difference between computation and measurement, remains unclear.

Concluding on the two half–equation models, Johnson–King and Johnson–Coakley, it may be underlined again that the Johnson–King model behaves much better in case of separation while the Johnson–Coakley model provides an increased predictive capability (for transonic flows) with respect to the location of shocks.

6 Figures

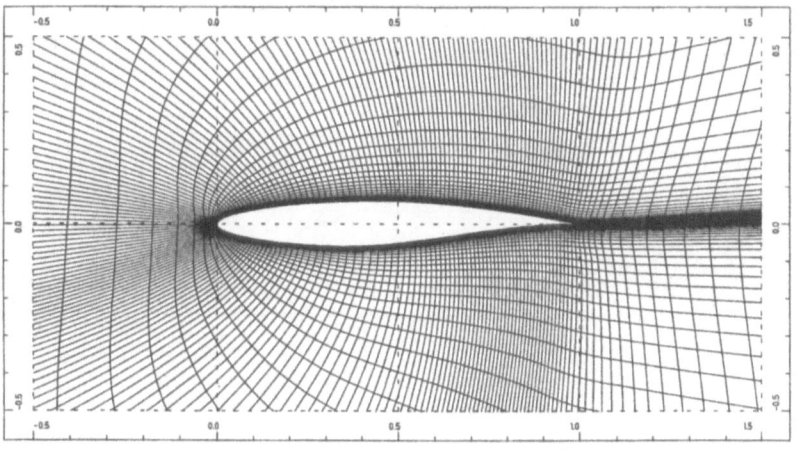

Figure 1 Mesh detail for RAE2822 airfoil, case 09 and 10, second mesh level (256x64)

507

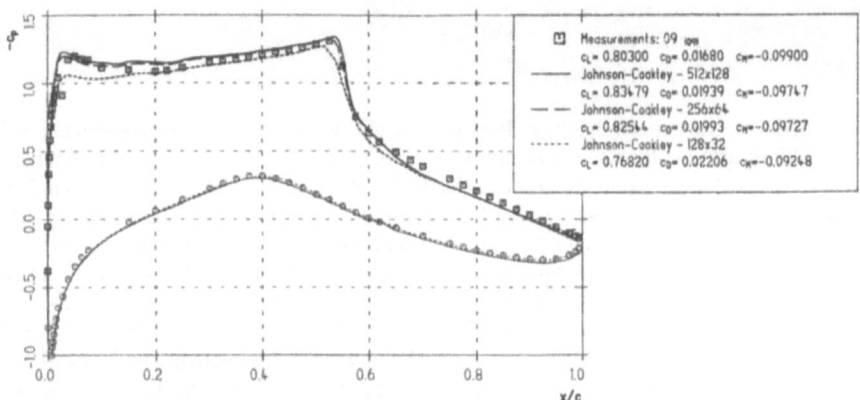

Figure 2 Pressure coefficient distribution: Grid–dependence study for
RAE2822 airfoil, case 09, using the Johnson–Coakley turbulence model

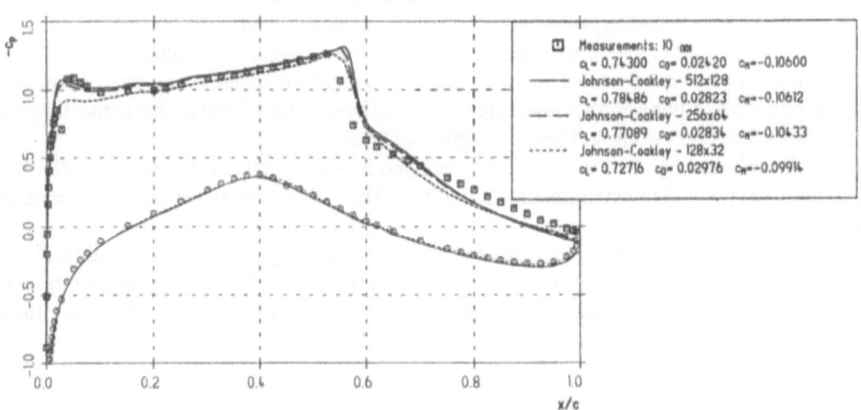

Figure 3 Pressure coefficient distribution: Grid dependence study for
RAE2822 airfoil, case 10, using the Johnson–Coakley turbulence model

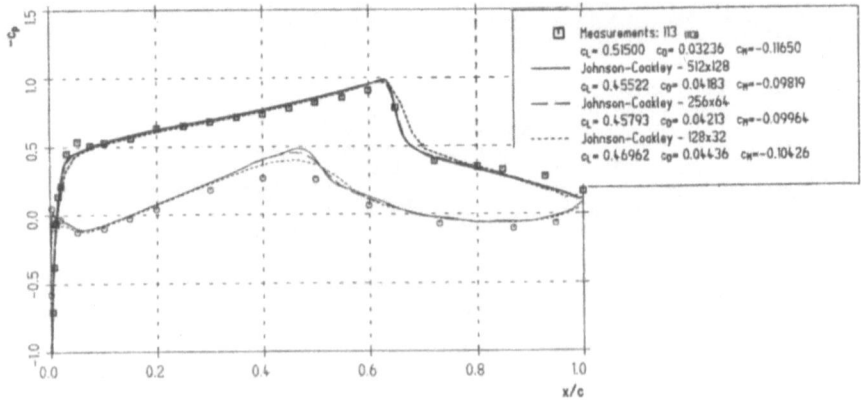

Figure 4 Pressure coefficient distribution: Grid dependence study for
MBB–A3 airfoil, case 113, using the Johnson–Coakley turbulence model

508

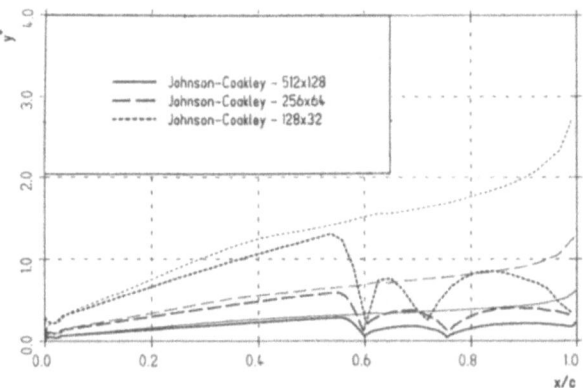

Figure 5 y^+_{Wall} distribution: Grid dependence study for
RAE2822 airfoil, case 10, using the Johnson–Coakley turbulence model;

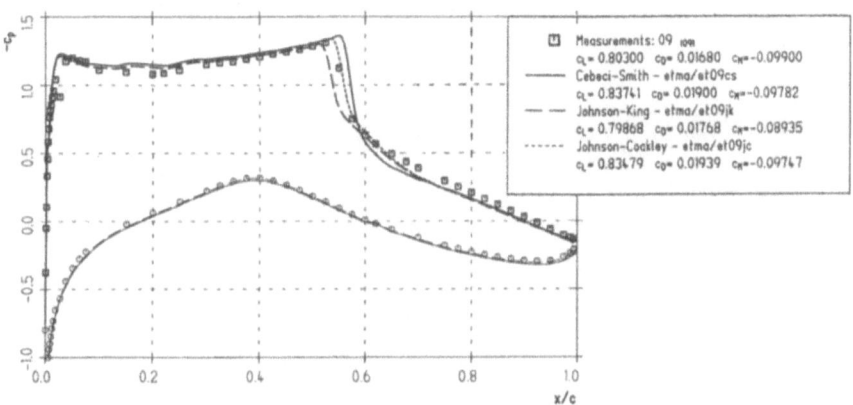

Figure 6 Pressure coefficient distributions for
RAE2822 airfoil, case 09, mesh–level 512x128

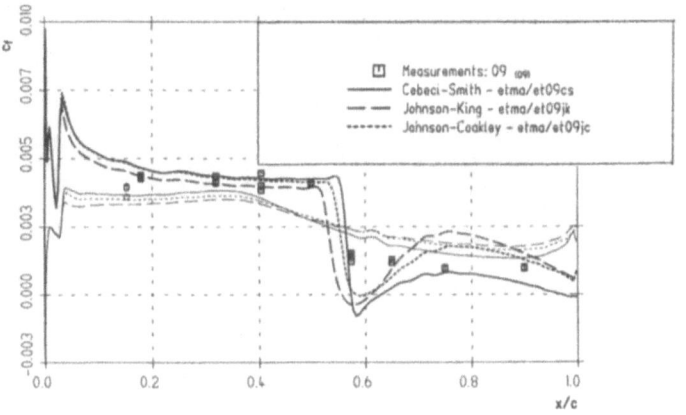

Figure 7 Skin friction distributions for RAE2822 airfoil, case 09, mesh–level 512x128

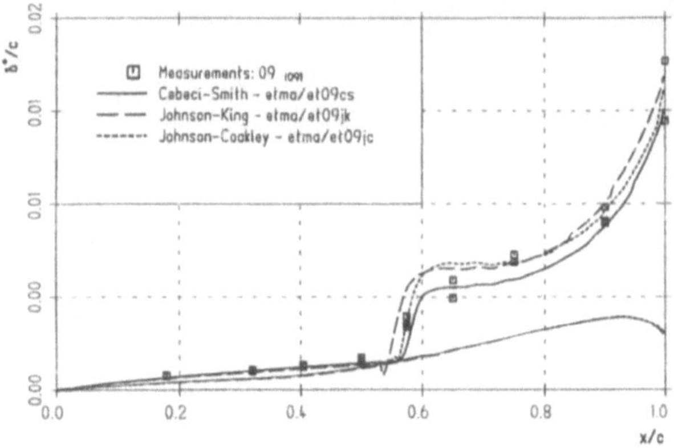

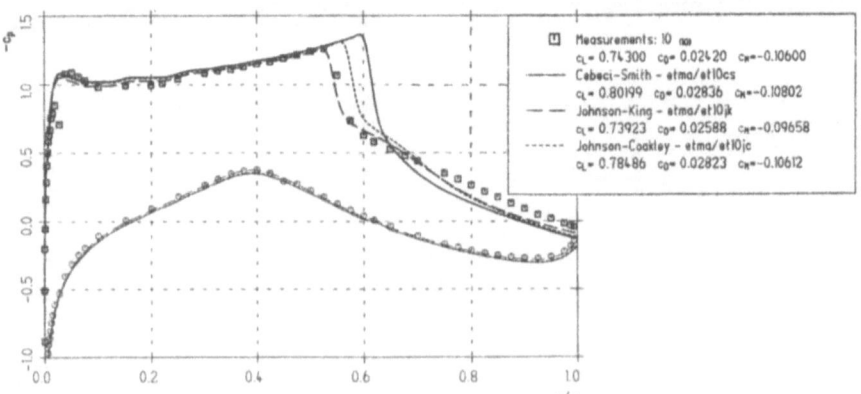

Figure 8 Displacement thickness distributions for RAE2822 airfoil, case 09, mesh–level 512x128

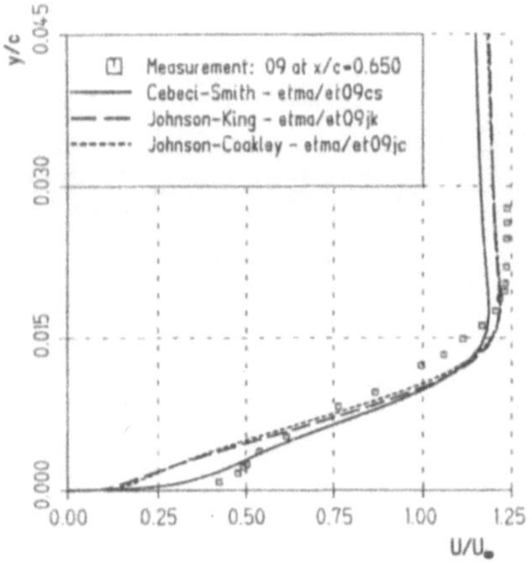

Figure 9 Velocity profiles at x/c=0.65 for RAE2822, case 09, 512x128

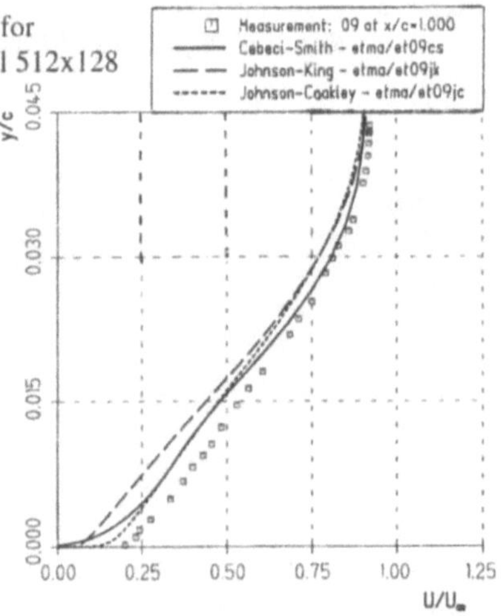

Figure 10 Velocity profiles at x/c=1.0 for RAE2822, case 09, 512x128

Figure 11 Pressure coefficient distributions for RAE2822, case 10, mesh–level 512x128

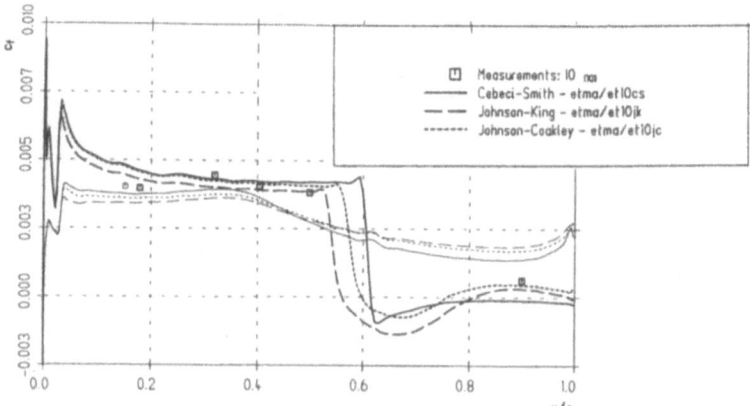

Figure 12 Skin friction distributions for RAE2822 airfoil, case 10, mesh–level 512x128

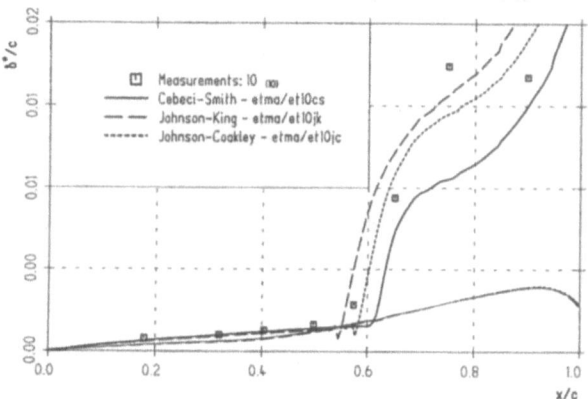

Figure 13 Displacement thickness distributions for RAE2822, case 10, mesh–level 512x128

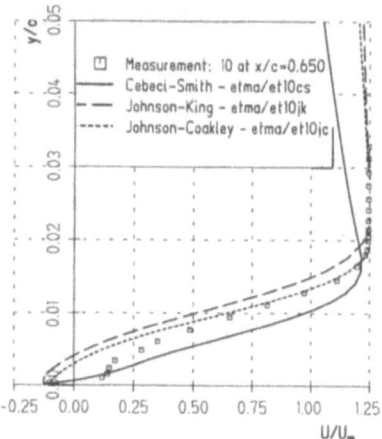

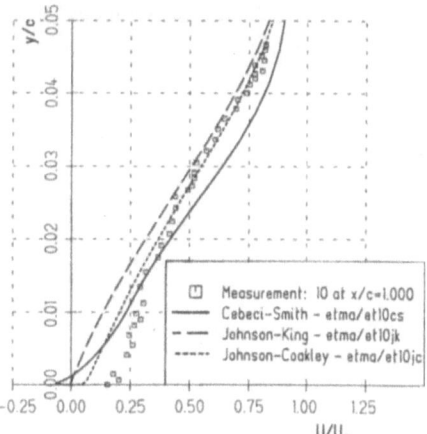

Figure 14 Velocity profiles at x/c=0.65 for RAE2822, case 10, 512x128

Figure 15 Velocity profiles at x/c=1.0 for RAE2822, case 10, 512x128

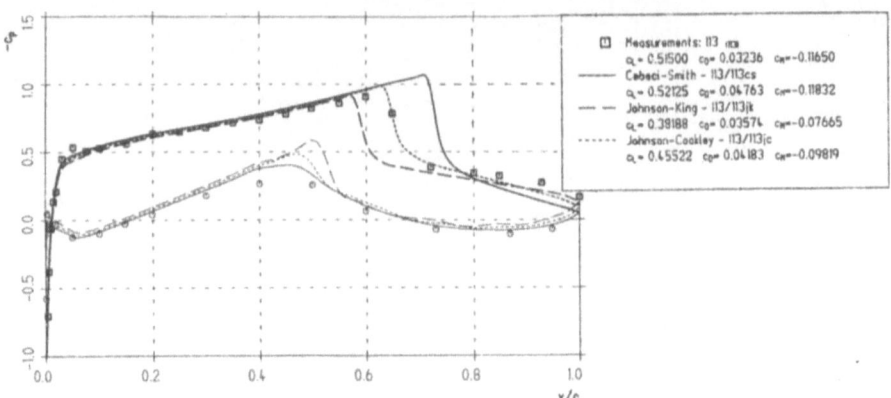

Figure 16 Pressure coefficient distributions for
MBB–A3 airfoil, case 113, mesh–level 512x128

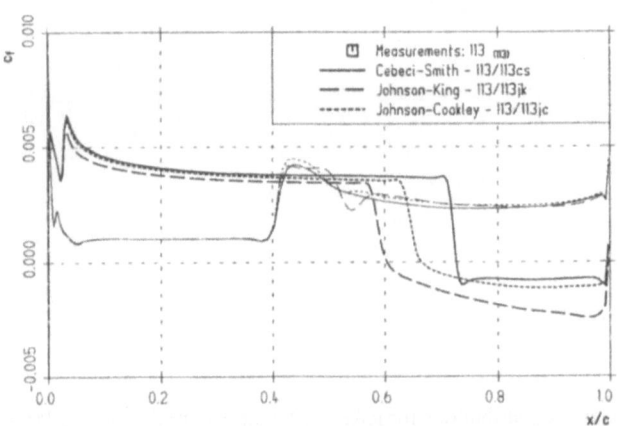

Fig. 17   Skin friction distributions for MBB–A3 airfoil, case 113, mesh–level 512x128

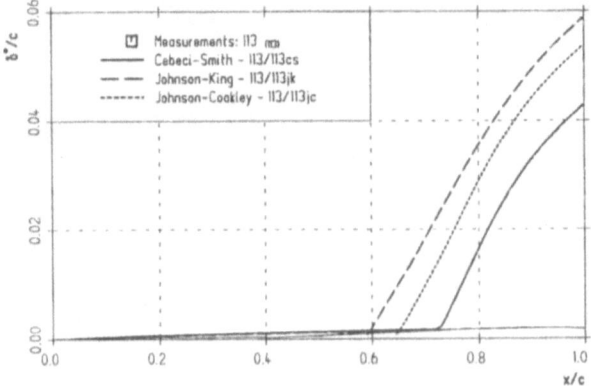

Figure 18 Displacement thickness distributions for MBB–A3 airfoil, case 113, mesh–level
512x128

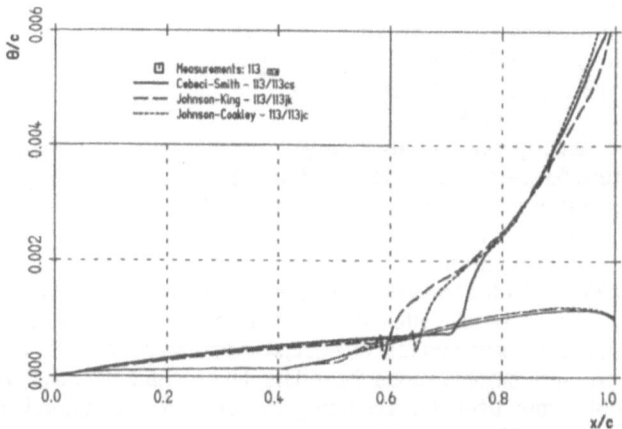

Fig. 19 Momentum thickness distributions for MBB-A3 airfoil, case 113, mesh-level 512x128

Flow Calculations Past RAE 2822 and MBB-A3 Airfoils for the ETMA Workshop Using the Navier Stokes Code ARC2D

M. Khalid
Aerodynamics Laboratory
Institute for Aerospace Research (IAR)
National Research Council of Canada (NRC)
Ottawa, K1A 0R6, CANADA

Summary

The paper contains CFD results for the airfoils RAE 2822 and MBB-A3 obtained through the use of the finite difference cell vertex implicit code, ARC2D. The exercise involved appropriate comparison of computed results against experimental data for the surface pressure distribution and other viscous boundary layer characteristics. A number of turbulence models were used for assessing the accuracy of numerical simulation particularly in the regions of separated flows.

Introduction

The 2D Navier Stokes code ARC2D [1] has been used to calculate flows past the RAE 2822 and MBB-A3 airfoils. The flow conditions used for the computations on both airfoils were taken directly from the specifications provided by the ETMA Workshop organizing committee. These specifications had asked for computations of case 9 ($M=0.734, \alpha=2.79$, Re = 6.5×10^6) and case 10 ($M=0.754$, $\alpha=2.81$, Re = 6.2×10^6) for the RAE 2822 airfoil and seven additional cases for the MBB-A3 airfoil. However for the MBB-A3 airfoil, as instructed later on, the main emphasis was on the last case number 7 ($M=0.84$, $\alpha=1.78$, Re = 6.08×10^6). For the RAE 2822 airfoil the transition was fixed at 3% of the chord for both upper and lower surfaces. On the MBB-A3 airfoil, the transition was specified to be at 3% of the chord from the leading edge for the upper surface and 40% of the chord for the lower surface.

The paper contains comparisons of computed results against experiment for the above three cases using four different turbulence models. The turbulence models investigated were Baldwin-Lomax, Johnson-King, Baldwin-Barth and k-ε two-equation model with Chien's near wall treatment.

Mostly, these comparisons involved the accuracy of Cp (pressure coefficient) prediction when judged against the measurement, as well as against the results from other participants. The comparison of results obtained from various groups was mainly carried out by the technical evaluator, Prof. L. J. Johnston. For more detailed comparisons and closer assessment of the turbulence models such parameters as skin friction distribution, velocity profiles and integrated loads on airfoils were also examined. Results from some computations showing the effects of the mesh coarseness are also included.

Computation

The grids for both airfoils were produced using the NASA AMES supplied hyperbolic code HYGRID. Most computations were performed on a mesh size of 369X65. The wall distance of the first grid line was kept at about 0.00005 of the unit chord. The far field grid line was placed at a distance of 25 chords away from the airfoil. For this mesh size, 32 points were located in the wake region while the remaining 305 points were placed on the surface around the airfoil. The grid spacing near the nose and tail regions on the airfoil surface was 0.001 and 0.002 respectively.

For the computational part of this investigation, the NASA code ARC2D which solves the thin layer Navier Stokes equations in two dimensions was used. The code was originally developed by Steger in 1978 and has since been modified and improved by Pulliam, Steger and Barth [2] and [3].

Within the framework of ARC2D, a number of computational schemes are identified for the solution of the Navier-Stokes equation matrices. For the current computations, the pentadiagonal implicit smoothing was used to solve the resulting factored form of the algorithm for the Navier Stokes equations. The values for the 2nd and 4th order dissipation terms were 1.0 and 0.64.

A number of measures were adapted to run this code in its most efficient mode. For example, the code was run in a grid cascade-sequencing mode where the solution is first obtained for coarser meshes before proceeding on to the fine mesh. The variable time step ($dt=\Delta t_{ref}/(1+\sqrt{J})$) used in the present calculations was geometrically based scaled by the Jacobian. Other code improvement measures inherent to ARC2D such as increased implicit treatment of dissipation terms, implementation of means to reduce inversion work and an ability to use scalar tridiagonal or pentadiagonal operators instead of block operators contribute towards robustness and efficiency of the code [3].

In its original form the code makes use of the Baldwin-Lomax [4] turbulence model (BL) which is known to be rather poor in regions of flow separation, and since some of the present cases involved separation and shock wave boundary layer interaction, higher order turbulence models were also investigated. The other three models investigated included: Johnson-King (JK) [5] which attempts to account for some flow history by continuously updating the chordwise maximum shear stress (τ_{max}), the Baldwin-Barth [6] one-equation turbulence model (BB) and a two-equation k-ε model with Chien's low Reynolds number adjustment close to the wall.

The Baldwin-Lomax turbulence model uses different algebraic expressions for turbulent viscosity in the inner ($v_{t\,in}=l^2|\omega|$, where l is the mixing length) and in the outer (v_{tout} =min(_max ,(U_max-Umin)^2/_max) viscous region with a blending formula in the overlap region. The function _ in the outer region is defined in terms of the vorticity ω, i.e., _= $y|\omega|[1-exp(-y^+/26)]$, see [4] for details. In fact use of Baldwin-Lomax model was also made when calculating the starting equilibrium conditions for the Johnson-King model, where a similar blending formula was also used for the overlap region between the inner and outer solution.

The Johnson-King turbulence model [5] also uses algebraic formulation for inner and outer eddy viscosities ($v_{t\,in}$= $D^2\kappa y(u_m'v_m')1/2$, v_{tout} = $0.0168Ue\delta^*/[1+5.5(y/\delta)^6]$) with a simple blending formula in the overlap region to evaluate the equilibrium conditions. The JK model then makes a novel use of the maximum shear stress development derived from the basic turbulent kinetic energy expression to obtain a formula which relates the 'instantaneous' maximum shear ($g=(u_m'v_m')1/2$) to the corresponding equilibrium conditions. This formula is given by:

$$\frac{dg}{dx} = \frac{a_1}{2U_m L_m}[(1-\frac{g}{g_{eq}})+\frac{C_{dif} L_m}{a_1\delta[0.7-(\frac{y}{\delta_m})]}|1-(\frac{v_{to}}{v_{to,eq}})^2|] \; .$$

Note that the subscript m corresponds to maximum shear stress conditions and all other parameters have the same definitions as in [5]. The eddy viscosity where the maximum shear stress occurs is calculated using v_t lm = $-u'v'_m/(\partial u/\partial y)$ lm which, when normalized with respect to the equilibrium conditions provides the ratio for scaling the eddy viscosity profile.

The Baldwin-Barth model is a one-equation turbulence model derived from the standard k-ε turbulence model equations. It is written in terms of the turbulent Reynolds number (R_t=vR_T):

$$\frac{D(R_t)}{Dt} = (v + 2.8554 v_T)\nabla^2(R_t) - 1.4277 v\vec{\nabla}(R_t) + (2 f_2 - 1.2)\sqrt{0.09 D_1 D_2 \Omega}(R_t)$$

where, $\mu_T = 0.09\rho(vR_T)D_1D_2$ and D_1,D_2 and f_2 are various damping functions well described in [6].

As for the Chien's low Reynolds number two-equation k-ε turbulence model which is documented in Reference 7 (also refer to Patel et al discussion in [8]), it is worth noting that he has added one additional term each, to the turbulent kinetic energy and dissipation transport equations to account for the wall effects. The new expressions for various damping terms appropriately relate to the y^+ value from the wall, and the standard coefficients ($c_\mu,c_{\varepsilon1},c_{\varepsilon2},\sigma_k$ and σ_ε) too have been modified.

Our experience with Baldwin-Lomax turbulence model is that, if pressure and viscous loads are integrated normally around the airfoil, it produces somewhat lower viscous drag, most probably caused by a thinner boundary layer prediction [9]. The viscous drag was thus recovered using the drag profile integration method [10] based on the momentum deficit considerations in the wake. For cases involving other turbulence models, lift, moment and drag coefficients were calculated using the usual surface forces' integration method.

Results and Discussion

For the RAE 2822 airfoil case 9 (M= 0.734, α = 2.79 and Re = 6.5 X 10^6) the pressure distribution obtained using the Baldwin-Lomax turbulence model is shown in Figure 1. It shows good comparison between experiment and fine mesh computations. As expected, the comparison deteriorates with the coarseness of the mesh. Within the thickness of the plotting symbol, the flow using the fine mesh seems to be well computed on the lower surface of the airfoil. Note that the computations also seem to pick up the flow expansion near the upper leading edge. Upper trailing edge comparison is not as good.

The pressure distribution comparisons on the more difficult case 10 are shown in Figure 2. This case seemed to be the one more suitable to study the performance of various turbulence models. This case is known to support regions of separation near the shock on the upper surface. Both Baldwin-Lomax and Johnson-King seem to be a little off in terms of the shock location prediction whereas, the Baldwin-Barth and Chien's two-equation k-ε models give a relatively better match. Baldwin- Barth turbulence model also shows a pressure decrease at the trailing edge, which is consistent with the flow separation. The pressure levels on lower surface are well computed by all the turbulence models whereas on the upper surface the k-ε

turbulence model shows a tendency to overpredict the pressures. Note once again that there is an attempt by nearly all the turbulence models to compute the suction peak near the leading edge on the upper surface of the airfoil.

The case 7 for the MBB-A3 airfoil shown in Figure 3 was found to be quite difficult to compute. Computations with Baldwin-Lomax model did not converge, as its empirical formulation is perhaps better suited for turbulent attached flows. On the other hand, satisfactory convergence was achieved using the Baldwin-Barth turbulence model. Results from Baldwin-Barth do resemble the experiment, in fact, they provide encouraging agreement near the leading edge and the upper trailing edge. The shock locations are not well predicted by computations and the lower surface prediction is quite different from the experiment.

The skin friction plots for the RAE 2822 airfoil cases 9 and 10 are shown in Figures 4 and 5 respectively. The measured values of the skin friction (supplied by ETMA) on the upper surface for both cases are also shown in each figure. It is evident from Figure 4, that the skin friction values for case 9 seem to be somewhat overpredicted on the forward upper surface before the shock. The computations then show a small region of separated flow immediately following the shock, beyond which point the flow reattaches, but the skin friction distribution remains less than the measured values. For case 10, in Figure 5, the results from the Baldwin-Barth and k-ε models are shown along with the five experimental points. It is interesting to note that on the forward portion of the airfoil, the two computed results are very similar, with the k-ε turbulence model, perhaps giving a better match with the experiment. Beyond shock, the Baldwin-Barth model predicts a completely separated flow, whereas the k-ε turbulence model shows a reattachment of the flow after a small separated region. Near the trailing edge, the k-ε turbulence model results, virtually pass through the lone measure point. On the lower surface, both turbulence models show completely attached flow, with the Baldwin-Barth turbulence model predicting notably lower skin friction distribution towards the latter half of the airfoil. The level of agreement between computed and measured data is rather similar to the results reported in [11].

For the RAE 2822 airfoil case 10, velocity profiles at axial station x/c = 0.65 and 1.0 are shown in Figures 6 and 7 respectively. Note that the experimental velocity distribution has been adjusted as in [11], to obtain U/U$_{INF}$. At station x/c = 0.65, all the turbulence models indicate a separated flow with the Baldwin-Barth model showing a rather pronounced separated velocity profile. Barth [12] (who had actually produced this velocity profile) recommends a wall distance of as low as 0.000009 of the chord for good viscous predictions in his updated model. Away from the sublayer, in the log-law region only the k-ε turbulence model seems to show a good comparison. In the outer edge of the viscous region, Baldwin-Barth turbulence model is noticeably the worst. At station x/c = 1.0, again Baldwin-Barth model continues to show the largest region of separated and reversed flow at the wall. Indeed, as observed in Figure 7, all turbulence models tend to underpredict the velocity in and close to the sublayer regions. In the main body of the boundary layer and in the outer regions, except for the k-ε turbulence model, which continues to under predict the velocity ratio, all other models are quite close to the measurement. The inability of the turbulence models to match the velocity profile measurements is not too dissimilar to the results reported in [10].

The viscous region for the RAE 2822 airfoil was further investigated in terms of the displacement thickness comparisons between computation and measurement. A set of displacement thickness curves for the RAE 2822 airfoil cases 9 and 10 are shown in Figures 8 and 9 respectively. For the case 9 in Figure 8, the comparison of displacement thickness between computation and experiment for most parts, both in trend and quality, is quite good. Only near the aft of the shock region, the computation shows some larger values than the measurement. For the case 10, results from four turbulence models are shown in Figure 9. In the forward portion of the airfoil, the match from Baldwin-Barth is notably good, otherwise the results from all four turbulence models are bunched fairly close together

Conclusion

The turbulence models investigated in this paper provided satisfactory agreement with experimental pressure distribution data for the RAE 2822 airfoil. The agreement for the MBB-A3 airfoil was not as good.

None of the present models can be relied upon to adequately predict flows with severe separation. This is particularly true for the velocity profile comparisons in viscous regions, where no one turbulence model could consistently provide a successful comparison with the experiment.

References

1. Pulliam, T., 'Euler and Thin Layer Navier-Stokes Codes: ARC2D, ARC3D', Notes for Computational Fluid Dynamics User's Workshop UTSI E02-4005-023-84, March 1984.

2. Pulliam, T.H. and Steger, J. L., 'Recent Improvements in Efficiency, Accuracy, and Convergence for Implicit Approximate Factorization Algorithms', AIAA-85-0360, AIAA 23rd. Aerospace Sciences Meeting January 14-17, 1985/Reno, Nevada.

3. Barth,T.J., and Steger, J.L., ' A Fast Efficient Implicit Scheme for the Euler and Navier - Stokes Equations Using Matrix Reduction Techniques', AIAA -85-0439 AIAA 23rd. Aerospace Sciences Meeting January 14-17, 1985/Reno, Nevada.

4. Baldwin, B. S., and Lomax, H., " Thin-Layer Approximation and Algebraic model for Separated Turbulent Flows," AIAA Paper 78-257, January 1978.

5. Johnson, D. A., and King, L. S., 'A Mathematically Simple Turbulence Closure Model for Attached and Separated Turbulent Boundary Layers', AIAA Journal, Vol. 23, Nov. 1985.

6. Baldwin, B. S., Barth, T. J., 'A One-equation Turbulence Transport Model for High Reynolds Number Wall-Bounded Flows', NASA TM 102847, August 1990.

7. Chien,K-U., 'Prediction of Channel and Boundary Layer Flows With a Low - Reynolds -Number Turbulence Model', AIAA Journal, January1982. Vol.20, No.1 pp. 33-38.

8. Patel, V.C., Rodi,W. and Scheuerer, G., 'Turbulence models for near-wall and low-Reynolds-number flows: A review', AIAA Journal, 1985., vol. 23, pp. 1308-1319.

9. Maksymiuk, C. M. and Pulliam, T. H., ' Viscous transonic airfoil workshop results using ARC2d', AIAA 25th. Aerospace Sciences Meeting, January 12-15, 1987.

10. Khalid, M and Jones, D. J. and Chan, Y. Y., 'Investigation of Viscous Flow Near the Blunt Trailing Edge of an Airfoil', 1993 European Forum - Recent Developments and Applications in Aeronautical CFD,' 1-3 Sep. 1993 Royal Aeronautical Society Conf. Bristol UK.

11. Haase, W., "Dornier Contribution to Task TC 8," Etma Workshop, November 14-17, 1994., UMIST, Manchester , UK.

12. Barth, T. J., 'Private Communication,' 13th, December 1994.

518

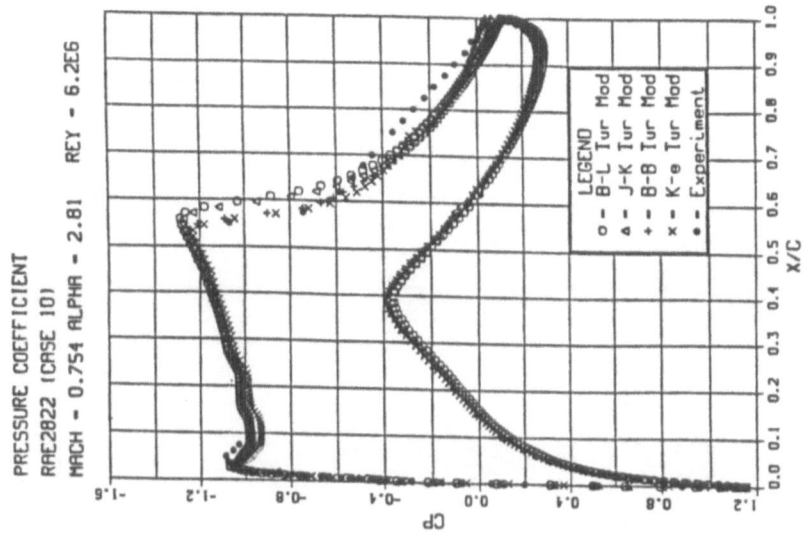

Fig. 1 Cp Distribution, RAE 2822 Airfoil Case 9 Fig. 2 Cp Distribution, RAE 2822 Airfoil Case 10

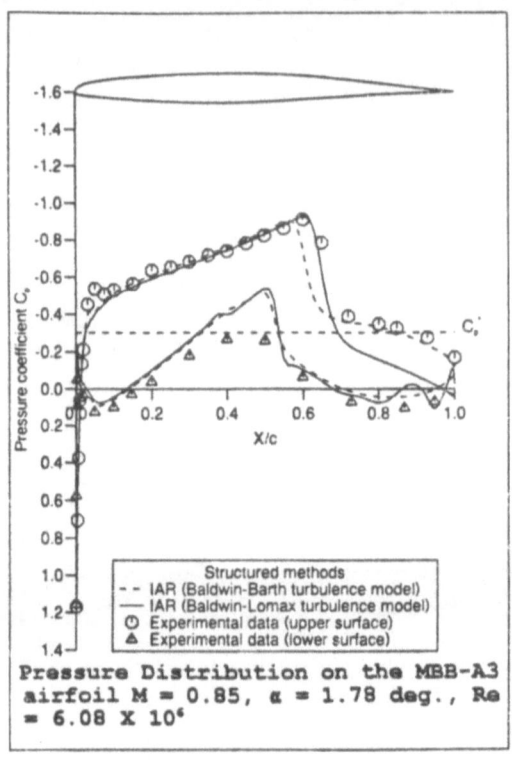

Fig. 3

Pressure Distribution on the MBB-A3 airfoil M = 0.85, α = 1.78 deg., Re = 6.08 X 10⁶

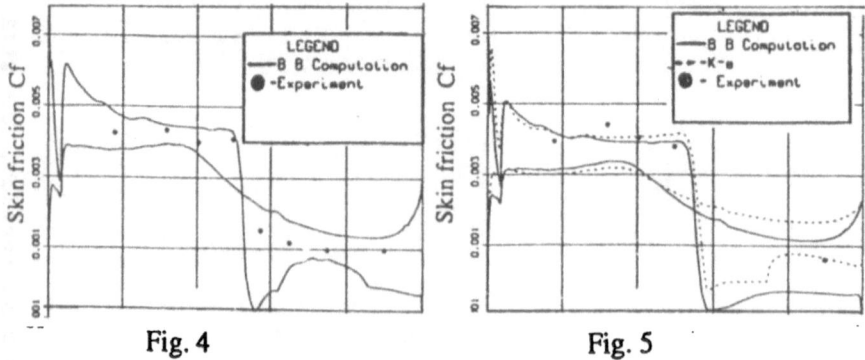

Fig. 4
Skin friction versus x/c, RAE 2822 AIRFOIL Case 9, M= 0.734, α = 2.79, Re = 6.5 X10⁶

Fig. 5
Skin friction versus x/c, RAE 2822 AIRFOIL Case 10, M= 0.754, α = 2.81, Re = 6.2 X10⁶

520

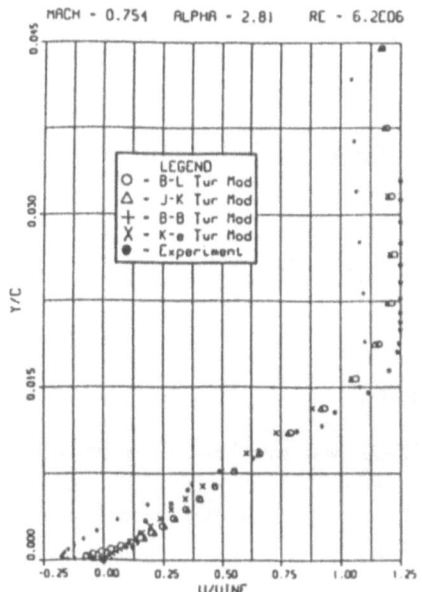

Fig. 6
Velocity Profile, RAE 2822 Airfoil
Case10 x/c = 0.65

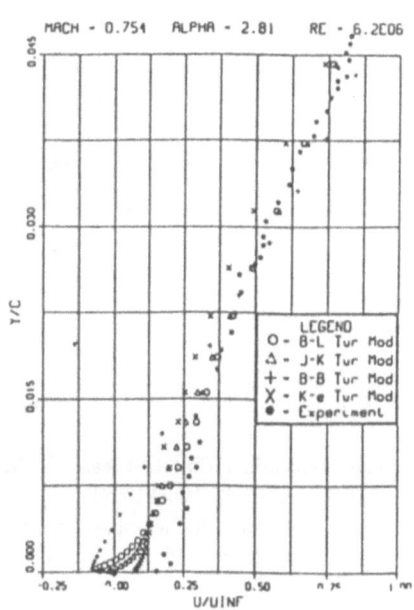

Fig. 7
Velocity Profile, RAE 2822 Airfoil
Case10 x/c = 1.0

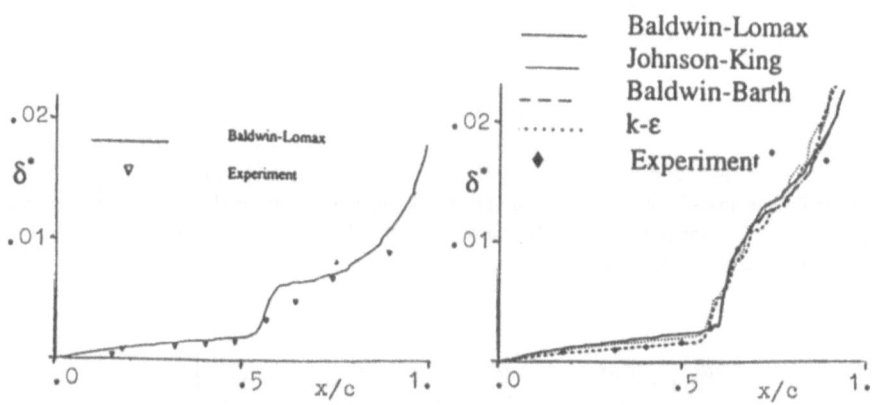

Fig. 8 δ* versus x/c. RAE 2822 Airfoil
Case 9, M = 0.734, α = 2.79,
Re = 6.5 X 10⁶

Fig. 9 δ* versus x/c. RAE 2822 Airfoil
Case 10, M = 0.754, α = 2.81,
Re = 6.2 X 10⁶

Calculation of the RAE2822 transonic airfoil using the $k - \tau$ model

Per-Åke Lindberg

Dept of Vehicle Engineering

Kungl Tekniska Högskolan

S-100 44 Stockholm, Sweden

Summary

The transonic airfoil test case, airfoil RAE2822, was calculated using the NSMB (Navier-Stokes Multi-Block) code. It is a finite-volume code for structured grids, solving the compressible Navier-Stokes equations. The calculations were performed using the $k - \tau$ model with two different sets of damping functions.

Introduction

The $k - \epsilon$ model has become a standard turbulence models for many aeronautical applications. However, it suffers from numerical problems and from the lac of a physically correct boundary condition for the dissipation rate at solid walls. With a variable transformation a similar model, based on the turbulent kinetic energy k and the turbulent time scale τ , can be formulated. It has been proven to give good and robust results for calculations of the A airfoil, Lindberg, Weber & Rizzi [3].

Turbulence models

The $k - \tau$ model, suggested by Speziale, Abid & Anderson [5] can be obtained from the standard $k - \epsilon$ model, Wilcox [6], by a variable transformation. If the turbulent time scale τ $(= k/\epsilon)$, is introduced in a low Reynolds number $k - \epsilon$ model the following equations are obtained after some manipulations.

$$\frac{\partial \rho k}{\partial t} + U_i \frac{\partial \rho k}{\partial x_i} = \tau_{ij} \frac{\partial U_i}{\partial x_j} - \frac{\rho k}{\tau} + \frac{\partial}{\partial x_i}\left[(\mu + \frac{\mu_t}{\sigma_k})\frac{\partial k}{\partial x_i}\right]$$

$$\frac{\partial \rho \tau}{\partial t} + U_i \frac{\partial \rho \tau}{\partial x_i} = (1 - C_{\epsilon 1} f_1)\frac{\tau}{k}\tau_{ij}\frac{\partial U_i}{\partial x_j} + \rho(C_{\epsilon 2}f_2 - 1) +$$

$$\frac{2}{k}(\mu + \frac{\mu_t}{\sigma_\epsilon})\frac{\partial k}{\partial x_i}\frac{\partial \tau}{\partial x_i} - \frac{2}{\tau}(\mu + \frac{\mu_t}{\sigma_\epsilon})\frac{\partial \tau}{\partial x_i}\frac{\partial \tau}{\partial x_i} +$$

$$\frac{\partial}{\partial x_i}\left[(\mu + \frac{\mu_t}{\sigma_\epsilon})\frac{\partial \tau}{\partial x_i}\right]$$

$$\tau_{ij} = \mu_t\left[\frac{\partial U_i}{\partial x_j} + \frac{\partial U_j}{\partial x_i}\right]$$

where f_1 and f_2 are damping functions and $C_{\epsilon 1}$, $C_{\epsilon 2}$, σ_k and σ_ϵ are model constants. The definition of the eddy viscosity also follows from the variable transformation, i.e.

$$\mu_t = \rho C_\mu f_\mu k \tau$$

where C_μ and f_μ are equivalent to the constant and function from the $k - \epsilon$ model.

Near wall treatment

In the present calculations the original damping functions suggested by Speziale et al. [5] are replaced by new functions. These new functions are deduced from the asymptotic behavior of the exact terms close to walls and from comparison with DNS data, Lindberg [2]. The DNS data used for the derivation of the damping functions are from Mansour, Kim & Moin [4].

The damping functions used in the present calculations are

$$f_1 = 1 + 2.05e^{-0.016(y^+ - 3)^2} - 0.6e^{-0.1y^{+2}}$$

$$f_2 = 1 - e^{-0.02y^{+2}} + 0.2ye^{-0.1(y^+ - 7)^2} +$$
$$0.04ye^{-0.11(y^+ - 12)^2}$$

$$f_\mu = \frac{\tanh 0.00039y^{+2}}{\tanh 0.0025y^{+3}}$$

and the model constants have the following values:

$$C_{\epsilon 1} = 1.44 \qquad C_{\epsilon 2} = 1.92 \qquad \sigma_k = 1.36 \qquad \sigma_\epsilon = 1.36 \qquad C_\mu = 0.09 \,.$$

An alternative set of damping functions based on $Re_k = y\sqrt{k}/\nu$ instead than y^+ are also used. The two sets of functions have the same shape when plotted using the DNS data from Mansour, Kim & Moin [4]

$$f_1 = 1 + 1.9e^{-0.12(\sqrt{Re_k} - 2.6)^2}$$

$$f_2 = 1 - e^{-0.07Re_k} + 0.32\sqrt{Re_k}e^{-0.4(\sqrt{Re_k} - 4.3)^2}$$

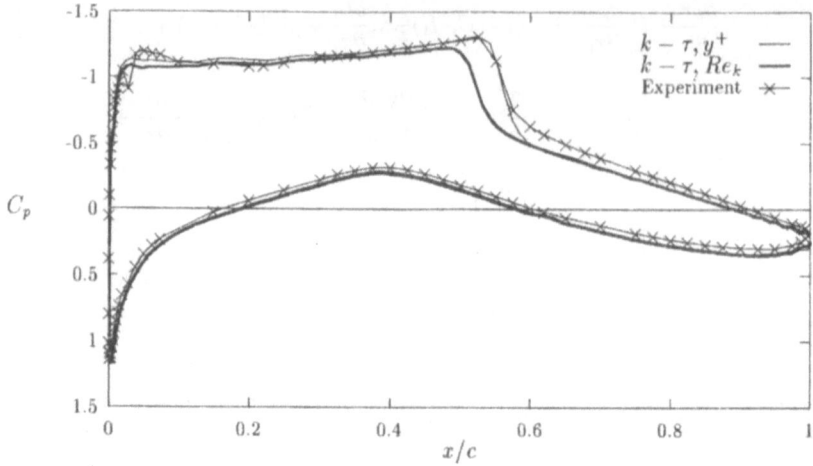

Figure 1: Pressure distribution, RAE2822 airfoil Case 9, $M = 0.73$, $\alpha = 2.79$, $Re = 6.5 \times 10^6$

$$f_\mu = \frac{\tanh 0.00005 \, Re_k^2}{\tanh 0.0004 \, Re_k^{2.5}}$$

and the model constants are the same as above.

Numerical method

The equations have been solved with a cell centered finite volume code, NSMB (Navier-Stokes Multi Block), using explicit Runge-Kutta time stepping. The equations for the turbulent quantities have been solved simultaneously with the other variables also using the explicit time stepping.

The transport equations have been discretized with standard second order methods. Which requires that node values are obtained from linear interpolation of cell center values. In most cases this is adequate if a fine enough mesh is used. However, k and τ are both $\sim y^2$ close to solid walls and the discretization of the k and τ equations must be able to resolve this quadratic behavior.

A standard Jameson 2nd and 4th order artificial dissipation have been used, Hirsch [1]. For the turbulent equations approximately a tenth of the artificial dissipation for the velocity was used, close to walls the artificial dissipation was damped by multiplying with an exponential function

$$1.0 - e^{-0.01y^{+2}}$$

If the artificial dissipation not was damped it was of the same order as the viscous terms and the shape of the mean velocity profile was distorted.

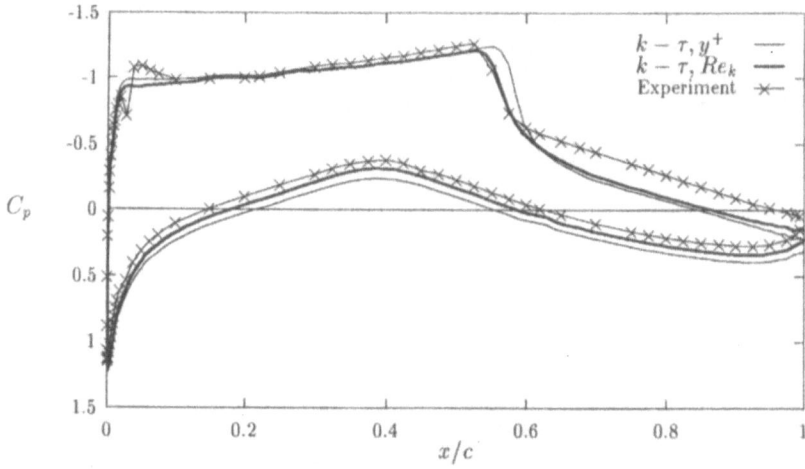

Figure 2: Pressure distribution, RAE2822 airfoil Case 10, $M = 0.75, \alpha = 2.81, Re = 6.2 \times 10^6$

Transition from laminar to turbulent flow was forced at $x/c = 0.03$ on both sides of the airfoil. The source terms and the eddy viscosity were multiplied by a factor which was zero in the laminar part and varied linearly to unity in the transition region. As free stream conditions $k = 0.001 * U_0^2$ and τ was given a value so that the eddy viscosity was 0.1% of the molecular viscosity. The free stream was corrected using the vortex correction as described by Hirsch [1].

The calculations were performed using a mesh with 256x64 grid points, which is the same mesh as used by Haase in this workshop.

Results for the RAE2822 airfoil

The damping functions based on y^+, $k - \tau_y$ model, have been used in several cases giving good results and robust calculations, see e.g. Lindberg, Weber & Rizzi [3]. However, in the present calculation where shock induced separation may occur the model might give some problems. The problems arise due to the fact that the friction velocity goes to zero at the separation point and the damping functions then are no longer valid. In order to come around this problem new damping functions based on the distance from the wall and the turbulent kinetic energy, $k - \tau_k$ model, was deduced. This model gives similar pressure distribution as the $k - \tau_y$ model but is more robust.

The overall agreement with the calculated pressure distribution and the experimental results is quite good. There are small differences between the results obtained with $k - \tau_y$ and the $k - \tau_k$ models, figures 1 & 2. The only remarkable difference is the shock position, which is further upstream in both cases when calculated with the $k - \tau_k$ model. Both models give a too high pressure right after the shock, especially for case 10.

The predicted skin friction from the $k - \tau_k$ model is too high compared with the experiment, figures 3 & 4. The $k - \tau_y$ model also gives higher C_f than the experiments, although the agreement is better than for the $k - \tau_k$ model. Downstream of the shock the calculated skin friction is too high for both models. In the experiment a small separated zone was noticed for case 10, this is not reproduced by neither of the models.

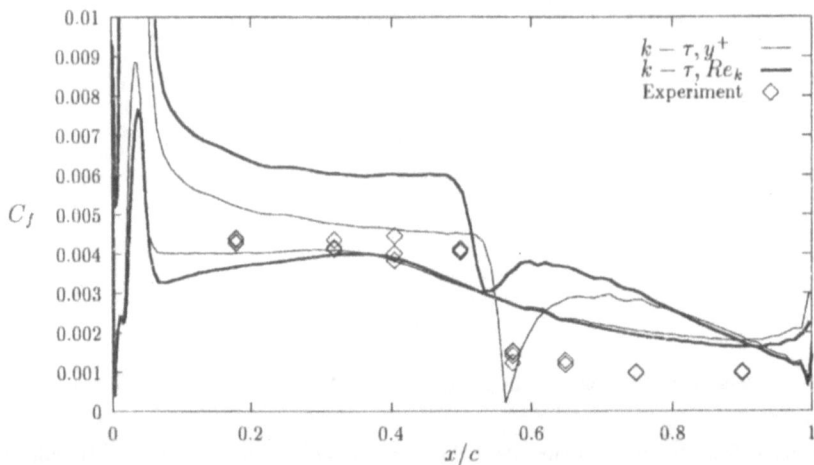

Figure 3: Wall friction distribution, RAE2822 airfoil Case 9, $M = 0.73, \alpha = 2.79, Re = 6.5 \times 10^6$

For the integrated loads the lift coefficient is about 5% off, generally predicted too high for with the $k - \tau_y$ model and too low with the $k - \tau_k$ model, table 1. The drag coefficient is predicted up to 25 % too high.

Table 1: Lift and drag coefficients, RAE2822 Case 9, $M = 0.734, Re = 6.5^6, \alpha = 2.79$ and Case 10, $M = 0.754, Re = 6.2^6, \alpha = 2.81$

Model	C_l, case 9	C_d, case 9	C_l, case 10	C_d, case 10
$k - \tau$, $f(y^+)$	0.832	0.0217	0.803	0.0261
$k - \tau$, $f(Re_k)$	0.775	0.0208	0.732	0.0265
Experiment	0.803	0.0168	0.743	0.0242

Conclusions

The calculations of the transonic airfoil have shown that the $k - \tau$ model is capable to compute the transonic flow fields with reasonable accuracy. The prediction of flow fields close to separation could be improved with use of modified damping functions. The present damping functions are based on results from direct numerical simulations (DNS) at low Mach numbers and they migth be improved if DNS for higher Reynolds numbers are used.

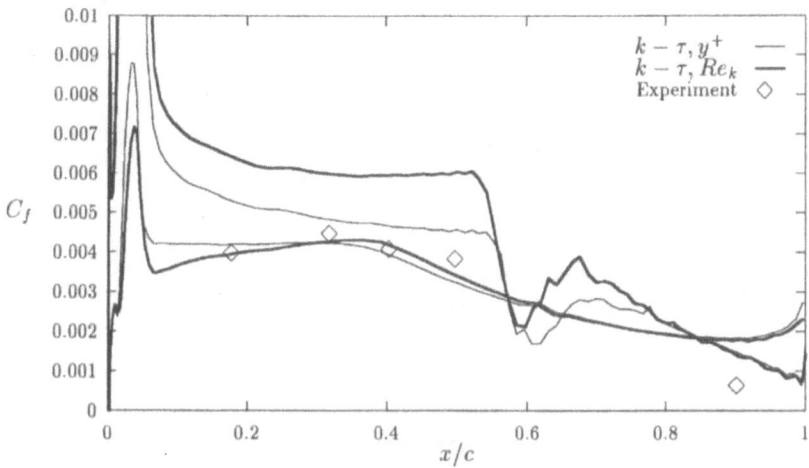

Figure 4: Wall friction distribution, RAE2822 airfoil Case 10, $M = 0.75, \alpha = 2.81, Re = 6.2 \times 10^6$

References

[1] C. Hirsch. *Numerical Computation of Internal and External Flows*. John Wiley & Sons, 1988.

[2] P.-Å. Lindberg. Near-wall turbulence models for 3D boundary layers. *Applied Scientific Research*, 53, 1994.

[3] P.-Å. Lindberg, C. Weber, and A. Rizzi. Calculation of the A-airfoil using the $k - \tau$ model. In *19th Congress of the International Council of the Aeronautical Sciences*, 1994.

[4] N.N. Mansour, J. Kim, and P. Moin. Reynolds-stress and dissipation-rate budgets in a turbulent channel flow. *Journal of fluid mechanics*, 194, 1988.

[5] C. Speziale, R. Abid, and C. Anderson. A critical evaluation of two-equation models for near-wall turbulence. *AIAA paper 90-1481*, 1990.

[6] D. Wilcox. *Turbulence Modeling for CFD*. DCW Industries, Inc., La Cañada, California, 1993.

APPLICATION OF AN UNSTRUCTURED GRID
FLOW SOLVER TO COMPRESSIBLE TURBULENT FLOWS

Luca Stolcis

Applied Mathematics and Simulation Group
CRS4, Via Nazario Sauro 10, I-09123 Cagliari (Italy)

Lars Davidson

Thermo and Fluid Dynamics
Chalmers University of Technology, S-41296 Gothenburg (Sweden)

SUMMARY

A numerical method for the prediction of turbulent compressible flows around complex configurations, which has been previously developed and validated [1], [2], has been employed for the computation of turbulent compressible flows of aeronautical interest. In order to allow a proper description of complex flow features, such as those resulting from shock-wave/boundary layer interactions, advanced turbulence models have been employed. Eddy-viscosity two-equation models as well as second-moment closure turbulence models have been adopted for this purpose. The results obtained for transonic flows around two-dimensional airfoils show that the usage of more sophisticated turbulence models such as the 'basic' Reynolds stress model can enhance the prediction capabilities.

NUMERICAL METHOD

The numerical method employed is based on that described in [1]. The approach adopted consists of solving the compressible Navier-Stokes equations in Reynolds-averaged form using unstructured computational grids. The governing equations for the mean-flow quantities are discretized in space by a cell-centred finite-volume technique. The method employs a central discretization for both convective and diffusive fluxes, with additional numerical dissipation introduced explicitly to damp oscillations due to odd-even decoupling and in proximity of shock-waves. An explicit multistage time-stepping scheme is employed to march the equations towards the steady-state solution. Acceleration techniques such as local time-step and residual smoothing are also utilised to increase the convergence rate.

Two different schemes can be used for the turbulence transport equations: an explicit scheme similar to that for the mean flow equations, and an implicit solver, which uses a point-by-point Gauss-Seidel relaxation to solve the discretized equations [5]. The choice of the implicit solver for the $k - \epsilon$ turbulence equations is due to the fact that this approach has proved to be very stable on structured grids [6]. Hybrid central/upwind differencing is used for the convective terms, and central differencing for the diffusive terms. The solver is written in such a general way, that it can handle control volumes of arbitrary number of cell-faces and grid points.

TURBULENCE MODELLING

Several turbulence models have been implemented in the present method, ranging from the standard high-Reynolds number $k - \epsilon$ model, up to the 'basic' Reynolds Stress Model.

a) Eddy viscosity models

The high-Reynolds $k - \epsilon$ turbulence model has been used in its standard form, which leads to the following two equations

$$\frac{\partial \rho k}{\partial t} + \frac{\partial \rho k u_j}{\partial x_j} + \frac{\partial}{\partial x_j}\left[-\left(\mu + \frac{\mu_t}{\sigma_k}\right)\frac{\partial k}{\partial x_j}\right] = P_k - \rho \epsilon \tag{1}$$

$$\frac{\partial \rho \epsilon}{\partial t} + \frac{\partial \rho \epsilon u_j}{\partial x_j} + \frac{\partial}{\partial x_j}\left[-\left(\mu + \frac{\mu_t}{\sigma_\epsilon}\right)\frac{\partial \epsilon}{\partial x_j}\right] = c_{\epsilon 1} P_k \frac{\epsilon}{k} - c_{\epsilon 2}\rho\frac{\epsilon^2}{k} \tag{2}$$

where u_j are the velocity vector components, ρ is the density and μ the viscosity. P_k is the production of turbulent kinetic energy, k, and μ_t is the turbulent eddy viscosity, defined as

$$\mu_t = c_\mu \rho \frac{k^2}{\epsilon} . \tag{3}$$

For a complete closure, the values of five constants c_μ, σ_k, σ_ϵ, $c_{\epsilon 1}$ and $c_{\epsilon 2}$ must be determined. The values of these constants for the two different approaches that have been used in the present work are shown in the following table.

Table 1: Coefficients for the $k - \epsilon$ models

	c_μ	σ_k	σ_ϵ	$c_{\epsilon 1}$	$c_{\epsilon 2}$
standard $k - \epsilon$	0.09	1.00	1.30	1.44	1.92
RNG $k - \epsilon$	0.085	0.7179	0.7179	equation(4)	1.68

The value of $c_{\epsilon 1}$ for the RNG [7] model is no longer constant and is determined as

$$c_{\epsilon 1} = 1.42 - \frac{\eta(1 - \eta/\eta_0)}{1 + \beta\eta^3} \tag{4}$$

where $\eta = Sk/\epsilon$, with $S = \sqrt{P_k/\mu_t}$, $\eta_0 = 4.38$ and $\beta = 0.015$.

b) Reynolds-stress model

The basic Reynolds stress model proposed by Gibson and Launder has been adopted here. The eddy viscosity assumption is removed and the Reynolds stresses are obtained directly from the following transport equations for the stresses

$$\frac{\partial}{\partial t}(\tau_{ij}) + \frac{\partial}{\partial x_k}(u_k\tau_{ij}) = P_{ij} + D_{ij} + \Phi_{ij} - \epsilon_{ij} \tag{5}$$

where P_{ij} indicates the production terms and is defined as

$$P_{ij} = -\left(\tau_{ki}\frac{\partial u_j}{\partial x_k} + \tau_{kj}\frac{\partial u_i}{\partial x_k}\right) \tag{6}$$

$$\epsilon_{ij} = \frac{2}{3}\rho\epsilon\delta_{ij} \tag{7}$$

$$\Phi_{ij} = \Phi_{ij1} + \Phi_{ij2} + \Phi_{ij}^w \tag{8}$$

with

$$\Phi_{ij1} = -c_1\frac{\epsilon}{k}\left(\tau_{ij} - \frac{2}{3}\rho k\delta_{ij}\right) , \quad \Phi_{ij2} = -c_2\left(P_{ij} - \frac{2}{3}\delta_{ij}P_k\right) \tag{9}$$

$$\begin{aligned}\Phi_{ij}^w = \ & [c_1'\frac{\epsilon}{k}(\tau_{km}n_kn_m\delta_{ij} - \frac{3}{2}\tau_{ik}n_kn_j - \frac{3}{2}\tau_{kj}n_kn_i) \\ & + c_2'(\Phi_{km2}n_kn_m\delta_{ij} - \frac{3}{2}\Phi_{ik2}n_kn_j - \frac{3}{2}\Phi_{kj2}n_kn_i)]\frac{k^{3/2}}{c_l\epsilon\eta_n} .\end{aligned} \tag{10}$$

Here, the subscript w indicates the terms arising from the presence of a solid wall, n_i are the wall-normal unity vectors along the co-ordinate directions x and y, and η_n is the dimensional distance from the wall. The diffusive terms are usually modelled using the so-called *generalize gradient hypothesis*, where a triple correlation is substituted by a combination of the turbulent stresses and their gradients. In particular, in the present work, an even more simplified expression based on an eddy viscosity concept, has been adopted

$$D_{ij} = \frac{\partial}{\partial x_k}(\mu_t/\sigma_t\frac{\tau_{ij}}{\partial x_k}). \tag{11}$$

As shown by Davidson and Rizzi [6] this more simplified approach for D_{ij} does not seem to particularly affect the quality of the computed flow-field. The standard set of constants has been adopted in present work

Table 2: Coefficients for the basic RSM model

c_1	c_2	c_1'	c_2'	c_l	σ_t
1.8	0.6	0.5	0.3	2.55	1.0

c) Near-Wall Treatment

In the viscous sub-layer near the wall, the turbulent Reynolds numbers become very low, and the above turbulence models do not describe correctly the physical phenomena that take place in that region. Therefore, particular methods have to be employed in order to deal with wall-bounded flows such as the ones under investigation during the present Workshop. The present method employed two different approaches: wall functions or a one-equation turbulence model.

When wall-functions are employed, the values of k and ϵ in the near-wall cells are not the ones obtained from the transport equation, but are fixed according to the universal law of the wall

$$u_T/u_\tau = ln(y^+E)/\kappa \,, \qquad u_\tau = \sqrt{\tau_{wall}/\rho_{wall}} \,, \qquad y^+ = y_n u_\tau \rho/\mu \qquad (12)$$

where u_T is the mean-velocity component parallel to the wall, u_τ is the friction velocity, y^+ is the non-dimensional wall distance, τ_{wall} is the wall shear stress, y_n the dimensional distance from the wall, κ the von Karman constant (≈ 0.41), and E the roughness parameter, which is equal to 9 for smooth walls.

Under the standard hypothesis of equilibrium between generation and dissipation of turbulence energy, if the length scale is assumed to be directly proportional to the distance from the wall, the relations become

$$k = u_\tau^2/\sqrt{c_\mu} \,, \qquad \epsilon = u_\tau^3/\kappa y_n \qquad (13)$$

where u_τ is obtained from the law-of-the-wall.

Instead of using Equation (13), one can solve the $k - \epsilon$ equations only in the high-Reynolds number regions and use a low-Reynolds number model in the near-wall region. The one-equation low-Reynolds number model of Wolfshstein, modified for compressible flows has been included in the present method. In regions adjacent to the surface, where $y^+ < 50$, the mean-flow equations and the equation for the turbulent kinetic energy are solved, whilst the characteristic length scales are determined via algebraic relations. In these regions, the eddy viscosity and the dissipation rate of k are defined as

$$\mu_t = c_\mu \rho k^{1/2} L_\mu \,, \qquad \epsilon = k^{3/2}/L_\epsilon \qquad (14)$$

with L_μ and L_ϵ being two algebraic length scales, defined as

$$L_\mu = c_1 y_n [1 - exp(-R_t/A_\mu)] \,, \qquad L_\epsilon = c_1 y_n [1 - exp(-R_t/2c_1)] \,. \qquad (15)$$

The model requires two constants which are usually taken as $A_\mu = 70$, $c_1 = \kappa c_\mu^{-3/4}$ where κ and c_μ take the usual values of 0.41 and 0.09. When y^+ becomes greater than 50, the standard high-Reynolds number $k - \epsilon$ model is employed as described before.

RESULTS

The present method has been extensively validated with the aid of several standard test cases, including the flow around the RAE 2822 airfoil, [1],[2],[3]. Some of the results obtained with the computational grids provided for the present workshop by UMIST are presented in Figures 1-4. The performance of the turbulence models for three reference cases (RAE 2822 and MBB A3 airfoils) can be detected by direct comparison of the results obtained. Figure 1 shows the surface pressure distribution for Case 9 ($M_\infty = 0.734$, $\alpha = 2.79$, Re= 6.5 million) obtained with the standard $k - \epsilon$ turbulence model with wall-functions. The computed data are in very good agreement with the experiments. This behaviour has been observed for all the turbulence models employed, which in fact produce results which are almost indistinguishable from each other. The surface skin friction coefficients are not as good as the pressure coefficients, mainly because of the use of wall-functions. In order to investigate the influence of the near-wall treatments upon the overall solution, we have compared the results obtained with wall-functions with those obtained with the two-layer model described in the previous paragraph. As expected, Figure 2, the skin friction coefficient levels computed with the two-layer model are much closer to the experiments than the ones computed with wall-functions. In spite of this, the pressure levels differ only in the supersonic region upstream- and in proximity of the shock. However, they are in very good agreement everywhere else, which is probably due to the fact that the low-Reynolds number grid is not fine enough. Therefore, the validation exercise on the more challenging Case 10 ($M_\infty = 0.754$, $\alpha = 2.81$, Re= 6.2 million) was performed using wall-functions. The surface pressures computed with the three models are shown in Figure 3. The two $k - \epsilon$ models give very similar results, whereas the Reynolds stress model shows an excellent agreement with experiments. However, despite the remarkable improvements on the shock location, the pressure levels in the recovery region downstream of the shock are still under-predicted. The skin friction coefficient plot confirms the difference between eddy-viscosity and second-moment closure models, with the $k - \epsilon$ result being a little closer to the experiments. The results obtained with the MBB A3 airfoil ($M_\infty = 0.85$, $\alpha = 1.78$, Re= 6.08 million) are presented in Figure 4. Also for this case, the usage of the RSM model has a remarkable influence upon the position of the shock-wave on the upper surface, which becomes closer to the experiments. However, there is still some disagreement regarding the pressure levels downstream of the shock and on the lower surfaces. The use of wall functions, as well as the uncertainties on the location of the transition points, seem to be the main reasons for this disagreement.

In general, we have noticed that for simple attached flows over airfoil there is little or nothing to gain by using higher-order models (indeed boundary-layer methods, or algebraic models perform well or even better). However, as the flow increases its complexity, only the use of more sophisticated models such as Reynolds stress models, can provide a better description of the flow-fields.

ACKNOWLEDGEMENTS

The work of the first author (LS) has been supported by the Sardinian Regional Authorities.

REFERENCES

[1] Stolcis, L. and Johnston, L.J. 'Computation of the viscous flow around multi-element aerofoils using unstructured grids', Notes on Numerical Fluid Mechanics, Vol. 35, pp. 311-320, Vos J.B, Rizzi A. and Ryhming I.L. Editors, Vieweg Verlag, 1992,

[2] Stolcis, L. and Johnston, L.J. 'Near-wall turbulence models and numerical solution of the Reynolds-averaged Navier-Stokes equations using unstructured grids', Lecture Notes in Physics, vol. 414, pp. 200-204, Napolitano M. and Sabetta F. Editors, Springer Verlag, 1993,

[3] Johnston, L.J. and Stolcis, L. 'Prediction of the high-lift performance of multi-element aerofoils using an unstructured Navier-Stokes solver', High-Lift System Aerodynamics, pp. 13.1-13.18, AGARD CP 515, AGARD, 1993

[4] Stolcis, L. 'Turbulent compressible flow computations using a second-moment closure model and unstructured grids', Report CRS4-APPMATH-94-3, CRS4, Cagliari (Italy), March 1994

[5] Davidson, L. and Stolcis, L. 'An efficient and stable solution procedure of compressible turbulent flow on general unstructured meshes using transport turbulence models', AIAA Paper 95-0342 , 33rd AIAA Aerospace Sciences Meeting Exhibit, Reno (USA), 9-12 Jan 1995

[6] Davidson, L. and Rizzi A. 'Navier-Stokes stall predictions using an Algebraic Stress Model', J. of Spacecraft and Rockets, Vol.29, pp. 794-800, 1992

[7] Speziale C.G. and Thangam S. 'Analysis of an RNG based turbulence model for separated flows', ICASE report No. 92-3, January 1992.

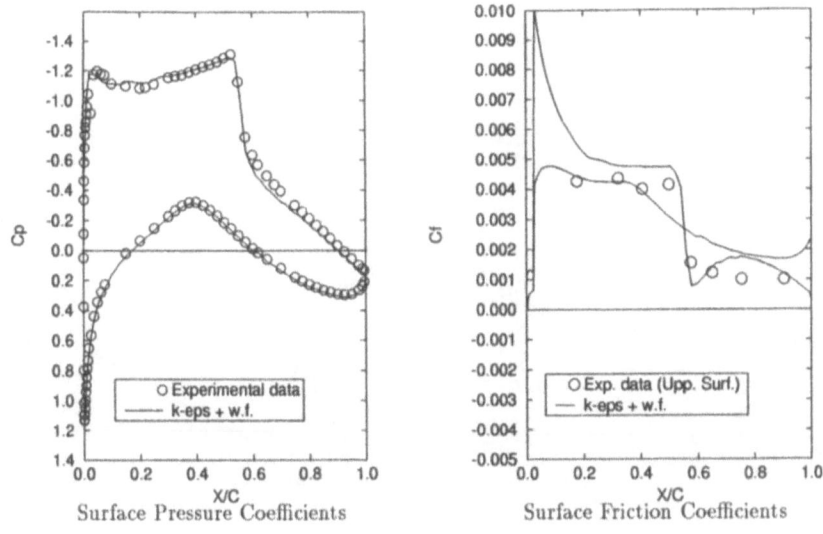

FIG 1 *RAE 2822 Airfoil - Case 9 -*

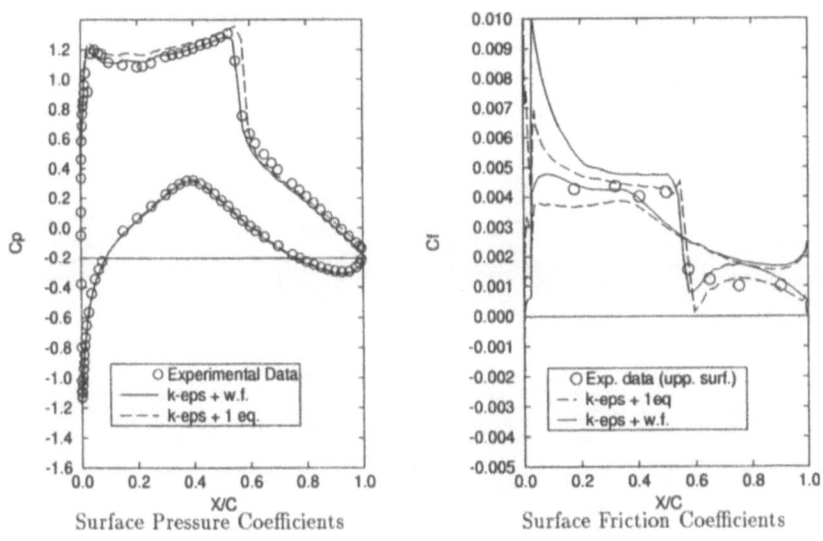

FIG 2 *RAE 2822 Airfoil - Case 9 -*

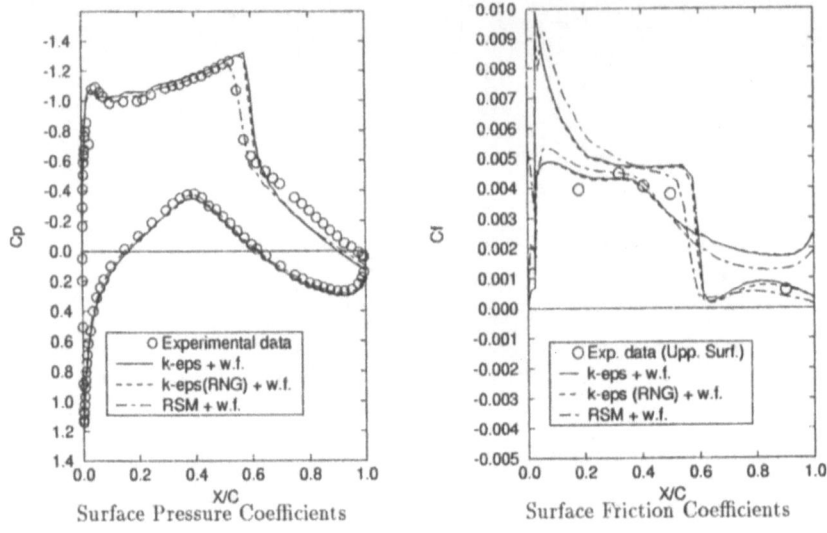

FIG 3 *Rae 2822 Airfoil - Case 10 -*

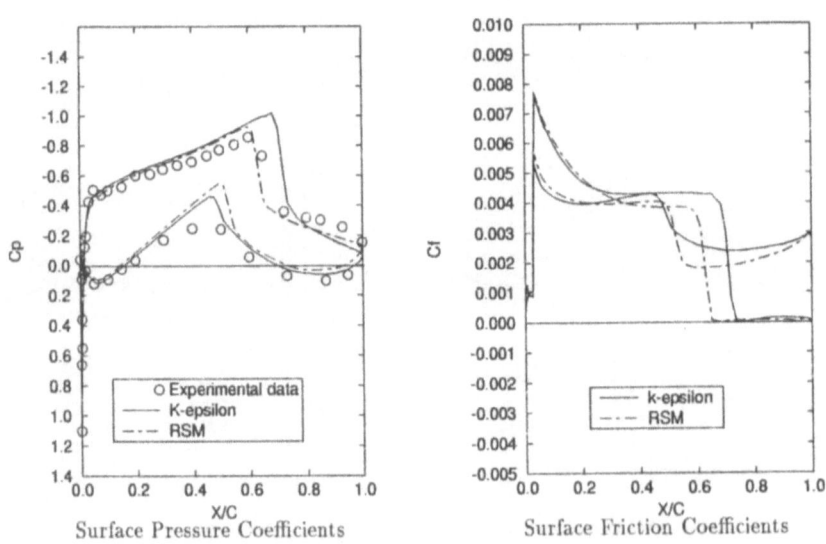

FIG 4 *MBB A3 Airfoil - Case 113 -*

534

CHAPTER 9 : UNSTEADY
AIRFOIL FLOW

PRESENTATION OF THE TEST-CASE TC8 bis

by

M. BRAZA[1] & S. TSANGARIS [2]

[1] Institut de Mécanique des Fluides de Toulouse, Unité Mixte de Recherche
C.N.R.S n° 5502, .Av. du Prof. Camille Soula
31400 Toulouse Cedex, France

[2]Laboratory of Aerodynamics, National Technical University of Athens
P.O. Box 64070, 15710 Zografos, Athens, Greece

The proposed test-case is the inherently unsteady flow over a 18% thick circular arc aerofoil at 0-deg angle of attack, Mach number 0.76 and Reynolds number 11×10^6, based on the chord length. Shock-induced separation is occurred at the foot of the shock wave, which is extended downstream beyond the trailing edge. Experimental data and numerical results for the time histories of the pressure, the mean velocity components and the mean flow direction on specific positions on the aerofoil's surface are available in ref. [1].

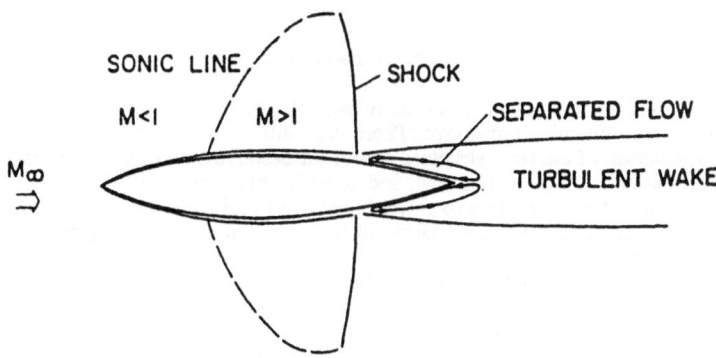

From the numerical point of view, we suggest the C-type grid of 273 x 41 points as a minimum size grid, with 203 points on the aerofoil's surface and 35 x 2 points along the wake. The outer boundaries are placed 10 chord lengths far from the aerofoil. The grid points are available in the data-base of INRIA. This grid is tentative, but it is proven to be able to seak the inherently unsteady flow. Of course, the participants are encouraged to use beyond this grid, other finer or coarser grids, in order to provide grid-independent results.

References

[1] SEEGMILLER, H.L., MARVIN, J.G. & LEVY, L.L. 1978 "Steady and unsteady transonic flow". AIAA J., Vol. 16, N° 12, pp1262, 1270.

PREDICTION OF THE UNSTEADY TRANSONIC TURBULENT FLOW AROUND A CIRCULAR-ARC AEROFOIL

M. Braza, F. Hanine, A. Bouhadji

Institut de Mécanique des Fluides de Toulouse, Unité Mixte de Recherche
C.N.R.S n° 5502, .Av. du Prof. Camille Soula
31400 Toulouse Cedex, France

Summary

In the present work we perform the prediction of the inherently unsteady turbulent flow around a circular-arc aerofoil, by using zero-equation turbulence models, especially modified to allow the appearance of the coherent structures in the mixing layer. It is shown that a good agreement with the experiment can be obtained by this category of models, concerning the aerodynamic parameters, the oscillation of the pressure coefficient and the frequency of the coherent eddies. The method described, classified as an Organised Eddy Simulation (OES) method, can be applicable in a general way, for external, separated and unsteady flows around aerofoils at high speed.

1. Introduction

The prediction of turbulent unsteady separated flows around aerofoils is a priority in the domain of aeronautics nowadays. There are numerous studies in this domain, but the accurate prediction of the flow structure and of the aerodynamic parameters, especially near stall conditions remain open questions, and considerable efforts have to be done in order to suggest an efficient methodology to solve this class of problems. The majority of the studies devoted to this topic use different classes of turbulence models employing the steady-state Reynolds averaged-equations. A widely used category of turbulence models for this flow are based in eddy-viscosity assumptions. The algebraïc models of turbulence remain popular for flows with complex geometries, owing to their simplicity in implementation and despite several limitations that they present in respect to the complex physics of this category of flows. Furthermore, they offer the opportunity to be a first step in modifying the scales of turbulence models, whenever flows in statistical non-equilibrium are considered, as the present class of flows with coherent structures. For these reasons we examine firstly this class of models for a high-speed flow around a circular-arc aerofoil at zero incidence, at Reynolds number 11×10^6 and Mach number 0.76, according to the physical experiment by Seegmiller, Marvin & Levy [1].

We discuss in the following sections the outlines of the numerical method and of the turbulence modeling implemented in the numerical code, as well as the choice of the optimum numerical parameters in order to ensure a *time-accurate* computation. In the last paragraph we present the results obtained in this study.

2. The governing equations and the numerical method

The equations of motion are written according to the Favre averaging, following the discussion presented for the TC1bis. These aspects are detailed in our contribution for the TC1bis problem. In the following, the averaging __ symbolizes the phase-averaging, as defined by numerous experimental [13] and numerical works [5], [6], [15], [16], [17], [18], in respect to the inherent unsteadiness. The details of the present unsteady flow modeling methodology are discussed in a following paragraph. The finally obtained equations are:

$$\frac{\partial \overline{\rho}}{\partial t} + \frac{\partial(\overline{\rho} \ \widetilde{u}_k)}{\partial x_k} = 0 \qquad (1)$$

$$\frac{\partial(\overline{\rho} \ \widetilde{u}_i)}{\partial t} + \frac{\partial(\overline{\rho} \widetilde{u}_i \ \widetilde{u}_k + \overline{\rho u''_i u''_k} - \overline{\sigma_{ik}})}{\partial x_k} = 0 \qquad (2)$$

$$\frac{\partial(\overline{\rho \widetilde{E}})}{\partial t} + \frac{\partial\left(\overline{\rho}(\widetilde{E}+\overline{p})\widetilde{u}_k + \overline{\rho u''_i u''_k}\widetilde{u}_i + \frac{\overline{\rho u''_i u''_i u''_k}}{2} + \gamma c_v \overline{\rho \theta'' u''_k}\right)}{\partial x_k} = \frac{\partial\left(\overline{\mu(\widetilde{S}_{ik}\widetilde{u}_i + \overline{S_{ik}u''_i} + \lambda \frac{\partial \widetilde{T}}{\partial x_k}}\right)}{\partial x_k} . \quad (3)$$

THE PHASE AVERAGING

It is recalled briefly that in the case of classical turbulence models, corresponding to the situations of flows in statistical equilibrium, the flow field is usually decomposed in two parts:

$$U_i = \overline{U_i} + u_i \qquad (4.a)$$

where $\overline{U_i}$ represents the statistical mean value which is independent of time and u_i is the random fluctuation. Often this statistical mean can be replaced by the temporal mean, under several assumptions (Hinze [11]), the temporal mean being easily measured.

However, in the present case of flows with a pronounced periodic component, a certain number of experimentalists used a technique to regroup the organized characteristics in an operator well known as " phase-averaged operator ". However, as the temporal mean is a widely measured quantity, they used to split the flow field in three parts, including explicitly the temporal mean. Hussain and Reynolds [12]) suggested to use a three-term decomposition of every physical quantity as follows:

$$U_i = \overline{U_i} + \widetilde{U_i} + u_i \qquad (4.b)$$

where $\overline{U_i}$ represents the steady averaged motion, $\widetilde{U_i}$ represents the periodic fluctuation part and u_i is the random fluctuation. This is also adopted by Cantwell and Coles [13], who further suggest a two-term decomposition. Indeed, the three-term decomposition applied to the Navier-Stokes equations leads to very complex transport equations, different from the Navier-Stokes ones, and moreover difficult to model.

The decomposition that we use therefore is a two-term decomposition defined as follows:

$$U_i = \langle U_i \rangle + u_i \qquad (4.c)$$

where $\langle U_i \rangle$ regroups $\overline{U_i} + \widetilde{U_i}$ and represents the phase-averaged organized motion, and u_i represents the random (chaotic) motion.

The first part which is represented by the phase-averaged operator $\langle \ \rangle$ yields equations similar to the Navier-Stokes ones, and in this way, the already available solvers can be used with certain modifications. The second part regroups chaotic characteristics, due to the random turbulence, which develops simultaneously with the organized motion. Owing to the non-linearity of the Navier-Stokes equations, following the decomposition (4.c), there appear new terms, which are the phase-averaged Reynolds stresses, in the same way as in the classic turbulence modeling. For this reason, the above terms have to be modeled by using

539

appropriate closures, which are not necessarily the same as in the case of classical turbulence models with steady temporal mean.

Beyond the practical interest of this methodology, there is a physical reason for this, since, according to the present decomposition the different kinds of structures in the unsteady wake are distinguished with a criterion based on their deterministic or chaotic character, rather than according to their size. This is a main difference of the present methodology, which is an Organised Eddy Simulation approach, comparing to the Large Eddy Simulation approach. Furhermore, the OES approach is not inherently three-dimensional and it involves a *physically existing and measurable* averaging: the phase-averaging [13], [26]. This decomposition has been advocated in our research group by Ha Minh *et al.* [15], [16], [17] and applied firstly for the investigation of a compressible boundary layer flow, by using a second-order closure models. In ref. (Braza, [27]), it has been shown that the phase-averaging operator can be extended to an ensemble-averaging one, which is able to regroup more than one predominant frequency effects due to different classes of the coherent structures, being in phase among them. Therefore, in the context of the Organized Eddy Simulation approach, the ensemble-averaging regroups all the organized characteristics of the flow system which are predictable by the complete system of the equations of motion. Braza and Noguès [5], [6] have also suggested this decomposition for turbulent wake flows and report that the shedding frequency due to the von Karman instability and the frequency related to the Kelvin-Helmholtz instability are correctly predicted by the OES approach. Franke *et al* [14] use the phase-average decomposition in wake flows and predict accurately the vortex shedding frequency due to the von Karman instability. In more recent studies, we use this methodology in combination with two-equation turbulence models for unsteady separated flows past aerofoils, (Jin & Braza [18]). In this methodology, the operator $\langle\ \rangle$ includes *all* the organized characteristics of the coherent structures predictable by the grid size and appearing in phase in the time-dependent evolution. In this context, the classical assumptions leading to different sets of constants in turbulent models, valid for steady-state cases, have to be completely reconsidered.

3. The turbulence model

In more recent studies of us, the OES approach is employed in combination with two-equation turbulence models [18] for incompressible flows around lifting bodies. Although the modifications suggested for two-equation models can be applied in high-speed flows, we examine in this paper the context of algebraïc models firstly. For simplicity, in this paper we replace the symbol $\langle\ \rangle$ by the symbol (__), but both designate *phase-averaged quantities*.

3.1 *THE ALGEBRAÏC MODEL*

In this category of models, the turbulent viscosity is given by a phenomenological law, based on theoretical developments and on turbulent experiments. Although the approaches available for algebraïc models suppose a conditioning of the variations of the turbulent stresses on those of the mean velocity gradients, these models merit to be examined, owing to their simplicity in implementation in flows of complex geometry and in three-dimensional applications, interesting the industrial domain.

The turbulent viscosity is expressed as a function of mean velocity gradients, through a length scale l_m which is the mixing length, defined by phenomenological laws. The mixing length in its physical meaning represents the characteristic length between two shear layers of the turbulent flow, along which the exchange of energy through the turbulent motion takes place.

In the seventies, more sophisticated algebraic models have been developed, taking into account two different expressions of the mixing length and of the eddy viscosity for internal and external regions of boundary layer configurations.

Cebeci, Smith and Mosinski [3] gave an extention of Cebeci-Smith model to compressible flows. They expressed the eddy viscosity as following :

* In the internal region the Prandtl hypothesis is used :

$$\mu_{ti}(y) = \overline{\rho}\, l_m^2 \left|\frac{\partial \overline{U}}{\partial y}\right| \quad \text{for}: 0 \le y \le y_c \tag{5}$$

where l_m is calculated with the van Driest approximation. This evaluation presents some restrictions for strong pressure gradients and mass transfer.

* In the external region :

$$\mu_{te}(y) = \alpha\, \overline{\rho}\, U_{max}\, \delta^* \left(1 + 5.5 \frac{y}{\delta}\right)^{-1} \quad \text{for}: y_c \le y. \tag{6}$$

α is the Clauser constant, δ and δ^* are respectively the boundary layer and momentum thicknesses, y_c is minimum value of y where $\mu_{ti}(y) = \mu_{te}(y)$.

In order to use this model, we must know the border of the boundary layer. This constraint is removed by the Baldwin-Lomax model.

Baldwin and Lomax [4] avoid the evaluation of δ by expressing:

$$\mu_t(y) = \begin{cases} \mu_{ti}(y) & \text{if } 0 \le y \le y_c \\ \mu_{te}(y) & \text{if } y_c \le y \end{cases} \tag{7}$$

where :

$$\mu_{ti}(y) = \overline{\rho}\, l_m^2 |\omega| \quad \text{and} \quad |\omega| = \left|\frac{\partial \overline{U}}{\partial y} - \frac{\partial \overline{V}}{\partial X}\right| \tag{8}$$

represents the dynamic eddy viscosity in the inner region and :

$$\mu_{te}(y) = \alpha\, \overline{\rho}\, C_{cp}\, F_{sillage}\, F_{Kleb}(y) \tag{9}$$

the outer region eddy viscosity .

$$F_{sillage} = \min \begin{cases} y_{max}\, F_{max} \\ C_{wk}\, y_{max}\, U_{dif}^2 / F_{max} \end{cases} \tag{10}$$

F_{max} is the maximum value of $F(y)$ and y_{max} the corresponding value. $F(y)$ is given by :

$$F(y) = y \left|\frac{\partial \overline{U}}{\partial y} - \frac{\partial \overline{V}}{\partial X}\right| \left(1 - \exp\left(\frac{y^+}{A}\right)\right) \tag{11}$$

$$F_{Kleb}(y) = \left(1 + 5.5 \left(\frac{C_{Kleb}\, y}{y_{max}}\right)^6\right)^{-1}. \tag{12}$$

F_{Kleb} represents the intermittency factor. U_{dif} is velocity scale given as following :

$$U_{dif} = \max\left(\sqrt{u^2 + v^2}\right) - \min\left(\sqrt{u^2 + v^2}\right). \tag{13}$$

For attached flows we have : $\min\left(\sqrt{u^2 + v^2}\right) = 0$.

The various constants used are :

A = 26	$C_{wk} = 0.25$
$\alpha = 0.0168$	$C_{Kleb} = 0.3$
$C_{cp} = 1.6$.	

The various models presented here have been largely used. Their success is due to the simplicity of their implementation, because they do not introduce additional equations. They give good results in case of equilibrium flows where production and dissipation rates are equal, concerning the kinetic turbulent energy $\tilde{k}$ and turbulent friction $- \overline{\rho}\, \overline{u''\, v''}$. We remind that an inconvenience of the algebraïc models remains in the fact that they present a loss of the flow history, in respect to the relaxation effects on the turbulent stresses.

In order to predict unsteady flows with coherent structures, we use the modification of the above model by Braza & Noguès [5], according to which the external region mixing length is multiplied by a function of the dimensionless vorticity ω, in its absolute value.

3.2 THE MODIFICATION OF THE LENGTH SCALE

In the case of the modeling of unsteady flows with coherent structures, there is a need to model differently the length scale in the outer region, owing to a different energy cascade from the external flow to the organized and random classes of vortices, unless the mixing length becomes excessively high; in fact, this is due to the fact that a considerable amount of the externally supplied energy is devoted to sustain not only the random motion but also the motion of the coherent structures. Following the analysis described for the TC1bis, we have introduced the dimensionles vorticity function in the outer law determining the turbulent viscosity. This yields:

$$\mu_{te}(y) = \alpha\, \overline{\rho}\, C_{cp}\, F_{sillage}\, F_{Kleb}(y)/\omega/. \tag{14}$$

In this study, we have implemented the phase-averaged equations for a turbulent, unsteady compressible flow, as explicited above. We have implemented the algebraïc models of Baldwin & Lomax and the one including the modification of Braza & Noguès, in the numerical code using the MUSCL scheme (van Leer [21], [22], [23]) in a second-order of accuracy and in a flux-vector-splitting implicit formulation.

The equations are writen in a general curvilinear coordinates system. A C type grid is used for the transformation. Zero-gradient boundary conditions are employed for the outlet boundaries.

542

4. The results

The computations are carried out according to two different grids. Firstly the one provided by partner 10/NTT:(273 x 41) points with 203 points on the aerofoil surface and 35 x 2 points along the wake. However, this grid, associated with the flux vector splitting scheme used, has been proven unsufficient to resolve the dynamics in the boundary layer, especially because this involves the appearance of the unsteady shock wave and separation. Certainly this grid size is more efficient with other categories of methods, as those introducing an artificial dissipation.

Secondly, we use a much finer grid, (500 x 120) which has been proven very satisfactory in a preliminary test-case that we have chosen, for a close value of the Mach number (M=0.70). We recall very briefly the performances of the method for this case, based to the physical experiment by Harris [24], at the NASA/Ames wind tunnel. The aerofoil configuration is a NACA0012, the Reynolds number Re=9 x 10^6 and the angle of attack a= 1°.86. The computations predict very well the appearance of the shock wave in the suction side, near the leading edge, as well as the mean values of the aerodynamic parameters:

Computational global parameters: C_L=0.279 C_D=0.0053

Experimental results: C_L=0.241 C_D=0.0079.

The pressure coefficient shows a very good agreement with the experimental results [24], figure 1.

For these reasons we adopted the same grid for the computation of the flow around a circular-arc. We have performed numerous tests for the optimum time-step to use. We have found that the optimum range for the time-step values is from 0.01 to 0.06 (dimensionles in respect to the chord).

The mean values of the aerodynamic parameters are found to be 0.085 for the drag coefficient and zero for the lift coefficient. The maximum and minimum amplitudes of the pressure coefficient (upper and lower sides) are shown on table 1. These values are in good agreement with the experiment and with the results by partners 1 and 10 (referenced in the synthesis of TC8bis). The turbulence model predicts the inherent unsteadiness (figure 3) and gives a fundamental frequency of the oscillating flow, equal to 168 Hz, whereas the experimental value is 188 Hz. The appearance of the inherent unsteadiness is due to the amplification of Tollmien-Schlichting waves developed in the separated shear layers, owing to the shape of the body. After the appearance of this phenomenon, the shock wave interacting with the separation region downstream, creates a synchronous mechanism which sustains the instability. A view of the field Mach number is presented on figure 6. A very good resolution of the boundary layer is obtained for all the thermodynamic quantities (figure 4).

5. Conclusions

The organized Eddy Simulation approach (OES), including a modified zero-equation turbulence model presented in this paper allows the appearance of the inherent unsteadiness in the flow around a circular-arc aerofoil, at Mach number 0.76. The predominant frequency predicted is in good agreement with the physical experiment, within the limits of errors unavoidably existing in both approaches. The phenomenon of shock-induced separation is also well predicted, although a rather simple turbulence model is used. However, the grid

543

required to achieve a time-accurate computation is considerably dense and this needs a rather long computational time to remove the residuals of the transitional phase.

Acknowledgements

We are thankful to Professor Seegmiller of NASA-AMES who provided us with additional results for the present study and also to Prof. S. Tsangaris and to Mr A. Pentaris for their valuable scientific discussions.

6. References

[1] SEEGMILLER, H.L., MARVIN, J.G. & LEVY, L.L. 1978 "Steady and unsteady transonic flow". AIAA J., Vol. 16, N° 12, pp1262, 1270.

[2] CEBECI, T. & SMITH, A. M. O. 1974 "Analysis of turbulent boundary layers". Applied Mathematics and Mechanics. Ed. Frankiel F. N. and Temple, G. Academic Press, New-York.

[3] CEBECI T., SMITH A. M. O., MOSINSKI S. G. 1970 "Calculations of compressible adiabatic turbulent boundary layer" AIAA Journal, N°8, pp 1974-1982.

[4] BALDWIN B. S., LOMAX H. 1978 "Thin layer approximation and algebraïc model for separated turbulent flows." AIAA Paper, 78-257, January 1978.

[5] BRAZA M., NOGUÈS, P 1991 "Numerical simulation and modeling of the transition past a rectangular afterbody". Proceedings of the VIIIth Turbulent Shear Flow Conference, Munich, 9-11 Sept. 1991 pp. I-2-1, I-2-2.

[6] NOGUÈS, P., BRAZA, M. 1992 " An organized-eddy simulation method for an unsteady wake flow in transition to turbulence". Internal Report IMFT.

[7] REYNOLDS O. 1883 "An experimental investigation of the circumstances which determine whether the motion of water shall be direct of sinuous and the law of resistance in parallel channels" Phys. Trans. Roy. Soc. London, 174, pp. 935-982.

[8] REYNOLDS O.1884 "On the dynamically theory of incompressible viscous fluid and the determination of the criterion." Phys. Trans. Roy. Soc. London, 186, pp. 123-161.

[9] FAVRE A. 1965 "Equations des gaz turbulents compressibles". Formes générales, Journal de Mécanique, Vol. 4, N°3, pp 361-390, Septembre 1965.

[10] FAVRE A. 1965 "Méthode des vitesses moyennées, méthode des vitesses macroscopiques pondérées par la masse volumique". Journal de Mécanique, Vol. 4, N°4, pp 390-421, Décembre 1965.

[11] HINZE, O. 1959 "Turbulence", McGraw-Hill and second edition 1975. New York.

[12] HUSSAIN, A. K. M. F. & REYNOLDS, W. C. 1975 "Measurements in fully developed turbulence channel flow". J. Fluid Eng. Vol. 97, 568.

[13] CANTWELL, B. J., & COLES, D. "An experimental study of entrainment and transport in the turbulent near wake of a circular cylinder" 1984 J. Fluid Mech. Vol. 136, 321-374.

[14] FRANKE, R. & RODI, W. 1991 Proceedings of the Eighth Symposium on Turbulent Shear Flows, Munich, Germany, Sept. 9-11. Vol. 2, 20-1.

[15] HA MINH, H., CHASSAING, P., BRAZA, M., KOURTA, A. & SEVRAIN, A. 1987 Rapport final, convention DRET / INPT n°84/190, Mars 1987, Institut de Mécanique des Fluides de Toulouse.

[16] HA MINH, H. & CHASSAING, P.1987 Rapport de convention DRET/INPT 87/131, Institut de Mécanique des Fluides de Toulouse.

[17] HA MINH, H., VIEGAS , J. R., RUBESIN, M. W., VANDROMME, D. D. & SPALART, P. 1989 Proceedings of the Seventh Symposium on Turbulent Shear Flows, Stanford, CA, USA, August 21-23.

[18] G. JIN & M. BRAZA (1994) "A two-equation turbulence model for unsteady separated flows around aerofoils". AIAA J., Vol. 32, N° 11, Nov. 94, pp. 2316, 2320.

[19] HANINE, F.1992. "Physique, Modélisation et Simulation des couches limites turbulentes compressibles". Thèse de Doctorat, I.N.P. Toulouse.

[20] BOUSSINESQ J. 1897 "Théorie de l'écoulement tourbillonnant et tumultueux des liquides dans les lits rectilignes à grande section." I-II- Gauthiers-Villars, Paris.

[21] VAN LEER B., THOMAS J. L., ROE P. L., NEWSOME R. W.1987 "A comparison of numerical flux formulas for the Euler and Navier-Stokes equations". AIAA Paper 87-1184.

[22] VAN LEER B. 1982 Flux vector splitting for the Euler equations. ICASE report 82-30.

[23] VAN LEER B. 1979 "Towards the ultimate conservative difference scheme. A second order sequel to Godunov's method". J. Comp. Phys. , Vol. 32, 101-136.

[24] HARRIS, C.D. 1981 "Two-dimensional Aerodynamic characteristics of the NACA0012 aerofoil in the Langley 8-foot transonic pressure tunnel". NASA TM 81927.

[25] LEVY, L.L., Jr.1978 "Experimental and computational steady and unsteady transonic flows about a thick aerofoil". AIAA Journal, VoL.16, June 1978, pp. 564-572.

[26] BOISSON, H.C. 1982 "Développement de structures organisées turbulentes à travers l'exemple du sillage d'un cylindre circulaire". Thèse de Docteur ès-Sciences, n°65. Institut National Polytechnique de Toulouse, France.

[27] M. BRAZA 1986 "Analyse physique du comportement dynamique d'un écoulement externe, décollé, instationnaire, en transition laminaire-turbulente. Application: Cylindre circulaire". Thèse de Doctorat d'Etat-ès-Sciences, I.N.P.T., Décembre 1986.

545

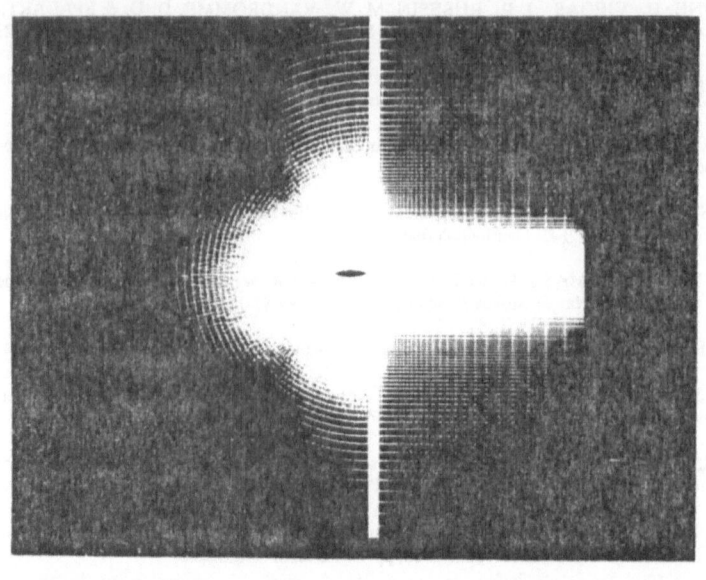

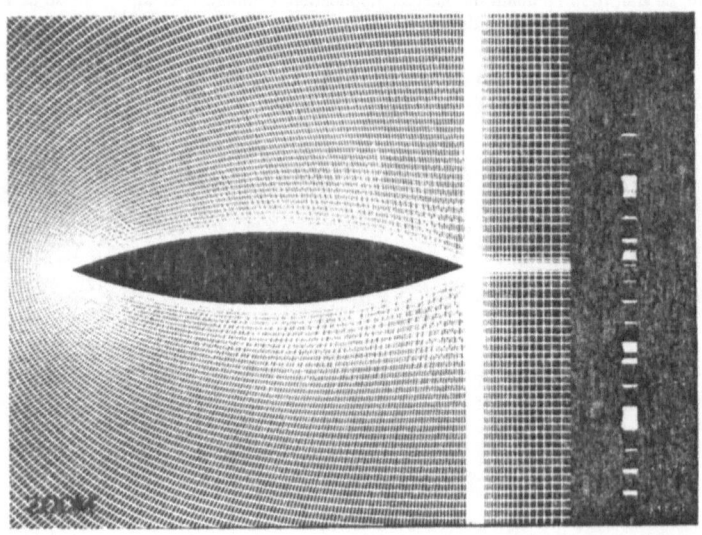

Fig. 1 Computational domain (a) and zoom (b) around the circular-arc aerofoil

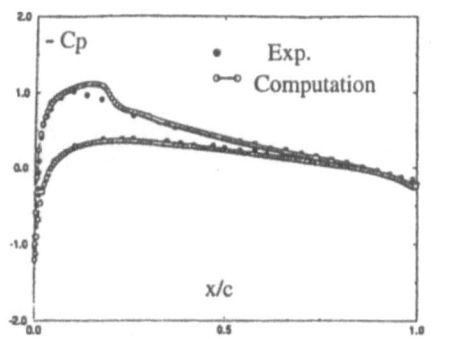 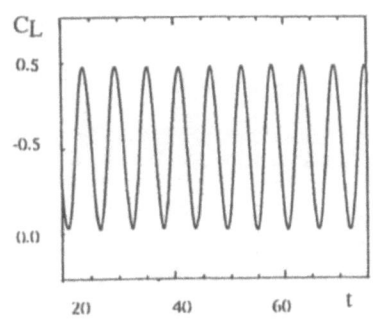

Fig. 2 Mean Pressure coefficient versus experiment.   Fig. 3 Lift coefficient versus time.

Table 1. Predominant frequency and maximum amplitudes of the pressure
coefficient oscillations. Comparison with the experiment.

	Frequency	Cp upper (max/min)	Cp lower (max/min)
Exp	188 Hz	0.1 -0.2	0.1 -0.21
This study	168	0.15 -0.2	0.13 -0.25

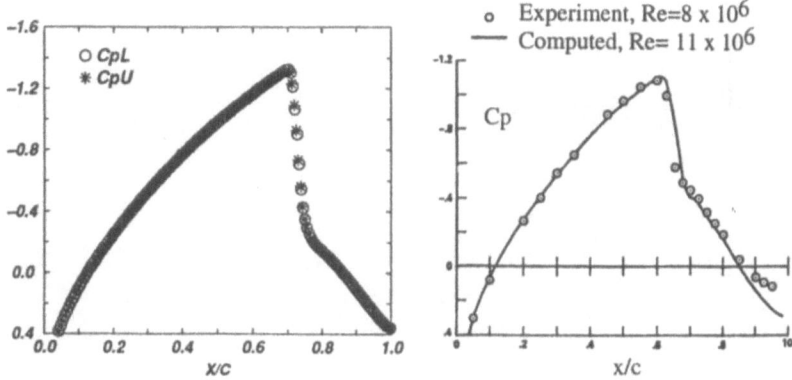

Fig. 5 Mean wall pressure coefficient ; Comparison with the experiment at Mach=0.754 ;
figure on the right, from ref [25].

Figure 6.   Iso-Mach number contours, t=240.

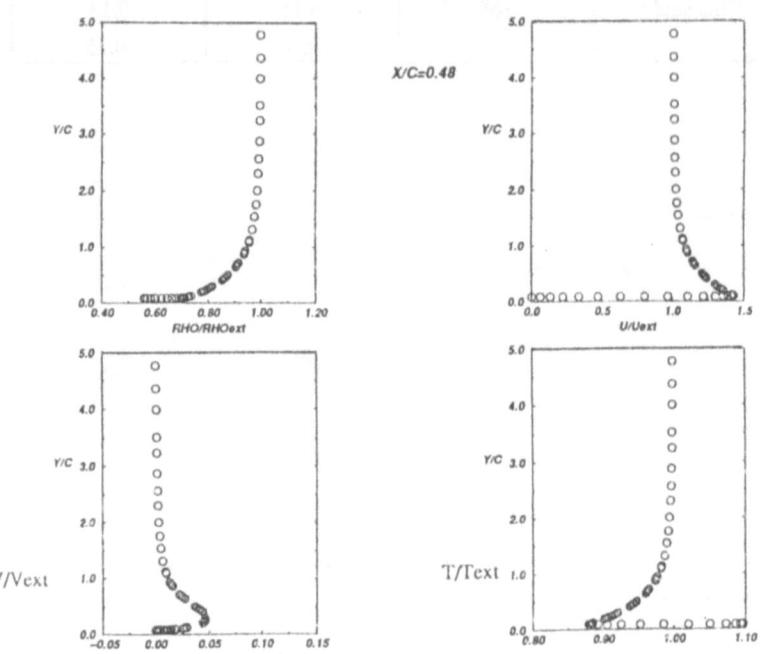

Figure 4.   Evolution of the mean velocity profiles (u,v), of density and temperature in the boundary layer.

Unsteady Separated Turbulent Flows Computation with Wall-Laws and $k - \varepsilon$ Model

Bijan Mohammadi

INRIA

Domaine de Voluceau B.P.105 - 78153 Le Chesnay, France

Abstract

This paper is devoted to the description of our experience in unsteady and separated flows simulations with the classical $k - \varepsilon$ model and wall-laws. The wall-laws we use are original and take into account for pressure and convection effects and are valid up to the wall.

1 Introduction

Statistical models like eddy viscosity or Reynolds stress models are expected not to be able to reproduce unsteady flows as contradictions between the averages in these models and unsteadiness can be pointed out. As discussed in [1] the Reynolds average $(\overline{f} = (\int_{t}^{t+T} f \, d\tau)/T)$ is not suitable for unsteady phenomena unless the problem has two disjoint scales. Indeed, if we apply it twice, the resulting signal will change $(\overline{\overline{f}} \neq \overline{f})$. This is of course incompatible with the basic requirement of the statistical approach (i.e. $\overline{\overline{f}} = \overline{f}$). Now, if we use a low-pass Fourier filter for instance, the averaged signal will remain unchanged through a second filtering. In other words, this means that the Reynolds average may work if the signal has clearly two disjoint parts in his spectra and that statistical models are able to capture mean unsteadiness if it has a spectra gap. Flows with 'Strouhal' frequencies are typical in this class.

In the same way, wall-laws are often used for attached and steady flows or flows where separation is induced by the geometry. But, wall-laws are expected to rather badly behave in situations where general separations and unsteadiness are involved. In our opinion, this often comes from the way these laws have been implemented and also the fact that too coarse a mesh has been used.

This paper is to show that interesting results can be obtained using wall-laws for general separation and unsteadiness. We use the standard $k - \varepsilon$ with wall-laws to reproduce mean unsteadiness and we compare the computed Strouhal number to its experimental value. Here, several difficulties are present coming from the question of the validity of $k - \varepsilon$ model for unsteady cases, wall-laws and of course implementation strategies.

The particular ingredients of our implementation related to wall-laws are:

1. Global wall-laws: they take into account for pressure and convection effects and are valid up to the wall (i.e. $\forall y^+ \geq 0$).

2. Weak formulation: this permits to easily take into account for pressure effects in the boundary integrals, something which is more difficult in a finite difference implementation for instance.

2. Small δ in wall-laws: this means that the computational domain should not be too far from the wall.

4. Fine meshes: this comes from the previous requirement and ask for meshes of 'laminar' type and not of 'Euler' type.

Indeed, if the computation domain is too far from the wall and the mesh too coarse in the boundary layer, the flow will keep its energy and will not separate. This is what often happen in computations with wall-laws.

By 'Euler' type mesh we mean that the first node is quite far from the wall (says at 10^{-2} m) for a Reynolds number of 10^7, 'Laminar' type mesh means that the first node is at about 10^{-4} m. For a low-Reynolds computation at the same Reynolds number (10^7), we would need a first node at about 10^{-7} m, something which is impossible to achieve for unstructured meshes in 3D for instance using Delaunay type mesh generator, even if we introduce adaptivity.

Another important and interesting feature of wall-laws is that they permit for explicit time integration. Something which is quite unrealistic if low-Reynolds modeling is introduced. To give an idea, if we use 'laminar' type meshes as defined above, this alows the wall-laws computation for time steps of at least 1000 times bigger than the one necessary for the low-Reynolds computation.

Models and numerics are described in sections 2 and 3. Section 4 is devoted to our wall-laws description and section 5 contains numerical examples.

2 Governing Equations

We split the variables into mean and fluctuating parts. We use a Reynolds average for density and pressure and a Favre average for other variables. We then consider the Reynolds averaged Navier-Stokes equations. Once the unknown correlations are modeled [1], we have:

$$\frac{\partial \rho}{\partial t} + \nabla \cdot (\rho u) = 0$$

$$\frac{\partial \rho u}{\partial t} + \nabla \cdot (\rho u \otimes u) + \nabla p = \nabla \cdot ((\mu + \mu_t)S) \tag{2.1}$$

$$\frac{\partial \rho E}{\partial t} + \nabla \cdot ((\rho E + p)u) = \nabla \cdot ((\mu + \mu_t)Su) + \nabla((\kappa + \kappa_t)\nabla T)$$

with

$$\kappa = \frac{\gamma\mu}{Pr}, \quad \kappa_t = \frac{\gamma\mu_t}{Pr_t},$$

$$\gamma = 1.4, \quad Pr = 0.72 \quad \text{and} \quad Pr_t = 0.9,$$

Where μ and μ_t are the inverse of the laminar and turbulent Reynolds numbers. In what follows, we call them viscosity. The laminar viscosity μ is given by Sutherland law but this is not very important for the cases presented here:

$$\mu = \mu_\infty (\frac{T}{T_\infty})^{1.5}(\frac{T_\infty + 110.}{T + 110.}), \tag{2.2}$$

where ∞ denotes reference quantities.

We do not take into account the turbulent kinetic energy contribution to the pressure and total energy and keep the usual laws for perfect gas.

The $k - \varepsilon$ model [2] we use is rather classical and is an extension to compressible flows of its incompressible version [1,3] and is defined by:

$$\frac{\partial \rho k}{\partial t} + \nabla.(\rho u k) - \nabla((\mu + \mu_t)\nabla k) = S_k, \tag{2.3}$$

and

$$\frac{\partial \rho \varepsilon}{\partial t} + \nabla.(\rho u \varepsilon) - \nabla((\mu + c_\varepsilon \mu_t)\nabla \varepsilon) = S_\varepsilon. \tag{2.4}$$

The right hand sides of (2.3)-(2.4) contain the production and the destruction terms for ρk and $\rho \varepsilon$:

$$S_k = \mu_t P - \frac{2}{3}\rho k \nabla.u - \rho\varepsilon, \tag{2.5}$$

$$S_\varepsilon = c_1 \rho k P - \frac{2c_1}{3c_\mu}\rho\varepsilon\nabla.u - c_2\rho\frac{\varepsilon^2}{k}. \tag{2.6}$$

The eddy viscosity is given by:

$$\mu_t = c_\mu \rho \frac{k^2}{\varepsilon}. \tag{2.7}$$

The constants $c_\mu, c_1, c_2, c_\varepsilon$ are respectively $0.09, 0.1296, 11/6, 1/1.4245$ and $P = S :$ ∇u. The constant c_2 and c_ε are different from their original values of 1.92 and $1/1.3$. The c_2 constant comes from the behaviour of k in isotropic turbulence:

$$k = k_0(1 + (c_2 - 1)\frac{\varepsilon_0}{k_0}t)^{\frac{-1}{c_2-1}},$$

which is consistent with the experimental results of Comte-Bellot [4] giving a decay of k in $t^{-1.2}$ if and only if $c_2 = 11/6$ while $c_2 = 1.92$ leads to a decay in $t^{-1.087}$ and therefore to an overestimation of k. This has also been reported in [5], where the author managed to compute the right recirculating bubble length for the backward step problem using the standard $k - \varepsilon$ model with $c_2 = 11/6$ and wall-laws but with $c_\varepsilon = 1/1.3$. In this work we have used the compatibility relation between the $k - \varepsilon$ constants to deduce c_ε:

$$c_\varepsilon = \frac{1}{\kappa^2 \sqrt{c_\mu}}(c_2 c_\mu - c_1), \quad \kappa = 0.41,$$

which comes from the requirement of a logarithmic velocity profile in the boundary layer.

3 Numerics

Spatial discretization of the Navier-Stokes equations is based on a Finite-Volume-Galerkin formulation. In this paper we use a Roe [6] Riemann solver for the convective part of the equations together with MUSCL reconstruction with Van Albada [7] type limitors. However, the limitors are only used in presence of shocks. The viscous terms are treated using a Galerkin Finite Element method on linear triangular elements.

We give a brief description of this technique applied to 2.1. Consider the following form of the Navier-Stokes equations:

$$\frac{\partial W}{\partial t} + \nabla.(F(W) - N(W)) = 0, \tag{3.8}$$

where $W = (\rho, \rho u, \rho v, \rho E)^t$ is the vector of conservation variables, F and N are the convective and diffusive operators.

Let $\Omega_h = \cup_j T_j$ be a discretization by triangles of the computational domain Ω and let $\Omega_h = \cup_i C_i$ be its partition in cells.

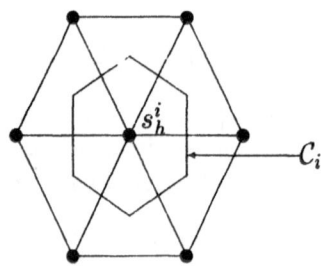

Thus, we can associate to each $w_h \in V_h$, where V_h is the set of the continuous affine functions on our triangulation, a w'_h piecewise constant function on cells by

$$w'_h | C_i = \frac{1}{|C_i|} \int_{C_i} w_h.$$

Conversely, knowing w'_h piecewise constant, w_h is obtained as $w_h(S_i) = w'_h | C_i$. The weak formulation of (3.8) is:
Find $W_h \in (V_h)^6$ such that, $\forall \phi_h \in V_h$

$$\int_\Omega \frac{\partial W_h}{\partial t} \Phi_h - \int_\Omega (F_h - N_h)(W_h) \nabla(\phi_h) \tag{3.9}$$

$$+ \int_{\partial \Omega} (F_h - N_h) \cdot n \phi_h = 0.$$

This is equivalent to the following weak formulation obtained by taking in the convective part of (3.9) for ϕ_h the characteristic function of C_i and by using an explicit time integration:

$$|C_i| (\frac{W_i^{n+1} - W_i^n}{\Delta t}) + \int_{\partial C_i} F_d(W^n) \cdot n = R.H.S. \tag{3.10}$$

We use a centered scheme to compute the right hand side:

$$R.H.S. = - \int_{\Omega_h} N(W^n) \nabla(\phi_h) + \int_{\partial \Omega} N(W^n) \cdot n \phi_h.$$

Moreover, $F_d(W_h^n) = F(W_{\partial \Omega})$ on $\partial C_i \cap \partial \Omega$ and elsewhere F_d is a piecewise constant upwinded approximation of $F(W)$ satisfying

$$\int_{\partial C_i} F_d \cdot n = \sum_{j \neq i} \Phi(W'|_{C_i}, W'|_{C_j}) \int_{\partial C_i \cap C_j} n. \tag{3.11}$$

After, writing $\tilde{B}$ for the jacobian of F at Roe's mean values, we take for Φ the Roe flux

$$\Phi_{Roe}(u, v) = \frac{1}{2}(F(u) + F(v)) - |\tilde{B}| \frac{(v - u)}{2}.$$

Spatial second order accuracy is obtained by using a MUSCL like extension involving a combinations of upwind and centered gradients. More precisely, let ∇W_i be an approximation of the gradient of W at node i. We define the following quantities on the segment $[i, j]$

$$W_{ij} = W_i + 0.5 \mathrm{Lim}(\beta(\nabla W)_i \vec{ij}, (1 - \beta)(W_i - W_j)),$$

and

$$W_{ji} = W_j - 0.5 \mathrm{Lim}(\beta(\nabla W)_j \vec{ij}, (1 - \beta)(W_j - W_i)),$$

with Lim being a Van Albada type limitor [7]:

$$\text{Lim}(a,b) = 0.5(1 + sgn(ab))\frac{(a^2 + \alpha)b + (b^2 + \alpha)a}{a^2 + b^2 + 2\alpha}$$

with $0 < \alpha << 1$ and β a positive constant containing the amount of upwinding $\beta \in [0,1]$ (here $\beta = 2/3$). Now, the second order accuracy in space is obtained by replacing W_i' and W_j' in (3.11) by W_{ij} and W_{ji}. These techniques have been successfuly extended to unstructured meshes in the past [8].

This approach does not guarantee the positivity of ρk and $\rho \varepsilon$. Therefore, the convective fluxes for the turbulent equations are computed using the PSI fluctuation splitting scheme [9] which is positive and linear preserving.

The boundary and initial conditions are classical. In particular, a Stegger-Warming [10] flux splitting scheme is used for in and outflow boundaries.

The spatial discretization (3.10) has been presented together with a first order scheme in time but as we are targeting unsteady computations, it is important to have a precise time integration scheme. In this paper, a low-storage four steps Runge-Kutta scheme has been used. Lets rewrite (3.8) as

$$\frac{\partial W}{\partial t} = RHS(W),$$

where RHS contains the nonlinear operators. The Runge-Kutta scheme we use is given by:

$$W^0 = W^n$$
$$W^k = W^0 + \alpha_k \Delta t RHS(W^{k-1}) \quad \text{for} \quad k = 1,..,4$$
$$W^{n+1} = W^4$$

with the following choices [11] for α_k:

$$\alpha_1 = 0.11, \alpha_2 = 0.2766, \alpha_3 = 0.5, \alpha_4 = 1.0.$$

4 Wall-Laws Implementation

In what follows, we describe our implementation of wall-laws. In weak form (finite element or finite volume approaches) the following boundary integrals appear in the momentum and energy equations in case of adiabatic walls ($(\vec{t}, \vec{n})$ denotes the local orthogonal basis for a wall node):

$$\int_{\Gamma_w} (\mathbf{S}.\vec{n})d\sigma,$$

$$\int_{\Gamma_w} (\vec{u}\mathbf{S})\vec{n}d\sigma,$$

where $S = (\mu + \mu_t)(\nabla u + \nabla u^t - \frac{2}{3}\nabla.uI)$ is the Newtonian strain tensor. We decompose $S.\vec{n}$ over $(\vec{t}, \vec{n})$:

$$S.\vec{n} = (S.\vec{n}.\vec{n})\vec{n} + (S.\vec{n}.\vec{t}).\vec{t}. \tag{4.12}$$

In our implementation, the first term (S_{nn}) in the right hand side of 4.12 is computed explicitely and the following wall-laws are used:

$$\vec{u}.\vec{n} = 0,$$

$$(S\vec{n}.\vec{t})\vec{t} = -\rho u_\tau^2 \vec{t},$$

$$\vec{u} S\vec{n} = -\rho u_\tau^2 \vec{u}.\vec{t},$$

where u_τ is the friction velocity, solution of $\vec{u}.\vec{t} = u_\tau f(u_\tau)$. We decompose $f(u_\tau)$ in two parts:

$$f(u_\tau) = f_r(u_\tau) + f_c(u_\tau),$$

with f_c being a new contribution when pressure and convection effects exist and $f_r(u_\tau)$ the nonlinear Reichardt equation:

$$f_r(y^+) = 2.5log(1 + \kappa y^+) + 7.8(1 - e^{-y^+/11} - \frac{y^+}{11}e^{-0.33y^+}),$$

with $y^+ = \frac{\rho u_\tau y}{\mu}$. Once u_τ is computed, k and ε are set to:

$$k = \frac{u_\tau^2}{\sqrt{c_\mu}}\alpha, \quad \varepsilon = \frac{u_\tau^3}{\kappa\delta}min(1, \alpha + \frac{0.2\kappa(1-\alpha)^2}{\sqrt{c_\mu}}),$$

where δ is the distance of the fictitious computational domain from the solid wall and $\alpha = min(1, \frac{y^+}{10})$ reproduces the behaviour of k when δ tends to zero. The distance δ is given a priori and is kept constant during the computation.

4.1 Convection and Pressure Gradient Correction

This is an attempt to take into account of the pressure gradient and convection effects in the classical wall-laws. We consider the following reduced form of the momentum equation in near-wall region (x and y denote the local tangential and normal directions):

$$\frac{\partial}{\partial y}((\mu + \mu_t)\frac{\partial u}{\partial y}) = \frac{\partial p}{\partial x} + \frac{\partial \rho u^2}{\partial x} + \frac{\partial \rho uv}{\partial y} \tag{4.13}$$

where

$$\mu_t = \kappa\rho y u_\tau(1 - e^{-y^+/70}),$$

is a classical expression for the eddy viscosity valid up to the wall. Suppose that the right hand side of equation 4.13 is known and keep it constant close to the wall:

$$C = \frac{\partial p}{\partial x} + \frac{\partial \rho u^2}{\partial x} + \frac{\partial \rho uv}{\partial y}.$$

We can then integrate this equation in y. We did not find the exact solution of 4.13, but after a first order developpment in y of the exponential, we found the following corrections:

$$f_c(y^+) = (\frac{35C\mu}{\kappa\rho^2 u_\tau^3})\log(1 + \kappa\frac{(y^+)^2}{70}) \quad \text{if} \quad y^+ \leq 5.26, \tag{4.14}$$

and

$$f_c(y^+) = \frac{C\delta}{\kappa\rho u_\tau^2}, \quad \text{if} \quad y^+ \geq 5.26. \tag{4.15}$$

The Reichardt law and our correction are shown in picture 1. Of course, this correction vanishes with C and we recover the Reichardt law.

5 ETMA TC8BIS

This computation has been done using NSC2KE fluid solver [12] which is in free access (anonymous ftp on piranha.inria.fr under pub/) on a HP workstation making 10 MFlops.

The mesh is unstructured and symmetric in y direction and has around 6000 nodes. No perturbation has been introduced for the flow to achieve unsteadiness.

We consider an unsteady 2D transonic turbulent flow over a 0.18 thick circular-arc airfoil. This is an unsteady separated turbulent test case where the separations are driven by the flow configuration and not by the airfoil geometry. The inflow Mach number is 0.754 and the chord Reynolds number is $Re_{\infty/c} = 11.10^6$. One difficulty in this case is that the periodic shock movement due to the oscillatory separations appears only in a small range of the Mach number [13]. For instance, for the same chord Reynolds number, at Mach number 0.73 the flow is steady with trailing-edge separations and at Mach number 0.76 the flow is again steady but with shock-induced separations.

The parameter δ in wall-laws is set up at 2.10^{-4} chord which corresponds to y^+ of less than 50. Approximately 300000 time steps were necessary to reach $T_{nondim} = \frac{tU_\infty}{Chord} = 50$. The computed reduced frequency is found to be around 0.512 (to be compared to the experimental value of 0.49 [13]). Results show iso-Mach contours for one cycle and the drag and lift coefficients histories as well as the pressure evolutions for four points at the wall. We can see that periodicity takes quite a long time to be established (about 20 chords traveled). The minimum drag agrees quite well with the experimental value of 0.085. The pressure evolutions have been measured for four points of the upper and lower surfaces respectively at 0.5 and 0.775 from the leading edge (the chord length being equal to 1). For the first location the maximum and minimum computed nondimensional pressure variations agree quite well with experimetal values of 0.15 and -0.04 while at the location 0.775 the agreement is not so good. Even for this complex configurations therefore, wall-laws gives surprisingly good results.

6 Concluding Remarks

Wall-laws are interesting as they remove most of the nonlinearity of the flow and therefore permit coarse meshes to be used together with explicit schemes. Something which is impossible when using low-Reynolds modeling where fine meshes and implicit schemes are necessary. However, both the classical $k - \varepsilon$ model and wall-laws are usually expected not to be able to predict unsteady separated flows. We have shown here that good numerics improve the situation considerably: weak formulation, global wall-laws, pressure and convection corrections and fine meshes. These results are encouraging us to continue in improving these wall-laws. More investigations and further computations are however necessary to make definitive conclusions.

REFERENCES

1. B. Mohammadi and O. Pironneau,*Analysis of the K-Epsilon Turbulence Model*, WILEY, 1994 (Book).

2. B.E. Launder and D.B. Spalding,*Mathematical Models of Turbulence*, Academic Press (1972).

3. D. Vandromme,*Contribution à la modélisation et la prédiction d'écoulements turbulents à masse volumique variable*, Ph.D. thesis, University of Lille, (1983).

4. G. Comte-Bellot, S. Corrsin,*Simple Eulerian Time-Correlation of Full an d Narrow-Band Velocity Signals in Grid-Generated Isotropic Turbulence*, JFM, vol.48, pp:273-33 7, 1971.

5. S. Thangam,*Analysis of Two-Equation Turbulence Models for Recirculating Flows*, ICASE report No. 91-61, 1991.

6. P.L.Roe, *Approximate Riemann Solvers, Parameters Vectors and Difference Schemes*, J.C.P. Vol.43, 1981.

7. G.D.Van Albada, B. Van Leer,*Flux Vector Splitting and Runge-Kutta Methods for the Euler Equations*, ICASE 84-27, June 1984.

8. A. Dervieux,*Steady Euler Simulations using Unstructured Meshes*, VKI lecture series, 1884-04, (1985).

9. R. Struijs, H. Deconinck, P. de Palma, P. Roe, G.G.Powel, *Progree on Multidimensional Upwind Euler Solvers for Unstrucutured Grids*, AIAA paper 91-1550, (1991).

10. J. Steger, R.F. Warming, *Flux Vector Splitting for the Inviscid gas dynamic with Applications to Finite-Difference Methods*, J. Comp. Phys. 40, pp:263-293. (1983).

11. M.H.Lallemand,*Schemas Decentrés Multigrilles pour la Résolution des Equations D'Euler en Eléments Finis*, Thesis, Univ. of Provence-Saint Charles, 1988.

12. B. Mohammadi ,*NSC2KE : an User Guide*, Technical report INRIA No.164, 1994.

13. L. Levy, *Experimental and Computational Steady and Unsteady Transonic Flows About a Thick Airfoil*, AIAA. J. Vol. 16. No. 6. pp. 564-572, 1982.

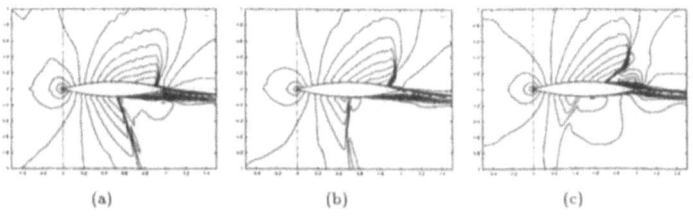

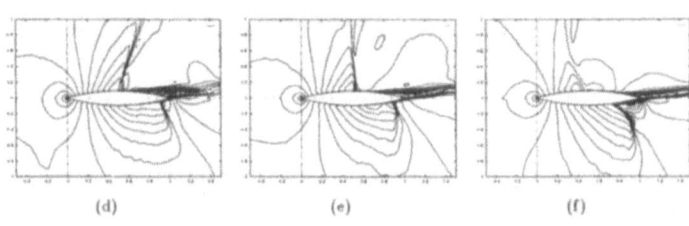

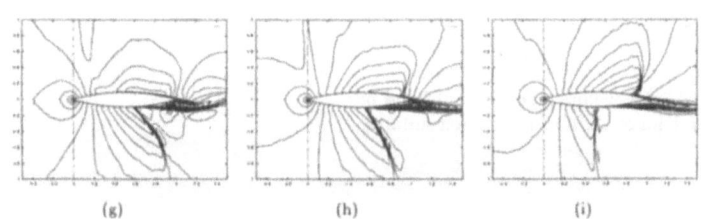

Figure 1: Iso-Mach evolution for one cycle.

559

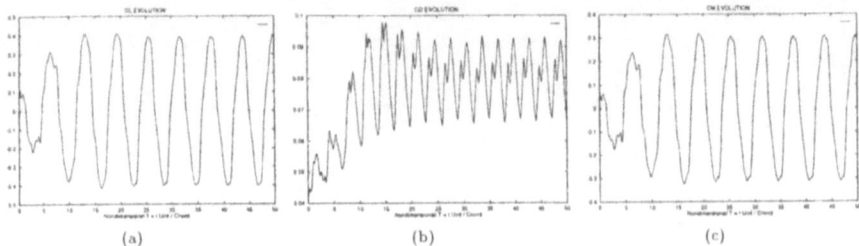

Figure 2: Lift (a), Drag (b) and Moment (c)
coefficients evolution.

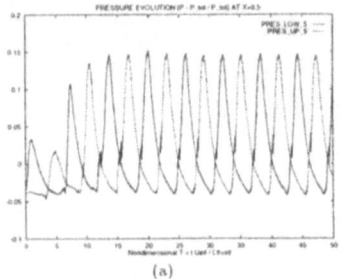 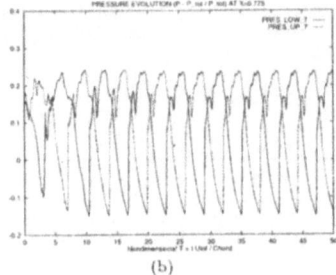

Figure 3: $\frac{P-P_{tot}}{P_{tot}}$ evolution at a: $x = 0.5$ and b:
$x = 0.775$ chord on the upper and lower surfaces.

UNSTEADY FLOW OVER A CIRCULAR ARC AIRFOIL

S. Tsangaris[1], A. Pentaris[1] and M. Thomadakis[2]

[1]National Technical University of Athens, Laboratory of Aerodynamics,

P.O. Box 64070, 15710 Athens

[2]Public Gas Corporation of Greece
DEPA SA

Mesogion 207, 11525 Athens

Summary

The unsteady flow over a circular arc airfoil of thickness ratio of 18% is numerically investigated. An implicit factored scheme, with second order accuracy in both space and time has been used. Two different grid sizes have been used and an algebraic turbulence model for the simulation of the viscous flow field. The unsteady type B oscillatory shock motion is accurately predicted. The computed reduced frequency of 0.54 is in good agreement to the experimental value of 0.49.

1. Introduction

As it has been experimentally observed, flow over a circular arc airfoil is either steady or unsteady, depending on the combination of Mach and Reynolds numbers. In a rather narrow range of free stream Mach number, the flow presents and unsteady oscillatory behavior. Various investigators [1,2,3,4], using various numerical methods and far field boundary conditions have managed to predict, in a sense, a similar behavior. It must be noted, though, that in most cases, the numerical solution is only qualitatively good agreement to the experimental data. Discrepancies have been observed either to the range of the unsteady flow regime, or to the reduced frequency of the unsteady motion. In addition, only rather coarse grids have been usually used, due to the increased CPU time required for the numerical resolution of the unsteady phenomena.

In the present paper, an adequately dense grid has been used. In addition, the use of a second order accurate scheme, permits the use of larger time steps, without the loss of accuracy, leading thus to reasonable CPU time requirements. The solution thus obtained is in better agreement to the experimental data than other numerical solutions for the same case.

561

2. Numerical method

The reader is referred to ref. [5] in the same volume for the description of the governing equations and the numerical solutions for the same test case.

3. Results

The flow over a circular arc airfoil at transonic speeds can be either steady or unsteady, depending on the combination of Mach and Reynolds numbers [4]. The combination of $M=0.76$ and $Re=11 \times 10^6$ gives, according to the experiment, a periodical unsteady flow. The numerical results for this case are presented bellow.

A coarse and a denser grid have been used for the simulation of the particular flow. The coarser grid consisted of 153 points in the streamwise direction (with 103 points on the body surface) and 41 in the normal direction. The denser grid had 273 points in the stream wise direction and the same point distribution in the normal to the body direction with the coarser grid. A close-up view of this denser grid is shown in figure 1. The maximum value of the y^+ was 20 and the minimum 0.5.

The computational domain extended 10 chord lengths in all directions. The boundary conditions implemented were imposed with the assumption of free air conditions. On the free boundaries the flow is considered inviscid and non reflecting boundary conditions are applied there to define the values of the conservative variables. The approximation of the locally one dimensional flow is considered to define the local Riemann invariants either by extrapolation from the interior, or by setting them equal to their free stream values. The conservative variables are then defined from the values of these Riemann invariants. Finally non-slip boundary conditions were imposed on the wall.

First and second order accurate computations have been performed. The time step value used was $dt=2 \times 10^{-3}$ for the 2nd order time accurate scheme and slightly lower (1.5×10^{-3}) for the 1st order scheme. In all cases, computations were initialized from free stream conditions. After a transient phase of 2000 time steps a periodically unsteady flow motion was established and remained unchanged and stable for the rest of the computation. Results obtained were qualitatively the same for both grid sizes and first and second order accurate schemes. Figure 2 presents a complete series of the instantaneous Mach number contours around the airfoil. In the beginning of the cycle ($t=0$) on the upper side, the flow is practically attached, with a series of weak shock waves above the rear part of the airfoil. These weak shock waves coalesce to form a strong oblique shock wave ($t=5T/16$), which then moves upstream, towards the midchord, weakens in strength, and finally disappears ($t=6T/16$ to $t=10T/16$). During this last phase of the upper shock motion, the shear layer on the upper side becomes thicker, and finally collapses. The same phenomena are repeated on the lower side of the airfoil, exactly half period out-of-phase. All these are in very good agreement to the experimental observation. For a small part of the cycle, shock waves exist on both sides, but the flow field can be characterized from a type-B unsteady shock motion Pressure time histories at two different positions, as obtained with the denser grid and the 2nd order accurate scheme, are presented in figure 3. It can be seen that the agreement to the experimental data -presented in the same figure- is very good. The power spectra density of all the pressure curves has given exactly the same main frequency of the unsteady phenomenon. Figure 4 presents the pressure PSD for one of the positions shown in fig.3. The main frequency deduced from the diagram is 168 Hz (figure 4), which is not much lower than the experimental value of 185Hz. Exactly the same value has been computed from the pressure and velocity signals at various positions of the domain. The frequency of the phenomenon was not influenced by the time accuracy of the computations, but it was affected by the density of the grid. For the coarser grid, the main frequency was found to be 155Hz, and this was the main difference between the numerical results obtained with the two grid densities.

The good behaviour of the numerical results is also verified by the comparison of the computed flow angle =arctan(v/u) with the corresponding experimental values and the computational results of [4], shown in figure 5. It can be clearly seen that the results obtained here are in quite good agreement to experiments and better than those of previous computations.

REFERENCES

[1] McDevitt J.B., Levy Jr L.L. and Deiwert G.S., AIAA J., Vol.14, No 5, pp. 606-613,1976.
[2] Deiwert G.S., AIAA J., Vol 14, No 6,1976.
[3] Levy Jr L.L., AIAA J., Vol 16, No 6,1978.
[4] Seegmiller H.L., Marvin J.G. and Levy Jr L.L., AIAA J., Vol.16, No12,1978.
[5] Tsangaris S., Pentaris A. and Thomadakis M.P., Vieweg-Verlag publications, present book.

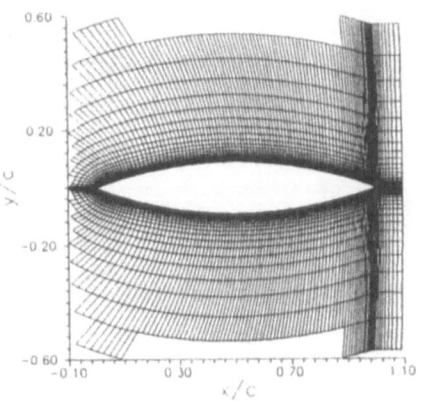

Figure 1. The 90x86 grid.

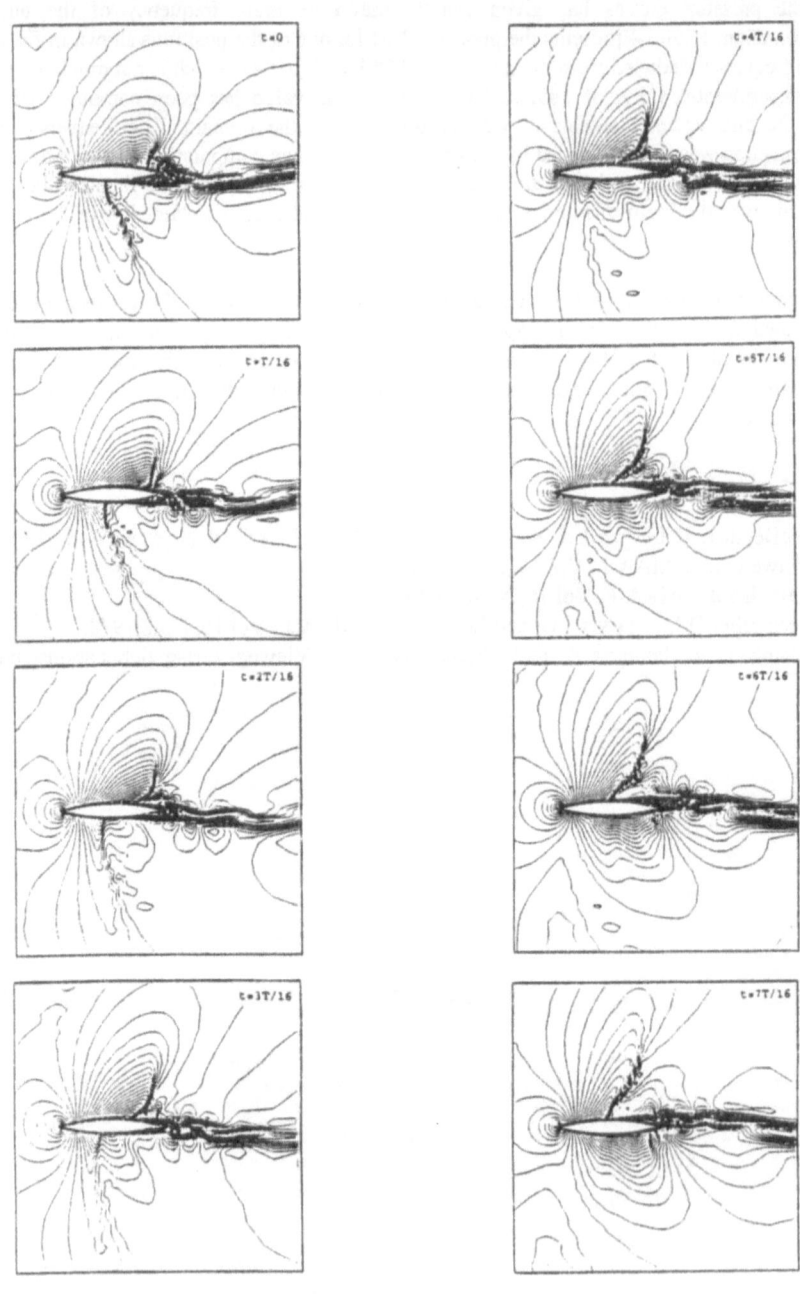

Figure 2. Instantaneous Mach number contours

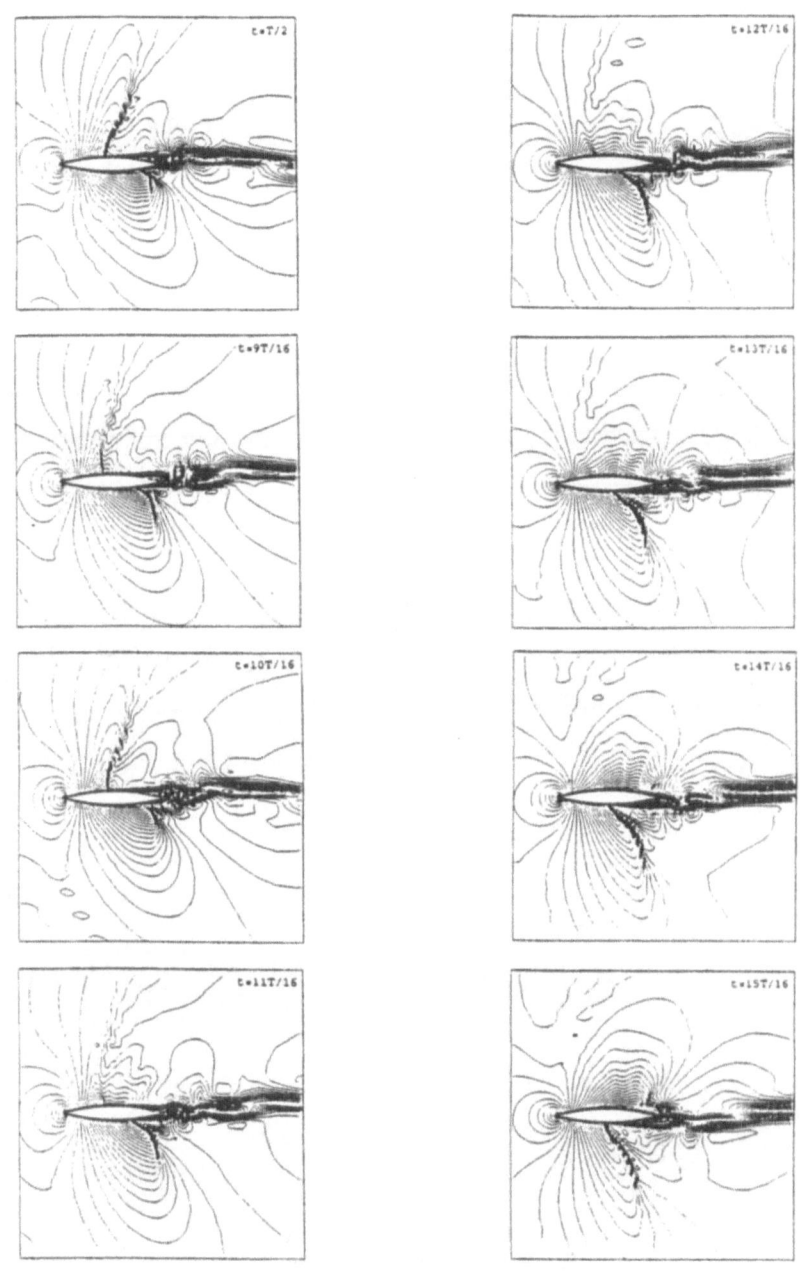

Figure 2. Instantaneous Mach number contours

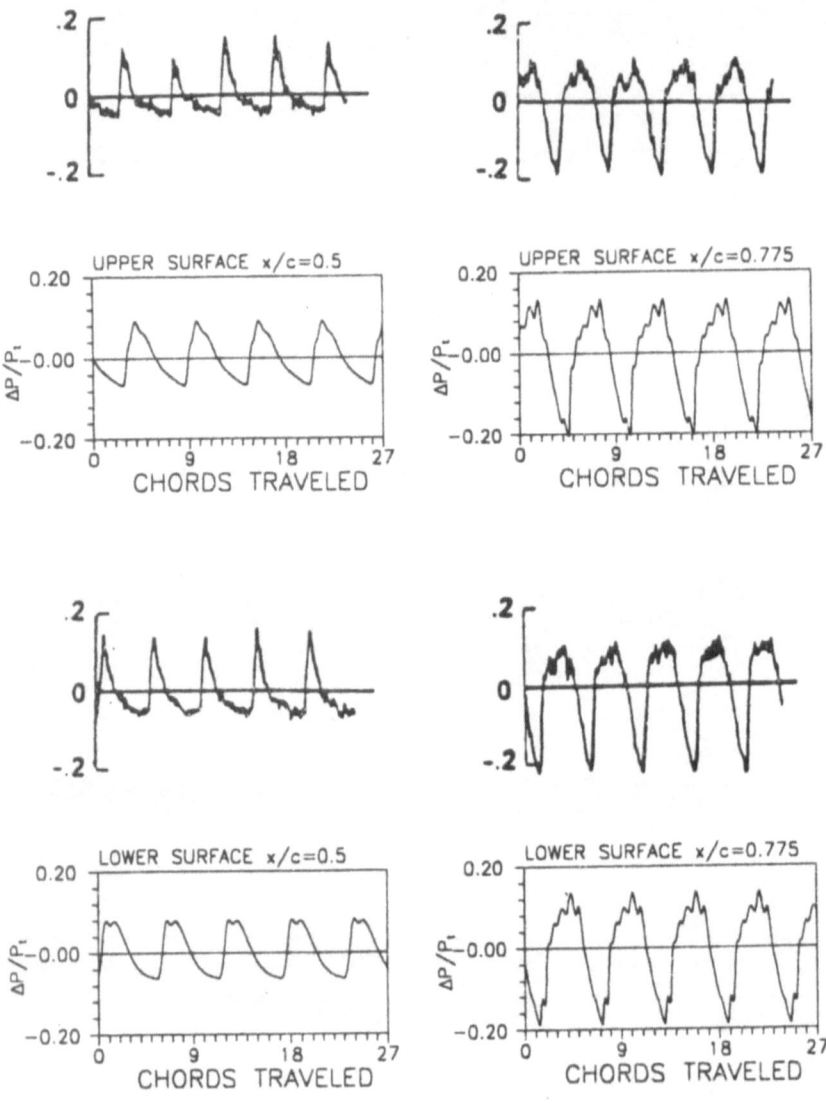

Figure 3. Pressure time histories

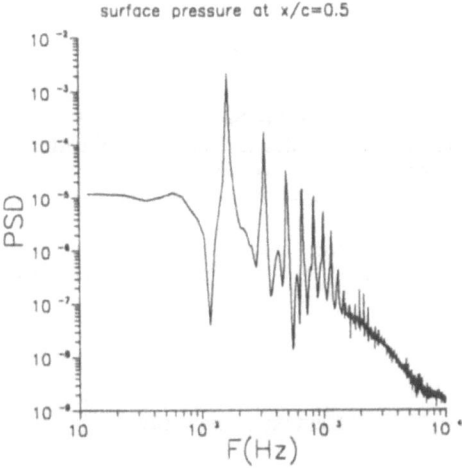

Figure 4. Power Spectra Density

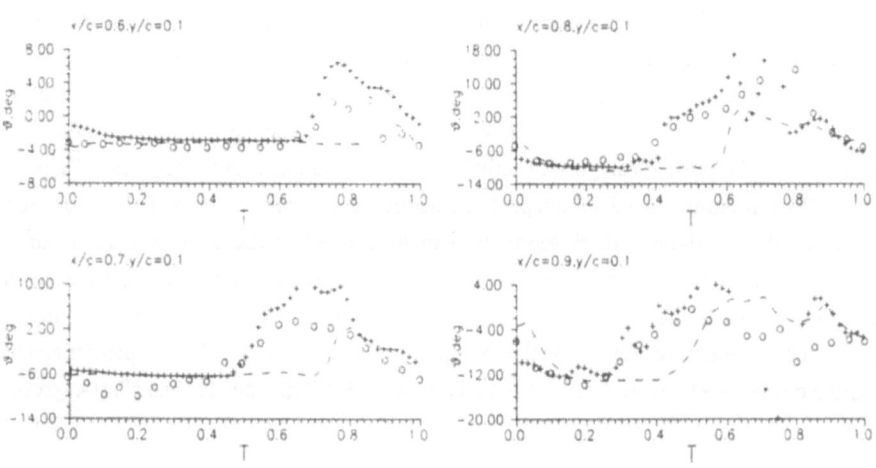

Figure 5. Flow angle

oooo Experimental data .... Num. Results ++++ Current results

Synthesis on the unsteady transonic flow around a circular-arc aerofoil (TC8bis)

M. BRAZA [1] & S. TSANGARIS [2]

[1] Institut de Mécanique des Fluides de Toulouse, Unité Mixte de Recherche
C.N.R.S n° 5502, .Av. du Prof. Camille Soula
31400 Toulouse Cedex, France

[2] Laboratory of Aerodynamics, National Technical University of Athens
P.O. Box 64070, 15710 Zografos, Athens, Greece

Context of the study

In the context of the ETMA research program, a special emphasis has been devoted in the prediction of unsteady high-speed flows *involving solid walls* and interesting the aeronautical applications. The inherently unsteady flow around an aerofoil in the transonic regime is examined. The present category of flows develop an inherently unsteady shock-induced separation, followed by the onset of periodic vortex shedding downstream. This is the consequence of the amplification of a von Karman instability downstream the separation point. Furthermore, a second kind of instability may occur in the separated shear layer: This is a shear-layer (Kelvin-Helmholtz) instability, which occurs under the amplification of Tollmien-Schlichting waves. These two possible modes are a typical pattern in many turbulent wakes characterized by a simultaneous development of coherent structures and of a random, fine-scale turbulence background [1]. Under the action of the mentioned mechanisms the present category of turbulence flows, examined in the test-case TC8bis of the ETMA program, need an adapted methodology in order to predict their unsteady dynamic characteristics, which represent a major interest in the domain of aeronautical applications. In this part of the study, essentially the prediction of the effect due to the unsteady shock-induced separation and of the von Karman flow pattern is achieved, according to the physical experiment of Seegmiller, Marvin & Levy [2]. The predictions are carried out by INRIA (partner 1), in cooperation with IMFT (partner 15) and NTUA (partner 10).

Methodology

The methodology adopted is the Organized Eddy Simulation approach [3], [6], [7], [8], [14], developed for flows characterized by pronounced periodicities. The principles of

568

this approach are based on the physics of the flow and are briefly recalled in the present synthesis: The distinction of the structures to be predicted and those to be modeled is based on the criterion of the organized or not character of the structures and not on their size. This is achieved by employing the phase-averaged decomposition, which is a *physically existing, measurable quantity* [5], [13] and not a mathematical averaging and which is proven able to regroup *all* the organized characteristics and predominant frequencies [14], of the coherent structures, appearing in phase in the dynamics of the flow system. This methodology is applied in the present study in two dimensions, because the OES approach has as an advantage to be not inherently 3-D. The mentioned elements constitute also the main differences of the OES, comparing to the LES approach. The derived phase-averaged time-dependent equations are closed in respect to the turbulent stresses, by using eddy-viscosity models, especially adapted for the present category of unsteady flows. It is noted that, although the implementation of the classic closure schemes is straightforward in the OES approach, owing to the same mathematical form of the equations, as in the Reynolds averaged equations [4], the hypotheses on which the existing closures are based, have to be reexamined and reconsidered, because they were derived for flows in statistical equilibrium, according to the statistical theory of Kolmogorov, which is not veryfied in the present category of unsteady flows with pronounced periodicities. In this respect, the studies carried out in the context of the ETMA program allowed the assessment of the behavior of standard algebraïc [12] and two-equation eddy-viscosity models and moreover, the study of the behavior of modified eddy-viscosity models [3], [9], [11]. The related models and the *three different numerical codes* used by the partners are presented on table 1. An important requirement for the numerical code in the context of OES methodology is to be *time-accurate*. This is respected by the present studies.

Conclusions and future developments

The main conclusions of this collaborative study are summarized as follows:

-The OES approach, applied in three fundamentally different numerical codes is proven promising for the prediction of a transonic flow around a circular-arc aerofoil, developing an inherent periodicity at Mach number 0.760 and Reynolds number 11 millions. The shock-induced separation, the vortex shedding pattern and the values of the predominant frequency are predicted satisfactorily. Furthermore, the maximum amplitudes of the oscillations are also in good agreement among all the approaches and with the physical experiment. These aspects are summarized on table 2.

- Concerning the turbulence modeling implemented in the OES methodology, it is found that both categories of algebraïc and two-equation models provide good results, without performing compressibility corrections. The standard Baldwin-Lomax model has the tendency to produce a higher rate of turbulent viscosity [11], [3], especially if it is used in a numerical code which does not employ artificial dissipation techniques. The shear-based two-equation model [9] suggested by Dr Mohammadi (INRIA), fundamentally similar to the one developped in refs. [7] and [8], is very promising to deal with the present flows. Especially, the use of wall functions, despite the complexity of the dynamics in the separation region, predicts the effect of the pressure gradient, governing this region and gives a satisfactory result, with reasonable CPU time, comparing to what is generally needed for the low-Reynolds number versions [7]. This aspect may be particularly interesting for the aeronautical applications, especially in reference to complex 3D configurations involving predominant unsteady effects.

- A general conclusion from this study is that even by using rather simple closure schemes, the principal flow characteristics are predictable, because a major part of the complex physical processes is contained and resolved in the phase-averaging approach.

-The innovative character and expected achievements set by the ETMA program concerning the present category of flows have been realized through the present studies.

- In order to complete this analysis, it is worthwhile to assess the effect of compressibility corrections in the two-equation model for unsteady flows, having as a starting point the developments in ETMA carried out for the test-case TC1 (steady mixing layer flow), under the responsibility of Dr Dussauge. Especially, comparisons of the near wall velocity and turbulence stress profiles are needed, without and with the compressibility corrections, which should also be improved in respect to their efficiency in the presence of solid walls, according to another general conclusion of ETMA (see synthesis by Dr Arina). Furthermore, a detailed study of modeling the dissipation equation is worthy to be examined, in order to increase the dissipation rate and hence to improve the general tendency of the two-equation model to produce an excessively high turbulent viscosity, especially for the unsteady flows. These aspects will be addressed in an on-progress study, in cooperation among INRIA, IMFT, IMST, POLITO, CRS4, IST and NTUA.

References

[1] M. BRAZA, P.CHASSAING, H. HA MINH 1990 "Prediction of large-scale transition features in the wake of a circular cylinder". Physics of Fluids A, Vol.2, N0 8, pp. 1461-1471.

[2] SEEGMILLER, H.L., MARVIN, J.G. & LEVY, L.L. 1978 "Steady and unsteady transonic flow". AIAA J., Vol. 16, N° 12, pp1262, 1270.

[3] BRAZA M., NOGUÈS, P 1991 "Numerical simulation and modeling of the transition past a rectangular

afterbody". Proceedings of the VIIIth Turbulent Shear Flow Conference, Munich, 9-11 Sept. 1991., pp. I-2-1, I-2-2.

[4] REYNOLDS O. 1883 "An experimental investigation of the circumstances which determine whether the motion of water shall be direct of sinuous and the law of resistance in parallel channels. "Phys. Trans. Roy. Soc. London, 174, pp. 935-982.

[5] CANTWELL, B. J., & COLES, D. 1984 "An experimental study of entrainment and transport in the turbulent near wake of a circular cylinder". J. Fluid Mech. Vol. 136, 321-374.

[6] HA MINH, H., VIEGAS , J. R., RUBESIN, M. W., VANDROMME, D. D. & SPALART, P. 1989 Proceedings of the Seventh Symposium on Turbulent Shear Flows, Stanford, CA, USA, August 21-23.

[7] G. JIN & M. BRAZA 1994 "A two-equation turbulence model for unsteady separated flows around aerofoils". AIAA Journal , Vol. 32, N0 11, Nov. 94, pp. 2316, 2320.

[8] G. JIN, M. BRAZA, P. CHASSAING, H. HA MINH 1995 "Prediction of unforced and forced flow around an aerofoil with two-equation turbulence models". Proceedings, 6th Asian Congress of Fluid Mechanics, Singapore, May 22-26.

[9] MOHAMMADI, B. 1995 "Unsteady transonic turbulent flow over an aerofoil using the k-epsilon model and wall-laws on unstructured meshes" Proceedings ETMA Workshop, Manchester, UMIST, Nov.1994., 2nd version March 1995.

[10] S. TSANGARIS, A. PENTARIS, M. THOMADAKIS 1994 "Unsteady flow over a circular-arc aerofoil" Proceedings ETMA Workshop, Manchester, UMIST, Nov.1994., 2nd version March 1995.

[11] BRAZA, M. HANINE, F. 1994 "Numerical simulation and modeling of an unsteady supersonic mixing-layer flow". Proceedings ETMA Workshop on Turbulence Modelling for Flows Arising in Aeronautics, UMIST, Manchester, Nov. 1994., 2nd version March 1995.

[12] BALDWIN B. S., LOMAX H. 1978 "Thin layer approximation and algebraïc model for separated turbulent flows." AIAA Paper, 78-257, January 1978.

[13] BOISSON, H.C. 1982 "Développement de structures organisées turbulentes à travers l'exemple du sillage d'un cylindre circulaire". Thèse de Docteur ès-Sciences, n°65. Institut National Polytechnique de Toulouse, France.

[14] M. BRAZA 1986 "Analyse physique du comportement dynamique d'un écoulement externe, décollé, instationnaire, en transition laminaire-turbulente. Application: Cylindre circulaire". Thèse de Doctorat d'Etat-ès-Sciences, I.N.P.T., Décembre 1986.

Table 1. Summarized characteristics of the numerical methods and
turbulence models used for TC8bis.

	Method	Turbulence Model
NTUA	Implicit Factored Fin-D,Beam-Warming 2000time steps/period	Baldwin-Lomax
INRIA	Explicit 4th order Runge-Kutta 10^6 time steps	two-equation model (shear based) wall functions
IMFT	Implicit 2nd order FVS-MUSCL 8192 time steps	Baldwin-Lomax $F(\omega)$ modif (Braza & Noguès)

Table 2. Synthesis on the results for TC8bis.

	Frequency	Cp upper (max min)	Cp lower (max min)	C_D
Exp	188 (Hz)	0.1 -0.2	0.1 -0.21	
NTUA	168	0.133 -0.2	0.135 -0.19	
INRIA	187	0.22 -0.15	0.22 -0.15	0.080
IMFT	168	0.15 -0.2	0.13 -0.25	0.085

ELEMENTS OF SYNTHESIS AND CONCLUSION

A. Dervieux (*), M. Braza (**), J.–P. Dussauge (***)
(*) Inria, B.P.93, F–06902 Sophia–Antipolis cedex
(**) IMFT, Av. du Prof. Camille Soula, 31400 Toulouse cedex
(***) IRPHE, 12 avenue Général Leclerc, F–13003 Marseille

Introduction

One of the objectives of the ETMA Project was to contribute to the improvement of nu-merical codes for turbulent high speed flows. Apart the studies on numerical method and on turbulence models, it was also planned to examine the influence of the turbulence model on the code itself from a numerical point of view: convergence properties, accuracy of the results. It is difficult and time consuming to test a large number of codes with a large number of turbulence models: so it was necessary to organize an international cooperative effort at European level. This implied sharing the same reference data, and also as far as possible, sharing some common elements in the numerical procedure, such as the same mesh generation. The other element was to compare, when feasible, various numerical methods, for example finite difference, finite elements and spectral methods. At last, some compressible turbulence models have appeared these last years.

These models are very often based on numerical simulations. They have provided a real improvement for the modelling and computation of some simple flows such as supersonic mixing layers, but have not really been widely tested in cases where it is suspected that there are significant effects related to compressible turbulence properties, for example in separated and reattaching zones or in shock wave/boundary layer interactions.

The natural way to compare the results obtained by all the partners was to organize a workshop. This workshop also provided an opportunity to control the behaviour of these models in some cases, and to compare them with models for incompressible turbulence.

At last, the workshop allowed to compare many recent numerical approximations and algorithms.

In the next sections, the results of the workshop will be briefly summarized, then more general informations on the advances made during the ETMA Project will be given, as well as the necessary ingredients which were used to reach the proposed objectives.

Main results of the workshop

Seven test–cases were examined. Three of them were "elementary test cases" in that respect that the geometry was very simple. These test cases can be categorized as follows: i) super-sonic turbulent mixing layer, ii) the boundary layer on flat plate (subsonic and supersonic) and iii) the shock reflection on a flat plate. More complex geometries were also examined, such as boundary layers around obstacles such as compression ramp or steps, and around a bump in transonic regime. The case of flows around airfoils was also considered: for these configurations, the statistically steady state was computed as well as the unsteadiness which can appear and produce flutter.

The conclusions about the turbulence models can be summed up as follows.

1) The modelling of compressible turbulence is still an issue. In the case of mixing layers, the overall picture suggests that models proposed by Zeman (1990) and by Sarkar et al. (1991) for the dilatation dissipation and also for pressure–divergence terms give a real improvement of the results for determining the rate of spread of the mixing layer and the mean fields: the convergence is faster and the accuracy better. However, the same models used in other flows (Boundary layer on flat plates or subjected to pressure variations) brought no improvement. Elements of an answer to this inconsistency are perhaps in the work by Huang, Coakley, Bradshaw (1994), who pointed out the difficulties to compute the equilibrium zone of a flat plate boundary–layer with variable density: For a k–ε model, problems come from the level of dissipation in the equilibrium layer, because se cannot be considered as a constant in presence of a density gradient. Improvements are needed on this point, and research work is still in progress.

2) Wall functions are still very good or still very bad, depending on the cases. They are very efficient when applied to cases for which they were designed, for example in layers with moderate pressure gradients. in this case, they work well, require a small memory space and lead to rapid computation. Their use in industrial codes is still interesting. They can behave rather poorly in separated flows for which low Reynolds number modelling has to be improved.

3) During the workshop, there was, once more, interest in such models as (q,ς) , (q,R) or (k,τ) which are less stiff than the classical k–ε model. These models use the same understanding of the physics of turbulent shear flows, but they lead to formulations more tractable than the k–ε equation: the formulation of the boundary condition for ς, R or τ at a wall is more direct, and the form of the equation for these variables is simpler, too. Generally, this results in a gain of accuracy and computing time.

4) It is likely that Algebraic Stress Modelling may improve the closure assumption in many distorted configurations. Although the results presented with this modelling were not totally convincing, it should be underlined that the potential improvements could not clearly appear in the present test cases. In these flows compressibility was probably a predominant factor while the models did not account for these effects but were designed more specifically for incompressible situations.

5) In the cases of transonic steady flow around airfoils and of subsonic flows around obstacles, it turned out that the Reynolds Stress Models give much better results than the k–ε model. This is not surprising for subsonic flows, since these models were primarily designed for this purpose. This is more surprising but very encouraging for the transonic flows. This will be discussed hereafter.

6) Good results have been obtained in transonic interactions with multiple scale turbulence models, suggesting that at moderate Mach numbers for which there is no by-pass of the energy cascade from large to smaller scales, the spectral transfers in accelerations and in decelerations follow the same physics as at low speeds.

7) The 24 degrees supersonic turbulent corner remains a difficult case. The final level of wall pressure can be poorly predicted with discrepancies of about 10% . This is probably due to a poor description of turbulent diffusion in the boundary layer downstream of the interaction. It may be noticed that the experimental values of

574

wall pressure correspond to a deviation rather less than 24 degrees, which suggests than along the rather short distance downstream of the interaction, the boundary layer has not had time enough to be adapted to the new conditions.

Some of the model pick up the right downstream level, but none of them capture the right separation length. The use of EARSM did not change the results significantly. It is not clear that all the possible traps of the numerical aspects including an optimum mesh grid definition have been totally explored in this difficult flow, where two directions, normal to the wall and to the shock wave, are strongly non homogeneous. The use of existing compressible turbulence models (Sarkar et al. 1990, for example) brought no improvement to the results.

8) Finally, very encouraging results have been obtained in the simulation of coherent eddies in the cases of supersonic mixing layers as well as flows around thick airfoils in transonic flows: the main properties of the flow are reproduced, suggesting that limited improvements of the models would give significantly better predictions.

9) From the point of view of numerical methods, it is interesting to note the large range of different methods covered by the workshop.

As far as approximation methods are concerned, structured and unstructured methods (i.e. methods applying to structured or unstructured meshes) were used; for structured methods, central differenced to upwind TVD shemes were applied; for unstructured meshes, central–differenced cell centered finite–volumes on triangles, upwind–TVD of MUSCL–type vertex centered finite–volume/elements (triangles), and Least–Square Galerkin (triangles, quadrangles) were used successfully. Some computations presented mesh–convergence study giving some better confidence to the results; it should be noted that mesh convergence is not obtained for difficult cases; from our opinion, in most cases, really assessed (mesh–converged) results would have demanded fine meshes involving at least 50,000 nodes.

At the same time, the use of fine enough meshes and accurate enough schemes in combination with wall laws seemed to brings a new light on the interest of wall laws for a larger set of flow than expected.

Many different solution algorithms have been also tested, such as explicit, explicit with residual averaging, implicit with linearisation, multi–grid. Although the workshop studies were not focused on the algorithmic issue, this issue had some impact on the choice of meshes and then on the workshop results.

The output of ETMA works for models improvement

An important part of the ETMA works was devoted to supersonic turbulent flows. It is therefore interesting to define what progress has been made in this field, and which new developments ETMA has made possible. The first obvious thing is that it would have been difficult to cover all the cases examined in ETMA in only one country. ETMA was probably well fitted to European objectives: to exchange ideas and expertise, broadcast the knowledge in the European Union, and also take advantage of the critical size necessary to test a large number of cases in a limited amount of time. With this stimulus, progress was possible in the understanding and the modelling of compressible turbulent flows. This is a result per se, but also constitutes a dynamic approach: by establishing a data base and writing a handbook, the ETMA Project has made this information available for the scientific and technical community in Europe, and has prepared the conditions for future progress in this field. It is therefore of some importance to define the improvements that ETMA has brought to the understanding and the modelling of supersonic turbulent flows.

The relationships between turbulence modelling in high – and low speed flows have been assessed in 1962 by Morkovin, by examining the few experimental data available for zero–pressure gradient turbulent boundary–layers in supersonic regime. The conclusion was: "We can expect with confidence that the essential of the dynamics of the supersonic shear flows will follow the incompressible pattern". This is expected to be true for equilibrium boundary layers for Mach numbers less than 5, and was used to justify the (successful) use of low speed closures such as mixing length or one–equation models for the prediction of wall layers.

If the Morkovin's hypothesis is extrapolated to the possible use of incompressible closures in supersonic flows, it is possible and useful to infer from the results of the two–years research ETMA program how far it holds, i.e. to what extent the incompressible turbulence models help to determine the compressible turbulent flows, and for which problems and flow cases specific modelling is needed.

Once more, it was checked that zero pressure gradient boundary–layers correspond well to this hypothesis. Recent work (Dussauge and Smits 1995) shows that such an interpretation can be justified by assuming that the relation between the time scales characteristic of the mean– and turbulent motions does not depend on Mach number. However, care should be taken when using two–equation models of turbulence. For example, Huang, Coakley, Bradshaw (1994) have shown that the standard k–ε model has difficulties reproducing the Couette approximation in supersonic boundary–layers. This is not an effect of compressibility, but only the consequence of density gradients. The k–ω model seems better behaved for such flow conditions. The conclusion on this point is that, although the overall physics of this case seems understood, the two–equation– and Reynolds stress models have still some aspects which should be examined for proper adaptation to high speed flows.

In the free shear flows, the effects of compressibility have been clearly identified, since there are consequences observable on macroscopic scales for moderate Mach numbers. The compressible turbulence models developed to cope with this point seem efficient, since they predict quite correctly the spreading rate of the supersonic mixing layer. The interaction with the numerical methods is good, since in general, better convergence is achieved. It seems therefore that the equation set has a better shape with a compressible turbulence model. In particular, it seems that the turbulent Mach number can be used as one of the relevant parameters to characterize compressible turbulence. However this sort of model, applied to the equilibrium boundary layers, leads to disappointing results, perhaps related to the remarks made previously on the modelling of the dissipation equation for variable density flows. It will be seen in flows with pressure gradients that some results are still more surprising.

Four high speed flows with pressure gradients were considered. All had a shock wave boundary layer interaction: two of them were transonic, and the two other ones supersonic. The computations with Reynolds stress models brought significant improvement to the computation of flows around airfoils, and the multiple scale model was very satisfactory for the computation of the transonic flow around the bump, while the supersonic shock wave interactions were rather poorly predicted.

This probably deserves some comments. We have already noticed that one of the relevant parameters of compressible turbulence for dilatation dissipation and probably for part of the

576

pressure–divergence terms is the turbulent Mach number $M_t=k^{1/2}/a$, where a is the speed of sound. In transonic flows, it can be expected that the main features of turbulence are incompressible: in most transonic interactions, $M_t \ll 1$, so that the previous effects are probably negligible.

Now, the interactions are also dominated by pressure gradient. In this case, turbulence is subjected to a distortion by the mean field, often a rapid distortion in the outer flow, always a slow one in the near–wall region. Recent rapid distortion problems have shown that another parameter, the gradient Mach number is probably also important. A mean distortion is rapid when applied to turbulence during a time short compared to a characteristic eddy time scale. In such conditions, the problem can be described by a linear formulation. Recent studies (Durbin and Zeman 1992, Jacquin, Cambon and Blin 1993) show that for an irrotational mean distortion, the nature of the potential part of the motion, which is also related to the pressure field, depends on a "gradient Mach number" $M_g=DL/a$, where D is a typical value of the mean velocity gradient, and L is a typical space scale of the eddies (for example an integral scale). If M_g is small, pressure behaves as at low speeds and limits the action of the mean velocity gradient; if M_g is large, pressure has no time to propagate and does not damp the amplification of the fluctuations due to the mean velocity gradient.

In the particular case of mean dilatations, no pressure fluctuations are developed, as was already noticed by Dussauge and Gaviglio (1987). If shock waves are very strong, they are very thin; D and M_g can be very large. However, very often for moderate nominal Mach numbers, and in the cases of the transonic airfoils and of the transonic bump, the value of M_g is probably not very large. Therefore, in the transonic interactions, and although the mean distortion is vortical, it is not completely surprising that the incompressible turbulence models behave quite well. Turbulent fluxes will follow processes (production of anisotropy by the distortion, role of pressure to limit the anisotropy variations, negligible production terms related to mean pressure gradients and spectral transfer bringing turbulence to an out of equilibrium state) similar to the subsonic ones.

In the supersonic separated interactions, the picture is rather different. None of the methods was really successful. However, it could have been expected that these flows would require particular care in the choice of the models, since many effects are present (effect of compression through the shock, possibly influence of the shock motion, separation of the layer, interactions of the structures generated by the separated zone with the boundary layer downstream of reattachment, etc.). This result is not totally surprising since the initial model is a simple k–ε model, where the effects of deceleration and normal strain are not taken into account. The overall picture is that the results are mediocre for wall pressure and still worse for wall friction: the level of pressure depends strongly on the wall deflection and is probably less sensitive to turbulence modelling than the turbulent quantities, the size and position of the separated bubble. To date, there are no turbulent closures taking into account the change in the nature of pressure according to M_g. Also, the present attempts to include compressibility effects used the available standard compressibility corrections. They were not efficient, and consequently, these attempts look like dead–ends. This suggest that a better analysis of the phenomena has to be performed in such flows before concluding on the appropriateness of compressible turbulence models. Lastly, since the models used in these predictions were not designed to predict these flows (most of them used assumptions valid only for a simple shear), it may be underlined that the results, even if not good, are not so far from the experimental observations. The computation of the 24 degrees corner flow which was made with the incompressible EARSM did not provide better result. This underlines perhaps the importance of compressibility effects on this particular configuration.

Finally, particular problems are found in the simulation of Large Eddies or of Organized Eddies. It is believed that a primary effect of compressibility is to promote the formation of Mach waves or possibly of shocklets in which energy is lost because of dilatation dissipation and of acoustic radiation. These perturbations are induced by large eddies, but produce very small scales, and numerical simulations indicate a significant increase of the spectra at high wave numbers. This may be a difficulty for LES in which a cut off in the spectrum is made at high wave numbers. Therefore two–point turbulence closures including compressible interactions between wave numbers have probably to be developped to make LES of compressible turbulence. The Organised Eddy Simulations may overcome this difficulty, since the fine– grained, "random" turbulence is globally represented in the shear layers, and that the external perturbations are explicitly computed. The first results are very promising, but it is clear that improvement have to be made, and that the prediction of flow unsteadiness related to eddies of long wave lengths remains a challenging question.

To conclude, it may be underlined that Morkovin's hypothesis can be used only in a limited number of flows. However, for many situations, the reference to low speed variable density flows has to be made before adjusting compressibility effects. A general formulation of one–point closure which is valid in free– and wall shear layers is still needed. Distorted flows and shock/ turbulence interactions are other problems in which the role of pressure has to be examined more deeply. The use of low speed models can probably be extended with some caution to the case of transonic interactions and also in the expansion of shear flows where dilatation effects are important, so that pressure fluctuations are not developped. A more global conclusion at this stage is that the major elements to develop efficient compressible turbulence models seem to have been identified.

ETMA contribution to numerical issues

The crucial role of numerics, especially for compressible flows has be again emphasized by the ETMA studies and the related workshop calculations.

Large errors can arise not only from inadequate approximations or meshes but also from insufficient iterative convergence.

While structured methods are yet rather well assessed, unstructured ones could find in these study the occasion of accurate evaluation and validation. Such rather novel works on unstructured meshes such as second moment airfoil calculations, fully low–Reynolds compressible flows have been done and contributed to the set of workshop results.

The calculation of low–Reynolds boundary layers remains a challenging issue; whatever be the approximation method, it seems difficult to use less that one dozen nodes normally to the wall for their prediction. Even in these conditions, more accurate approximations are required and it is now expected that more–than–second–order schemes will at last be derived for compressible turbulent flows.

Concerning the solution methods, the new birth of explicit methods (Residual–averaged Runge–Kutta, smart alternated directions) is an interesting event first by itself, and also because the natural evolution is to encapsulate the explicit solver into a multi–grid method.

Implicit methods were also successfully applied, sometimes with only a small benefit as compared to explicit, due to the strong nonlinear stiffness of the turbulence models; we believe that the accumulation and diffusion of experience and knowledge will help making progress in that direction.

A few extra words about multi–grid solvers. The ETMA project was the opportunity of developing a new implicit unstructured multi–grid method for a two–equation model, with wall law. Mesh as heavy as 48,000 nodes could be used for airfoil flows, with reasonable workstation times. From our point of view, this constitutes a determinant step for reaching the main issue in today's engineering problems solution, the prediction of flows with boundary layers. The difficulty is then to compute these layers on stretched meshes. Recently, the above work was extended by Francescatto and Dervieux to anisotropic (directionally coarsened) multigrid treatment of low–Reynolds 2D flows with convergence rates-rather insensitive to aspect ratio.

From a general point of view, it appears that these nonlinear problems are still of costly resolution in terms of number of floating point operations by unknown and much extra effort is necessary for reaching reasonably expensive and enough robust simulations of some of the more sophisticated models used for complex turbulent flow simulation.

As concluding remarks, some overall prospects can be outlined.

An objective is to have rather simple formulations avoiding the full Reynolds stress modelling; rather general but simple formulations may be derived from the EARSM. Another problem to be examined is the determination of the proper time scale of turbulence: this is particularly important for flow out of equilibrium: in equilibrium shear flows, this scale is adapted to the scale of the mean flow This is no longer verified in distorted flows or in presence of pressure gradients. The multi–scale modelling is an interesting example of an attempt to address this point. A better modelling of this phenomenon would produce probably more efficient computations.

Two other problems need further development: the modelling of compressible turbulence, for which it seems that many pending questions have been identified, and the computation of unsteady flows, for which more exploratory work has to be performed.

Finally, a more general issue can be expected. For flow sensitive to boundary conditions, of for applications like turbulent flow control, it is important to predict the large scale structures, so that "Large Eddy Simulation" or " Ordered Eddy Simulations" could be the answer. Improved resolution techniques to solve the Navier–Stokes equations for calculating explicitly unsteady part of the motion related to large scales, and efficient turbulence models for the turbulent closure on the fine scales will be of course necessary to make these computations effective and attractive enough for industrial applications, and will certainly necessitate further fundamental research work.

References

Durbin P.A., Zeman O., "Rapid distortion theory for homogeneous compressed turbulence with application to modelling", J. Fluid Mech., 242, 349–370, 1992.

Dussauge J.P., Gaviglio J., "The rapid expansion of a supersonic turbulent flow: role of bulk dilatation", J. Fluid Mech., 1987.

Dussauge J.P., Smits A.J., "Characteristic scales for energetic eddies in turbulent supersonic boundary layers", Accepted for presentation, Turbulent Shear Flow Symposium, Penn State, Pa, U.S.A., August 1995.

Francescatto J., Dervieux A., " Anisotropic Agglomeration multigrid for low–Reynolds 2D turbulent flows", INRIA Report, 1996, to appear.

Huang P.G., Bradshaw P., Coakley T.J., "Turbulence models for compressible boundary layers", AIAA J., Vol. 32, N 4, 1994.

Jacquin L., Cambon C., Blin E., "Turbulence amplification by a shock wave and rapid distortion theory", Phys. Fluids A, 5, (10), October 1993.

Morkovin M.V., "Effects of compressibility on turbulent flows", Mécanique de la turbulence, A. Favre ed., C.N.R.S., Paris 1962.

Sarkar S., Erlebacher G., Hussaini M.Y., Icase Report 92–6, 1992.

Sarkar S., "The stabilizing effect of compressibility in turbulent shear flow", J. Fluid Mech., Vol. 282, 1995.

Zeman O., "Dilatation dissipation: the concept and application in modelling compressible mixing layers", Phys. Fluids A, Vol. 2, N 2, 1990, pp. 178–188.

Addresses of the Editors of the Series "Notes on Numerical Fluid Mechanics"

Prof. Dr. Ernst Heinrich Hirschel (General Editor)
Herzog-Heinrich-Weg 6
D-85604 Zorneding
Germany

Prof. Dr. Kozo Fujii
High-Speed Aerodynamics Div.
The ISAS
Yoshinodai 3-1-1, Sagamihara
Kanagawa 229
Japan

Prof. Dr. Bram van Leer
Department of Aerospace Engineering
The University of Michigan
3025 FXB Building
1320 Beal Avenue
Ann Arbor, Michigan 48109-2118
USA

Prof. Dr. Michael A. Leschziner
UMIST-Department of Mechanical Engineering
P.O. Box 88
Manchester M60 1QD
Great Britain

Prof. Dr. Maurizio Pandolfi
Dipartimento di Ingegneria Aeronautica e Spaziale
Politecnico di Torino
Corso Duca Degli Abruzzi, 24
I-10129 Torino
Italy

Prof. Dr. Arthur Rizzi
Royal Institute of Technology
Aeronautical Engineering
Dept. of Vehicle Engineering
S-10044 Stockholm
Sweden

Dr. Bernard Roux
Institut de Recherche sur les Phénomènes Hors d'Equilibre
(IRPHE)
Technopole de Chateau-Gombert
F-13451 Marseille Cedex 20
France

Brief Instruction for Authors

Manuscripts should have well over 100 pages. As they will be reproduced photomechanically they should be produced with utmost care according to the guidelines, which will be supplied on request.

In print, the size will be reduced linearly to approximately 75 per cent. Figures and diagrams should be lettered accordingly so as to produce letters not smaller than 2 mm in print. The same is valid for handwritten formulae. Manuscripts (in English) or proposals should be sent to the general editor, Prof. Dr. E. H. Hirschel, Herzog-Heinrich-Weg 6, D-85604 Zorneding.